Sustainable Aquaculture: Management and Techniques

Sustainable Aquaculture: Management and Techniques

Editor:

Bruno Augusto Amato Borges

Delve Publishing

www.delvepublishing.com

Sustainable Aquaculture: Management and Techniques

Editor: Bruno Augusto Amato Borges

708 3rd Avenue, 6th Floor
New York, NY 10017
United States of America
www.delvepublishing.com

ISBN: 978-1-68095-856-0

Library of Congress Control Number: 2016951609

About the Editor

Bruno Augusto Amato Borges

Bruno Augusto is Aquaculture Engineer from Federal University of Santa Catarina. His focus is on writing projects and development of new technologies in aquaculture. He is currently project service provider, consultant and develops research in the field of Biofloc Technology (BFT) for freshwater fish farming.

Contents

List of Contributors

John Bostock
Institute of Aquaculture, University of Stirling, Stirling FK9 4LA, UK

Alistair Lane
European Aquaculture Society, Slijkensesteenweg 4, 8400 Ostend, Belgium

Courtney Hough
Federation of European Aquaculture Producers, rue de Paris 9, 4020 Liege, Belgium

Koji Yamamoto
Institute of Aquaculture, University of Stirling, Stirling FK9 4LA, UK

Emily M. Trentacoste
Scripps Institution of Oceanography, University of CaliforniaSan Diego, 9500 Gilman Dr., La Jolla, CA 92093, USA

Alice M. Martinez
A. M. Martinez T. Zenk Sapphire Energy, Inc., 3115 Merryfield Row, San Diego, CA 92121, USA
Algae Biomass Organization, 125 St. Paul Street, Preston, MN 55965, USA

Tim Zenk
Algae Biomass Organization, 125 St. Paul Street, Preston, MN 55965, USA

Zahira Yaakob
Department of Chemical & Process Engineering Faculty of Engineering & Built Environment, University Kebangsaan Malaysia, 43600 Bangi, Malaysia

Ehsan Ali
Centre for Energy Systems, National University of Sciences and Technology, Sector H-12, Islamabad, Pakistan

Afifi Zainal
Department of Chemical & Process Engineering Faculty of Engineering & Built Environment, University Kebangsaan Malaysia, 43600 Bangi, Malaysia

Masita Mohamad
Department of Chemical & Process Engineering Faculty of Engineering & Built Environment, University Kebangsaan Malaysia, 43600 Bangi, Malaysia

Mohd Sobri Takriff
Department of Chemical & Process Engineering Faculty of Engineering & Built Environment, University Kebangsaan Malaysia, 43600 Bangi, Malaysia

Matthew W. Fields
Department of Microbiology and Immunology, Montana State University, 109 Lewis Hall, Bozeman, MT 59717, USA
Center for Biofilm Engineering, Montana State University, Bozeman, MT, USA

Adam Hise
Department of Environmental Sciences and Engineering, University of North Carolina, Chapel Hill, NC, USA

Egan J. Lohman
Center for Biofilm Engineering, Montana State University, Bozeman, MT, USA
Department of Chemical and Biological Engineering, Montana State University, Bozeman, MT, USA

Tisza Bell
Department of Microbiology and Immunology, Montana State University, 109 Lewis Hall, Bozeman, MT 59717, USA
Center for Biofilm Engineering, Montana State University, Bozeman, MT, USA

Rob D. Gardner
Center for Biofilm Engineering, Montana State University, Bozeman, MT, USA

Luisa Corredor
Department of Microbiology and Immunology, Montana State University, 109 Lewis Hall, Bozeman, MT 59717, USA
Center for Biofilm Engineering, Montana State University, Bozeman, MT, USA

Karen Moll
Department of Microbiology and Immunology, Montana State University, 109 Lewis Hall, Bozeman, MT 59717, USA
Center for Biofilm Engineering, Montana State University, Bozeman, MT, USA

Brent M. Peyton
Center for Biofilm Engineering, Montana State University, Bozeman, MT, USA
Department of Chemical and Biological Engineering, Montana State University, Bozeman, MT, USA

Gregory W. Characklis
Department of Environmental Sciences and Engineering, University of North Carolina, Chapel Hill, NC, USA

Robin Gerlach
Center for Biofilm Engineering, Montana State University, Bozeman, MT, USA
Department of Chemical and Biological Engineering, Montana State University, Bozeman, MT, USA

Krystal K Lum
Department of Animal Science, Cornell University, Ithaca, NY 14850, USA

Jonggun Kim
Department of Animal Science, Cornell University, Ithaca, NY 14850, USA

Xin Gen Lei
Department of Animal Science, Cornell University, Ithaca, NY 14850, USA

T. Catalina Adarme-Vega
Algae Biotechnology Laboratory, School of Agriculture and Food Sciences, the University of Queensland, Brisbane, QLD 4072, Australia.

David K Y Lim
Algae Biotechnology Laboratory, School of Agriculture and Food Sciences, the University of Queensland, Brisbane, QLD 4072, Australia.

Matthew Timmins
Centre for Metabolomics, School of Chemistry and Biochemistry, the University of Western Australia M313, 35 Stirling Highway, Crawley, WA 6009, Australia.

Felicitas Vernen
Algae Biotechnology Laboratory, School of Agriculture and Food Sciences, the University of Queensland, Brisbane, QLD 4072, Australia.

Yan Li1
Centre for Metabolomics, School of Chemistry and Biochemistry, The University of Western Australia M313, 35 Stirling Highway, Crawley, WA 6009, Australia.

Peer M Schenk
Algae Biotechnology Laboratory, School of Agriculture and Food Sciences, The University of Queensland, Brisbane, QLD 4072, Australia

Céline Rebours
Bioforsk, Norwegian Institute for Agricultural and Environmental Research, Frederik A. Dahlsvei 20, 1430 Ås, Norway

Eliane Marinho-Soriano
Department of Oceanography and Limnology, Federal University of Rio Grande do Norte, Via Costeira, Praia de Mãe Luiza, s/n, Natal, RN 59014-002, Brazil

José A. Zertuche-González
Instituto de Investigaciones Oceanológicas, Universidad Autónoma de Baja California, Km 107 Carretera Tijuana-Ensenada, 22860 Ensenada, BC, Mexico

Leila Hayashi
Aquaculture Department, Federal University of Santa Catarina, Rod. Admar Gonzaga 1346, Itacorubi, Florianópolis, SC, Brazil

Julio A. Vásquez
Facultad de Ciencias del Mar, Centro de Estudios Avanzados en Zonas Aridas, Universidad Católica del Norte, Coquimbo, Chile

Paul Kradolfer
PSW SA, Av. Paul Poblet Parcela D19 Lote 1, Lurín, Lima 16, Peru

Gonzalo Soriano
Soriano SA, JC Evans 40, 9105 Gaiman, Chubut, Argentina

Raul Ugarte
Acadian Seaplants Limited, 30 Brown Ave., Dartmouth, NS B3B-1X8, Canada

Maria Helena Abreu
ALGAPlus Lda, Travessa Alexandre da Conceição 3830-196, Ílhavo, Portugal

Ingrid Bay-Larsen
Nordland Research Institute, Bodø, Norway

Grete Hovelsrud
Nordland Research Institute, Bodø, Norway

Rolf Rødven
Bioforsk, Norwegian Institute for Agricultural and Environmental Research, Frederik A. Dahlsvei 20, 1430 Ås, Norway

Daniel Robledo
Department of Marine Resources, CINVESTAV-IPN, Km 6 Carretera Ant. Progreso, 97310 Mérida, Yucatán, Mexico

Adam D Hughes
Scottish Association for Marine Science, Scottish Marine Institute Department of Ecology

Maeve S Kelly
Scottish Association for Marine Science, Scottish Marine Institute Department of Ecology

Kenneth D Black
Scottish Association for Marine Science, Scottish Marine Institute Department of Ecology
Scottish Association for Marine Science, Scottish Marine Institute Marine Bio Energy Scotland

Michele S Stanley
Scottish Association for Marine Science, Scottish Marine Institute Department of Ecology
Scottish Association for Marine Science, Scottish Marine Institute Marine Bio Energy Scotland

Mukundh N. Balasubramanian
INRA, UR1067 NUMEA Nutrition, Métabolisme et Aquaculture, Pôle d'Hydrobiologie INRA, 64310 Saint Pée-sur-Nivelle, France.

Stephane Panserat
INRA, UR1067 NUMEA Nutrition, Métabolisme et Aquaculture, Pôle d'Hydrobiologie INRA, 64310 Saint Pée-sur-Nivelle, France.

Mathilde Dupont-Nivet
INRA, UMR1313 GABI Génétique Animale et Biologie Intégrative, 78350 Jouy-en-Josas, France

Edwige Quillet
INRA, UMR1313 GABI Génétique Animale et Biologie Intégrative, 78350 Jouy-en-Josas, France

Jerome Montfort
INRA, UR 1037 Laboratoire de Physiologie et Génomique des Poissons (LPGP), Rennes, France

Aurelie Le Cam
INRA, UR 1037 Laboratoire de Physiologie et Génomique des Poissons (LPGP), Rennes, France

Francoise Medale
INRA, UR1067 NUMEA Nutrition, Métabolisme et Aquaculture, Pôle d'Hydrobiologie INRA, 64310 Saint Pée-sur-Nivelle, France.

Sadasivam J. Kaushik
INRA, UR1067 NUMEA Nutrition, Métabolisme et Aquaculture, Pôle d'Hydrobiologie INRA, 64310 Saint Pée-sur-Nivelle, France.

Inge Geurden
INRA, UR1067 NUMEA Nutrition, Métabolisme et Aquaculture, Pôle d'Hydrobiologie INRA, 64310 Saint Pée-sur-Nivelle, France.

Gary S Burr
ARS National Cold Water Marine Aquaculture Center, USDA, 25 Salmon Farm Road, Franklin, ME 04634, USA.

William R Wolters
ARS National Cold Water Marine Aquaculture Center, USDA, 25 Salmon Farm Road, Franklin, ME 04634, USA.

Frederic T Barrows
Agricultural Research Service, Fish Technology Center, USDA, Bozeman, MT 59715, USA.

Alan W Donkin
Northeast Nutrition Inc, 494 Willow St, Truro, Nova Scotia B2N 6X8, Canada

Chika Yoshida
Frontier Research Labs, Institute for Innovation, Ajinomoto Co., Inc., 1-1 Suzuki-cho, Kawasaki-ku, Kanagawa 210-8681, Japan
Present Address: Strategy Implementation Group, Business Strategy and Planning Department, Ajinomoto Animal Nutrition Group, Inc., Tokyo 104-0031, Japan.

Mayumi Maekawa
Material Development and Application Labs, Research Institute for Bioscience Products and Fine Chemicals, Ajinomoto Co., Inc., 1-1 Suzuki-cho, Kawasaki-ku, Kanagawa 210-8681, Japan.

Makoto Bannai
Frontier Research Labs, Institute for Innovation, Ajinomoto Co., Inc., 1-1 Suzuki-cho, Kawasaki-ku, Kanagawa 210-8681, Japan

Takeshi Yamamoto
Feed Group, Tamaki Laboratory, National Research Institute of Aquaculture, Fisheries Research Agency, Tamaki, Mie 519-0423, Japan.

GUI Jian-Fang
State Key Laboratory of Freshwater Ecology and Biotechnology, Institute of Hydrobiology, Chinese Academy of Sciences, University of Chinese Academy of Sciences, Wuhan 430072, China

Faouzi Ben Rebah
Laboratoire de Biochimie et de Ge´nie Enzymatique des Lipases, ENIS, BPW 1173-3038 Sfax, Tunisia

Nabil Miled
Laboratoire de Biochimie et de Ge´nie Enzymatique des Lipases, ENIS, BPW 1173-3038 Sfax, Tunisia

Christopher Marlowe A. Caipang
School of Applied Science, Temasek Polytechnic, Singapore 529757, Singapore

Han Xuan Choo
School of Applied Science, Temasek Polytechnic, Singapore 529757, Singapore

Zemin Bai
School of Applied Science, Temasek Polytechnic, Singapore 529757, Singapore

Huilin Huang
School of Applied Science, Temasek Polytechnic, Singapore 529757, Singapore

Clara M. Lay-yag
School of Applied Science, Temasek Polytechnic, Singapore 529757, Singapore

Ngo Thi Phuong Lan
University of Social Sciences and Humanities, Vietnam National University, Ho Chi Minh City, Vietnam

Suwat Tanyaros
Faculty of Science and Fisheries Technology, Rajamangala University of Technology Srivijaya, Trang 92150, Thailand

Woraporn Tarangkoon
Faculty of Science and Fisheries Technology, Rajamangala University of Technology Srivijaya, Trang 92150, Thailand

Tossaporn Klomkleing
Aquaculture Technology Program, Faculty of Science and Fisheries Technology, Rajamangala University of Technology Srivijaya, Trang 92150, Thailand

Rini Kusumawati
Rural Development Sociology Group, Wageningen University, Wageningen, The Netherlands.

Simon R. Bush
Environmental Policy Group, Wageningen University, Wageningen, The Netherlands.

Isao Tsutsui
Fisheries Division, Japan International Research Center for Agricultural Sciences (JIRCAS), Tsukuba, Ibaraki, Japan
Department of Animal Production and Fisheries, Faculty of Agricultural Technology, King Mongkut's Institute of Technology Ladkrabang (KMITL), Bangkok, Thailand

Jaruwan Songphatkaew
Department of Animal Production and Fisheries, Faculty of Agricultural Technology, King Mongkut's Institute of Technology Ladkrabang (KMITL), Bangkok, Thailand

Chonlada Meeanan
Shrimp Co-culture Research Laboratory (SCORL), KMITL, Bangkok, Thailand

Dusit Aue-umneoy
Department of Animal Production and Fisheries, Faculty of Agricultural Technology, King Mongkut's Institute of Technology Ladkrabang (KMITL), Bangkok, Thailand

Halethichanok Sukchai
Shrimp Co-culture Research Laboratory (SCORL), KMITL, Bangkok, Thailand

Piyarat Pinphoo
Department of Animal Production and Fisheries, Faculty of Agricultural Technology, King Mongkut's Institute of Technology Ladkrabang (KMITL), Bangkok, Thailand

Sirimas Klomkling
Shrimp Co-culture Research Laboratory (SCORL), KMITL, Bangkok, Thailand

Monthon Ganmanee
Department of Animal Production and Fisheries, Faculty of Agricultural Technology, King Mongkut's Institute of Technology Ladkrabang (KMITL), Bangkok, Thailand

Hiroyuki Sudo
Fisheries Division, Japan International Research Center for Agricultural Sciences (JIRCAS), Tsukuba, Ibaraki, Japan

Kaoru Hamano
Research Center for Marine Invertebrates, National Research Institute of Fisheries and Environment of Inland Sea, Onomichi, Hiroshima, Japan

Tanamy Paul
Department of Microbiology, Vidyasagar University, Midnapore 721102, West Bengal, India

Suman K. Halder
Department of Microbiology, Vidyasagar University, Midnapore 721102, West Bengal, India

Arpan Das
Department of Microbiology, Vidyasagar University, Midnapore 721102, West Bengal, India

Kuntal Ghosh
Department of Microbiology, Vidyasagar University, Midnapore 721102, West Bengal, India

Arpita Mandal
Department of Microbiology, Vidyasagar University, Midnapore 721102, West Bengal, India

Pijush Payra
Department of Microbiology, Vidyasagar University, Midnapore 721102, West Bengal, India

Prasenjit Barman
Department of Microbiology, Vidyasagar University, Midnapore 721102, West Bengal, India

Pradeep K. Das Mohapatra
Department of Microbiology, Vidyasagar University, Midnapore 721102, West Bengal, India

Bikas Ranjan Pati
Department of Microbiology, Vidyasagar University, Midnapore 721102, West Bengal, India

Keshab C. Mondal
Department of Microbiology, Vidyasagar University, Midnapore 721102, West Bengal, India

Eranga K Galappaththi
Natural Resources Institute, University of Manitoba, 70 Dysart Road, Winnipeg, Manitoba, Canada R3T 2N2

Fikret Berkes
Natural Resources Institute, University of Manitoba, 70 Dysart Road, Winnipeg, Manitoba, Canada R3T 2N2

Preface

Currently, we see a growing demand for aquatic foods and as a consequence the growth of aquaculture production worldwide. It is imperative that this growth is properly managed in a sustainable way, supporting the environmental, economic and social aspects. For this, several works have been made to improve management and create new techniques that allow for the sustainable development of the activity.

As any activity of the branch of production, aquaculture has positive and negative impacts. Despite the social and economic benefits related to job creation and technological development, the activity is considered to be impacting the environment mainly because nutrient amount wasted during the processes involved and the protein source used in the formulation of commercial feed. Seeking to reduce these impacts, some management techniques are already used in aquaculture ventures. The use of recirculation systems with biological filter, the use of effluent for vegetable production and cultivation systems with zero water exchange are examples of nutrient management seeking responsible use of resources. One of the main current discussions that live aquaculture production chain is the partial or total replacement of fish meal by ingredients of plant origin. However, it has been difficult to find the ingredient to replace fish meal and keep the zootechnic rates and feed conversion. To that must be fetched alternatives for composition and formulation of diet with the goal of finding the ideal combination and balance of nutrients. In addition, aquaculture presents a huge potential in the search for alternative energy sources through the development of research for use of biofuel from microalgae and macroalgae biomass. The most important thing when you want to start an aquaculture production is planning. To seek sustainability several factors should be analyzed before any decision, since the volume of water to be used up to what will be done with the waste. The techniques used successfully by a producer may not always apply to all because every project has its peculiarities. The aim of this book is to present the current aquaculture situation related to aspects of sustainability, showing different techniques and managements that are used in search of intelligent use of available resources.

Editor
Bruno Augusto Amato Borges

INTRODUCTION

Aquaculture is projected to be the prime source of seafood by 2030, as demand grows from the global middle class and wild capture fisheries approach their maximum take. When practiced responsibly, fish farming can help provide livelihoods and feed a global population that will reach nine billion by 2050. But for an aquaculture system to be truly sustainable, it must have:

- **Environmental sustainability** — Aquaculture should not create significant disruption to the ecosystem, or cause the loss of biodiversity or substantial pollution impact.
- **Economic sustainability** — Aquaculture must be a viable business with good long-term prospects.
- **Social and community sustainability** — Aquaculture must be socially responsible and contribute to community well-being.

Sustainable aquaculture is a dynamic concept and the sustainability of an aquaculture system will vary with species, location, societal norms and the state of knowledge and technology.

Several certification programs have made progress in defining key characteristics of sustainable aquaculture. Some essential practices include:

- **Environment practices**: Mangrove and wetland conservation; effective effluent management and water quality control; sediment control and sludge management; soil and water conservation; efficient fishmeal and fish oil use; responsible sourcing of brood stock and juvenile fish; control of escapes and minimizing biodiversity and wildlife impact.
- **Community practices**: Establish well-defined rights, aquaculture zones and responsibilities for aqua culturists; regulatory compliance and effective enforcement; community involvement; worker safety, fair labor practices and equitable compensation.
- **Sustainable business and farm management practices**: Effective biosecurity and disease control systems; minimal antibiotic and pharmaceutical use; microbial sanitation; maintain global standards for hygiene; efficient and humane harvest and transport; accountable record-keeping and traceability; profitability.

Is sustainable aquaculture possible?

The EU's Blue Growth Strategy1 identifies aquaculture — the farming of fish, shellfish and aquatic plants — as a sector which could boost economic growth across Europe and bring social benefits through new jobs. The reformed Common Fisheries Policy2 also aims to promote the sector and EU Member States are currently developing national aquaculture strategies. Presently, a quarter of seafood products consumed in the EU (including imports) are produced on farms; in 2011, 1.24 million tons of aquaculture goods were produced in the EU, worth €3.51 billion (European Commission, 2014a). There are over 14 000 aquaculture enterprises in the EU, directly employing 85 000 people in total. In contrast with other regions of the world, aquaculture production is stagnating in the EU, while imports are rising. At the same time, there is a growing gap between the amount of seafood consumed in the EU, and the amount caught from wild fisheries. The European Commission calls for this gap to be partly filled with environmentally responsible aquaculture (European Commission, 2013). Aquaculture thus has an important role to play in Europe's food security as well as its economic growth.

In its expansion, aquaculture must continue to respect environmental legislation. This report outlines research into a selection of water and ecological impacts of aquaculture, alongside information on existing and forthcoming measures to mitigate negative impacts. It also highlights the possibility that filling certain knowledge gaps could help to further improve aquaculture's sustainability. In some cases, aquaculture may have positive effects on nature and water quality, and it has been suggested that certain types of farming could help meet the goals of environmental legislation. The importance for the industry of water quality in the wider environment is also considered. Aquaculture is a hugely diverse industry and it should be emphasized that environmental impacts cannot be generalized across the sector. Impacts vary with species, farming methods and management techniques, precise location and local environmental conditions and wildlife.

Five ways to make aquaculture more sustainable

Aquaculture, or the rearing of fish in captivity, is the world's fastest-growing protein-producing activity, with nearly 50 percent of all seafood being farmed rather than caught in wild fisheries.

This rapid growth has provoked questions of sustainability in the global aquaculture industry, including how to handle the massive amounts of salt water being imported inland for fish farms. While researchers warn of dangerous overfishing and decline in the world's wild fish population, aquaculture stands as a potentially sustainable alternative, and recent innovations promise to enhance the efficiency, safety, and sustainability of aquaculture while improving the lives of its fish farmers.

Today, Nourishing the Planet examines five innovations that are improving the sustainability of aquaculture around the world.

1. ***Integrating rice-and-fish farming:*** In many parts of Asia, rice farming provides a major source of income. Rice paddies and fish have long coexisted incidentally, since many fish species find their way into flooded rice fields and actually prefer the fields for reproduction and habitation. But, recently farmers have intentionally imported fish into their rice fields. The advantages of integrated rice-fish farming include a more productive and nutrient-rich rice crop, because fish increase the availability of phosphorous and nitrogen in soils; a reduction in disease-carrying aquatic weeds and algae, which compete with rice for nutrients but are a favored food among fish; and an extra source of income for farmers who can find markets for their fish.
2. ***Combating salmon lice with wrasse fish:*** The spread of disease in aquaculture poses a serious threat not only to farmed fish, but also to wild fisheries. Although one such disease, salmon lice, occurs naturally in the wild, salmon lice has been intensified by aquaculture because of its high concentrations and varieties of species – in some areas of Norway, for example, wild salmon and sea trout had 3-5 times more lice than what is considered to be a "fatal dose." Furthermore, the lice can be transmitted from fish to fish or across large distances via currents, making the disease very difficult to contain. If aquaculture contributes to the incidence of a potentially fatal disease in wild habitats, then it may contribute to the collapse of global wild fisheries. For these reasons, scientists from Stirling University in Scotland are studying the effect of wrasse, a family of fish that cleans other fish of parasites and has been shown to help control lice in farmed salmon. If wrasse can effectively control the incidence of salmon lice, fish farms can reduce their use of medicines and other inputs, and limit their environmental impact.
3. ***Recirculating aquaculture systems:*** A form of aquaculture that has gained popularity in the last few years is called recirculating aquaculture systems, or RAS. These systems recirculate the water used in the fish tank after flowing through a treatment tank, so they use up to 99 percent less water than other aquaculture systems. Because they are maintained in controlled environments, RAS can reduce the discharge of waste and the need for antibiotics or chemicals used to combat disease, as well as prevent fish and parasite escapes. RAS can also incorporate hydroponics, or the water-based cultivation of plants, because the plants thrive in the nutrient-rich water and actually help purify it for reuse. In addition, RAS are less damaging to the environment than many other aquaculture systems, such as open-ocean farms, because of their limited pollution and low demands for space.
4. ***Using locally caught fish as feed:*** The question of how to feed fish raised in aquaculture operations is controversial. Many researchers, such as Rosamund Naylor and Marshall Burke from Stanford University, now estimate large-scale, industrial aquaculture to be a "net drain" on the world's fish supply, meaning that farms raising larger fish such as tuna actually consume more fish in the form of ground-up feed than they produce for human consumption. In addition, farmers are increasingly cutting costs by feeding fishmeal to traditionally herbivorous fish. Aquaculture that relies on local supplies of fish to feed their fish stock could reduce the inputs of industrial operations.

5. ***Involving women in aquaculture:*** Women in developing countries can have a large role in small-scale, sustainable aquaculture systems because they are often charged with managing their family's land while the men seek work in cities. Commercial aquaculture often replaces paddy fields or other agricultural activities in which women are traditionally involved. Because there is often bias against employing women in these larger aquaculture operations, the involvement of women in home-based aquaculture systems, such as backyard ponds, would provide them with a reliable source of income. These operations would also provide nutritional, monetary, and social benefits for the family and community.

Sustainable aquaculture development

Aquaculture is currently playing, and will continue to play, a big part in boosting global fish production and in meeting rising demand for fishery products. A recent session of the FAO Committee on Fisheries (COFI) stressed the increasingly important and complementary role of aquaculture and inland capture fisheries in fish production for human nutrition and poverty alleviation in many rural areas.

Aquaculture, in common with all other food production practices, is facing challenges for sustainable development. Most aqua-farmers, like their terrestrial counterparts, are continuously pursuing ways and means of improving their production practices, to make them more efficient and cost-effective. Awareness of potential environmental problems has increased significantly. Efforts are under way to further improve human capacity, resource use and environmental management in aquaculture. COFI emphasized enhancement of inland fish production through integrated aquaculture-agriculture farming systems and integrated utilization of small and medium-size water bodies.

Integrated aquaculture has a variety of benefits for farmers in addition to the production of fish for consumption or sale. In Asia, for example, rice farmers use certain species of fish to fight rice pests such as the golden snail. With rice-fish farming, they boost their rice yields and harvest the fish. Under FAO's Special Programme for Food Security (SPFS), farmers in Zambia are introducing small ponds into their home gardens for irrigation and aquaculture. Mud from the bottom of fish ponds is also an organic mineral-rich fertilizer.

In traditional, extensive aquaculture, fish can be bred in open waters such as lakes, estuaries or coastal bays, where they feed on naturally available nutrients, or in farm ponds, where they can be fed with by-products from the farm. Traditionally in China, more than five species of carp are bred together to make the best use of feeds and ponds.

The promotion of sustainable aquaculture development requires that "enabling environments", in particular those aimed at ensuring continuing human resource development and capacity building, are created and maintained. The FAO Code of Conduct for Responsible Fisheries contains principles and provisions in support of sustainable aquaculture development. The Code recognizes the Special Requirements of Developing Countries, and its Article 5 addresses in particular these needs, especially in the areas of financial and technical assistance, technology transfer, training and scientific cooperation.

Importance of Sustainability

The sustainability of aquaculture is crucial if the industry is not to go the way of the fisheries sector. About 75 percent of the world's most valuable marine fish stocks are either fished to the limits or over-fished. At the same time world fish consumption has increased from 45 million tonnes in 1973 to more than 130 million in 2000 and the FAO estimates an additional 40 million tonnes of seafood will be required by 2030, just to maintain current levels of consumption. In order to serve this increasing demand in the long run, sustainable alternatives have to be strengthened. The most promising of these is the aquaculture industry. With a growth rate of 8% per year since the 1980's, aquaculture is probably the fastest growing food-production industry, that today accounts for almost half the fish consumed globally, up from 9% in 1980.

Limits of the system

To keep the practice of "sustainability and aquaculture" manageable and practicable, it is important to define the limits of the system for which sustainability is defined.

1. "Farm level": includes the factors that can be directly influenced by the farmer, for instance water quality, nutrient and energy management, fish health, etc.
2. "Second level": addresses the factors directly linked to the farm processes for which the farmer does not have direct influence, but on which he could potentially have an influence if he/she wanted or needed to. For instance: fish feed quality, how fish feed is composed/ processed, distance of transportation for the feed, the kind of energy the farmers uses (renewable or non-renewable), markets for the products (distant markets – requiring long transport distance, local markets – requiring short transport distance), etc. The farmer might also "transfer" some factors of the second level into the "farm level", e.g. by producing fish feed on the farm, using energy produced on the farm or by selling the products directly from the farm. The first two circles are the most relevant for the project.
3. The "Third level": contains factors that are indirectly linked to the farm processes but which can normally not be influenced by the farmer. These are factors like sustainability of the packaging material (production, material, etc.), the type of fuel for the transportation of the fish, etc.

Figure 2: Three levels of system limits for which sustainability is defined.

Sustainability indicators and certification

The limited availability of natural resources coupled with increasing energy prices emphasise the need to move forward in aquaculture to become more sustainable. The aquaculture industry is already working on this demanding task, but there is still a long way to go. Compared to other animal production systems, aquaculture is put under special pressure to become more sustainable, because of their use of important natural resources, such as freshwater, wetlands, coastal areas and also the wild catch of fish for fish feed production or stock recruitment. The sustainability of an activity and its measurement is not a static topic as, by definition, it incorporates economic, environmental and social considerations (see Figure 3). Each approach to sustainability as well as being based on indisputable facts contains some level of attached societal values and value judgements, which may be under discussion or may change over time. This means it is not always possible to decide unambiguously whether a process is sustainable or not. Often there are transitions between non-sustainable to sustainable processes.

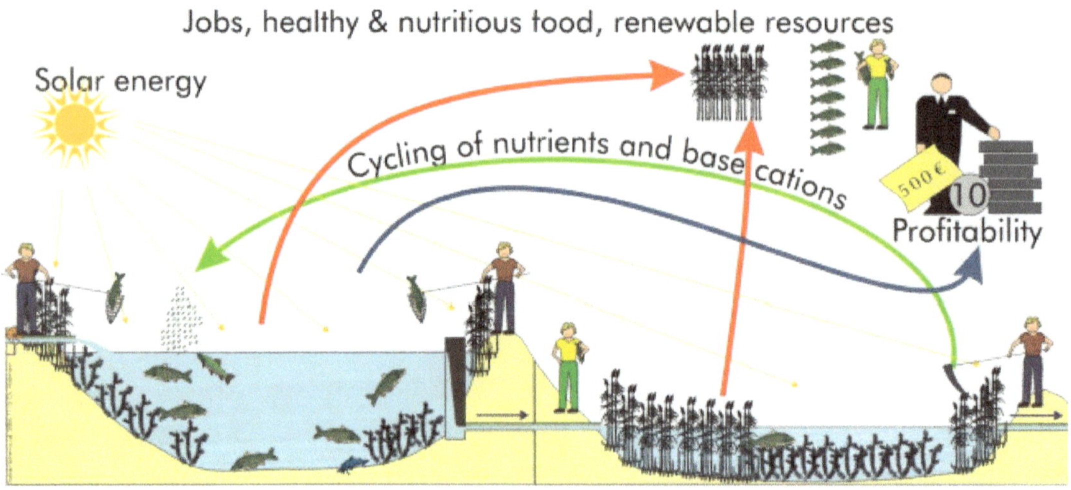

Figure 3: Sustainable freshwater aquaculture combines ecological, economic and social aspects.

The different Codes of conducts and criteria systems mentioned earlier aim to resolve this issue of how to achieve sustainability and are intended to support a sustainable cultivation of aquaculture products. But up until now there have been no complete and practicable European criteria, indicators, and related labelling systems which are really able to certify the sustainability status of a fish product. The project intents to contribute to the development of criteria which are currently being developed by various initiatives (see above). They are primarily designed to give a measurable orientation to the transferability and practicability of the research carried out in the five case studies in order to develop applicable methods and technologies for more sustainable aquaculture production in Europe. It is not the task of judge, if a certain freshwater aquaculture farm is sustainable or not, but to provide an unambiguous direction, on what can be done in a case study or at a specific farm to improve sustainability.

Application of Sustainable Principles to Aquaculture

In the following paragraphs, the principles of each sustainability area will be introduced in detail. In addition, general suggestions are made on how to make an aquaculture system more sustainable by considering these principles.

Improving Ecological Sustainability

Water, nutrients, the area used for the farm, and energy are the most important topics related to the ecological sustainability of aquaculture farms. Regarding water, both the amount needed and the quality are important aspects. Freshwater may be obtained from surface sources, such as lakes or rivers, or from the ground (aquifers) by the use of wells. One important goal in all systems is to reduce the amount of freshwater needed to relieve the natural ecosystems. An equally essential objective is (as in most cases the outflow of an aquaculture contains a lot of nutrients which may eutrophicate the natural systems), to reduce the amount of wastewater and to optimise the effluent treatment. The best management practice naturally depends on the type of aquaculture. Traditional carp ponds, for example, need water only for replacing the evaporation and seepage; the outflow is limited to the harvest. Recirculation aquaculture systems, like the Danish model trout farms, are another example of how to substantially reduce the amount of water needed. In the latter case, they use for instance plant lagoons to retain the nutrients of the outflow. An efficient use of the required nutrients is also essential for environmental sustainability. Reducing feed losses by an advanced feeding regime and the selection of appropriate feeds is the first step. The additional use of the remaining nutrients is again a site-specific task. The use of periphyton, as in the Hungarian case study, is one possibility. The use of different fish species in the same ponds, a polyculture, may raise the nutrient efficiency because of the different ecological niches of the fish species, which are used e.g. in the Polish case study. However, it should be considered in this case, to avoid the use of alien and exotic species. If sufficient area is available, renewable resources like reed or willow (one example is the Hungarian case study), or garden plants, as in the Danish case study, are further examples of how to increase nutrient utilization efficiency. The origin of the feed used is a further task to contribute to ecological sustainability, for instance to use fishmeal produced from by-catch originating from sustainable fisheries (e.g. MSC certified). Sustainability with regard to the area used for the aquaculture farm depends greatly on the local circumstances. In general, the need to produce renewable resources in addition to food puts more pressure on the land use. The decreased land used per unit of fish produced in some recirculation aquaculture systems can offer a contribution. On the other hand, the pond area of the aquaculture farm can also contribute to local climate stabilization by increased evapotranspiration. Ponds can also provide excellent ecologically valuable areas. Regarding the use of energy, this is a particularly major topic in recirculation aquaculture systems, as in the case of the Netherlands (see chapter 'Netherlands'). Also in other aquaculture systems, it is possible and important to reduce the amount of energy by increasing the energy efficiency, e.g. through the use of more efficient pumps. With regard to energy use, the aim is to produce at least the same amount of fish with less energy or more fish with the same amount of energy.

Improving Economic Sustainability

An aquaculture is economically sustainable and viable, if the farm is profitable, the farm revenue is reliable and the farm system and products are accepted by the consumer. In many cases, improving environmental sustainability can be connected to the optimization of economic sustainability. For instance, a more efficient use of feed and nutrients or the reduction of the use of freshwater is not only positive for the environment, it can also reduce costs. Depending on national laws, reducing wastewater contributes also to the lowering of production costs. The same is true for all energy dependent processes. A more local or regional distribution of products will decrease the transport costs, which are partly energy costs. The diversification of the aquaculture can buffer market fluctuations. Production of traceable high quality products can both increase realized prices and consumer confidence. Last, but not least, fully endorsing sustainability (and not just adopting under duress as a necessary chore) can be a valuable argument to increase consumer acceptance. However, all these aspects need to be evaluated very individually, because the availability of all resources needed for an aquaculture (water, land, nutrients, and energy) vary greatly between the different European countries and regions. In the vicinity of a big city, for instance, a highly intensive recirculation system might be very much sustainable, especially if it can be heated by waste heat; whereas in rural areas, as is the case in many areas of Hungary, it might be economically much more sustainable to run a large extensive carp pond, as land and water is relatively cheap and available.

Improving Social Sustainability

The issue of social sustainability is also very complex. It includes employment opportunities in the sector, the conditions of employment on the aquaculture farm (hygiene, safety, training), but also the general public in connection with e.g. recreation, health and nutritional issues. Important aspects are also the attractiveness of aquaculture to the younger generation or in which way an aquaculture system preserves culture and traditions, for example with pond fish farming in Eastern Europe. Social sustainability was not a primary focus, which concentrated more on technical solutions to directly increase economic and environmental sustainability, but which, nonetheless, if achieved support social sustainability (securing jobs, ensuring functional environment for recreation, contributing to high-quality and healthy nutrition, etc.).

Pond Fish Farming

Production of freshwater fish in artificial ponds is often considered as the oldest fish farming activity in Europe, dating back to medieval times. Ponds were built in areas where water supply was available and the soil was not suitable for agriculture. The wetlands of Central and Eastern Europe are good examples of this. The total European production from pond farming is approximately 475 000 tons per annum. About half of this production is cyprinid fish, such as common carp, silver carp and bighead carp. The main producer countries are the Russian Federation, Poland, the Czech Republic, Germany, Ukraine and Hungary. Typical fish ponds are earthen enclosures in which the fish live in a natural-like environment, feeding on the natural food growing in the pond itself from sunlight and nutrients available in the pond water. In order to reach higher yields, farmers today introduce nutrients (organic manure) and additional food (grain). This is accompanied by the stocking of fingerlings. Fish pond production, however, remains 'extensive' or 'semi intensive' (with supplementary feeding) in most countries, where semi-static freshwater systems play an important role in aquaculture. Chemicals and therapeutics are not usually used in such ponds. Chemicals and therapeutics are not usually used in such ponds. Hence, the main environmental issue is the use of organic fertilizers, which may cause eutrophication in the surrounding natural waters. The use of organic fertilizers is regulated at national levels. Extensive fish ponds are usually surrounded by reed belts and natural vegetation, thus providing important habitats for flora and fauna. They play a growing role in rural tourism. Many pond fish farms have been turned into multifunctional fish farms, where various other services are provided for recreation, maintenance of biodiversity and improved of water management.

CHAPTER 1

AQUACULTURE PRODUCTION AND POLICIES FOR SUSTAINABLE DEVELOPMENT

John Bostock, Alistair Lane, Courtney Hough, Koji Yamamoto

Keywords: Aquaculture EU Europe Economics Policy Systems Competitiveness

INTRODUCTION

Fish and shellfish have been produced using aquaculture techniques for hundreds of years. However, a new phase of growth commenced in the 1970s–1980s as production from capture fisheries reached a plateau, whilst demand for aquatic food products continued to rise. Total EU output increased throughout the 1990s but stabilized in the early 2000s although consumption and imports maintained an upward trend. This review paper examines the structure of the EU aquaculture sector, the contribution it makes to the EU economy and the policy environment for past and future development. The paper is based on work carried out by the authors for the European Parliament in 2014 (Lane et al. 2014) which in particular featured a preliminary (i.e. based on expert knowledge rather than survey-based farm data) disaggregation of production and value data by system type. Whilst this cannot be taken as definitive, it provides another view on the industry which should help inform future analysis and policy development.

Citation: John Bostock Alistair Lane Courtney Hough Koji Yamamoto, aquaculture production and policies for sustainable development, DOI 10.1007/s10499-016-9992-1

THE SCALE AND STRUCTURE OF THE EUROPEAN AQUACULTURE INDUSTRY

This section provides a short overview of production of the key European species. The authors have identified the main technologies used and have developed a new classification of 'production by technology' based on expert knowledge of the sector.

Aquaculture Development in European Countries

Whilst global aquaculture production expanded at an average annual rate of 6.2 % in the period 2000–2012 (9.5 % in 1990–2000) (FAO 2014a) to reach 90.4 million tonnes (live weight equivalent), growth in Europe was well below this at an average of 2.9 % and total production is only 4.3 % of the global total at 2.88 million tonnes. Most of the positive growth was due to Norway, whilst production in the EU actually declined after 1999 (See Fig. 1).

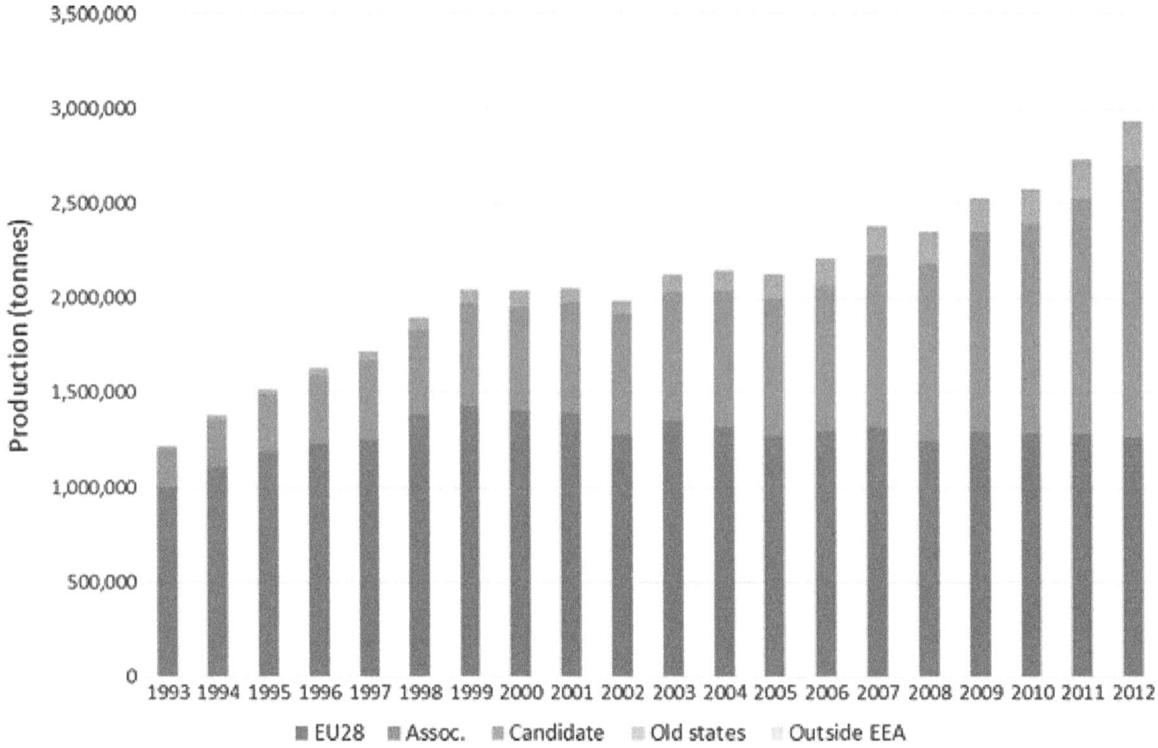

Figure 1: Development of European aquaculture production in relation to EU membership (*Source:* developed from FAO FishStat database)

Aquaculture is practised in almost every European country with the largest producers being Spain, France, UK, Greece and Italy (Fig. 2).

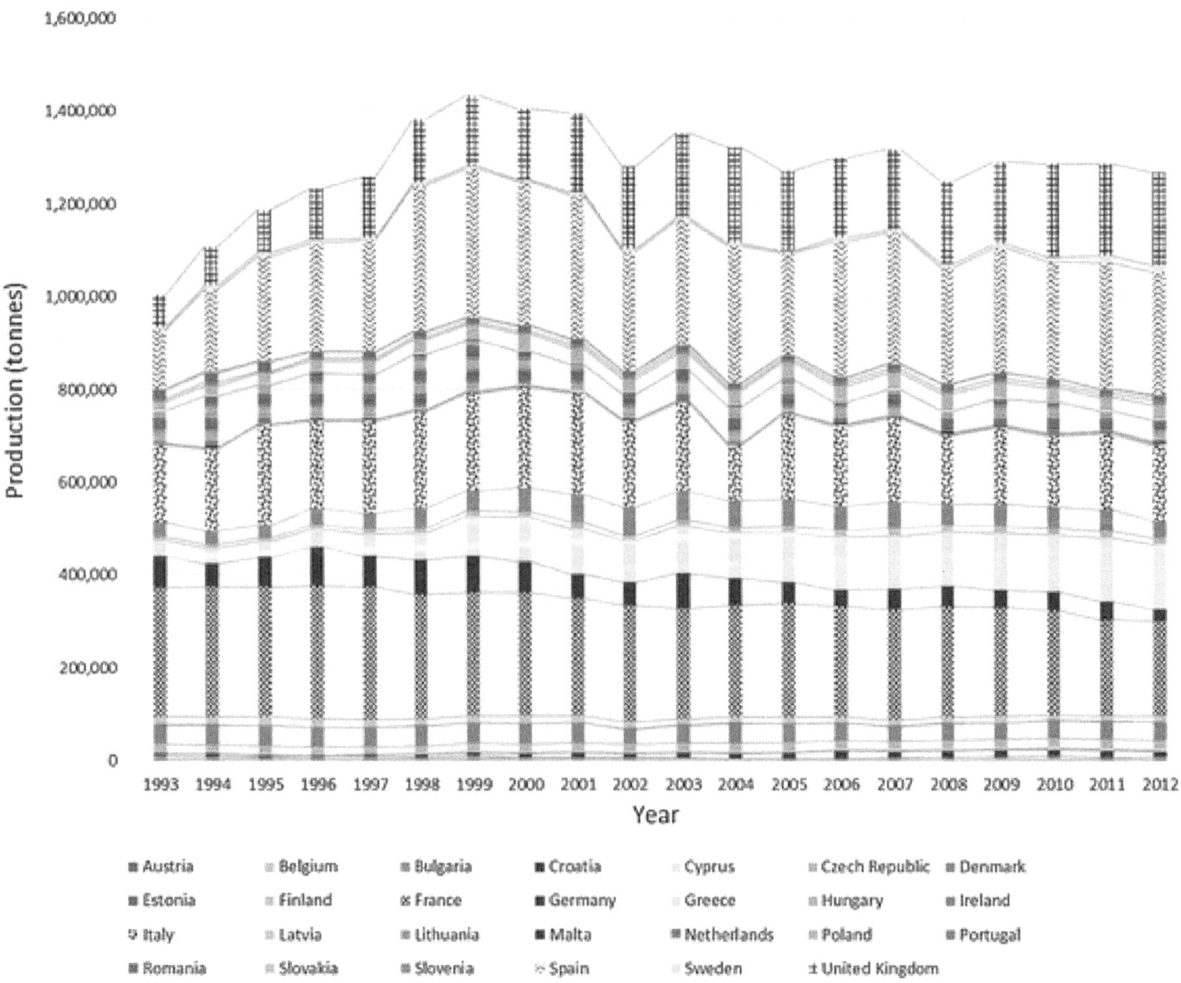

Figure 2: Geographic distribution of EU production (*Source:* developed from FAO FishStat database)

Production by Species of the Principal Sub-Sectors of European Aquaculture

In volume terms, the largest segment of EU aquaculture is shellfish—mainly of mussels and oysters. In value terms, however, fish production is more significant. More than 70 different fish species are cited for aquaculture in the EU (FAO 2014b), but production (in 2012) was dominated by rainbow trout, Atlantic salmon, gilthead sea bream, European sea bass and common carp. These five species make up 90 % of all fish production in the region.

For the purpose of this study, we consider the EU aquaculture industry as comprising five separate segments, namely: cold-water marine fish, warmwater marine fish, freshwater fish, shellfish and algae and other aquatic products (Fig. 3).

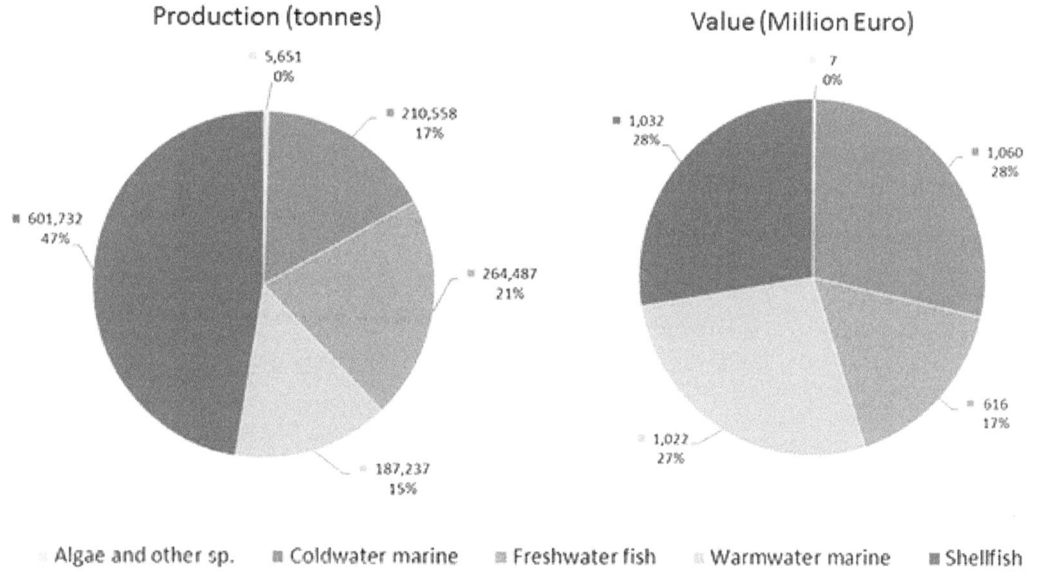

Figure 3: EU aquaculture production by major group (tonnes and €million in 2012) (*Source:* FAO FishStat data)

Coldwater Marine Fish

The cultivation of Atlantic salmon dominates this segment and is seen as the major success story in this sector. Production rose from only 900 tonnes in 1980 (FAO 2014b) to 168,000 in 2013 (FEAP 2014), reared mainly in Scotland (UK) with 11,500 coming from Ireland. The EU production is, however, significantly overshadowed by Norway (1.2 million tonnes) in 2013 (FEAP 2014). This is supplemented by the rearing of large rainbow trout in marine/brackish water conditions (Scotland, Denmark, Finland and Sweden) of 24,000 tonnes (FEAP 2014) (Fig. 4).

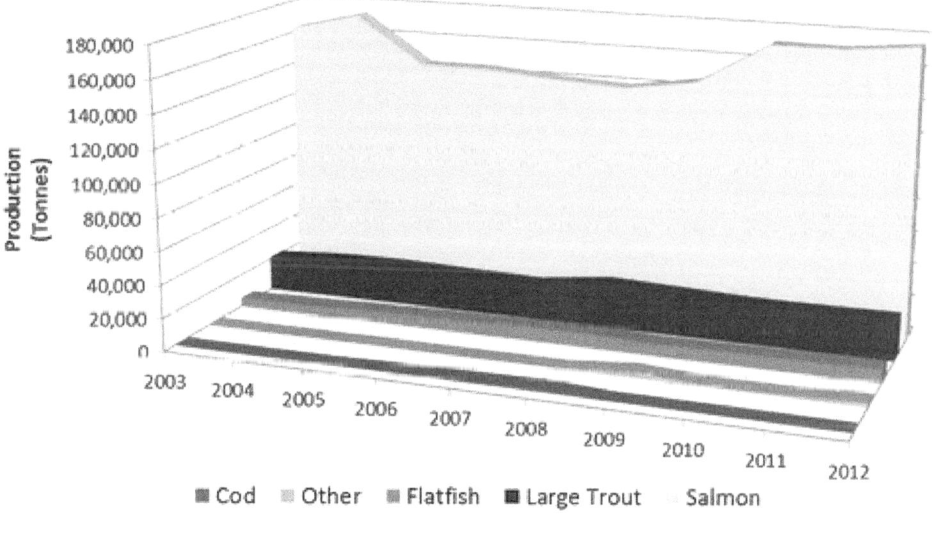

Figure 4: EU production of marine coldwater fish (*Source:* FAO FishStat data)

Efforts on diversification have focused on cod, sole and halibut—but so far with a small number of operations servicing mostly niche markets. Norway invested heavily in cod production, resolving technical challenges and reaching annual production levels exceeding 20,000 tonnes. However, resurgent wild stocks and market/price competition led to a crash and suspension of this activity (Jensen et al. 2014; Norwegian Directorate of Fisheries 2015).

Warmwater Marine Fish

This sector covers fish farming in warmer marine conditions, notably in the Mediterranean and Southern Europe (France, Portugal, Spain). The dominant species are gilthead sea bream and European sea bass, with turbot and meagre leading the remainder. EU production totalled 218,000 tonnes in 2013 (FAO 2014b). As with salmon farming, the total production of these 3 species in 1980 was less than 500 tonnes. Growth started in the 1990s, when hatcheries were able to produce juveniles more readily for stocking in cages (Fig. 5).

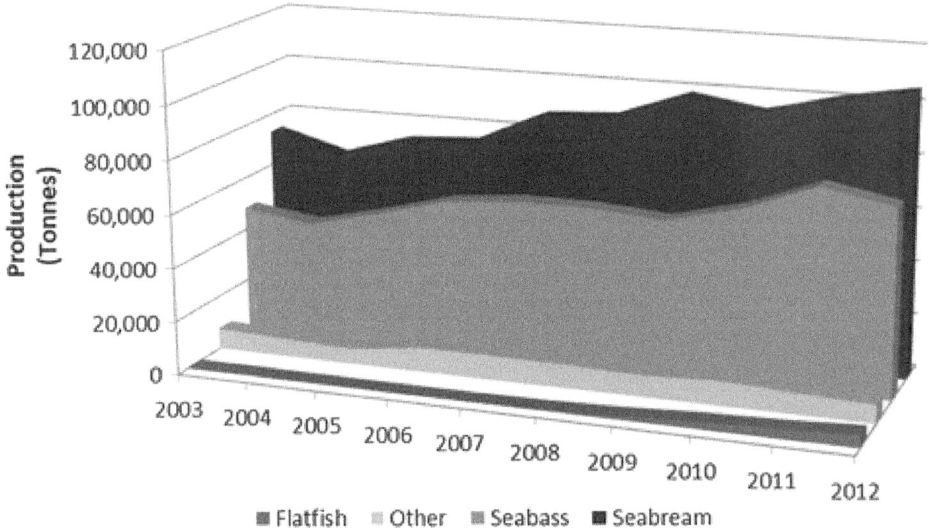

Figure. 5: EU production of warmwater marine fish (*Source:* FAO FishStat data)

The growth in Turkish aquaculture (of sea bass and sea bream) has also been significant, especially as it exports mainly to the EU. From 25,000 tonnes of sea bass/sea bream in the late 1990s, Turkey produced 78,000 tonnes in 2010 and is estimated to be over 85,000 tonnes in 2013 (FEAP 2014).

Freshwater Fish

Whilst the freshwater sector is by far the oldest in European fish farming, it is divided into 2 main components, those of rainbow trout and common carp. This distinction principally reflects climatic conditions where rainbow trout is more suitable for temperate environments with carp being better in the more extreme continental conditions seen in Central Europe (hot summers, cold winters) (Fig. 6).

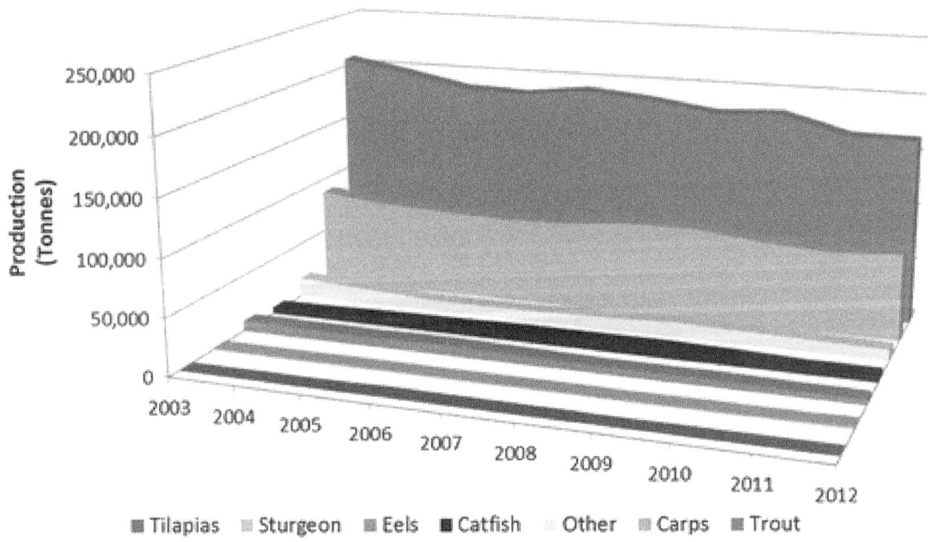

Figure 6: EU production of freshwater fish (*Source:* FAO FishStat data)

Rainbow trout production expanded from the 1980s, moving from 100,000 tonnes to a peak of approximately 250,000 tonnes in 2000–2001. This has since fallen to around 165,000 tonnes in 2012, although combined with the large trout reported under cold water marine fish, production still exceeded that of salmon in 2012. It is produced not only for consumption (principally as a portion-size fish (±250–350 g.) that is produced in one growing season) but also for stocking lakes and rivers for sport fishing. The decline in production is attributed to operational and licensing difficulties, in relation to environmental legislation, and the difficulties faced by smaller companies to deal with evolving market conditions and competition, notably with salmon. Significant reductions have been seen in Italy, Germany, Denmark, France and Spain (FEAP 2014).

Common carp attained 90,000 tonnes of production in the 1970s; major producers were Poland, Romania, Hungary, Germany, Czech Republic and Poland (each >10,000 tonnes). Smaller production levels were reported for France, Croatia and Lithuania. Carp is seen as a 'traditional' species, being a cultural fish dish for many inland countries that did not have access to marine seafood. EU carp production is now around 60,000 tonnes (FEAP 2014). This reduction is attributed to different causes; predation from wild birds (cormorants, herons) is a major factor and disease (koi herpes virus) is an additional issue. In addition, with adhesion of Central European States to the EU, supermarkets have replaced traditional markets and access to other seafood and fish products has ensued (e.g. marine fish, salmon, pangasius catfish). Carp markets have remained traditional, and added-value products are relatively few (STECF2013a, b).

Eel farming reached a maximum of 11,000 tonnes in 2000 but is entirely dependent on the wild catch of young eels (glass eels) for its initial stock; the reductions in wild catch availability and high price competition for the stock with Asia have led to a fall in production, which was around 6000 tonnes in 2012 (FAO 2014b).

Many other freshwater species have been cultured at smaller scale in the EU, including Arctic char and other trout species, sturgeon—mainly for caviar, perch, pike-perch, roach and tench, African catfish—reared in warmwater, recirculation systems (mainly in the Netherlands) and tilapia and barramundi—tropical species reared in warmwater and recirculation systems. With the exception of African catfish and sturgeon, these all remain very minor components of EU aquaculture due to technical and/or marketing issues. Sturgeon rearing for caviar has expanded significantly in recent years, assisted by restrictions on the exploitation of wild sturgeon and trade in wild caviar under CITES advice and agreements (Bronzi et al. 2011; Doukakis et al.2012).

Shellfish

European shellfish production is mainly of oysters (the Pacific cupped oyster, *Crassostrea gigas*, and the European flat oyster, *Ostrea edulis*) and mussels (the blue mussel, *Mytilus edulis*, and the Mediterranean mussel, *Mytilus galloprovincialis*). Together, oyster and mussel culture represent 93 % of the total European cultivated mollusc production (Eurostat 2011). The third element of shellfish production is of clams, cockles and arkshells (Fig. 7).

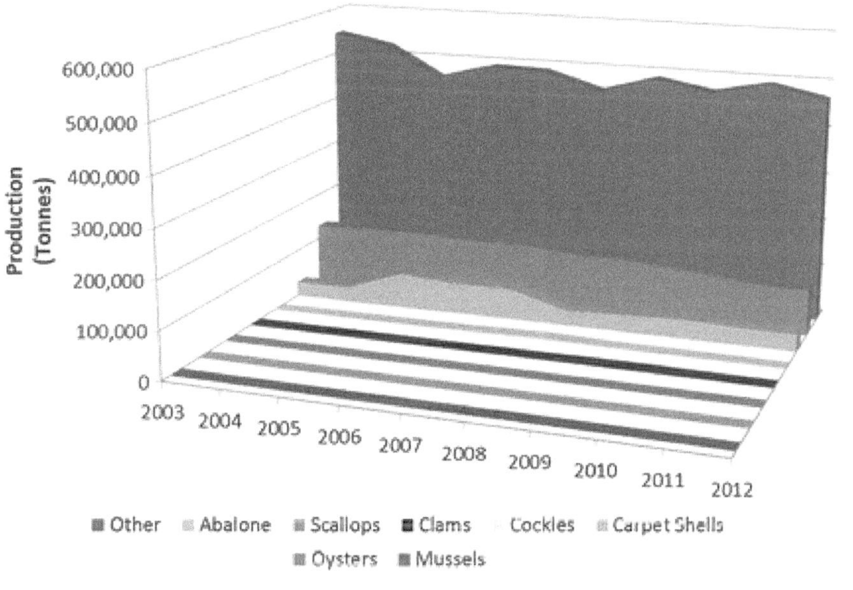

Figure 7. EU production of shellfish (*Source:* FAO FishStat data)

France is by far the leading producer of oysters (± 85,000 tonnes in 2011), Spain of mussels (±209,000 tonnes in the same year) and Italy of clams (± 32,000 tonnes in the same year).

Over recent years, the production of oysters has declined with abnormal mortality events of*Crassostrea gigas* adults reported in most French oyster production areas, generally linked with oyster herpes virus (Martenot et al. 2011). It is now thought that the bacterial pathogens*Vibrio splendidus* and *V. aestuarianus* may play a more significant role in the mortality events, with interactions between pathogens and the development of some resistance by previously infected stocks playing a role in the severity of losses (Vezzulli et al. 2015; Petton et al. 2015). The UK, Jersey, Ireland and the Netherlands have all suffered recent similar mortalities (Renault 2011).

Algae and other Aquatic Products

European production of algae and other species has only really started to evolve since 2007. Of these, aquatic plants (including brown seaweeds of the genus *Phaeophyceae*) have seen the most growth and this has been reported for Denmark (FAO 2014b).

Other species groups include crustaceans (mostly freshwater crayfish species in Central and Eastern European countries, but also several prawn species including the kuruma prawn,*Penaeus japonicus* in Southern Europe).

Mollusc species include mainly octopus and the miscellaneous aquatic animal group such as sea urchins and non-specified aquatic animals from inland waters.

EUROPEAN AQUACULTURE TECHNOLOGIES AND ESTIMATION OF PRODUCTION BY TECHNOLOGY

Classification of Systems and Overview of Production

European aquaculture is a very diverse activity. Not just in terms of species produced, but also in the technologies that are used for rearing them. Aquaculture can be classified according to various primary characteristics such as scale (Lazard et al. 1991), intensity of production and feeds used (e.g. Edwards 1993), and finally by a matrix of environment and/or species (or species group).

A technology-based classification primarily considers the nature of the containment system, and a primary distinction is made between land-based and water-based systems. For the former, there are potential benefits both to the farm and to the environment if water is recycled and treated on the farm. 'Recirculated aquaculture systems' (RAS) are normally defined by replenishing less than 5 % of the system volume per day. This approach allows systems to be effectively isolated from the environment and water quality conditions fully controlled. Water-based containment systems are either floating structures that are moored in position or static constructions usually placed in the intertidal zone or shallow water and fixed into the substrate.

A recent study (Lane et al. 2014) performed for the Committee on Fisheries of the European Parliament selected 11 production technologies (see Table 1) as being the most dominant technologies for EU production and, using data from FAO, characterized by country and by species by assigning the 2012 production to the most dominant technology used for (ongrowing) production (Fig. 8). Hatchery and nursery technologies were therefore not included as their impacts are small and included in overall production calculations.

Table 1: Key to system types used in analysis

EP code	System type
EP01	Freshwater pond aquaculture (extensive to semi-intensive)
EP02	Coastal pond aquaculture (mostly semi-intensive)
EP03	Intensive freshwater flow-through and partial recirculation systems (mostly tanks, raceways and small ponds)
EP04	Intensive marine flow-through and partial recirculation systems (mostly large tanks)
EP05	Indoor land-based recirculated aquaculture systems (freshwater)
EP06	Indoor land-based recirculated aquaculture systems (marine)
EP07	Small cage systems—freshwater
EP08	Small cage systems—sheltered marine
EP09	Large cage systems—marine in exposed sites, using mechanized systems
EP10	Marine bottom culture (non-fed sedentary and attached animals and plants)
EP11	Marine supported and suspended culture (non-fed sedentary and attached animals and plants)

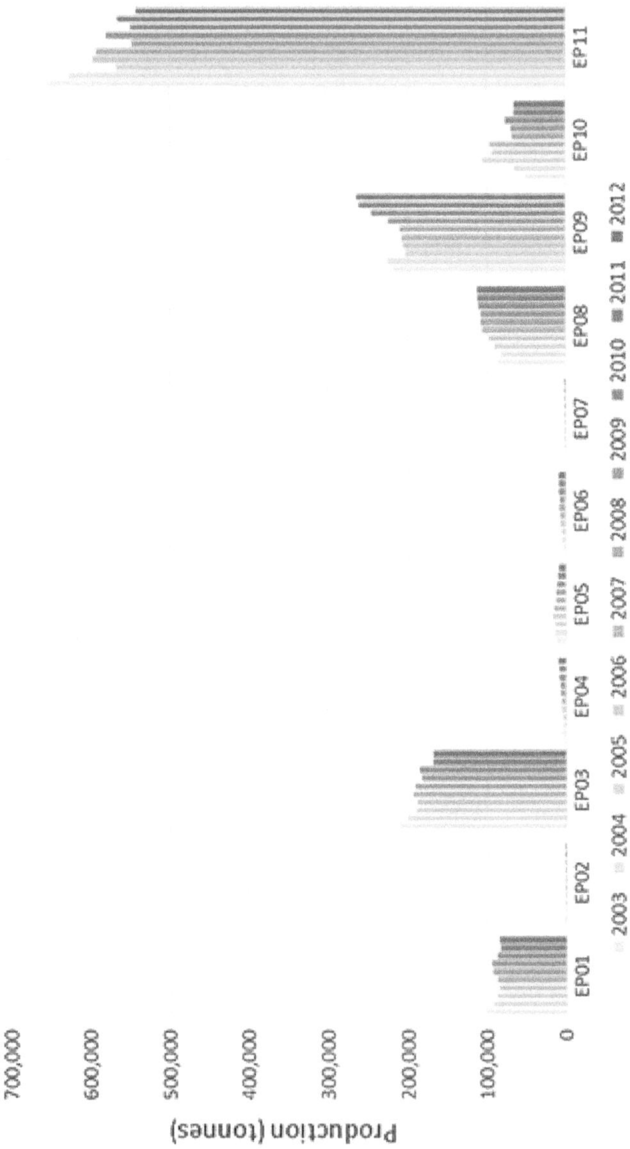

Figure 8: EU aquaculture production by system type (*Source:* Lane et al. 2014)

Analysis of main segments by production system

Coldwater marine fish

Cage/net production is the primary ongrowing technology (Table 2), and there has been a gradual expansion of unit size accompanied by automatically controlled feed distribution, remote-monitoring and mechanized movement/harvesting of fish. This has significantly reduced labour costs as well as providing safer working conditions in the marine environment (Bostock et al. 2010).

Table 2: Coldwater marine fish—estimated production quantity by system type

Technology	Estimated 2012	
	Production (tonnes)	**Value (million euro)**
Large cage systems—marine in exposed sites, using mechanized (automated) systems	190,090	870.21
Small cage systems—sheltered marine	11,000	37.46
Indoor land-based recirculated aquaculture systems (marine)	9180	51.85
Coastal pond aquaculture	180	0.75
Intensive marine flow-through and partial recirculation systems (mostly large tanks)	100	0.87
Total	210,550	961.14

Source: Lane et al. (2014) and present study

The vast majority of cage-based marine fish farming is conducted in relatively sheltered coastal areas that provide protection from storms and ease of access by personnel (for maintenance work, stocking, feeding and harvesting, etc.). However, the number of such sites that have been licenced for development over the past 15 years has been limited due to planning authorities responding to expressed concerns over the potential and perceived impacts of the waste produced by farms, influence of escapees (e.g. from storm damage) on wild stocks, visual impact (e.g. to coastal residences and hotel businesses) or interactions with wildlife and conservation interests. For these reasons both 'offshore' farming (further away from the coastal zone) and closed containment system (e.g. onshore RAS) have been promoted as being a panacea for the future (Olsen et al. 2008; Klinger and Naylor 2012).

Warmwater Marine Fish

As for the coldwater marine segment, ongrowing is usually done in cages, although generally smaller than used for salmon (Table 3). It is anticipated, however, that production in larger cage sizes will be prevalent in the future and that production will develop more in exposed or high energy sites.

Table 3: Warmwater marine fish—estimated production quantity by system type

Technology	Estimated 2012	
	Production (tonnes)	**Value (million euro)**
Small cage systems—sheltered marine	102,420	558.69
Large cage systems—marine in exposed sites, using mechanized (automated) systems	74,330	369.88
Intensive marine flow-through and partial recirculation systems (mostly large tanks)	9910	58.47
Coastal pond aquaculture and valliculture	900	4.18
Indoor land-based recirculated aquaculture systems (marine)	340	3.57
Total	187,900	1021.80

Source: Lane et al. (2014) and present study

There are a few specialized pump-ashore marine tank systems mostly for flatfish, and some use of coastal ponds, particularly in Italy and Portugal, but the contribution to total marine fish production is low.

Freshwater fish

Freshwater aquaculture—compared to marine aquaculture—is fixed to the site where the installations are made and is largely dependent on adequate access to water for its production potential (Table 4). Traditionally, incoming water provides the environment for growing the fish and site selection is usually based on a combination of clean water and temperatures appropriate for the species.

Table 4: Freshwater fish—estimated production quantity by system type (tonnes)

Species	Freshwater pond	Intensive freshwater flow-through	Indoor land-based recirculated	Small cage	Total production
Trout	310	164,390	0	250	164,950
Carps	77,860	0	0	0	77,860
FW other	6640	2580	290	0	9510
Catfish	1140	400	3790	0	5330
Eels	0	0	4960	0	4690
Sturgeon	400	1910	0	0	2310
Tilapias	0	0	450	0	450
Salmon	0	5	0	0	5

Total (t)	86,350	169,285	9220	250	265,105
Total (€M)	184.22	464.63	60.66	2.07	711.58

Source: Lane et al. (2014) and present study

Both trout and carp were originally reared in earthen ponds that require considerable maintenance (banks, floor, inlets/drains, etc.). Trout farming expanded as a result of increased productivity, due to the development of compound feeds adapted to nutritional requirements and the use of concrete tanks/raceways and latterly oxygenation allowing higher stocking levels and greater mechanization and automation for feeding, grading and harvesting fish.

Carp and coarse fish are generally reared in large freshwater ponds with minimal extra feeding (extensive production); the use of specific compound feeds is much rarer. This means that the aquaculture process to provide fish of market size is longer (2–3 years) and less direct stock management (e.g. grading and vaccination) is practised. Nonetheless, the infrastructure of extensive pond farming is recognized as providing environmental services—providing flood protection, habitats for wild birds and animals, and water treatment (outlet water is often cleaner than inlet) (Szücs et al. 2008).

Sturgeon production for caviar often uses water recirculation systems—particularly for younger fish—and uses advanced monitoring procedures of individual fish, so as to follow ovary development for caviar harvesting.

Recirculated aquaculture systems (RAS) have been in use since the 1980s as a means of reducing water usage (and thereby increasing potential production for any given site/water supply), maintaining stable and more biosecure environments, and in some cases to assist the conservation of heat energy. Installations and monitoring equipment have evolved considerably, and large-scale units have been installed for different freshwater species (notably for salmon juvenile production, eels, African catfish, barramundi and tilapia). State-of-the-art systems use approximately 500 litres of water per kg production, but with increased capital and energy costs (Verdegem et al. 2006; Martins et al. 2010).

Shellfish Production

European oyster production is usually a three-year process that starts with the collection of small oysters on a support from which they can be easily removed (spat collector, e.g. white-painted roofing tiles, as practised in the Bassin d'Arcachon) or on a substrate such as mussel shells. The newly settled oyster juveniles are then transferred to intertidal or sub-tidal bottom culture plots, or in intertidal bags on trestles, or suspended from frames.

There are three different culture techniques for mussel culture—using poles ('bouchot'), suspended ropes or bottom culture. A 'bouchot' is a wooden pole, placed upright into the sand. Mussel seed, collected (usually around March) either on poles (placed further out to sea) or on ropes, is transplanted onto the growing poles and harvested after 12–15 months. Suspended rope culture: Ropes covered with mussel seed are suspended from either frames, floating structures (rafts), or longlines with surface buoys. Frames are built from metallic poles, placed upright into the ground, at water depths ranging between three and nine metres. Bottom culture: Based on the harvesting of naturally producing mussel beds—a technique that is widely practised in the Netherlands and Germany and to some extent in UK and Ireland. Since the 2000s mussel seed is also derived from suspended seed collectors and then relayed on bottom culture plots, particularly in the Netherlands and Germany.

Scallop, clam and cockle species are generally produced using bottom culture techniques, with juveniles from the wild or in some cases, hatcheries (Table 5).

Table 5: Shellfish—estimated production quantity by system type

Technology	Estimated 2012	
	Production (tonnes)	**Value (million euro)**
Marine supported and suspended culture (non-fed sedentary and attached animals and plants)	537,050	837.21
Marine bottom culture (non-fed sedentary and attached animals and plants)	65,440	204.01
Total	602,490	1041.23

Source: Lane et al. (2014) and present study

Algae and other Aquatic Products

The majority of reported production volume is of brown seaweeds using suspended culture techniques. Other species groups, such as crayfish and shrimps, are generally produced in ponds, whether freshwater or brackish water (Table 6).

Table 6: Algae and other aquatic products—estimated production quantity by system type

Technology	Estimated 2012	
	Production (tonnes)	**Value (million euro)**
Marine supported and suspended culture (non-fed sedentary and attached animals and plants)	5352	0.01
Coastal pond aquaculture	247	2.39
Freshwater pond aquaculture	38	0.31
Marine bottom culture (non-fed sedentary and attached animals and plants)	9	2.95

Small cage systems—sheltered marine	5	0.02
Intensive freshwater flow-through and partial recirculation systems	1	0.007
Indoor land-based re-circulated aquaculture systems (marine)	1	0.01
Total	5653	5.70

Source: Lane et al. (2014) and present study

Small cage marine systems are used for the culture of cephalopods, mainly octopus species.

THE ECONOMIC VALUE OF EU AQUACULTURE

Overview of Value Chain Structures

The EU Fish Processors and Traders Association estimated the total turnover of the seafood sector in the EU to be around €27 billion (AIPEC-CEP 2015). The contribution of EU aquaculture producers and capture fisheries to this was, respectively, €4 billion and €7 billion (STECF 2013a, b). The remainder of the €16 billion mostly comprises the value of imports. This value is doubled, however, once retailing and food service are taken into account. The European Market Observatory for Fisheries and Aquaculture Products estimated total expenditure on fish and seafood products in the EU in 2011 to be €52.2 billion (EUMOFA 2014). Their figures also suggest the value of imports to be €19.2 billion in 2012. These figures can best be understood through value chain analysis.

The value of any product at the point of consumption can be considered to be the sum of the value added through many intermediate activities needed to produce it and deliver it to the consumer. Such value chains can be quite complex, but with respect to aquaculture production can be considered to consist of farming itself and associated upstream (suppliers) and downstream (processing and distribution) activities. The price of a product at any point is a combination of the cost of all the inputs required to reach that point in the value chain plus a contribution to company profits. Where each step is carried out by separate companies, there will be an element of profit taken at each step. If all the activities are carried out by a single company (vertical integration), the profit margin may only be applied at the final stage (Fig. 9).

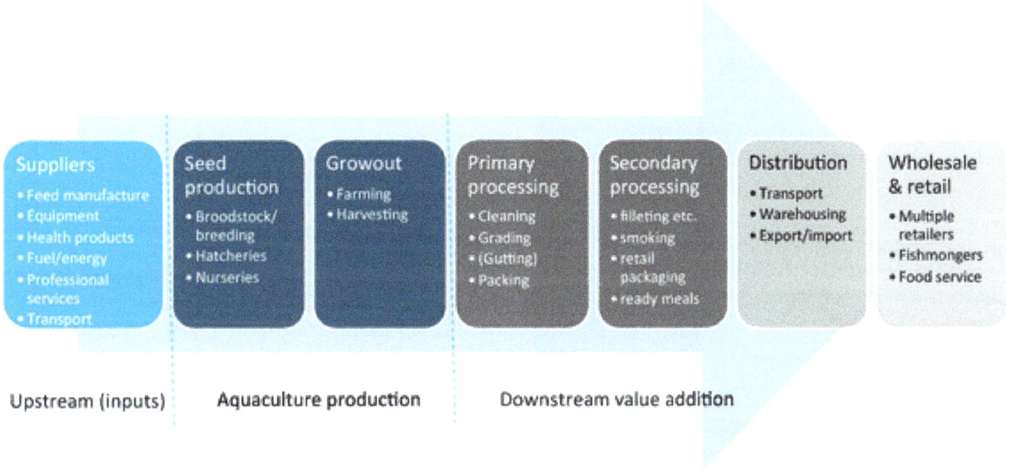

Figure 9: Generic value chain (simplified)

The value chain for aquaculture products varies regionally depending on the nature of the production process and the degree of processing prior to market. The simplest will be for shellfish that are marketed without substantive processing and packaging, whilst the more complex will be for salmon or other species that have substantial upstream inputs and downstream processing and distribution. Based on limited studies of seafood value chains (e.g. Gudmundsson et al. 2006; EUMOFA 2013) raw material (e.g. output from fishing or farming) accounts for between 15 and 40 % of the final retail price. Processing and distribution adds between 10 and 30 % to the final price, whilst retail costs and margins adds between 25 and 40 % to the final price. A specific example for farmed fish is a study by EUMOFA (2014) on fresh sea bream in Italy (i.e. no secondary processing) which found ex farm values to be around 40 % of the final retail price. Care needs to be taken in such an analysis to take account of wastage (e.g. in gutting and filleting) although where value is obtained from by-product that should also be considered. In the case of food service (e.g. institutional catering and restaurants), the ingredient cost will only represent 30–35 % of the price charged per meal,[1] so the cost of primary production (e.g. a farmed fish) may only represent 10–15 % of the final value of the product as sold to a consumer. For this reason, it is important to consider the whole value chain as changes in output from the production sector, for instance, can have a larger economic impact in the downstream activities than in the production sector itself. For instance, an analysis of the Scottish salmon farming industry (IMANI/SRSL 2014) estimated the turnover from primary production to be £537 (€660) million in 2012 with downstream processing, distribution, retail and catering adding a further £270 (€332) million within Scotland and over £860 million (€1.06 billion) across the whole UK value chain (total £1.4 (€1.72) billion). On the upstream side, feed is generally the most significant input for aquaculture. In the case of the Scottish salmon industry, this was estimated to account for between 50 and 60

% of production costs (IMANI/SRSL 2014). Other activities supplying aquaculture producers include engineering, transport and equipment supply companies, fuel supply and professional services.

Economic Assessment at the EU Level

Since many companies involved in both the upstream and downstream activities are also servicing other sectors, they are generally not fully reliant on aquaculture production for their turnover. In some cases, they could (and often do in the case of secondary processor, retailers and food service) substitute local aquaculture products with fisheries products or imported products. For this reason, a formal analysis of economic impact of an activity (e.g. aquaculture production) is considered in terms of direct impact (e.g. hatcheries, farms, primary processing and sales operations), indirect impact (suppliers and wider business transactions) and induced (expenditure of company employees in the broader economy, etc.). Economic input–output models are used to quantify these effects, usually within defined geographic boundaries. Few examples of this approach are available, but a study by Fisheries and Oceans Canada (2010) quotes a direct output value of the Canadian aquaculture industry in 2007 to be C$1.026 billion, whilst the gross output (including calculated indirect and induced impacts) was C$2.1 billion (i.e. over double). This multiplier effect is industry and region dependent and can be applied to other measures of economic value including employment and value added (discussed below). In the case of the aquaculture sector in Scotland for instance, the output multiplier for indirect effects is 1.7, whilst the multiplier for both indirect and induced effects is 1.9 (Scottish Government 2015—data for 2012).

Whilst a full value chain approach or the use of input–output models provides an important overall perspective on the value of aquaculture activities, obtaining the appropriate economic data can be difficult or expensive. Many studies (e.g. STECF 2014) focus only on direct impacts for which data are more readily available. The main economic indicator used so far is output which is the total value of the products from aquaculture—essentially the same as turnover. However, it is more meaningful in economic terms to consider the value that is added by a particular activity (e.g. the farming process) as gross value added (GVA). This is the value of the output minus the cost of purchased inputs (e.g. feed, medicines, fuel, services, etc.). The value that is added is therefore comprises employee costs, the cost of capital employed and the profit that is made. GVA values are therefore substantially lower than output values. Hence, the total GVA for the EU aquaculture sector in 2012 was reported to be €1.294 billion, whilst total turnover (output) was €4.365 billion (i.e. the multiplier from turnover to GVA is around 0.3) (STECF 2014). However, this varies by sector with shellfish averaging 0.54 and sea bass and sea bream less than 0.1. Shellfish provides a greater GVA in relation to output because there are no costs for feed. The GVA for sea bass and sea bream is low due to negative profitability in the sector at this time (2012).

A further economic indicator is profitability (measured as earnings before interest and taxes (EBIT) albeit with some variation in definition) and ratios of EBIT to turnover and capital employed. These are particularly useful for assessing the economic health of the sector and potential for reinvestment or tax revenues. STECF (2014) using data from 18 countries (excluding Greece and Poland) found the overall ratio of EBIT to turnover was around 8 % in 2012 and return on investment (EBIT divided by total assets) was 7 % (down from 9 % in 2011). Further analysis and perspectives on EU aquaculture sector profitability are provided by Guillen et al. (2015) using data from submitted company accounts and somewhat different means of calculation. This analysis by country and segment illustrates significant variability over time suggesting the need for relatively long-term financial planning on the part of companies.

The social economic value of the sector is generally measured in terms of the number of people employed and the normalized measure of FTE (full-time equivalent). The employment generated by aquaculture is mostly in rural and coastal regions with fewer economic opportunities and, hence, may have a greater social impact than an equivalent number of jobs in cities. A more precise measure of economic value is labour income, which is a measure of earnings by the people employed in the industry. STECF (2014) estimates total employment in EU aquaculture to be around 80,000 people and around 40,000 FTE (indicating a fairly high degree of part-time and seasonal work). Average wages per FTE (based on data from 19 countries) were €22,100 per annum in 2012, but this varied from €3100 in Bulgaria to €70,000 in Denmark. Total labour earnings in EU aquaculture were therefore around €884 million.

Economic Value by Sub-Sector

A detailed analysis of the economic performance of the EU aquaculture sector by species/system and country has been carried out by the EU Scientific, Technical and Economic Committee for Fisheries (STECF 2014) as far as data are available. Data from that report have been combined with the analysis of FAO FishStat data in order to provide a perspective using the species groupings and system types defined earlier.

Five countries accounted for around 78 % of the direct output value of EU aquaculture in 2012, the UK, France, Greece, Italy and Spain. For the UK, the main component is Atlantic salmon culture, whilst for France the output is dominated by shellfish production. Greece is predominantly sea bass and sea bream, whilst both Italy and Spain have more diverse aquaculture sectors involving shellfish and both marine and freshwater finfish (Fig. 10).

The importance of the UK (Scottish) salmon production by output value carries through to an analysis by system type with large marine cage systems (supplemented by some production of sea bass and sea bream in several Mediterranean countries) providing the greatest direct output value (€1.26 billion). This is followed by the marine shellfish sector and in particular suspended systems (e.g. mussel culture in Spain) (€0.84 billion). Interestingly, this achieves a higher GVA (€436 million) than large cage culture (€360 million) due to the lower purchased inputs for shellfish farming. Conversely, smaller marine cage systems for sea bass and bream and intensive systems for trout had lower GVAs (around 20 %) in relation to output value due to poor profitability in 2012 (Fig. 11).

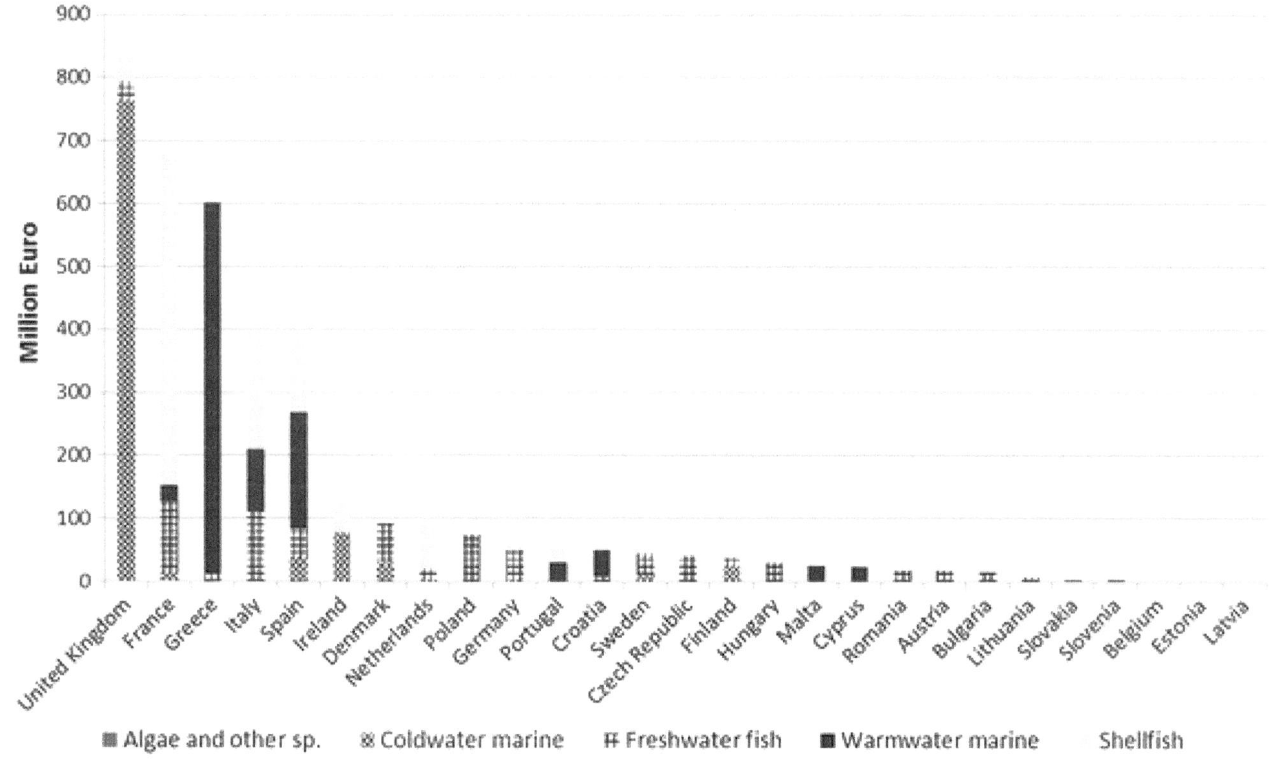

Figure 10: Direct output value of EU aquaculture production (2012) [*Source:* analysis of FishStat (FAO 2014b) data]

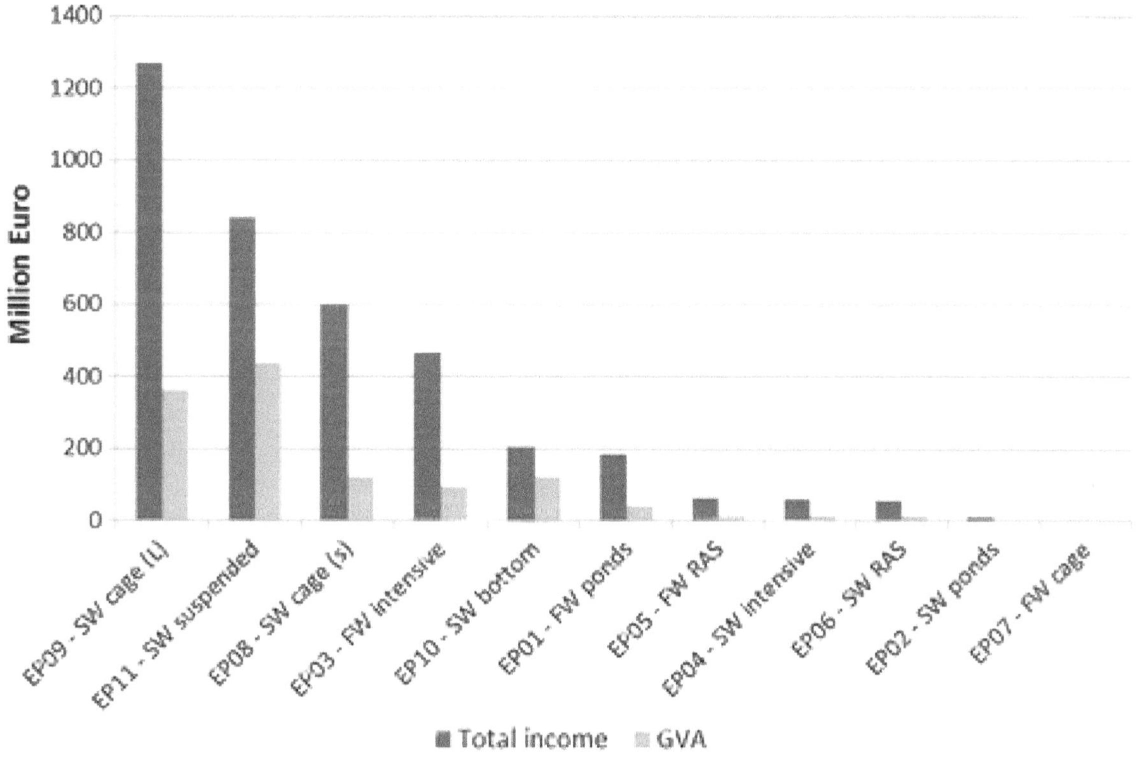

Figure 11 Output value and estimated GVA of EU aquaculture production by production system type (2012) [*Source:* based on FishStat (FAO 2014b) and STECF (2014) data]

Some analysis of employment patterns is possible using data from STECF, although data are not available for all countries or species/system combinations. Dividing output value by FTE employment provides an indication of labour productivity and can also be used to estimate the FTE employment by sub-sector. The highest productivities are seen in the Scottish salmon industry at around €490,000 output per FTE employee. On the basis of available data, the lowest productivity can be found in the freshwater pond sector with an output of only €14,000 per FTE employee. The shellfish sector is second lowest at around €70,000 per FTE. This means that these sectors provide more employment in relation to output and hence will have greater social impact. However, in practice it can mean low wages which can prompt the local population to seek higher returns from alternative occupations and increased dependency on immigrant labour from economically weaker countries (Fig. 12).

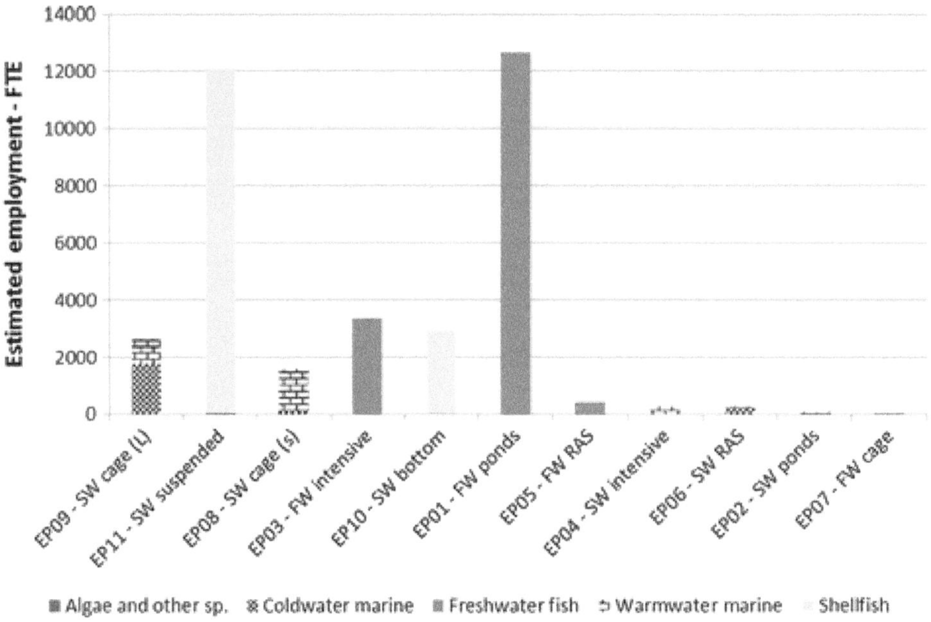

Figure 12: Estimated employment (FTE) in EU aquaculture by system type (2012) [*Source:*calculated using FishStat (FAO 2014b) value data and STECF data on employment. Note that limited data mean that the same employment ratios are applied to all shellfish systems and to minor fish systems]

The labour productivity figures are best in the sectors which are most invested in technology and larger-scale farming operations. Low productivity is associated with smaller-scale enterprises with relatively low mechanization and use of capital. Analysis carried out by STECF (2014) confirmed the EU production sector in general to be highly fragmented with 12,466 companies recorded (14–15,000 estimated) with aquaculture as their main activity in 2012. These have an average of 5.5 employees with only 9 % of companies having more than ten. In contrast, the salmon industry is increasingly consolidated. The number of active companies in Scotland has fallen from 132 in 1993 to 15 in 2013. These employ an average of 72 people (Marine Scotland Science 2014). The sea bass and bream sector in Southern Europe are following a similar trend.

Competitiveness of Aquaculture

The outputs from European aquaculture enterprises find a place on the market alongside products from capture fisheries and imported seafood (both captured and farmed) from around the world. This implies competition on price, quality and other attributes determined by consumer perceptions. Price is the dominant factor as conventional supply and demand economics predicts that demand falls as prices rise, whilst the incentive and then ability of producers to supply the market fall as prices fall. Price is the mechanism that balances supply and demand. As consumers have choice over their expenditure, and particularly in a commodity market such as unprocessed fish and seafood, potential market volumes for aquaculture producers are increasingly limited as production costs rise. This is illustrated in Fig. 13 which shows sales volumes of EU capture fish and aquaculture production in relation to wholesale price band (2012). It must be noted that the capture fisheries data include industrial fisheries for reduction to fishmeal (approximately 50 % of the €0–€1/kg price band) and that sales values recorded for aquaculture products will be higher than capture fisheries products due to most aquaculture prices including primary processing and distribution, whilst capture fisheries prices are generally based on landed auction prices. Note that higher price categories are included in the following charts as there are species falling into most of these categories, although production is sometimes too small to register on the chart.

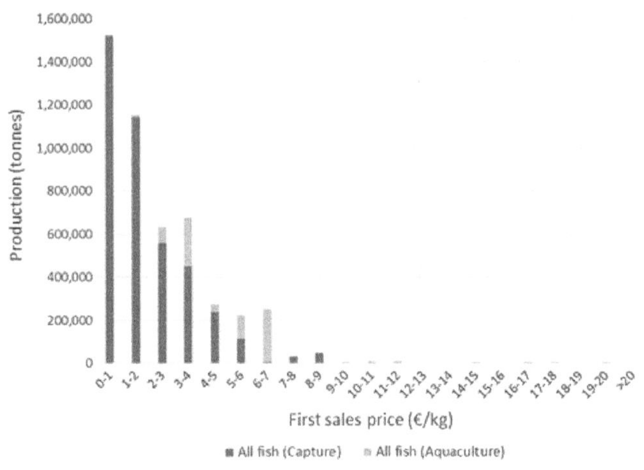

Figure 13: EU capture finfish landings and EU aquaculture fish production (2012) by price band [*Source:* calculated using FishStat production data and value data for aquaculture (FAO2014b) and capture fisheries prices from EUMOFA (http://www.eumofa.eu)]

Although the prices are not strictly comparable, it is interesting to note that over 50 % of EU finfish production has a first sale price below €2/kg, which is below the cost of production for most European fish species using existing aquaculture systems. Aquaculture starts to make a contribution to supplies as prices rise above €2/kg and particularly over €3/kg. It is dominant in the €6–7/kg category. This picture represents around 40 % of total EU supplies (due to imports), but gives some indication of overall market structure.

Breaking this down by sub-sector shows that freshwater aquaculture and shellfish farming are more competitive with capture fisheries products than finfish (Figs. 14, 15, 16, 17).

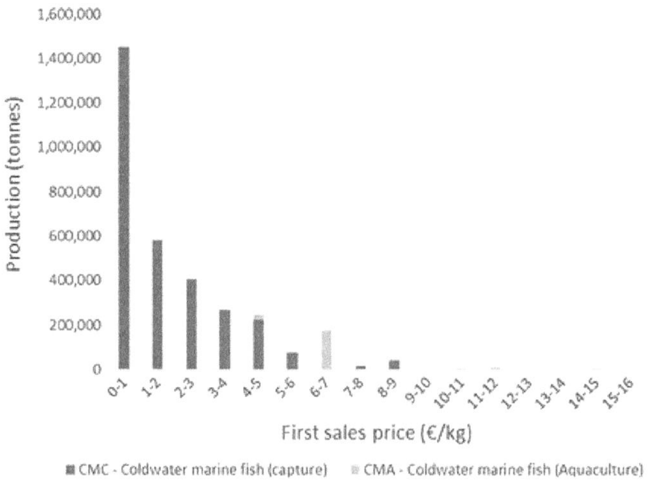

Figure 14: Price structure of EU capture fisheries and aquaculture production of coldwater marine fish species (2012). [*Source:* calculated using FishStat production data and calculated prices for aquaculture (FAO 2014b) and capture fisheries prices from EUMOFA (http://www.eumofa.eu)]

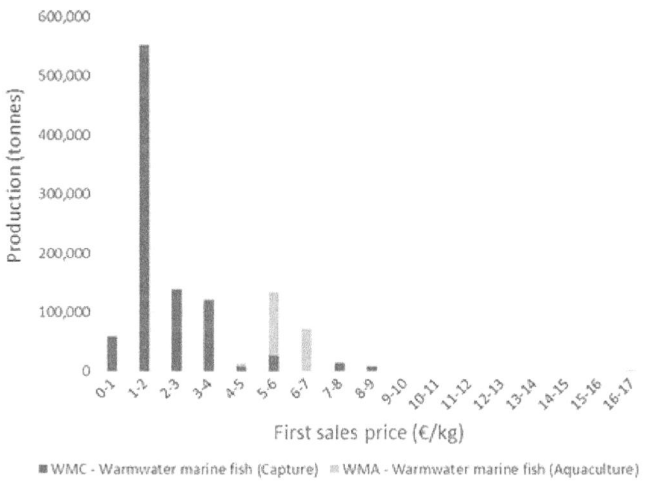

Figure 15: Price structure of EU capture fisheries and aquaculture production of warmwater marine fish species at first sale (2012) [*Source:* calculated using FishStat production data and calculated prices for aquaculture (FAO 2014b) and capture fisheries prices from EUMOFA (http://www.eumofa.eu)]

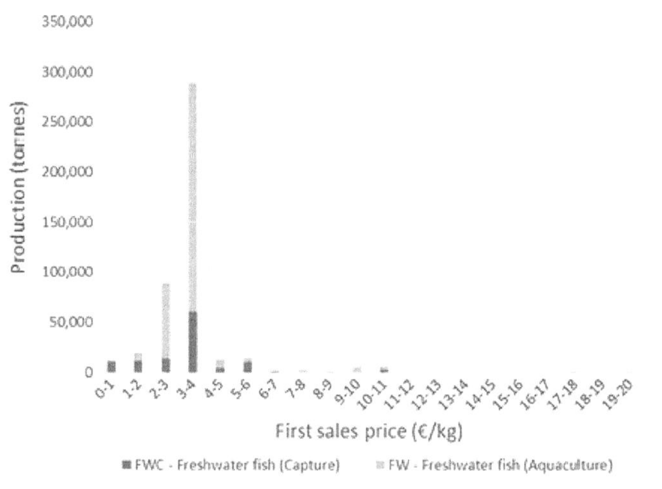

Figure 16: Price structure of EU capture fisheries and aquaculture production of freshwater fish species at first sale (2012) [*Source:* calculated using FishStat production data and calculated prices for aquaculture (FAO 2014b) and capture fisheries prices from EUMOFA (http://www.eumofa.eu)]

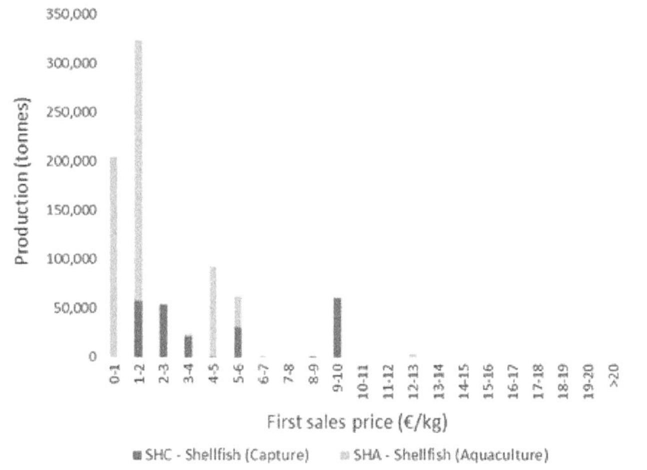

Figure 17: Price structure of EU capture fisheries and aquaculture production of shellfish species at first sale (2012) [*Source:* calculated using FishStat production data and calculated prices for aquaculture (FAO 2014b) and capture fisheries prices from EUMOFA (http://www.eumofa.eu)]

If products from aquaculture and capture fisheries readily substitute for each other, the volume of aquaculture production within the EU can be expected to be strongly influenced by its competitiveness on price in relation to other sources (capture fisheries and imports). Technical and management innovations that reduce the cost of production would be expected to increase competitiveness and increase the potential for EU production, whilst higher input and regulatory compliance costs should reduce the scope for EU production (Bostock et al. 2009). On this basis, there is the potential for aquaculture to be more competitive and eventually replace some of the capture fisheries supplies. This hypothesis was examined by Asche et al. (2001) which found little evidence of market interactions between species other than where both farmed and wild species were available together. Overall, it was considered that aquaculture production expanded the market for seafood. This conclusion was further strengthened in a later study (Asche et al. 2009) on the European whitefish market. However, stronger market interactions have been found in some cases. For instance, whilst Norman-Lopez (2009) found no interactions between farmed catfish and tilapia in North America, fresh farmed tilapia fillets were found to compete with wild whole red snapper wild fresh fillets of sea dab, and blackback flounder. Regnier and Bayramoglu (2014) found greater evidence of differentiation in the French fresh fish market between farmed and wild sea bass than between farmed and wild sea bream. They attributed this to sea bass being a higher value product with consumers being more willing to pay higher prices (in this case for wild caught fish) to make ethical choices with higher priced products. As aquaculture production overtakes capture fisheries for human consumption, a greater diversity of market segments and interactions can be anticipated particularly with respect to ethical issues relating to production methods (Natale et al. 2013).

POLICY SUPPORT MEASURES AND MECHANISMS FOR EUROPEAN AQUACULTURE AND THEIR INFLUENCE

Studies and Policy Development

Identifying challenges for sustainable development of aquaculture, and hence growth, has to refer to an extensive combination of economic, social and technical issues. Whilst some of these apply specifically to aquaculture, many are more general in nature, covering livestock rearing conditions, water use, environmental considerations and food supply in general. The complex interweaving of legislative and societal approaches that cover these issues tends to show that, until recently, aquaculture has been an 'add-on' that remained outside core policy considerations.

This is most clearly demonstrated by the history of the Common Fisheries Policy (CFP) where aquaculture was included primarily because its products are in the same market(s) as those of fisheries. This is reflected in the requirements of the oldest component of the CFP, the Common Organisation of the Markets for Fisheries and Aquaculture Products, which has existed since 1970. As aquaculture grew in the 1980s–1990s, both in Europe and globally, its higher level of visibility meant that both markets and legislation had to adapt to this new developing activity.

After different reforms, the CFP[2] in 2014 recognized aquaculture to be a key component of its scope, alongside traditional fisheries and seafood processing.

Nonetheless, whilst the CFP is the legislative instrument used for the management of European fish stocks that are a common resource/good, **aquaculture is predominantly an economic activity of the private sector, where the stock belongs to the operator**. This gives different responsibilities to both the operator and suppliers, which include:

1. Environmental respect and, in many cases, management, covering:

a. Water use and waste management

b. Adaptation to: for freshwater environments the Water Framework Directive, River Basin Management principles and for marine environments the Marine Strategy Framework Directive (2008), the Directive for Maritime Spatial Planning (2014) and, more generally, Natura2000 and the Habitats Directive.

2. Health and welfare of livestock (mainly covered by EFSA advice)

a. Standards for the protection of animals bred or kept for farming purposes (including fish)[3]

b. Aquatic Animal Health Directive[4]

c. Live Animal Transport Directive[5] (for stock movements [e.g. from hatcheries])

d. Stunning and killing

3. Feed composition

a. Awareness of the finite nature of fishmeal and fish oil sources

b. Ban on ruminant protein sources (following TSE and dioxin crises) although non-ruminant PAPs were reintroduced in 2013

c. Promotion of plant-based and alternative protein/oil sources

4. Processing standards

a. European food processing standards

b. Adherence to market-oriented quality and/or certification labels

Complementary policies (non-exhaustive) that influence aquaculture include the Integrated Maritime Policy, Blue Growth, the Circular Economy and the Bioeconomy.

Challenges to European aquaculture development are numerous, and the most important ones identified by Lane et al. (2014) are:

Strong market competition within the EU seafood market, particularly with imported seafood (fisheries and aquaculture), which has kept market prices down for several years.

Administrative burdens, particularly slow times to licensing (and the very number of licences/permits needed to operate), that restrict investments and expansion potential.

High labour and employment costs and working conditions (e.g. 35 h week).

Adequate access to bridging finance for working capital (investment in stocks).

These challenges are at the core of the European Commission's strategic guidelines for the sustainable development of European aquaculture,[6] where they were described in detail and specific targets for each of these were set for compliance by Member States and also for the Commission.

The European aquaculture industry agreed that the main challenges to progress are (EATIP2012):

Competition in the marketplace, principally from imports

Access to and competition for space for coastal and inland aquaculture

Maintaining health and welfare of livestock

Improving resource use (husbandry, feeds, farm technology)

Governance within the Common Fisheries Policy

Support for Research and Technological Development

The EU aquaculture sector has been supported through both national and EU-level research funding. EU funding is mainly through the RTD framework programmes, usually through primary research themes but also included in various cross-cutting programmes which have supported networking, mobility, access to research infrastructures and SME participation, etc. (Table 7).

Table 7: Major EC funding for aquaculture research and technological development

Framework programme	Main theme including aquaculture	Aquaculture priorities	Funding allocation
FP4 (1994–1998)	FAIR	The impact of environmental factors; ecological impact of aquaculture; biology of species and optimization of aquaculture; improved methodology for data collection and analysis	103 M ECU for fisheries and aquaculture
FP5 (1998–2002)	Quality of live (QoL)—key action 5, sustainable agriculture, fisheries and forestry and integrated development in rural areas including mountain areas	Sustainable production systems with the reduction of the impact on ecosystems and diversification of cultivated species (both plant and animals); improvement of production techniques, genetic improvement; disease resistance and control	520 M EUR for action 5 (sustainable agriculture and fisheries and forestry….)
FP6 (2002–2006)	Scientific support to policy (SSP—linked with the Common Fisheries Policy); Priority 5—'Food Quality and Safety' and Priority 6—'Sustainable Development'	Integration of environmental requirements into the CFP; sustainable aquaculture production; synthesis and dissemination of results to end-users; to support a better seafood production, improving the knowledge about diseases, health conditions, processing, etc	159 fisheries and/or aquaculture-related projects were funded with a total budget of 292.9 million euro. Of these, 75 projects (98 million euro) were aquaculture specific
FP7 (2007–2013)	Food, agriculture and biotechnology	Sustainable production and management of biological aquatic resources; fork to farm—seafood, health and well-being	1760 M euro on food, agriculture, fisheries and biotechnology
Horizon 2020 (2014–2020)	Food security, sustainable agriculture and forestry, marine, maritime and inland water research and the bioeconomy (under societal challenges pillar)	Domestication of new species; interactions with the environment; health and nutrition of farmed aquatic organisms; reproduction and breeding; social aspects of the seafood sector for the cohesion of the social fabric in the European coastal areas	4152 M euro for all food security including funding of the European Institute of Innovation and Technology

Sources: Impact Fish (2005), EC (2008),http://ec.europa.eu/research/horizon2020/pdf/press/horizon_2020_budget_constant_2011.pdf and Cordis (http://cordis.europa.eu/)

Research funding under the main RTD programmes has been increasingly directed by the European Commission to ensure it properly supports strategic and policy objectives. In responding to the needs for consultation on research needs in aquaculture, a multistakeholder technology platform was created in 2008, the European Aquaculture Technology and Innovation Platform (EATiP). Recognized as an official European Technology Platform (ETP)s, EATiP has to have a strategy, mobilization and dissemination function, encompassing.

- The preparation of an industry-focused **strategic research and innovation agendas**including technology roadmaps and implementation plans;
- encouraging **industry participation in Horizon 2020**, the EU's framework programme for research and innovation, and cooperating with networks in Member States;
- fostering **networking opportunities** with other ETPs and other partners along the value chain to address cross-sectoral challenges and promote the move

towards more open models of innovation;

- identifying **opportunities for international cooperation;**
- acting as one of the channels of external **advice** for the programming and implementation of Horizon 2020; notably, ETPs have been a key driving force behind the launch of high profile public–private partnerships under the programme.

EATiP provided a detailed Vision document, and a strategic research and innovation agenda (SRIA) was made in 2012, which identified specific sectoral needs—in terms of both legislation and research requirements. Its value for policy was recognized within the EC strategic guidelines for the sustainable development of European aquaculture, and several of the topics identified in its SRIA have become topics for the subject of calls in recent work programmes (Fig. 18).

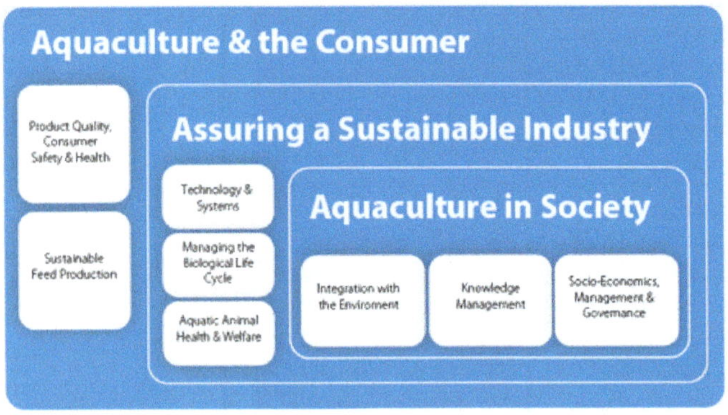

Figure 18: Thematic priorities identified by EATiP

Further inputs on strategic research priorities are provided by the European Fisheries and Aquaculture Research Organisation (EFARO—http://www.efaro.eu) and the COFASP project (Cooperation in Fisheries, Aquaculture & Seafood Processing—http://www.cofasp.eu).

Support for Aquaculture Industry Development

Direct support for aquaculture industry investment is provided through EU structural funds. The first of these was the Financial Instrument for Fisheries Guidance (FIFG) which ran over the period 1994–1999 and 2000–2006. These provided grants that were part-funded by national governments within a framework of multiannual support programmes with differing levels of contribution depending on country region. The total programmed expenditure under FIFG 1994–1999 was ECU 2.676 billion and for FIFG 2000–2006 €3.746 billion.[7] In the latter programme, €0.317 billion was provided to support a total investment of €1.199 billion in the aquaculture sector (EC 2010).

FIFG was followed by the European Fisheries Fund (EFF), which ran from 2007 to 2013. With a budget of €4.3 billion, its aim was to improve the competitiveness of the sector and help it become environmentally, economically and socially sustainable. Aquaculture, processing, marketing and inland fisheries made up Axis 2 of the fund, with aid being available for diversification into new aquaculture species with good market prospects, environmentally friendly aquaculture, public and animal health measures, processing and marketing and lifelong learning. Special provision existed for inland fishing, reflecting its importance in Central and Eastern Europe.

In its interim evaluation report of the EFF (Ernst and Young 2011), Ernst and Young reported that of a total of €1.24 billion programmed under Axis 2, €518 million (43 %) had been committed to 3556 projects across all Member States. Delays in implementation were (in general) stated to have been due to limited co-financing in an unfavourable economic environment, and the late launch of the programme, mostly due to delays in validation of the operating programmes.

Specifically under Axis 2, measure 2.1 (aquaculture) represented 27 % of the EFF spend, measure 2.2 (inland fishing) less than 1 % and measure 2.3 (fish processing and marketing) the vast majority with 72 %. Projects under measure 2.1 were principally focussed on productive investments, although many were reported as being constrained by environmental impact assessment, requirements. Investments in 'environment measures' were observed to be well implemented, but animal and public health measures were considered by many Member States to be unsuitable and not adapted to the reality in both the fish and shellfish sectors. Under measure 2.3, the majority of investments were in increasing production capacity, improving systems and improving hygiene and working conditions in processing. These reportedly produced positive impacts on employments, but the national evaluations did not enable an assessment on their effects on competitiveness and sustainability.

In its 6th annual report on EFF (COM (2013) 921 final), the European Commission reported that the expenditure in aquaculture measures had leveraged additional national public contribution of €183 million and a further €538 million of private funding. Hence €1 of EFF funding had a leveraging effect of €1.68.

The EFF has been superseded by the European Maritime and Fisheries Fund (EMFF) which will run from 2014 to 2020. This has a total EC budget of €6.4 billion (Table 8).[8] Each country is required to draw up an operational programme[9] which takes account of both the EU fisheries and aquaculture policies and national policies and priorities (in the case of aquaculture expressed through a National Aquaculture Plan). Applicants are required to show compliance with the relevant national operational programme, and as with previous programmes, there will be an element of national co-funding. A new element is Fisheries and Aquaculture Monitoring and Evaluation (FAME) which will be carried out by an expert support unit based in Brussels. This has the remit to (i) assist in monitoring and evaluating the implementation of the EMFF and provide the Commission with regular updates and analysis of the state of play of the implementation of the EMFF, and (ii) build capacity across the Member States and in the Commission on evaluation and monitoring methodologies, indicators and good practices.[10]

Table 8: Total EU allocations of European Maritime and Fisheries Fund 2014–2020 (unit €, current prices)

MS	2014	**2015**	**2016**	**2017**	**2018**	**2019**	**2020**	**Total**	**%**
BE	5,722,130	5,795,229	5,848,204	5,942,991	6,081,279	6,122,861	6,233,357	41,746,051	0.7
BG	12,071,289	12,225,498	12,337,253	12,537,214	12,828,942	12,916,663	13,149,763	88,066,622	1.5
CZ	4,263,975	4,318,446	4,357,922	4,428,555	4,531,602	4,562,588	4,644,927	31,108,015	0.5
DK	28,559,270	28,924,111	29,188,510	29,661,596	30,351,790	30,559,328	31,110,815	208,355,420	3.6
DE	30,100,054	30,484,577	30,763,242	31,261,850	31,989,281	32,208,016	32,789,256	219,596,276	3.8
EE	13,840,012	14,016,816	14,144,946	14,374,205	14,708,679	14,809,253	15,076,507	100,970,418	1.8
IE	20,231,798	20,490,256	20,677,561	21,012,701	21,501,645	21,648,669	22,039,349	147,601,979	2.6
EL	53,289,776	53,970,543	54,463,896	55,346,644	56,634,503	57,021,756	58,050,796	388,777,914	6.8
ES	159,223,336	161,257,387	162,731,468	165,369,007	169,216,972	170,374,037	173,448,682	1,161,620,889	20.2
FR	80,594,423	81,624,003	82,370,140	83,705,190	85,652,923	86,238,597	87,794,897	587,980,173	10.2
HR	34,629,786	35,072,176	35,392,777	35,966,420	36,803,321	37,054,974	37,723,684	252,643,138	4.4
IT	73,642,561	74,583,332	75,265,111	76,485,002	78,264,728	78,799,884	80,221,941	537,262,559	9.3
CY	5,443,762	5,513,306	5,563,703	5,653,880	5,785,440	5,824,999	5,930,119	39,715,209	0.7
LV	19,167,006	19,411,862	19,589,309	19,906,810	20,370,021	20,509,307	20,879,427	139,833,742	2.4
LT	8,694,653	8,805,725	8,886,220	9,030,247	9,240,371	9,303,555	9,471,451	63,432,222	1.1
HU	5,358,928	5,427,387	5,477,000	5,565,770	5,695,280	5,734,223	5,837,705	39,096,293	0.7
MT	3,101,540	3,141,162	3,169,876	3,221,253	3,296,208	3,318,746	3,378,637	22,627,422	0.4
NL	13,915,788	14,093,559	14,222,391	14,452,906	14,789,211	14,890,336	15,159,053	101,523,244	1.8
AT	954,693	966,888	975,727	991,541	1,014,613	1,021,551	1,039,987	6,965,000	0.1
PL	72,814,233	73,744,422	74,418,532	75,624,702	77,384,410	77,913,547	79,319,610	531,219,456	9.2
PT	53,797,969	54,485,229	54,983,288	55,874,453	57,174,593	57,565,539	58,604,393	392,485,464	6.8
RO	23,085,512	23,380,425	23,594,150	23,976,562	24,534,471	24,702,232	25,148,019	168,421,371	2.9
SI	3,400,584	3,444,026	3,475,509	3,531,839	3,614,022	3,638,734	3,704,400	24,809,114	0.4
SK	2,163,649	2,191,290	2,211,321	2,247,162	2,299,451	2,315,174	2,356,953	15,785,000	0.3
FI	10,197,069	10,327,335	10,421,739	10,590,653	10,837,087	10,911,188	11,108,097	74,393,168	1.3
SE	16,469,779	16,680,178	16,832,654	17,105,477	17,503,503	17,623,188	17,941,225	120,156,004	2.1
UK	33,327,114	33,752,863	34,061,403	34,613,468	35,418,887	35,661,073	36,304,629	243,139,437	4.2
EU 27[a]	788,060,689	798,128,031	805,423,852	818,478,098	837,523,233	843,250,018	858,467,679	5,749,331,600	100

Source: DG MARE

[a]LU is excluded, because it is not a recipient of EMFF

Future directions

At the present, the principle EU policy for aquaculture development remains the Common Fisheries Policy, supported by the Common Organisation of the Markets for Fisheries and Aquaculture Products and the financial instrument, the EMFF.

The submission of the National Aquaculture Plans foreseen in the strategic guidelines for the sustainable development of European aquaculture has been delayed, principally due to late finalization of the EMFF and the need to combine these multiannual plans with the operational programmes of each Member State. It is anticipated that these will be published in 2016.

From the consultation within EATiP, during the preparation of its Vision document, the following forward production projections were prepared from expert inputs and assessments for individual species and sectors (Table 9).

Table 9: EATiP exercise and forward production projections (a) coldwater marine assumptions, (b) warmwater marine assumptions, (c) fresh water marine assumptions, (d) shellfish assumptions, (e) summarized assumptions

	Forecast growth (tonnes)				Value 2010	Value 2030	Jobs		
	APR (%)	2010	**2020**	**2030**	**M€**	**M€**		**2010**	**2030**
(a) Coldwater marine assumptions									
Cod	4.0	1	1	2	€ 0.00	€ 0.01	Upstream	143	191
Flatfish	9.6	189	473	1,079	€ 1.21	€ 5.39	On-farm	951	1273
Large trout	2.0	30,220	36,838	44,025	€ 75.55	€ 110.06	Added-value (processing)	951	1273

Salmon	4.0	159,912	236,709	336,910	€ 495.73	€ 1044.42	Downstream	95	127
Subtotal		190,322	274,021	382,016	€ 572.49	€ 1159.88	Total jobs	2139	2865
Total increase	101.0			191,694		587.40	Feeds (tonnes)	285,561	458,419
(b) Warmwater marine assumptions									
Flatfish	5.0	10,102	16,455	25,527	€ 89.30	€ 225.66	Upstream	912	1358
Other	4.0	16,453	24,355	34,664	€ 151.70	€ 319.60	On-farm	6080	9051
Sea bass	4.0	74,907	110,881	157,818	€ 346.07	€ 729.12	Added-value (processing)	6080	9051
Sea bream	4.0	111,322	164,784	234,539	€ 445.29	€ 938.15	Downstream	608	905
Subtotal		212,784	316,474	452,548	€ 1,032.36	€ 2,212.54	Total jobs	13,679	20,365
Total increase	112.0			239,764		€ 1,180.18	Feeds (tonnes)	383,011	543,057
(c) Fresh water marine assumptions									
Carps	1.5	67,484	78,318	89,548	€ 131.59	€ 174.62	Upstream	1175	2259
Catfish	6.0	7,279	13,036	22,023	€ 10.41	€ 31.49	On-farm	12,653	15,060
Eels	5.0	6,561	10,687	16,579	€ 99.73	€ 252.01	Added-value (processing)	7,592	10,583
Other	3.0	1,988	2,672	3,486	€ 10.73	€ 18.82	Downstream	1265	1506
Sturgeon	8.0	1,928	4,162	8,320	€ 15.19	€ 65.56	Total jobs	22,686	29,409
Tilapias	25.0	130	1,211	9,021	€ 0.22	€ 15.33			
Trout	1.5	246,498	286,071	327,090	€ 601.45	€ 798.10			
Subtotal		331,868	396,156	476,068	€ 869.33	€ 1355.94	Total jobs	17,473	25,925
Total increase	43.0			144,200		€ 486.61	Feeds (tonnes)	305,658	367,977
(d) Shellfish assumptions									
Abalone	25.0	2	19	139	€ 0.10	€ 6.83	Upstream		
Carpet shells	4.0	37,800	55,953	79,639	€ 138.60	€ 292.02	€ On-farm		
Clams	10.0	1700	4,409	10,397	€ 3.92	€ 23.97	€ Added-value (processing)		
Cockles	5.0	2000	3,258	5,054	€ 5.47	€ 13.82	€ Downstream		
Mussels	1.0	490,000	541,265	591,973	€ 434.70	€ 525.17	Total jobs		
Oysters	1.5	121,800	141,354	161,623	€ 434.49	€ 576.54			
Scallops	25.0	16	149	1,110	€ 0.09	€ 6.42			
Subtotal		653,318	746,407	849,935	€ 1017.37	€ 1444.77	Total jobs		
Total increase	30.0			196,617		€ 427.39			

(e) Summarized assumptions	Growth	Forecast growth (tonnes)			**Value 2010**	**Value 2030**	**Jobs**		
	%	**2010**	**2020**	**2030**	**M€**	**M€**		**2010**	**2030**
Coldwater marine	101.0	190,322	274,021	382,016	€ 572.49	€ 1159.88	Coldwater marine	2139	2865
Freshwater	43.0	331,868	396,156	476,068	€ 869.33	€ 1355.94	Freshwater	22,686	29,409
Mediterranean	112.0	212,784	316,474	452,548	€ 1032.36	€ 2212.54	Mediterranean	13,679	20,365
Shellfish	30.0	653,318	746,407	849,935	€ 1017.37	€ 1444.77	Shellfish	–	–
Subtotal		1,388,292	1,733,058	2,160,566	3491.55	€ 6173.13	Total jobs	38,504	52,638
Total increase	55.6			772,275		€ 2681.58	Feeds (tonnes)	974,230	1,369,453

The total increase in volume and value (constant prices) of for 2030 compared to 2010 data

The job data are based on estimates taking into account technology developments and automation

The feeds required take into account expected advances in food conversion ratio (FCR)

The AquaInnova forecasts used here did not have complete data for job developments in the shellfish sector

Source: AquaInnova (http://www.eatip.eu/default.asp?SHORTCUT=405)/EATIP forecasts for scenarios in each sub-sector, re-formatted by the study authors

Based on our categorization of the sector by production technology (Tables 2, 3, 4, 5, 6) and by the outcomes of the consultations within EATiP and directly with the professional sector, an estimation on the trends (for 2030) of the use of technologies for each of the production sub-sectors was prepared. This is shown in Table 10. The arrow indicates the trends (upwards, stagnant and downward).

Table 10: Trends in prevalence of production technologies

Technology	Estimated 2012 production (tonnes)	Trend in use of technology 2030
Coldwater marine finfish		
Coastal pond aquaculture	180	↘
Indoor land-based recirculated aquaculture systems (marine)	100	↗
Intensive marine flow-through and partial recirculation systems (mostly large tanks)	9800	↗
Small cage systems—sheltered marine	11,000	↘
Large cage systems—marine in exposed sites, using mechanized (automated) systems	190,090	↗
Warmwater marine finfish		
Coastal pond aquaculture and valliculture	900	→
Indoor land-based recirculated aquaculture systems (marine)	9910	↗
Intensive marine flow-through and partial recirculation systems (mostly large tanks)	340	↗
Small cage systems—sheltered marine	102,420	↘
Large cage systems—marine in exposed sites, using mechanized (automated) systems	74,330	↗
Freshwater fish		
Freshwater pond aquaculture (extensive to semi-intensive)	86,350	↗
Intensive freshwater flow-through and partial recirculation systems (mostly tanks, raceways and small ponds)	169,285	↘
Indoor land-based recirculated aquaculture systems (freshwater)	9220	↗
Small cage systems—freshwater	250	→
Shellfish		
Marine bottom culture (non-fed sedentary and attached animals and plants)	65,440	→
Marine supported and suspended culture (non-fed sedentary and attached animals and plants)	537,050	↗

Source: estimated by the authors of this study

The main conclusions of this foresight exercise were as follows:

Productivity/competitiveness drive towards larger cages, particularly in offshore locations, seen as increasing location trend for both Mediterranean and coldwater farming

Coastal pond aquaculture will decline or stagnate, principally because of lower yields and competition for space (availability and licences)

Indoor marine recirculation systems will increase for hatcheries, but less likely for ongrowing (costs of investment, energy costs) due to cost comparison with cage production (Mediterranean and coldwater production) although may play a role in the development of some higher value species such as sole and shrimp

Freshwater pond production, stable or increasing, dependent on a combination of market demand, diversification activities and recognition of environmental services

Intensive flow-through systems for freshwater will probably decline, dependent on a combination of market demand, water availability and diversification towards specialized/niche markets (e.g. organic label) where lower intensity demanded

Freshwater recirculation systems will increase, notably for high-value production (sturgeon, pike-perch) and potentially for warmwater species that can be produced at high density (e.g. African catfish, eel, tilapia)

Shellfish production will continue to be dominated by supported/suspendedcultivation systems

It is anticipated by many stakeholders that an increasingly proportion of the expected increase in large cage culture for fish species and supported and suspended culture for shellfish species will be done in integrated multitrophic aquaculture (IMTA) systems, where species are combined (e.g. salmon, seaweeds, mussels) within a complementary area so as to best use space and to mitigate environmental impact. At this time, it is not possible to estimate the percentage production from such systems, since practicality, legal frameworks, licensing conditions and economic viability need to be elaborated by region or coastal area.

CONCLUSIONS

EU aquaculture production contributes the equivalent of 2.66 kg (live weight) of fish and seafood per EU citizen with a first sale value of €4.76 billion contributing a GVA of €1.29 billion and employing around 80,000 people (40,000 FTE) (STECF 2014). Whilst the value of EU aquaculture industry increased by 70 % between 2000 and 2012, actual volumes declined slightly. As the overall market for fish and seafood products in the EU is in the region of €27 billion (AIPEC-CEP 2015), the lack of output growth has been widely attributed to competitiveness issues including access to suitable sites (EC 2013). Recognizing this, and based on widespread recognition of the potential contributions that the sector could provide to European seafood supplies, the 2014 reform of the CFP provided a new impetus for the sustainable development of European aquaculture.[11] This was accompanied by a number of different studies and position papers on issues and challenges that affect directly or indirectly the potential for development. Consultations with industry stakeholders presented above indicate potential to responsibly increase output volume by 55 %, value by 77 % and employment by 40 % by 2030 if suitable social and economic conditions prevail. The overall strategy to achieve this sustainable development of (European) aquaculture is by maximizing the quality and health benefits of farmed products, whilst improving resource efficiency and minimizing impacts. Based on this strategy, and within the framework of the CFP, EU Member States are preparing national multiannual plans for aquaculture development. These will, for the first time, provide an overview of Member State intentions to grow aquaculture and the approaches that are prioritized to achieve this.

Monitoring progress is another challenge, and the STECF (2013a, b) study underlined the difficulties of obtaining accurate and up-to-date statistics on the performance of the EU Aquaculture Sector. However, the Data Collection Framework (DCF) continues to be implemented and should be adapted to include all forms of aquaculture, be it marine or freshwater. This was also a clear recommendation to the Commission from the European Court of Auditors (2014) when considering the effectiveness of the EFF support for aquaculture. With improved data it will be easier to understand developments within the sector and develop appropriate policies to support social and economic development.

There is increasing recognition that aquaculture can contribute to additional EU policies and strategies, including Blue and Green Growth and the Bio economy, the use of sustainable resources, food security and public health—by providing sustainably produced, high quality and healthy food. Clarification of the position of aquaculture within the existing environmental legislation has been initiated with recommendations existing for Natura2000 and is under development for the EU Water Framework Directive and for the Marine Strategy Framework Directive. This is essential so as to boost the potential integration of the activity and the expansion of site availability. Assessment and agreed quantification of environmental services provided by aquaculture requires to be integrated, alongside potential incentives.

Finally there is a need to better understand market interactions and competitiveness for EU aquaculture products to guide future investment and support policies. The establishment of EUMOFA[12] is an important step in this direction together with the PrimeFish Horizon 2020 project[13] which is studying several key species and developing models that will help understand boom and bust cycles, the success factors for new product development and the interactions between key aquaculture, capture and imported whitefish species.

Taken together, these policy positions, strategies and project initiatives, particularly the requirement for National Aquaculture Plans, will form the basis of the planning required to move towards the next stage of EU aquaculture policy realization and implementation.

Footnotes

1 http://restaurants.about.com/od/menu/a/foodcost.htm.

2 Regulation (EU) No 1380/2013.

3 Directive 98/58/EC.

4 Directive 2006/88/EC.

5 Directive 95/29/EC.

6 COM(2013) 229.

7 http://www.eurocbc.org/page355.html.

8 http://ec.europa.eu/fisheries/documentation/publications/2015-cfp-funding_en.pdf.

9 Seehttp://ec.europa.eu/fisheries/cfp/emff/country-files/index_en.htm.

10 http://ec.europa.eu/fisheries/cfp/emff/fame/index_en.htm.

11 http://ec.europa.eu/fisheries/cfp/index_en.htm.

12 http://www.eumofa.eu.

13 www.primefish.eu.

REFERENCES

1. AIPCE-CEP (2015) Finfish study 2015. EU Fish Processors and Traders Association, Brussels.http://aipcecep.drupalgardens.com/content/white-fish-study
2. Asche F, Bjørndal T, Young JA (2001) Market interactions for aquaculture products. Aquac Econ Manag 5(5–6):303–318. doi:10.1080/13657300109380296CrossRef
3. Asche F, Roll KH, Trollvik T (2009) New aquaculture species—the whitefish market. Aquac Econ Manag 13(2):76–93. doi:10.1080/13657300902881641CrossRef
4. Bostock J, Murray F, Muir J, Telfer T, Lane A, Papanikos N, Papergeorgiou P, Alday-Sanz V (2009) European aquaculture competitiveness: Limitations and possible strategies. Report for the European Parliament, Directorate-General for Internal Policies, Policy Department B, Structural and Cohesion Policies, Fisheries.http://www.europarl.europa.eu/RegData/etudes/etudes/join/2009/431569/IPOL-PECH_

ET(2009)431569_EN.pdf

5. Bostock J, McAndrew B, Richards R, Jauncey K, Telfer T, Lorenzen K, Little D, Ross L, Handisyde N, Gatward I, Corner R (2010) Aquaculture: global status and trends. Philos Trans R Soc B 365(1554): 2897–2912.http://rstb.royalsocietypublishing.org/content/365/1554/2897.full
6. Bronzi P, Rosenthal H, Gessner J (2011) Global sturgeon aquaculture production: an overview. J Appl Ichtyol 27(2):169–175. doi:10.1111/j.1439-0426.2011.01757.xCrossRef
7. Doukakis P, Pikitch EK, Rothschild A, DeSalle R, Amato G, Kolokotrionis S-O (2012) Testing the effectiveness of an international conservation agreement: Marketplace forensics and CITES caviar trade regulation. Plos One 7(7) e40907, 1–9. doi:10.1371/journal.pone.0040907
8. EATIP (2012) The future of European aquaculture—Our Vision: a strategic agenda for research & innovation. European Aquaculture Technology and Innovation Platform.http://www.eatip.eu/default.asp?SHORTCUT=92
9. EC (2008) Synopsis of fisheries & aquaculture research projects in the 6th Framework Programme. Directorate General for Maritime Affairs and Fisheries & Directorate General for Research. European Commission, Brussels. http://cordis.europa.eu/food/docs/2008-0699 en.pdf
10. EC (2010) Ex-post evaluation of the Financial Instrument for Fisheries Guidance (FIFG) 2000–2006. Executive summary. European Commission, Brussels. http://ec.europa.eu/fisheries/documentation/studies/fifg_evaluation/fifg_evaluation_executive_summary_en.pdf
11. EC (2013) Strategic Guidelines for the sustainable development of EU aquaculture. Communication from the Commission to The European Parliament, The Council, The European Economic and Social Committee and the Committee of the Regions. COM (2013) 299 final.http://ec.europa.eu/fisheries/cfp/aquaculture/official_documents/com_2013_229_en.pdf
12. Edwards P (1993) Environmental issues in integrated agriculture-aquaculture and wastewater fed systems. In: Pullin RSV, Rosenthal H, Maclean JL (eds) Environment and aquaculture in developing countries. Worldfish, pp 139–170
13. Ernst and Young (2011) Interim Evaluation of the European Fisheries Fund (2007 2013) Final report. http://ec.europa.eu/fisheries/documentation/studies/eff_interim_evaluation_en.pdf
14. EUMOFA (2013) Case study: price structure in the supply chain for fresh cod in the United Kingdom. European Market Observatory for fisheries and aquaculture products. European Commission.http://www.eumofa.eu/documents/20178/0/Price+structure+Cod+in+the+United+Kingdom/707cbc82-e17c-4331-b9a1-463fc5e77bb1
15. EUMOFA (2014) The EU fish market, 2014th edn. European Market Observatory for fisheries and aquaculture products, European Commission
16. European Court of Auditors (2014) The effectiveness of European Fisheries Fund support for aquaculture. Special report. Publications Office of the European Union, Luxembourg. doi:10.2865/24917.http://www.eca.europa.eu/Lists/ECADocuments/SR14_10/QJAB14010ENC.pdf
17. Eurostat (2011) Fishery statistics. European Commission. http://ec.europa.eu/eurostat/statistics-explained/index.php/Fishery_statistics
18. FAO (2014a) FishStat—Software for fishery statistical time series. United Nations Food and Agriculture Organisation, Rome. http://www.fao.org/fishery/statistics/software/fishstatj/en
19. FAO (2014b) The state of world fisheries and aquaculture—opportunities and challenges. Food and Agriculture Organization of the United Nations. UN FAO, Rome. http://www.fao.org/3/a-i3720e.pdf
20. FEAP (2014) European aquaculture production report 2001–2013. Federation of European Aquaculture Producers. http://www.feap.info/shortcut.asp?FILE=1274
21. Fisheries and Oceans Canada (2010) Socio-economic impact of aquaculture in Canada. Report prepared by GardnerPinfold Consulting Economists Ltd. Fisheries and Oceans Canada, Ottawa, DFO/2010-1645. http://www.dfo-mpo.gc.ca/aquaculture/ref/aqua-es2009-eng.pdf
22. Gudmundsson E, Asche F, Nielsen M (2006) Revenue distribution through the seafood value chain. FAO Fisheries Circular No. 1019, FIIU/C1019 (En). Food and Agriculture Organization of the United Nations, Rome. http://www.fao.org/docrep/009/a0564e/a0564e00.htm
23. Guillen J, Natale F, Fernández Polanco JM (2015) Estimating the economic performance of the EU aquaculture sector. Aquacult Int 23(6):1–14. doi:10.1007/s10499-015-9891-xCrossRef
24. IMANI/SRSL (2014) An assessment of the benefits of aquaculture to Scotland. Marine Scotland, Government of Scotland. http://www.gov.scot/Resource/0045/00450799.pdf
25. Impact Fish (2005) Impact assessment of the FP4 and FP5 research programmes on fisheries, aquaculture and seafood processing research area and the fisheries industry. EC FP6 Project (FP6-2003-SSP3-513651) involving the European Aquaculture Society and Oceanic Development. http://www.easonline.org/component/content/article/1/7-impact-of-eu-research
26. Jensen F, Nielsen M, Nielsen R (2014) Increased competition for aquaculture from fisheries: Does improved fisheries management limit aquaculture growth? Fish Res 159:25–33. doi:10.1016/j.fishres.2014.05.004CrossRef
27. Klinger D, Naylor R (2012) Searching for solutions in aquaculture: charting a sustainable course. Annu Rev Environ Resour 37:247–276. doi:10.1146/annurev-environ-021111-161531CrossRef
28. Lane A, Hough C, Bostock J (2014) The long-term economic and ecologic impact of larger sustainable aquaculture. Study for the European Parliament, Directorate General for Internal Policies, Policy Department B, Structural and Cohesion Policies—Fisheries.http://www.europarl.europa.eu/RegData/etudes/STUD/2014/529084/IPOL_STU(2014)529084_

EN.pdf

29. Lazard J, Lemcomte Y, Stomal B, Weigel JY (1991) Pisciculture en Afrique Subsaharienne: Situations et Projects dans les Pays Francophones. Propositions d'Action, Ministere de Cooperation et de Developpement, Paris

30. Marine Scotland Science (2014) Scottish Fish Farm Production Survey 2013. The Scottish Government. http://www.gov.scot/Publications/2014/10/7776/0

31. Martenot C, Oden E, Travaillé E, Malas JP, Houssin M (2011) Detection of different variants of Ostreid Herpesvirus 1 in the Pacific oyster, Crassostrea gigas between 2008 and 2010. Virus Res 160(1-2): 25–31. doi:10.1016/j.virusres.2011.04.012

32. Martins CIM, Eding EH, Verdegem MCJ, Heinsbroek LTN, Schneider O, Blancheton JP, Roque d'Orbcastel E, Verreth JAJ (2010) New developments in recirculating aquaculture systems in Europe: a perspective on environmental sustainability. Aquac Eng 43(3):83–93. doi:10.1016/j.aquaeng.2010.09.002CrossRef

33. Natale F, Hofherr J, Fiore G, Virtanen J (2013) Interactions between aquaculture and fisheries. Mar Policy 38:205–213. doi:10.1016/j.marpol.2012.05.037CrossRef

34. Norman-Lopez A (2009) Competition between different farmed and wild species: the US tilapia market. Mar Resour Econ 24(3):237–251. doi:10.5950/0738-1360-24.3.237CrossRef

35. Norwegian Directorate of Fisheries (2015) Aquaculture Statistics.http://www.fiskeridir.no/English/Aquaculture/Statistics

36. Olsen Y, Otterstad O, Duarte CM (2008) Status and future perspectives of marine aquaculture, Chap 10. In: Holmer M, Black K, Duarte CM, Marbà N, Karakassis I (eds) Aquaculture in the ecosystem. Springer, Netherlands, pp 293–319. doi:10.1007/978-1-4020-6810-2CrossRef

37. Petton B, Bruto M, James A, Labreuche Y, Alunno-Bruschia M, Le Roux F (2015) Crassostrea gigas mortality in France: the usual suspect, a herpes virus, may not be the killer in this polymicrobial opportunistic disease. Frontiers Microbiol 6:686. doi:10.3389/fmicb.2015.00686CrossRef

38. Regnier E and Bayramoglu B (2014) Competition between farmed and wild fish: the French sea bass and sea bream markets. In: Presented at 16th annual BIOECON conference on biodiversity, ecosystem services and sustainability. 21–23 September 2014, Kings College, Cambridge, UK. http://www.bioecon-network.org/pages/16th_2014/Bayramoglu.pdf

39. Renault T (2011) A review of mortality outbreaks in the Pacific oyster, *Crassostrea gigas*, reported since 2008 in various European Union Member States and the related implementation of Council Directive 2006/88/EC. Bulletin de l'OIE 2011 No 3: 51–52.http://www.oie.int/doc/ged/D11038.PDF

40. Scottish Government (2015) Input–Output Tables 1998–2012—Latest Year (2012).http://www.gov.scot/Topics/Statistics/Browse/Economy/Input-Output/Downloads/IO1998-2012Latest

41. STECF (2013) The economic performance of the EU aquaculture sector (STECF 13-29). JRC Scientific and Policy Reports. Scientific, Technical and Economic Committee for Fisheries (STECF). Publication Office of the European Union, Luxembourg. EUR 26366 EN, JRC 86671. doi:10.2788/41898

42. STECF (2013a) The 2013 Annual Economic Report on the EU Fishing Fleet (STECF-13-15). 2013. European Commission, Joint Research Centre, Scientific, Technical and Economic Committee for Fisheries (STECF). Publications Office of the European Union, Luxembourg, EUR 26158 EN, JRC 84745. doi:10.2788/23331

43. STECF (2014) The economic performance of the EU aquaculture sector (STECF 14-18). JRC Scientific and Policy Reports. Scientific, Technical and Economic Committee for Fisheries (STECF). Publication Office of the European Union, Luxembourg. EUR 27033 EN, JRC 93169. doi:10.2788/15501

44. Szücs I, Stündl L, Váradi L (2008) Carp farming in Central and Eastern Europe and a case study in multifunctional aquaculture, Chap 26. In: Leung PS, Lee CS, O'Bryen PJ (eds) Species and system selection for sustainable aquaculture. Wiley, New York, pp 389–413. doi:10.1002/9780470277867.ch26

45. Verdegem MCJ, Bosma RH, Verreth JAJ (2006) Reducing water use for animal production through aquaculture. Int J Water Resour Dev 22(1):101–113. doi:10.1080/07900620500405544CrossRef

46. Vezzulli L, Pezzati E, Stauder M, Stagnaro L, Venier P, Pruzzo C (2015) Aquatic ecology of the oyster pathogens *Vibrio splendidus* and *Vibrio aestuarianus*. Environ Microbiol 17(4):1065–1080. doi:10.1111/1462-2920.12484/pdf

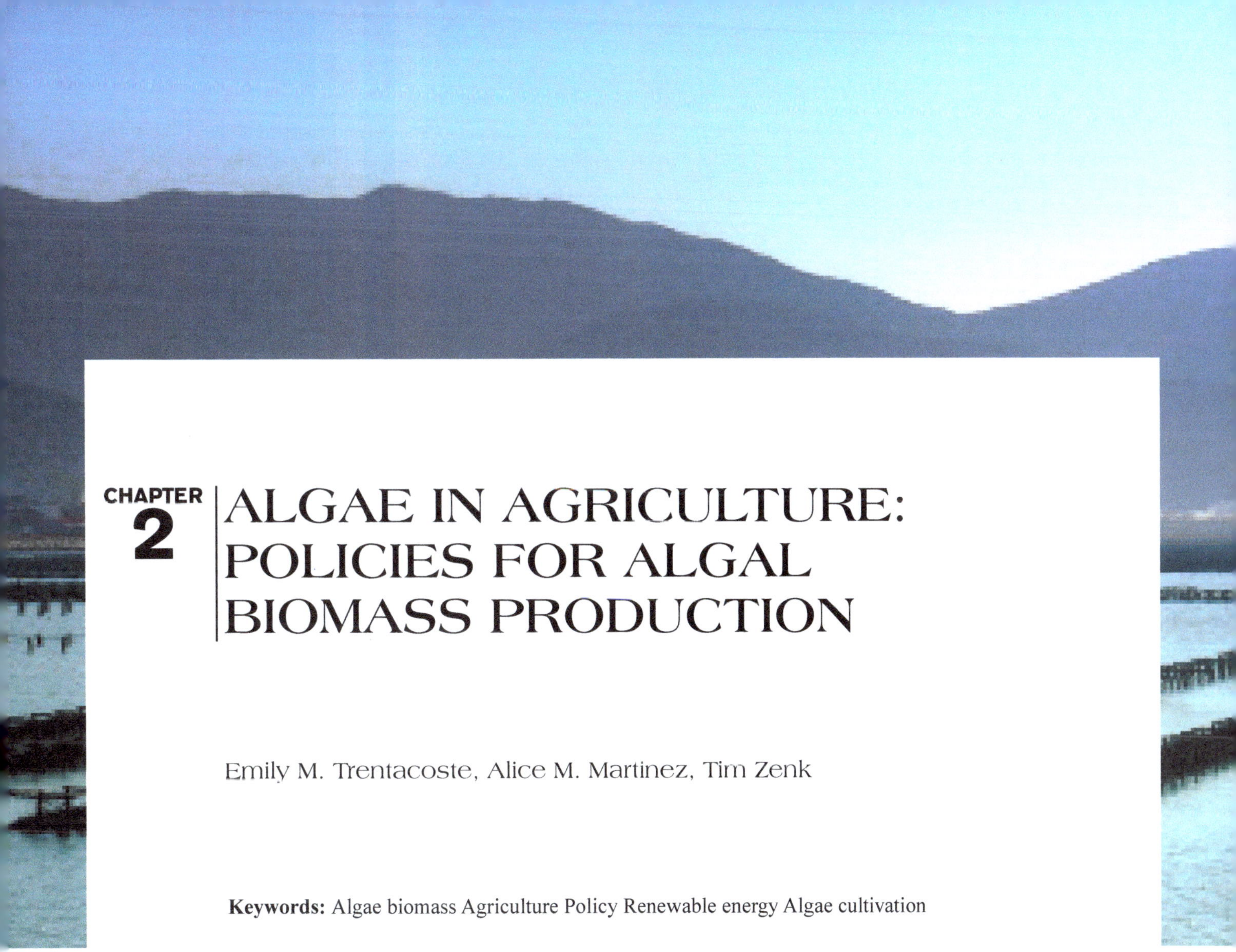

CHAPTER 2 ALGAE IN AGRICULTURE: POLICIES FOR ALGAL BIOMASS PRODUCTION

Emily M. Trentacoste, Alice M. Martinez, Tim Zenk

Keywords: Algae biomass Agriculture Policy Renewable energy Algae cultivation

INTRODUCTION

Algae are simple, photosynthetic, generally aquatic organisms that, like plants, use energy from sunlight to sequester carbon dioxide (CO_2) from the atmosphere into biomass through photosynthesis. Plants evolved from ancient algae ancestors, and the photosynthetic machinery in both plants and algae originally came from the same source: cyanobacteria (Falcón et al.2010; Fehling et al. 2007). Although algae and plants differ in many ways, the fundamental processes, such as photosynthesis, that make them so distinguished among Earth's organisms and valuable as crops, are the same.

Certain strains of algae have been used for anthropogenic purposes for thousands of years, including as supplements and nutraceuticals (Kiple and Ornelas 2000) and in the fertilization of rice paddies (Tung and Shen 1985). As early as the 1940s, other strains were identified as possible fuel sources (Borowitzka 2013a) because of their ability to produce fuel or fuel precursor molecules. Large-scale production and cultivation systems, including photobioreactors and outdoor open ponds, were developed in the early 1950s in the U.S., Germany, Japan, and the Netherlands (Borowitzka 2013b; Tamiya 1957). By the onset of the U.S. Department of Energy's (DOE) aquatic species program (ASP) in the U.S. in 1980, various species of microalgae and cyanobacteria were being produced and farmed on commercial scales around the world, and had been for over 20 years, mostly for the health food and nutritional supplement industries (Borowitzka 2013b).

Microalgae have evolved to be practically ubiquitous throughout the globe, and their varied distributions and evolutionary histories (Fehling et al. 2007) are reflected in extremely diverse metabolic capabilities between species (Andersen 2013). These diverse metabolisms produce a myriad of compounds with anthropogenic relevance including nutraceuticals, such as the carotenoids produced by *Dunaliella* and *Haematococcus* (Borowitzka 2013a, b), the polyunsaturated fatty acids (PUFAs) produced by various species (Ratledge 2004), and the high-value proteins and carbohydrates available in whole-cell supplements of *Spirulina*

Citation: Emily M. Trentacoste, Alice M. Martinez, Tim Zenk, algae in agriculture: policies for algal biomass production, DOI 10.1007/s11120-014-9985-8

and*Chlorella* (Görs et al. 2010; Khan et al. 2005). Some microalgae produce compounds of biotechnological interest including fluorescent compounds, such as phycoerythrin, and many produce isoprenoid molecules that can be used in food and over-the-counter products (Andersen 2013).

Microalgae have also been identified as attractive sources of biofuel because different species can produce a variety of fuel products. Various microalgal species have the ability to produce large quantities of lipid while sequestering CO_2, particularly neutral lipids in the form of triacylglycerol (TAG), which can be converted to fatty acid methyl esters (FAMEs), the main components of biodiesel (Hossain et al. 2008), through trans-esterification, or refined into other fuel constituents (Pienkos and Darzins 2009). Total lipids and other biomass constituents can be converted into crude oil alternatives through thermochemical processes such as hydrothermal liquefaction (Barreiro et al. 2013). Microalgal carbohydrates can be fermented into ethanol, and some species can produce biohydrogen (Radakovits et al. 2010). In addition to their diversity of products, microalgae are attractive as fuel sources because many species grow relatively fast compared to terrestrial plants and can be grown on brackish or saline water, thus avoiding the use of unsustainable quantities of freshwater, an increasingly limited resource (Dismukes et al. 2008). Table 1 provides an overview of some commercial algal products and potential sources.

Table 1: Commercial products from algae

Product	Use	Example source	Reference
β-Carotene	Supplement	*Dunaliella*	Lamers et al. (2008)
Astaxanthin	Supplement	*Haematococcus*	Lorenz and Cysewski (2000)
Whole-cell nutra-ceuticals	Supplement	*Spirulina*	Khan et al. (2005)
		Chlorella	Görs et al. (2010)
Aquaculture feed	Animal feed	*Tetraselmis*	Gladue and Maxey (1994)
		Isochrysis	Gladue and Maxey (1994)
Polyunsaturated fatty	Supplement	*Crypthecodinium*	Jiang et al. (1999)
acids (PUFAs)		*Shizochytrium*	Spolaore et al. (2006)
Phycoerythrin	Biotechnology	Red algae	Pulz and Gross (2004)
Fuel molecules	Energy	*Botryococcus*	Ashokkumar and Rengasamy (2012)
		Scenedesmus	Mandal and Mallick (2009)
		Neochloris	Gouveia et al. (2009)
Anticancer drugs	Pharmacueticals	*Symploca*	Coates et al. (2013)

Algaculture, or the farming of algae (Savage 2011), merges the requirements of traditional terrestrial plant agriculture such as sunlight, water, CO_2, nutrient inputs, and harvesting systems with additional aquaculture requirements such as self-contained aquatic systems, water quality, and waste disposal/recycling (Fig. 1). Because of their capability to produce commodities that span multiple markets, including those of health food, nutraceuticals, pharmaceuticals, animal feed, chemicals and energy, algae are uniquely versatile crops (Rosenberg et al. 2008). These diverse metabolic capabilities are due, in part, to the diversity of strains found within the algal lineage. Algae strains grown for food purposes, such as *Spirulina*, have a starkly different metabolic profile from strains grown for energy, such as *Scenedesmus*. The diversity of their end products, and their cultivation using both agriculture and aquaculture practices make algae unique among other agricultural products.

Figure 1: Algaculture in the U.S. Algaculture can take place in closed photobioreactors, like those of Algenol in Florida (a) and Solix Biosystems in Colorado (b), or in open ponds like those of Sapphire Energy, Inc. in New Mexico (c). Like agriculture, algae cultivation requires growth as well as harvesting infrastructure, such as that of Sapphire Energy Inc. (d)

Despite significant overlap with both traditional agriculture and aquaculture (which Congress has defined as agriculture, including that of aquatic plants) (Food and Agriculture Act of 1977,1977), algaculture has not yet been afforded an official position within Title 7 of the U.S. Code (USC) for Agriculture. There are currently a number of other crops that share commonalities with algae in their cultivation practices or diversity of end-use markets, but these have all been designated a place within Title 7. For example, the commercial cultivation of aquatic plants, such as seagrass, is eligible for a diverse array of agricultural programs. Similarly, the farming of terrestrial crops for renewable energy, which shares the same end market and purpose as many algal-farming operations, benefits from its definition as agriculture.

Funding for research and development of algal biomass cultivation has increased over the last decade, and has led to the emergence of research programs, private projects, demonstration- and commercial-scale facilities across the U.S. (Fig. 2). The increase is primarily due to the growth of the algal biofuel industry in response to the demand for alternative fuel sources driven by the renewable fuel standards (RFS) (Tyner 2013). While the use of algae as functional food or feed ingredients is also on the rise (Ibañez and Cifuentes 2013), there are currently few federal program resources focused in this area. The production of algae for any end product is a two-phase process involving the farming and cultivation of algal biomass followed by processing of the harvested biomass. The ability of the algal biomass industry to access federal programs that support the agricultural phase is imperative for future growth. This report analyzes the place of algae in the current agricultural policy and funding landscape, and the opportunities and pitfalls that exist for algae within this policy framework.

Figure 2: Algae projects in the U.S. Algal biomass projects exist in almost every state in the U.S.*Blue pins* denote a research institution, *green* denote a private project or company

AGRICULTURAL PROGRAMS

Congress has legislated a number of renewable energy programs that can be applied to algae such as the Bioenergy Program for Advanced Biofuels, the Rural Energy for America Program, the Biomass Research and Development Initiative and various grants and loans established in the 2008 Farm Bill in section 9003 of the USC (Food, Conservation, & Energy Act of 2008, 2008). These programs, however, focus on research and development of algae for fuels at smaller scales. While this initial investment in research & development (R&D) is essential to build knowledge, expertise, and technology around algae, the industry is now entering the formative stage of large-scale commercialization, which requires broader coordination among federal agencies and support infrastructure to gain proper alignment at the federal and state level required for a successful industry.

Biomass Crop Assistance Program

The Biomass Crop Assistance Program (BCAP) was established in the 2008 farm bill (Food & Conservation Act of 2008, 2008) to financially assist farmers wishing to establish, produce, and deliver biomass feedstocks. BCAP's purpose is to promote farming of bioenergy crops. The program provides either one-time establishment payments, annual payments, or matching payments to help with harvest, storage, and transportation of biomass. Proposals for BCAP funding are submitted to the FSA and can come from either producers or conversion facilities (Schnepf 2011). While many traditional biofuel crops are currently eligible for BCAP funding, such as switchgrass and most non-food biomass, the 2008 farm bill specifically excluded algae from participation in the matching payment side of BCAP but qualifies algae for establishment payments through BCAP (Food & Conservation Act of 2008, 2008).

Support Programs

Congress has appropriated numerous federal agencies, such as the USDA and DOE, funds and authorization to implement programs that aid and support development of agriculture and aquaculture resources (Table 2). Since the passage of the original Agricultural Adjustment Act of 1933, each subsequent farm bill has evolved to address rising relevant issues in agriculture. This frequently involves drafting new programs or expanding existing programs to the new developing technologies. The 1977 farm bill (Food & Agriculture Act of 1977, 1977) expanded the definition of agriculture to include aquaculture, thus spurring the development of industry in the U.S. The 2002 farm bill was the first to include a title (9003) on energy (Farm Security & Rural Investment Act of 2002, 2002), enabling the initial research and development of biofuels and bioenergy and set the stage for bio-based energy standards in the 2005 and 2007 energy bills.

Table 2: Overview of federal support programs

Program	Description	Program Crops	Specialty Crops	Aquaculture	Algae
Farm Service Agency					
Commodity Operations	Price and market support programs to purchase, deliver, dispose of designated commodities for domestic and foreign markets	●	●	●	○
Conservation Programs	Conservation related programs	●	●	●	◐
Direct/Counter Cyclical	Income support	●	●	●	◐
Disaster Assistance	Disaster assistance insurance	●	●	●	◐
Economic and Policy	Market development assistance	●	●	●	○
Biomass Crop Assistance	Production incentive	●	●	●	◐
Farm Loan Program	Operating and capital Loans	●	●	●	◐
Energy Assistance					
Biorefinergy Assistance	Loan guarantee/grant program to assist in the development of Advanced Biofuels	○	●	●	●
Bioenergy Program for Advanced Biofuels	Producer payments to expand production of Advanced Biofuels	○	●	○	●
Rural Energy For America	Loan and Grant program for individual farmers	●	●	●	◐
Biomass Research and Development Initiative	Research and development, demonstration of biofuels and biobased products	●	●	●	◐

Agricultural and energy support program provided by the Farm Service, USDA and DOE. Shaded circles signify all feedstocks within that crop group are eligible for a particular service, empty circles signify no feedstocks within that crop group are eligible, and half-shaded circles signify only certain feedstocks within that crop group are eligible. For example, farm service programs are only available for algal biomass feedstocks that are used to produce food or feed commodities

The current farm bill, primarily through the arm of the USDA and associated agencies, funds a large number of assistance programs for agriculture and aquaculture (Agricultural Act of 2014,2014). All of the major farm price and income support programs comprising the farm safety net are available only to the "program crops" of corn, cotton, wheat, tobacco, peanuts, rice, and some new oil crops such as sunflower and oilseed. The main farm safety net programs restricted to program crops include the Marketing Assistance Loan, Price Loss Coverage, and Agriculture Risk Coverage. Additional programs, such as the Feedstock Flexibility Program for sugar, also instill price control while simultaneously attempting to bridge the gap with biofuel producers looking to meet RFS standards. These programs ensure that market prices for program crops never fall below a certain limit and provide direct income support or revenue assistance. Farmers of specialty crops, such as fruits and vegetables, aquaculture crops, horticulture crops, and livestock are eligible for a range of support programs outside of the safety net. These programs provide extension services, loans, crop insurance, and incentives for improving environmental quality of farms (Mercier 2011).

Extension Services

Some of the most important benefits allotted to agriculture and aquaculture in the U.S. are research, teaching, and extension services. Extension services are some of the oldest programs in U.S. agriculture, dating back to the Smith-Lever Act of 1914 that established a link between universities and the USDA (Smith-Lever Act 1914). The purpose of the programs has always been to (1) develop applications for agricultural research and (2) provide instruction on agricultural technologies to farmers. Today, the Cooperative Extension Service program of the USDA provides funding through the National Institute of Food and Agriculture to support programs that connect scientific agricultural research with local farmers. Extension services are administered through regional offices that bring expertise from land-grant universities to local levels to instruct farmers in emerging technologies that can increase productivity.

Extension services are essential for disseminating information about innovative research and technologies throughout the agricultural industry. They also play an extremely important role in providing more immediate assistance to issues faced by local farmers and in developing plans that address regional problems. The application of USDA's extension services to aquaculture in the 1981 farm bill was instrumental in expanding the industry and coordinating research and commercialization efforts (Agriculture and Food Act of 1981).

Federal Crop Insurance Programs

The additional support programs available for all farmers are important for the continuing success of non-program crops. These programs provide assistance for the development, commercialization, and continuation of farms and provide incentives for environmentally sound farming practices. The largest of these programs, in which all farmers (including those of aquaculture and livestock) can participate, is the crop insurance program. The original crop insurance program began in 1938 and only covered major crops (Agricultural Adjustment Act of1938, 1938), but the passing of the Federal Crop Insurance Act of 1980 expanded the program to be universal (Federal Crop Insurance Act of 1980, 1980). Crop insurance is run by the USDA Risk Management Agency (RMA) and paid for by the separate Federal Crop Insurance Corporation (FCIC).

Over 100 crops are currently eligible for the Federal Crop Insurance (FCI) program, in which farmers pay a subsidized premium for insurance delivered by private companies. While program crops are eligible for revenue-based loss insurance, specialty crops typically only participate in physical crop-loss insurance. If a crop is ineligible for the program, then it can still be insured through the Non-insured Crop Disasters Assistance program, established in the 1996 farm bill and run by the Farm Service Agency (FSA), which functions similarly to FCI (Federal Agriculture Improvement & Reform Act of 1996, 1996). Sea grass, a similar crop to algae that requires a blend of agriculture and aquaculture, is eligible for Non-Insured Crop Disasters Assistance (FSA 2011). Additional insurance support is available for all farmers to cover losses from natural disasters under the Supplemental Revenue Assurance Program. This program provides additional assistance beyond crop insurance to farmers who experience a decrease in revenue due to natural disasters and is only available for crops that are enrolled in one of the crop insurance programs.

The expansion of crop insurance programs to specialty crops, aquaculture, and livestock was important for the development and protection of these industries. Farms of these commodities are all affected by the same environmental factors as those of program crops, such as lower-than-expected production due to droughts, natural disasters, soil quality, water availability, etc. The farming of algae is equally susceptible to different but similar factors that affect biomass and crop yields.

Farm Loan Programs

Farm loans are essential in successful agriculture as up-front capital is needed to make purchases of inputs such as fertilizer, equipment, land, etc. Most farm loans are authorized by the Consolidated Farm and Rural Development Act (1961) and can be in the form of direct loans, guaranteed loans or emergency loans. Direct loans cover input purchases and farmland purchases, require farmers to complete financial training courses and are given preferentially to beginning farmers. Guaranteed loans are available in coordination with banks and emergency loans can help cover natural disasters.

Environment and Conservation Programs

Agriculture, aquaculture, and livestock farms have traditionally been eligible for a number of federal programs that incentive environmentally friendly practices and resource conservation. Most notable, the Environmental Quality Incentives Program (EQIP), introduced in the 1996 farm bill, provides technical and financial assistance to farmers to increase the environmental quality of their farmland. EQIP funds are distributed by states in competitive programs that focus either on innovation of novel conservation practices or water enhancement, including enhancing water quality and conservation. EQIP also works in partnership with farms to aid in farm design that promotes environmental quality and resource conservation.

The Conservation Stewardship Program (CSP) awards funds to farmers that have adopted uncompensated practices across their entire operation for overall conservation. To be eligible for CSP funds, farmers must be sustaining conservation of a certain resource and must demonstrate improvement and maintenance of conservation practices. Farmers can receive both EQIP support and CSP rewards. The final environmental program, the Agricultural Management Assistance (AMA) Program was established in the Agricultural Risk Protection Act of 2000 to address the fact that crop insurance is heavily concentrated among program crops in only a few states. The AMA provides assistance for conservation practices in a select 16 states.

The algae industry, which has most recently been associated with renewable energy production with the added constraints of reducing greenhouse gas emissions and being cost-competitive with fossil fuels, has already made substantial technological advances in freshwater conservation and nutrient recycling for commercial-scale production. In order to be categorized as advanced biofuel, the overall process of algal fuel production must represent a 50 % decrease in GHG emission compared to fossil fuels (Energy Independence & Security Act of 2007, 2007). A study conducted by the University of Virginia found that commercial scale production of algae-to-energy can result in a 68 % reduction in overall greenhouse gas emissions when compared to traditional fossil petroleum (Liu et al. 2013). Additionally, to increase economic feasibility, algae can be grown on non-potable saline or wastewater and nutrients can be recycled, drastically mitigating freshwater use and fertilizer inputs. The company BioProcess Algae, for example, has successfully utilized waste outputs of water, heat, and CO_2 from corn ethanol fermentation to cultivate algal biomass for various end products. Coupling algae cultivation with waste outputs from other industrial processes provides cost-effective and sustainable solutions to cultivation barriers.

Marketing Services

Agricultural products are frequently subjected to market analyses by the USDA such as economic and census reports. As the commercialization of algae progresses, market analyses will be advantageous to assess the strengths and weaknesses of the industry, the interplay between the agricultural and energy aspects of algae, and the outlook of the industry. The USDA also provides marketing assistance to farmers through financial assistance, research and promotion (AMS 2013). To successfully break into the agricultural market, algae would benefit from the marketing services available from the USDA.

State Programs

Defining the commercial cultivation of algae as agriculture provides opportunities at the state level as well. Many states offer additional loan and financing programs, especially for first-time farmers, such as "Aggie Bonds" that encourage private lenders to loan to beginning farmers (CDFA 2005). Beyond financial assistance, states can control laws associated with agricultural property and zoning. For example, the Ohio state legislatures recently defined algaculture as agriculture to allow use value assessments of algae cultivation land for tax purposes, thus lowering property taxes for land used for commercial algaculture (OH-H.R. 2012). The law additionally limits the authority of zoning laws to restrict algae cultivation on lands. Although decisions on specific investments in algae development are made at the regional and local levels, a federal initiative is still imperative to establish and influence direction and focus for the industry, as well as to develop guidance for new algae programs.

APPLICATION OF AGRICULTURAL PROGRAMS TO ALGAE

Opportunities currently exist for algae cultivation to expand commercialization within agriculture if it were defined as such. The most notable is the potential to fill a large void in agriculture of the use of non-arable land to produce renewable hydrocarbons and high value protein. Unlike terrestrial crops, algae do not require fertile soil or arable land for growth, thus expanding the areas of the country in which algae can be cultivated. Algae do require other inputs such as salt or freshwater, nutrients, and consistent year-round sunlight. Taking all of these factors into account, a recent study by the Pacific Northwest National Laboratory (PNNL) identified ~90,000 sites in the U.S. that would be suitable for algaculture, comprising ~5.5 % of the contiguous U.S. land mass and consisting predominantly of shrub/scrub landscape. These sites exclude any cropland, urban land, protected lands, wetlands, wilderness, or significantly sloping landscapes (Wigmosta et al. 2011). To compare, agricultural land currently utilizes over 40 % of the total U.S. land mass.

The USDA currently asserts jurisdiction of algae as an agricultural crop, and can potentially offer agricultural safety net programs to algal biomass companies. Despite the role of the USDA in overseeing agricultural programs for algae, barriers still exist to the application of these programs. Many of these barriers exist at the federal and state levels, and stem from lack of an overall national plan for the development of algaculture, from the overlapping jurisdictions of other federal agencies over different aspects of algae cultivation, (Fig. 3), and from the diverse end products generated by algae.

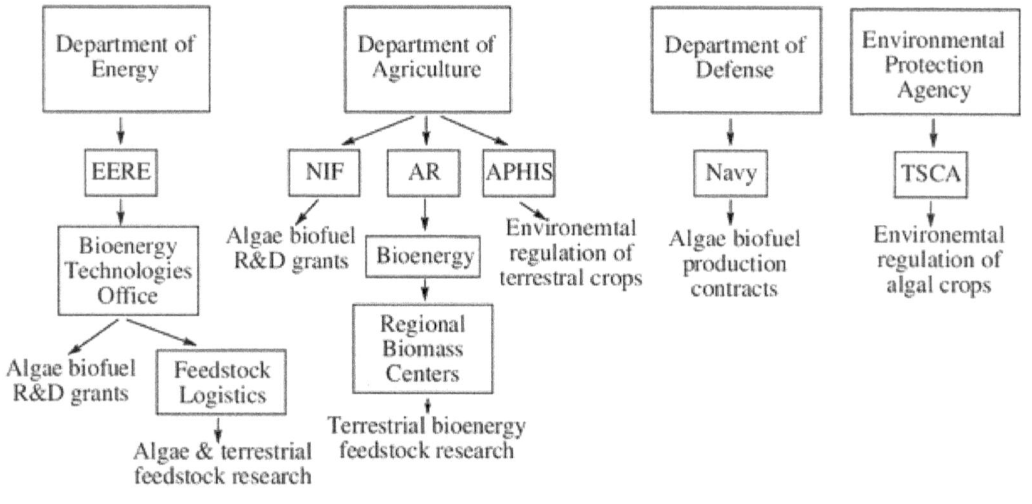

Figure 3: Federal agency jurisdiction over algae versus terrestrial crops. Four different federal departments hold jurisdiction over various aspects of algae cultivation, research, and products. *EERE* energy efficiency & renewable energy, *NIFA* National Institute of Food & Agriculture,

ARS Agricultural Research Service, *APHIS* Animal & Plant Health Inspection Service, *TSCA* toxic substance control act

Agencies that currently hold some responsibility over algae are the DOE, USDA, DOD, and EPA. The DOE has been involved in algae biofuel research since the onset of the 25-year long ASP in 1980 and has done extensive research on both algal biology and large-scale cultivation under its Biomass Program (Sheehan et al. 1998). Findings have been reported in both the ASP close-out report and the National Algal Biofuels Technology Roadmap (U.S. DOE 2010). The DOE also appropriates funding for grants and loans to industry and academic partners doing algae biofuel R&D. The DOD appropriates R&D grants and participates in demonstrations for algal biofuel use. It has currently entered contracts for developing commercial-scale production. While the USDA is responsible for regulatory oversight and approval, biotechnology and environmental regulation of genetically modified crops, the EPA has asserted jurisdiction for the permitting of genetically engineered algae varieties under its Toxic Substance Control Act, further supporting the notion of uncoordinated and overlapping federal support and regulation of the algae industry. There are also statutory limitations for the USDA's support of algae. Existing law, although not defined well and left open to individual programs for interpretation, may have the ability to support algae when used to produce a feed or food; the same standard, however, is not applied to algae if the end product is used to produce energy. None of these inconsistencies exist for the program crops (e.g., corn); they qualify for the vast array of USDA assistance no matter what products they support.

The USDA asserts responsibilities for agricultural policies pertaining to algae, but the end-use of algae as an energy source has created uncertainty in the applicability of these policies to algae cultivation. While a clear case can be made for expanding these programs for algal biomass used for food and nutraceutical purposes, there are still holes in the existing framework to accommodate algal biomass grown for bioenergy purposes. Because algae are such unique crops in their diversity of end product potential, no precedent exists to determine if a particular algae cultivation facility is eligible for agricultural programs or not. The USDA currently has no clear methodology for evaluating algal biomass producers within the agricultural landscape.

The uncertainty in algae's eligibility under agriculture is further exacerbated by insufficient communication about algal policies between the USDA's national leadership and its state and regional offices. The USDA's work, including decisions on application of policies to various USDA state offices, is primarily carried out in the field through more local offices, but while the national office claims jurisdiction over algae, there is again no precedent for state offices to follow. For example, the USDA's five Regional Biomass Centers, which are designed to lead research in sustainable biomass production, currently specifically exclude algae to avoid DOE overlap (Steiner 2011). Extension services, such as those provided under the Smith-Lever Act, would be appropriate to link regional USDA centers with local institutions and algae cultivators to develop methodology for evaluating algal biomass production under the agricultural framework. Another notable barrier is the lack of an overall algae-specific plan to move algae past R&D and into the formative stages of commercialization. The DOE has written an algae-specific roadmap, but this is primarily a summary of technologies that were available at the time and directions for R&D, without specific suggestions for moving into development and commercial stages (U.S. DOE 2010). Since then, a number of reports have been published agreeing that commercialization of algae, particularly for biofuels, is feasible given certain improvements in the production process (NRC 2012; ANL et al. 2012). Furthermore, since these reports, many of these improvements have been made and technologies have been developed that successfully demonstrate the ability to sustainably cultivate and harvest algae on large scales. While continued R&D is imperative to maintain and drive such improvements in the overall production process, it is now more important than ever for federal agencies to map out the next stage of the scale-up process.

The overlapping jurisdiction of algae, lack of a national plan, and specifically the assumption of major responsibility by the DOE, has caused the focus of algal policies to primarily revolve around its downstream use for energy, and to overlook expansion of policies that would support its most basic properties as a crop. Consistent, long-term federal policies are essential for scaling up biomass production of algae for energy, carbohydrates, protein and many other products (U.S. DOE 2012). The farming of algae requires biology, cultivation, harvest, and biomass processing practices, modeled after agricultural systems, which require independent and unique support networks for commercialization from those required for the downstream conversion of biomass into fuel (such as extraction, conversion, and biorefining processes).

Looking Forward

While we have discussed the successes for algae in the U.S. agricultural framework and the pitfalls that still exist, we can also identify areas of progress. Individual states have taken initiative to pave the way in recognizing algae cultivation as agriculture. In 2012 two states, Arizona and Ohio, specifically amended their laws to define algaculture as part of agriculture. While these changes had different specific effects in each state, they were both carried out with the purpose of increasing investment in algaculture and attracting the industry to those states. In Ohio, the recognition of algae farming as agriculture allows land used for algae cultivation to be eligible for the same land use valuation as agriculture, thus allowing lower property taxes for algae farms. It also limits the authority of zoning laws to restrict algaculture on lands. The Ohio legislation was proposed with widespread support from many factions including the Farm Bureau, the Poultry Association and the Soybean Association (OH-H.R. 2012). In Arizona, state trust lands can now be leased for algaculture, and algae farmland is eligible for lower property taxes afforded to traditional farmland (AZ-HR 2012a, b). In 2013, Iowa also passed a similar bill defining land used for algal cultivation as agricultural (IA-H.R. 2013).

Arizona's bills have allowed for the development of a national test bed for algal biomass production, led by Arizona State University. This multi-regional private and public partnership, funded by the DOE, focuses on developing algae cultivation on large, economically relevant scales and involves coordination between facilities in Arizona, Ohio, California,

Hawaii, and Georgia. Other public–private partnerships include the California Center for Algal Biotechnology, which coordinates and promotes research, commercialization and public education projects.

CONCLUSIONS

Large-scale cultivation of algae, or algaculture, has existed for over half a century. More recently, algaculture for food and fuel purposes has begun the transition from R&D and pilot-scale operations to commercial-scale systems. It is crucial during this period that institutional frameworks (i.e., policies) support and promote development, and commercialization. While the U.S. government has supported the R&D stage of algaculture for biofuels over the last few decades, it is imperative that policies anticipate and stimulate the evolution of the industry to the next level.

Large-scale cultivation of algae merges the fundamental aspects of traditional agriculture and aquaculture. Despite this overlap, algaculture has not yet been afforded an official position within agriculture or the benefits associated with it. Recognition of algaculture as part of agriculture under the USDA at national, regional, and local levels will expand agricultural support and assistance programs to algae cultivation, thus encouraging progression of the industry. The U.S. is currently the world leader in algal biomass technology and hosts a disproportionate number of companies devoted to the industry (Fig. 4). Continued federal support and initiatives will provide the spark needed to drive algaculture into the next stage of commercialization.

Figure 4: The global algal biomass industry. Locations of algal biomass projects, production, and companies around the world

ACKNOWLEDGMENTS

Thanks to L. Purpuro for providing information. Thanks to S. Whitaker, W. Gerwick and M. Hildebrand for support. This work was performed while ET was supported by NIH Marine Biotechnology Training Grant Fellowship 5T32GM067550.

REFERENCES

1. Agricultural Act of 2014, Pub. L. no. 113-79, 128 Stat. 649 (2014)
2. Agricultural Adjustment Act of 1938, Pub. L. no. 75-430, 52 Stat. 31 (1938)
3. Agricultural Marketing Service (AMS) (2013) Commodity Areas. USDA Agricultural Marketing Service. http://www.ams.usda.gov/AMSv1.0. Accessed 7 April 2013
4. Agriculture & Food Act of 1981, Pub L. no. 97-98, 95 Stat. 1213 (1981)
5. Andersen RA (2013) The microalgal cell. In: Richmond A, Hu Q (eds) Handbook of microalgal culture: applied phycology and biotechnology, 2nd edn. Wiley, Oxford, pp 1–20
6. Argonne National Laboratory (ANL), National Renewable Energy Laboratory (NREL), Pacific Northwest National Laboratory (PNNL) (2012) Renewable Diesel from Algal Lipids: An integrated baseline for cost, emissions, and resource potential from a harmonized model. ANL/ESD/12-4; NREL/TP-5100-55431; PNNL-21437. Argonne, ANL; Golden, CO: NREL; Richland, WA: PNNL
7. Ashokkumar V, Rengasamy R (2012) Mass culture of Botryococcus braunii Kutz. under open raceway pond for biofuel production. Bioresour Technol 104:394–399
8. AZ-HR 2225, 50th Legislature, 2nd Sess, (2012a)

9. AZ-HR 2226, 50th Legislature, 2nd Sess (2012b)
10. Barreiro DL, Prins W, Ronsse F, Brilman W (2013) Hydrothermal liquefaction (HTL) of microalgae for biofuel production: state of the art review and future prospects. In: 20th Eur Biomass Conf. vol 53 pp 113–127
11. Borowitzka MA (2013a) High-value products from microalgae—their development and commercialisation. J Appl Phycol 25:743–756
12. Borowitzka MA (2013b) Energy from microalgae: a short history. In: Borowitzka MA, Moheimani NR (eds) Algae for biofuels and energy. Springer, Houten, pp 1–15
13. Coates RC, Trentacoste EM, Gerwick WH (2013) Bioactive and novel chemicals from microalgae. In: Richmond A, Hu Q (eds) Handbook of microalgal culture: applied phycology and biotechnology, 2nd edn. Wiley, Oxford, pp 504–531
14. Consolidated Farm & Rural Development Act of 1961, Pub. L. No. 87-128, 75 Stat. 294 (1961)
15. Council of Development Finance Agencies (CDFA) (2005) Aggie Bonds Fact Sheet. CDFA.http://www.cdfa.net/cdfa/cdfaweb.nsf/ordredirect.html?open&id=cdfafactsheetaggiebonds.html. Accessed 7 April 2013
16. Dismukes GC, Carrieri D, Bennette N, Ananyev GM, Posewitz MC (2008) Aquatic phototrophs: efficient alternatives to land-based crops for biofuels. Curr Opin Biotechnol 19:235–240. doi:10.1016/j.copbio.2008.05.007
17. Doe US (2012) Biomass multi-year program plan. Office of Energy Efficiency and Renewable Energy, Washington, DC
18. Energy Independence & Security Act of 2007, Pub. L. no. 110-140 (2007)
19. Falcón LI, Magallón S, Castillo A (2010) Dating the cyanobacterial ancestor of the chloroplast. ISME J 4:777–783
20. Farm Security & Rural Investment Act of 2002, Pub. L. no. 107-171, 116 Stat. 134 (2002)
21. Farm Service Agency (FSA) (2011) Noninsured Crop Disaster Assistance Program (NAP) for 2011 and Subsequent Years. USDA Farm Service Agency.http://www.fsa.usda.gov/Internet/FSA_File/nap_august_2011.pdf. Accessed 7 April 2013
22. Federal Agriculture Improvement & Reform Act of 1996, Pub. L. 104-127, 110 Stat. 888 (1996)
23. Federal Crop Insurance Act of 1980, Pub. L. No. 96-365, 94 Stat. 1312 (1980)
24. Fehling J, Stoecker D, Baldauf SL, Falkowski PG, Knoll AH (2007) Photosynthesis and the eukaryote tree of life. In: Falkowski PG, Knoll AH (eds) The evolution of primary producers in the sea. Academic Press, New York, pp 76–107
25. Food and Agriculture Act of 1977, Pub. L. No. 95-113, 91 Stat. 913 (1977)
26. Food, Conservation, & Energy Act of 2008. Pub. L. No. 110-234, 122 Stat. 923 (2008)
27. Gladue R, Maxey J (1994) Microalgal feeds for aquaculture. J Appl Phycol 6:131–141
28. Görs M, Schumann R, Hepperle D, Karsten U (2010) Quality analysis of commercial Chlorella products used as dietary supplement in human nutrition. J Appl Phycol 22:265–276
29. Gouveia L, Marques A, da Silva T, Reis A (2009) Neochloris oleabundans UTEX#1185: a suitable renewable lipid source for biofuel production. J Ind Microbiol Biotechnol 36:821–826
30. Hossain ABM, Salleh A, Boyce AN, Chowdhury P, Naqiuddin M (2008) Biodiesel fuel production from algae as renewable energy. Am J Biochem Biotechnol 4:250–254
31. IA-H.R. 632, 85th General Assembly (2013)
32. Ibañez E, Cifuentes A (2013) Benefits of using algae as natural sources of functional ingredients. J Sci Food Agric 93:703–709
33. Jiang Y, Chen F, Liang SZ (1999) Production potential of docosahexaenoic acid by the heterotrophic marine dinoflagellate *Crypthecodinium cohnii*. Process Biochem 34:633–637
34. Khan Z, Bhadouria P, Bisen PS (2005) Nutritional and therapeutic potential of spirulina. Curr Pharm Biotechnol 6:373–379
35. Kiple KF, Ornelas KC (2000) The Cambridge world history of food. Cambridge University Press, Cambridge
36. Lamers PP, Janssen M, De Vos RCH, Bino RJ, Wijffels RH (2008) Exploring and exploiting carotenoid accumulation in Dunaliella salina for cell-factory applications. Trends Biotecnol 26:631–638
37. Liu X, Saydah B, Eranki P, Colosi LM, Mitchell GB, Rhodes J et al (2013) Pilot-scale data provide enhanced estimates of the life cycle energy and emissions profile of algae biofuels produced via hydrothermal liquefaction. Bioresour Technol 148:163–171
38. Lorenz RT, Cysewski GR (2000) Commercial potential for Haematococcus microalgae as a natural source of astaxanthin. Trends Biotechnol 18:160–167
39. Mandal S, Mallick N (2009) Microalga Scenedesmus obliquus as a potential source for biodiesel production. Appl Microbiol Biotechnol 84:281–291
40. Mercier S (2011) Review of U.S. Farm Programs. AGree, Washington, DC
41. National Research Council (NRC) (2012) Sustainable development of algal biofuels in the United States. The National Academies Press, Washington, DC
42. OH-H.R. 276, 129th General Assembly (2012)
43. Pienkos PT, Darzins A (2009) The promise and challenges of microalgal-derived biofuels. Biofuels Bioprod Biorefin 3:431–440
44. Pulz O, Gross W (2004) Valuable products from biotechnology of microalgae. App Microbiol Biotechnol 65:635–648
45. Radakovits R, Jinkerson RE, Darzins A, Posewitz MC (2010) Genetic engineering of algae for enhanced biofuel production. Eukaryot Cell 9:486–501
46. Ratledge C (2004) Fatty acid biosynthesis in microorganisms being used for single cell oil production.

Rec Adv Lip Metab Relat Disord 86:807–815

47. Rosenberg JN, Oyler GA, Wilkinson L, Betenbaugh MJ (2008) A green light for engineered algae: redirecting metabolism to fuel a biotechnology revolution. Tissue Cell Pathw Eng 19:430–436

48. Savage N (2011) Algae: the scum solution. Nature 474:S15–S16

49. Schnepf R (2011) Renewable energy programs and the farm bill: status and issues. Congressional Research Service, 7-5700

50. Sheehan J, Dunahay T, Benemann J, Roessler P (1998) A Look Back at the U.S. Department of Energy's Aquatic Species Program—Biodiesel from Algae. National Renewable Energy Laboratory, Golden

51. Smith-Lever Act, ch. 79, 38 Stat. 372, 7 USC. 341 et. seq. (1914)

52. Spolaore P, Joannis-Cassan C, Duran E, Isambert A (2006) Commercial applications of microalgae. J Biosci Bioeng 101:87–96

53. Steiner JJ (2011) USDA Biomass Research Centers & ARS Contributions. USDA Agricultural Research Service. http://www.csrees.usda.gov/nea/technology/pdfs/11_bioenergy.pdf. Accessed 7 April 2013

54. Tamiya H (1957) Mass culture of algae. Annu Rev Plant Physiol 8:309–334

55. Tung HF, Shen TC (1985) Studies of the *Azolla pinnata*—Anabaena azollae symbiosis: concurrent growth of Azolla with rice. Aquat Bot 22:145–152

56. Tyner WE (2013) Policy update: the US renewable fuel standard up against the wall. Biofuels 4:475–477

57. US DOE (2010) National algal biofuels technology roadmap. Office of Energy Efficiency and Renewable Energy, Washington, DC

58. Wigmosta MS, Coleman AM, Skaggs RJ, Huesemann MH, Lane LJ (2011) National microalgae biofuel production potential and resource demand. Water Resour Res 47:W00H04

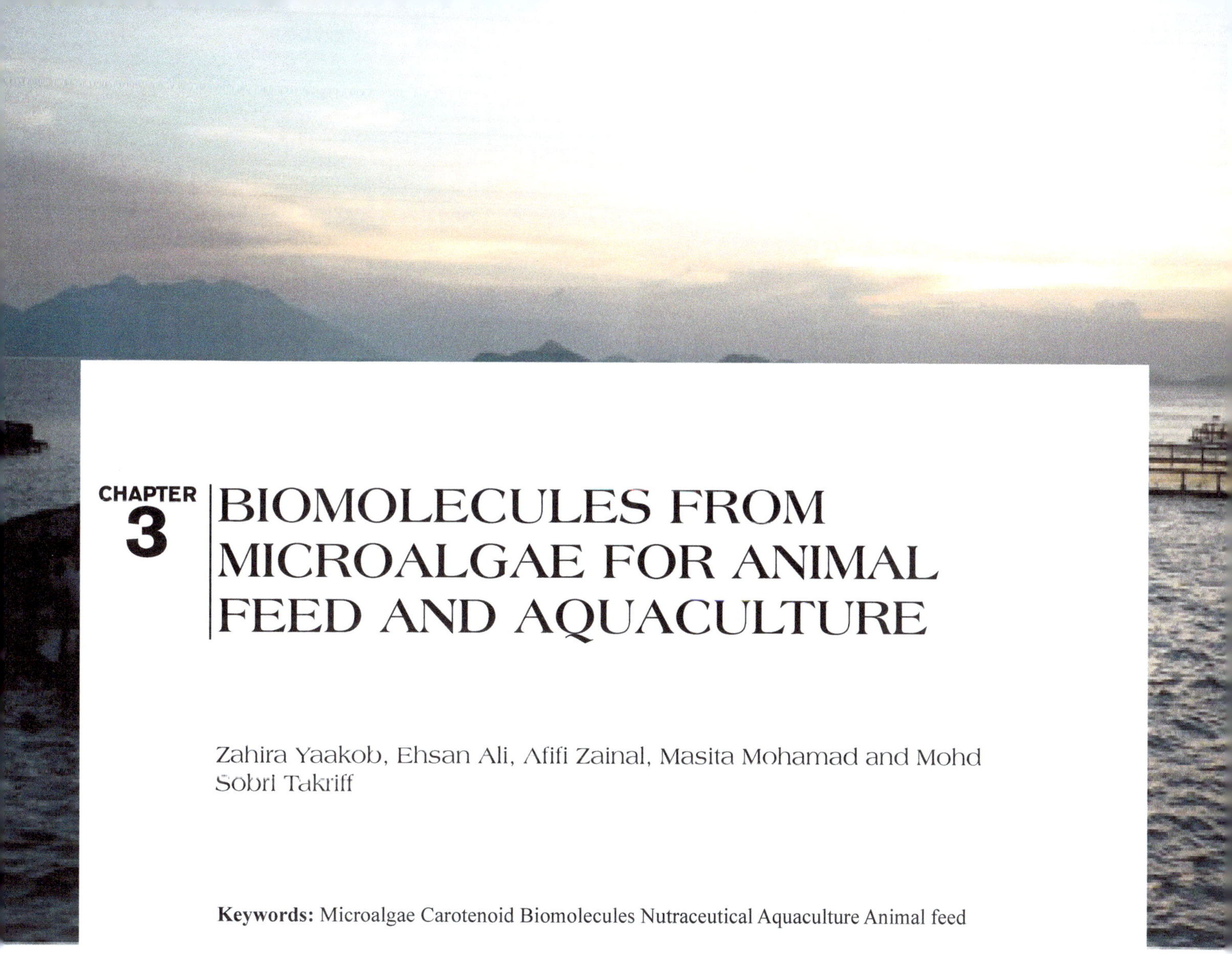

CHAPTER 3

BIOMOLECULES FROM MICROALGAE FOR ANIMAL FEED AND AQUACULTURE

Zahira Yaakob, Ehsan Ali, Afifi Zainal, Masita Mohamad and Mohd Sobri Takriff

Keywords: Microalgae Carotenoid Biomolecules Nutraceutical Aquaculture Animal feed

INTRODUCTION

Nutritional composition of algae is well documented and it mainly contains proteins, carbohydrates, lipids and trace nutrients, including vitamins, antioxidants, and trace elements. These all algal components have the characteristics to be a natural supplement in human and animal feed to replace synthetic components or meet the increasing demands of such ingredients. Increased consumption of animal proteins is creating a towering demand of high quality fish and animal feed to generate healthy proteins for humans. Fish meal is the protein source traditionally used in aquaculture diets but it is considered as limited and expensive. The nutritional contents of algae are rapidly gaining importance as a renewable source to substitute the conventional ingredients in the aquaculture/animal feed. Algae is defined as heterogeneous assemblage of organisms that range in size from tiny cells to giant seaweeds, thus algae mostly are photosynthetic species which includes eukaryotes and prokaryotic Cyanobacteria (blue-green algae) and live in aquatic habitats. Algae can be differentiated (by the body size and structure) either to microalgae (with algal bodies that need microscope to be observed) or to macroalgae (large enough to be seen with unaided eye) [1]. Algae culturing has been adopted since the late of 1800s and early 1900s, while microalgae mass culturing begin to develop since the achievement of Allen and Nelson in 1910 by cultivating *Chlorella* sp. for aquaculture purposes in Berlin, Germany [2]. To date, it seems that microalgae have the potential to play pivotal roles in order to remedy the energy, environment and food crisis prevailing in the world. This review encompasses the roles of algae in biomolecules production including astaxanthin, lutein, beta carotene, chlorophyll, phycobiliprotein, Polyunsaturated Fatty Acids (PUFAs), beta-1,3-glucan, and pharmaceutical and nutraceutical compounds. This review also emphasizes on the utilization of these biomolecules in livestock, poultry and aquaculture (and especially shrimp farming) feed.

Citation: Zahira Yaakob, Ehsan Ali, Afifi Zainal, Masita Mohamad and Mohd Sobri Takriff, biomolecules from microalgae for animal feed and aquaculture, DOI: 10.1186/2241-5793-21-6

REVIEW

Production of Value Added Products from Algae

The global carotenoids market is projected to reach US$1.2 billion by the year 2015, because of the rising consumer awareness about health benefits offered by various carotenoids, and the shift towards healthy and natural food products (Global Industry Analysts, Inchttp://www.strategyr.com/Carotenoids_Market_Report.asp/). According to a recently published report, global carotenoids market tends to reach US$1.3 billion by 2017 (http://www.prweb.com/pdfdownload/8849957.pdf). The nutraceuticals boom has also integrated carotenoid mainly on the claim of their proven antioxidant properties; carotenoids are usually considered important for industrial use in food products, cosmetics as vitamin supplements, health food products, and as feed additives for poultry, livestock, fish, and crustaceans [3]. In contrast, carotenoid in fish oil comes from microalgae, therefore it is logical to cultivate microalgae for carotenoids production [4]. Del Campo *et al.* [3] have reported that structure of more than 600 different carotenoids have been derived from a 40-carbon polyene chain. This paper will focus on carotenoids such as astaxanthin, lutein, beta-carotene and other high-valued molecules which are produced by microalgae (or in some case by macro algae) and Cyanobacteria and can be industrialized. Moreover, microalgae play a key role in high grade animal nutrition food from aquaculture to farm animals. Comprehensive nutritional and toxicological evaluations have demonstrated suitability of algal biomass as a valuable feed supplement or substitute for conventional animal feed sources [5]. Therefore, the usage of microalgae biomass for feed productions will be discussed in this review which might be helpful to encourage the trends of mass cultivation of microalgae for commercial and beneficial purposes.

Astaxanthin

Astaxanthin, a carotenoid equipped with two asymmetric carbon located at the 3 and 3′ position of the benzenoid rings on either end of the molecule, has been proposed as super vitamin E due to its natural antioxidant activities [6–8]. As a natural pigment, the major market for astaxanthin is as pigmentation source in aquaculture [9]. Natural astaxanthin can be produced by *Haematococcus* microalgae, *Chlorella zofingiensis*, *Chlorococcum* sp. and the yeast *Xanthophyllomyces dendrorhous* [10–13]. Studies regarding effects on lipid metabolism and antioxidant defense mechanisms revealed that astaxanthin is responsible for lowering the plasma lipid concentrations and enhancing antioxidant defense in apolipoprotein E knockout mice [14]. The increasing growth of salmonid farming has created an enormous demand for natural astaxanthin as pigment. Out of total carotenoid (astaxanthin, cantaxanthin), the astaxanthin was determined as the most important carotenoid in salmons and rainbow trouts [15]. Salmonid are unable to synthesize astaxanthin *de novo*, therefore carotenoid pigments must be supplied in their artificial aquaculture diet [16]. The carotenoid pigment astaxanthin has important applications and is routinely being used in the nutraceutical, cosmetics, food and feed industries [17]. In aquaculture feed, it has been used to culture salmon, shrimp, ornamental fish and sea bream. A recent report showed that *Haematococcus pluvialis*especially in dose of 3 g kg^{-1} feed administration may effectively enhance the antioxidant system and some biochemical parameters in rainbow trout [18]. Astaxanthin is not responsible only to provide pigmentation but also has been attributed with extraordinary potential for protecting the organism against a wide range of diseases, and has considerable potential and promising applications in human health [13]. Natural astaxanthin was also found effective to treat the patients with functional dyspepsia, caused by *Helicobacter pylori* [7]. UV radiation from sunlight is known as the most potent environmental risk factor in skin cancer pathogenesis, astaxanthin has been demonstrated as a protective tool against UVA-induced DNA alterations in human skin fibroblasts (1BR-3) [19].

Epidemiological studies have demonstrated a correlation between increased carotenoid intake and reduced incidence of coronary heart disease, certain cancer types, macular degeneration, and increased resistance to viral, bacterial, fungal and parasitic infection [20]. The price for astaxanthin varies in the market and depends on the percentage contents in algal source; however the overall maintained price of 5% astaxanthin is about US$1900 kg^{-1}(http://www.herbridge.com/bencandy.php?fid-55-id-23446-page-1.htm).

Lutein

Lutein is another important carotenoid which can be found naturally in algae. It is prominent in human serum and foods [21]. Lutein has been used for the pigmentation of animal tissues and products, and it is considered important for the natural coloration of foods, drugs and cosmetics. It has been documented that it plays an active role in delaying chronic diseases, stimulating the immune response, hampering the development of cataracts and the progression of early atherosclerosis, and preventing blindness or decrease in vision caused by age-related macular degeneration [22–24]. Global lutein market in 2004 accounted for US$139 million and was considered as the fastest projected growth in individual carotenoids sales [3]. The recent market value of lutein is ~ US$233 million (2010), and is expected to reach US$309 million by 2018 with a compound annual growth rate of 3.6% (http://www.companiesandmarkets.com/Market/Food-and-Drink/Market-Research/The-Global-Market-for-Carotenoids/RPT988273). French marigold (*Tagetes patula*) is currently the most widely applied source for lutein production in the world. In the USA, two lutein containing products, Aztec Marigold and Tagetes have been commercialized. Lutein is usually produced from three different marigold flowers (*Tagetes patula* L.) (marigold orange, marigold yellow, and marigold red) but the marigold orange (MGO) variety contains the maximum amount of lutein [25]. However, mass plantation of marigold occupies a large land area and it is easily influenced by seasonal and climatic change [24]. Although an established commercial system for the production of lutein from microalgae does not exist yet, the basis for outdoor production of lutein-rich cells of strains of *Muriellopsis* sp. and *Scenedesmus* sp. at a pilot scale have already been set up [3]. Recently, there has been remarkable interest in dietary carotenoids with respect to their antioxidant properties and their ability to reduce the chances of some chronic diseases which occurred because of the free radicals. Possibly, carotenoids protect cells from oxidative stress by quenching singlet oxygen damage with a variety of mechanisms. Therefore, algae derived carotenoids could be a leading natural resource in the search of potential functional ingredients [26]. The microalgae biomass from chlorophycean is also considered as the major source of lutein, especially *Chlorella* sp., *Scenedesmus* sp. and*Muriellopsis* sp. [3, 24]. Laboratory studies showed that heterotrophically cultivated *Chlorella*sp. produce considerable amount of lutein [27]. The ability of heterotrophic growth in fermenters makes *Chlorella* sp. a potential alternative resource for commercial production of lutein [24]. Studies showed that more lutein will be produced by *Chlorella* sp. using Basal Medium compared to Kuhl Medium because of the high concentration of EDTA and minerals (500 mg l^{-1}: 9.3 mg l^{-1}) [22]. On the other hand, Wu *et al.* [24] have modeled lutein production by heterotrophic *Chlorella* sp. in batch and fed-batch cultures as a potential alternative source for commercial production.

Beta-carotene

Beta-carotene is a pigment of increasing demand and a wide variety of market applications: as food coloring agent, pro-vitamin A (retinol) in food and animal feed, additive to cosmetics and multivitamin preparations, and as a health food product under the antioxidant category. The market value of beta-carotene was estimated about US$253 million in 2009 [3]. According to a report on the global market for carotenoids (September 2001), the market value of beta-carotene was estimated at around US$261 million in 2010 and this market is expected to grow to US$334 million by 2018 at a compound annual growth rate of 3% (http://www.companiesandmarkets.com/News/Chemicals/Beta-carotene-the-Largest-Carotenoid-Sector-is-Forecast-to-be-Worth-US-334-Mn-by-2018/NI2439).

Because of limited supply from natural sources, a number of companies have introduced synthetically produced carotenoids to meet the high demands in the global market. Beta-carotene, a provitamin A is the only carotenoid which has the potential to form two molecules of vitamin A (retinol) [28]. Natural production of beta-carotene can be produced by the microalgae genus *Dunaliella*, strains that have shown to change their color from green to orange may produce and accumulate large amount of cellular beta-carotene [29]. *Dunaliella* sp. occurs in oceans, brine lakes, salt marshes, salt lagoon and salt water ditches near sea, predominantly in water bodies containing more than 2 M salt and high level of magnesium. *Dunaliella* sp. is recognized as being the most halotolerant eukaryotic photosynthetic organism, showing a remarkable degree of adaptation to variety of salt concentrations from as low as 0.1 M to salt saturation about 4 M [30]. The unicellular halotolerant alga *Dunaliella bardawil*was shown to contain high concentration of beta-carotene, composed of the all-*trans* and the 9-*cis* isomers [28]. Beta-carotene is a structural analogue of vitamin A and its topical application can effectively be used to treat skin melasma [31]. It has been reported that a daily dose of*Dunaliella* sp. beta-carotene exerts a protective effect against exercise induced asthma probably through *in vivo* antioxidative effect [32]. Age-related macular degeneration is one of the most common eye diseases causing severe or permanent loss of vision, an antioxidant therapy using combinations of high dosage antioxidant vitamins C, E, beta-carotene, and zinc are employed [33, 34].

The price for natural beta-carotene depends on demand varying from US$300 to US$3000 kg^{-1}. Purified beta-carotene is sold in vegetable oil form ranging from 1% to 20% of concentration depending on type of food products. Naturally produced purified beta-carotene is usually accompanied with other *Dunaliella* sp. carotenoid (carotenoid mix) which includes lutein, neoxanthin, zeaxanthin, violaxanthin, cryptoxanthin, and α-carotene. Variety of natural beta-carotene can be found for sale as health food and supplement under vitamin section. In powder form, natural beta-carotene is used for colorization and pro-vitamin A for animal (cattle and poultry) and aquaculture (shrimp and fish) feed [30].

Chlorophyll

Chlorophyll is a green pigment and is an extremely important biomolecule, critical in photosynthesis, which allows plants to absorb energy from light for carbon dioxide fixation. The most green leaf material contains about 0.3% chlorophyll and on extraction using conventional methods yields about only 5%. Cheap sources, like grass and lucerne (alfalfa), are usually selected as chlorophyll source [35]. In more esoteric extractions, leaves of nettles and elder are preferred, whilst algae and silk worm droppings have also been used commercially [36].

Structurally, chlorophyll is essentially two parts: a substituted porphyrin ring known as an excellent chelating ligand and a long carbon chain phytol [37]. Dietary chlorophyll is predominantly composed of lipophilic derivatives including chlorophyll a and b. The structural modifications of the chlorophyll pigments mainly the de-esterification of phytol ring may significantly increase (65-fold) its transfer from the food matrix to the intestinal epithelial cells during digestion [38]. Sodium/copper derivatives of chlorophyll are frequently used as food additives or in beverages [39]. Sodium copper chlorophyllin is an approved food color in the USA (http://www.accessdata.fda.gov/scripts/cdrh/cfdocs/cfCFR/CFRSearch.cfm?fr=73.125). Besides that, it repairs cells, increases hemoglobin in blood and promotes the cell growth [5]. Regarding the cytotoxity of chlorophyll derivatives, the application of pheophorbide and pheophytin against tumor cell revealed that the cellular uptake and inhibition of myeloma cell multiplicity was greater for pheophorbide than pheophytin, hence it indicates that chlorophyll derivates may play a role in cancer prevention [28]. Patients with chronic relapsing pancreatitis can be treated with chlorophyll-α without any unfavorable side-effects, such as of allergic, or photosensitive, or hepatotoxic nature [40].

Chlorophyll has also been investigated as potential source of pigments in cosmetics; the brown and red algae are mostly used in the cosmetic industries [5]. Furthermore, in food industry, chlorophyll is used as natural pigment ingredient in processed foods. Because of its strong green pigment and consumer demand for natural foods, chlorophyll is gaining importance as food additive. This in turn is encouraging food processors to switch from artificial pigments to chlorophyll-based natural coloring. However, a downstream process needs to be developed to purify chlorophyll a and b from algae [5].

Phycobiliproteins

Phycobiliprotein, a high-potential molecule, has been utilized commercially as natural dyes and has a variety of applications in pharmaceutical industry. It is classified according to UV-visible absorption maxima as phycocyanin (blue pigment), phycoerythrins (red pigment), and allophycocyanin (pale-blue pigment). The annual market of phycocyanin was around US$5-10 million [41]. The Cyanobacterium *Spirulina* sp. is a source of c-phycocyanins and allophycocyanins and contains relatively low nucleic acid content; is composed of 55%-70% protein, 6%-9% fat, and 15%-20% carbohydrate, and is rich in minerals, vitamins, fibres, and pigments [42]. Another type of algae that can be a source for phycobiliprotein is*Aphanizomenon flos-aquae* [28, 43]. *Arthrospira* sp. is a group of Cyanobacteria characterized by loosely spiral shaped trichomes arranged in an open helix enclosed in a thin mucilaginous sheath, usually found in alkaline, brackish and saline waters. They usually become predominant species and form massive blooms [44]. The phycobiliprotein may be utilized for labeling antibodies in clinical tests (such as immunofluoresscence or flow cytometry) by binding to biologically active molecules such as immunoglobulin, biotin or protein [28, 45]. Scientific studies have shown that the synergistic action of a wide spectrum of antioxidants is better than the activity of a single antioxidant, and that antioxidants from natural sources (primarily foods) have a higher bioavailability and therefore higher protective efficacy than synthetic antioxidants [43]. It is reported that phycocyanin acts as antioxidant against free radicals and selenium and phycocyanin have been reported to show potent cancer chemopreventive activities [46, 47]. It is proven that phycocyanin is 20 times more efficient than ascorbic acid against hemolysis induced by peroxyl radicals in human erythrocytes [28].

Polyunsaturated Fatty Acids (PUFAs)

Polyunsaturated Fatty Acids (PUFAs) include docosahexaenoic acid (DHA), eicosapentaenoic acid (EPA), arachidonic acid (AA), γ-linolenic acid (GLA) and have been widely recognized as beneficial towards human health [48]. According to a new technical market research report "Global Market for Oleochemical Fatty Acids" (CHM062A) from BCC Research (http://www.bccresearch.com), the naturally produced fatty acid market was valued at US$7.2 billion in 2011 and should reach US$6.9 billion in 2012. The total market value of fatty acids expected to reach US$13 billion in 2017 (http://www.slideshare.net/bccresearch).

Natural PUFAs can be obtained from fish or extracted from fish oil. However, there is report on the possibility of accumulated toxin in the PUFAs derived from fish sources, furthermore the unpleasant smell, taste and poor oxidative stability has limits the application of fish oil as food additive [4]. Fish oil derived PUFAs have low oxidative stability and the oxidative stability of DHA containing algae oils varies widely depending on some physical parameters. However, EDTA can be used to inhibit oxidation of foods containing long-chain PUFA from either fish or algae and fortified with iron [49]. Moreover, a recent study has shown that PUFAs (450 mg day^{-1}) are not being met by the diet in the majority of the population; the current mean intake of PUFAs by adults is estimated to be about 282 mg day^{-1} (EPA and DHA contributing about 244 mg day^{-1}) [48]. It is undeniable that microalgae can be the alternative natural source for PUFAs; PUFAs found in fish are actually originated from microalgae. However, DHA is the only algal PUFA commercially available as an omega-3 fatty acid and has been documented as a major structural fatty acid in the grey matter of the brain [50]. DHA is important for correct brain and eye development in infants and has been shown to support cardiovascular health in adults [4, 51]. DHA helps body to fight against diseases and recent studies explore the significant role of PUFAs in the immune system through peroxisome proliferator-activated receptors (PPAR-α and PPAR-γ), and favorable effects of n-3 PUFA supplementation on allergic diseases and novel therapeutic strategies to treat eosinophilic disorders have also been reported [52].

Marine microalgae have significantly higher DHA contents compared to fresh water microalgae [5]. DHA is found in a limited selection of foods such as fatty fish and organic meat; it also occurs naturally in breast milk and its concentration in breast milk might be influenced by environmental factors, however it is not available in cow's milk [53]. Since 1990, a number of health and nutrition organizations have specifically recommended the inclusion of DHA in infant formula for preterm and full term infants. DHA is the characteristic PUFA of the marine dinoflagellates. *Crypthecodinium cohnii* is a non-photosynthetic, marine dinoflagellate producing DHA predominantly [54]. Nearly 30-50% of its constituent fatty acids are C22:6 (n-3) fatty acid and no other PUFAs are present in excess of 1%. It is, thus easy to separate DHA from the fatty acid mixtures. For this reason, *Crypthecodinium cohnii* represents a promising source for the commercial production of DHA [55]. The world wholesale market for infant formula was estimated to be about US$10 billion per year [56]. A number of food items containing PUFAs (including DHA) are being sold in market [4]. Some algae-related heterotroph strains that have high DHA such as *Schizochytrium mangrove*, reported to have main component of DHA in a range of 33-39% of total fatty acid while *Amphidium caryerea*and *Thrautocytrium aureum* contain 17% and 16.1%, respectively [5].

Nannochloropsis sp. has been proposed as a source of PUFAs due to its high contents of EPA [57]. Thus, *Nannochloropsis sp.* production has been used to address the rotifer requirements of 8 million gilt-head sea bream fingerlings production annually [58]. Being suitable for aquaculture industry, EPA too has been used as nutritional supplement. The diatoms*Phaeodactylum tricornutum* and *Nitzschia laevis* are algae strains to be utilized as sources of EPA [4, 59]. *Chaetoceros muelleri* var. *subsalsum* and *Isochrysis galbana* have record percentage of DHA and EPA; they present the highest output of EPA (3.5% and 4.8% on dry basis, respectively) using vertical plate glass reactor [58, 60]. EPA is an n-3 PUFA that plays an important role in the regulation of biological functions and prevention and treatment of a number of human diseases such as heart and inflammatory diseases [61]. The eicosanoids are hormone-like substances including prostaglandins (PG), thromboxanes (TX) and leukotrienes (LT). AA and EPA are precursors of eicosanoid compounds. However, the eicosanoids from these two fatty acids are different structurally and functionally, and are sometimes even antagonistic in their effects. A balanced uptake of EPA/AA can prevent eicosanoid dysfunctions and may be effective in treating a number of illnesses and metabolic disorder [59].*Porphyridium cruentum*, a red alga, has been used for massive cultures due to its relatively high AA and EPA concentration. Nevertheless, *Porphyridium cruentum* biomass is high in protein (34.1%), carbohydrates

(32.1%) and essential minerals [62]. AA is potentially used for infant formulas (either for full-term or preterm infants) and as nutritional supplement [4].

Aside from phycobiliprotein, *Arthrospira* sp. is known to be the richest algal source of GLA. GLA is important to the infant formulas for full-term infant and has the potential to lower low-density lipoproteins in hypercholesterolemic patients, alleviate the premenstrual syndrome and treat atopical eczema [4, 44]. GLA may attenuate the signs and symptoms of inflammatory diseases such as rheumatoid arthritis and atopic dermatitis [63, 64]. From both animal and human trials, it has been revealed that dietary supplementation with GLA can modulate inflammation [63].

Beta-1,3-Glucan

Beta-1,3-glucanis considered responsible to initiate host defense reactions in response to pathogen surface molecules. The most important substance other than oil in *Chlorella* sp. is beta-1,3-glucan. The world annual sales of *Chlorella* sp. are in excess of US$ 38 billion [65]. Initially, beta-1,3-glucan was named as Zymosan in the 1940s in Norway. It was found in the cell wall of baker's yeast. Further studies have been conducted by Riggi & Di Luuzio in 1961 [66] in Italy, and since then the bio-active compound was renamed as beta-1,3-glucan. Further studies on beta-1,3-glucan has been done in Japan around 1970s. Beta-1,3-glucan is important to human health. Feeding of immunostimulant beta-1,3-glucan to healthy fish may raise the non-specific and specific immunity level and the protection against bacterial infection compared with the control [67]. *Chlorella* sp. is a major source of beta-1,3-glucan [68]. Beta-1,3-glucan had significant effect as a prophylactic treatment to reduce the anthrax infection in mice and the same type of treatment can also inhibit the growth of cancer cells *in vivo* by involving stimulation of three important cytokines [69, 70]. Positive effects were also found in patients after cardiopulmonary bypass, and inhibition of antiviral activity has been found in HIV-infected patients, moreover beta-1,3-glucans are routinely used in patients for tumor immunotherapy [71]. The use of beta-1,3-glucan is of special interest in the cancer patient undergoing chemotherapy and/or radiation treatment, since beta-glucans have shown a remarkable ability to accelerate hematopoetic recovery in both sublethally and lethally irradiated mice. It can also stimulate recovery of the bone marrow following chemotherapy, which is vital to restricting tumor growth and preventing infectious complications during treatment. Amparyup *et al.* [72] showed that pattern recognition proteins [including lipopolysaccharide and beta-1,3-glucan binding protein (LGBP)] could enhance the phenoloxidase activity in shrimp.

Pharmaceutical and Nutraceutical Compounds

Algae have the potential to be a source of compounds having antibiotic and anticancer activity. A wide variety of Cyanobacteria produce compounds including sulfolipids that were active against herpes virus, pneumonia virus and HIV [73, 74]. Cyanobacteria can produce tolytoxin (6-hydroxy-7-O-methylscytophycin B), an anti-fungal antibiotic and highly toxic to mice [75,76]. The marine cyanobacterium *Lyngbya majuscula* produces aurilides B and C having cytotoxicity against human lung tumor and mouse neuroblastoma cell line [77]. Moreover the macroalgae *Laphocladia* sp. is capable of producing lophocladine alkaloids that exhibit cytotoxicity to NCI-H460 human lung tumor and MDA-MB-435 breast cancer cell lines [78]. Therefore, once the algal sources of useful medicinal compound are identified, the algae can be cultivated in mass scale and the material can be extracted, purified, and marketed [1]. Some microalgae species are established in the skin care market, the main ones being *Arthrospira* sp. and *Chlorella* sp. A number of cosmeticians have even invested in their own microalgae production (http://www.fao.org/docrep/012/i1704e/i1704e01.pdf). Two examples of commercially available products and their properties claimed by their companies are a) a protein-rich extract from *Arthrospira* sp. that repairs the signs of early skin aging, exerts a tightening effect and prevents stria formation [4] (Protulines, Exsymol S.A.M., Monaco) and b) an extract from *Chlorella vulgaris* that stimulates collagen synthesis in skin, thereby supporting tissue regeneration and wrinkle reduction (Dermochlorella, Codif, St. Malo, Francehttp://www.coptis.com/site/desc.php?RefFab=1CDF). Recently, two new products have been launched by Pentapharm (Basel, Switzerland): a compound from *Nannochloropsis oculata* is used for excellent skin-tightening properties (short and long-term effects) (Pepha-Tight) and an ingredient from *D. salina* is known to have the ability to markedly stimulate cell proliferation and turnover to positively influence the energy metabolism of skin (Pepha-Ctive) [4].

Livestock Feed

Spirulina has been frequently used in feed supplements due to its excellent nutrient compounds and digestibility because it has small amount of carbohydrates. The livestock feed is another useful product which may be obtained from the algae, since many of algae species have been examined for their biochemical compositions to evaluate their suitability as a substitute or primary livestock feed [79]. Spirulina has a unique blend of nutrients containing nutrients that include B-complex vitamins, minerals, proteins, gamma-linolenic acid and super anti-oxidants (such as beta-carotene, vitamin E, trace elements) and a number of unexplored bioactive compounds [80]. Compared to other microorganisms, *Spirulina* sp. can be cultivated in high saline water and alkaline conditions which give an advantage to function as a feedstock for livestock feed [81]. Recently, studies regarding growth and body conformation responses of genetically divergent Australian sheep to Spirulina (*Arthrospira platensis*) supplementation revealed that Spirulina can increase live weight, growth and body conformation significantly [82]. Feeding lipid-encapsulated algae supplements may increase n-3 content in milk fat without adversely affecting milk fat yield [83].

An algal supplementation level of about 10 g kg^{-1} of dry matter intake proved to be effective in reducing milk fat content and in modifying the milk fatty acid composition toward increased conjugated linolenic acid (CLA) and DEA concentrations [84]. A related study demonstrated that dietary supplementation with fish meal or n-3 PUFAs in early-lactating dairy cows significantly increased milk yield with no change in milk composition [85]. In addition, some sea weeds like red algae (e.g. *Porphyra* sp., *Gracilaria* sp. and *Kappaphycus* sp.) and brown algae (particularly *Laminaria* sp.,

Undaria sp. and *Hizikia fusiforme*) were used in human food exhibiting potential to be used for livestock feed. *Porphyra* sp., *Gracilaria* sp., *Grateloupia* sp., and *Kappaphycus* sp. have also been genetically transformed for enhanced production of respective products [86]. The annual *Porphyra* sp. harvest worldwide has been estimated to be worth of US$2.5 billion. *Porphyra* sp. primarily contains vitamin A, vitamin B, vitamin C, beta-carotene, as well as essential minerals including iodine. Besides that, a study on *Laminaria digitata* suggested that algae supplemented feed has increased the pig weight up to 10% on a daily basis [1, 5].

Poultry Feed

There are more chickens in the world than any other bird; in fact, more than 50 billion chickens are reared annually as a source of food, for both their meat and eggs (http://www.ciwf.org.uk/farm_animals/poultry/). Comprehensive analyses and nutritional studies have demonstrated that algae contain proteins and these proteins are of high quality and comparable to conventional vegetable proteins [87]. Ginzberg *et al.* [88] studied the role of algae *Porphyridium* sp. as feed supplement on metabolism of chicken. It was found that cholesterol of egg yolk was reduced about 10% and color of egg yolk became darker, indicating the production of higher carotenoids [5, 89]. In poultry rations, algae up to a level of 5-10% can be used safely as partial replacement for conventional proteins. Prolonged feeding of algae at higher concentrations may cause adverse effects. The yellow color of broiler skin and shanks as well as of egg yolk is the most important characteristic that can be influenced by feeding algae. Moreover, the Institute für Getreideverarbeitung (Bergholz-Rehbrücke, Germany) produces a natural feed with the algae *Chlorella* sp. and *Arthrospira* sp. called Algrow [90].

Aquaculture Feed

Demand for fish for human consumption has risen due to the environmental concerns over Open Ocean fishing [91]. The world market for farmed fish and aquatic edible plants was valued in 2008 at US$106 billion. Products from fisheries and aquaculture combined are supplying the world with 142 million tonnes of protein every year (http://www.blueeconomy.eu/m/articles/view/Fish-without-Feed-Part-1-The-Market).

According to another report by United Nations Food and Agriculture Organization (http://www.fao.org/docrep/016/i2727e/i2727e.pdf), in the last three decades (1980-2010), world food fish production of aquaculture has increased by almost 12 times, at an average annual rate of 8.8% and the world aquaculture production attained another all-time high in 2010 at 60 million tonnes (excluding aquatic plants and non-food products), with an estimated total value of US$119 billion (http://aquabounty.com/aquaculture-market/). Non-toxic marine microalgae, including the stramenopiles *Isochrysis* sp., *Pavlova* sp. and*Nannochloropsis* sp. as well as various diatoms, represent the primary food source at some stages in the life cycle of most cultivated marine animals [1]. Therefore, algal biomass will be in high demand for the fish-food and aquaculture markets in the future and provide ample revenues for the algae industry [91].

A challenge in fish nutrition is to generate end products with high levels of health-promoting long chains of omega-3 fatty acids for the consumer, while reducing the use of fish oils. Omega-3 oils have a high market demand in both the human nutraceuticals and animal feed industries. This growing trend is an additional driving force for the marketing of non-marine-based omega-3 oils and alternative feed ingredients. Algae is a rich source of high quality protein, vitamins, micronutrients (trace elements), and carotenoids, Long-chain PUFAs, especially of n-3 and n-6 series such as eicosapentaenoic (EPA), docosahexaenoic (DHA), and arachidonic (AA) are considered pharmacologically important for dietetics and therapeutics [90]. Algal meal is a rich source of high-quality protein, vitamins, micronutrients (trace elements) and carotenoids, which can be used directly in aqua feeds [85]. Furthermore, fish phosphoglycerides generally contain 50% of their total fatty acids as n-3 PUFA with a ratio of 22:6 (n-3):20:5 (n-3) of about 2:1 and it can be seen most clearly in the phosphoglycerides of fish eggs [59]. In addition, better growth weight and protein efficiencies ratio of tilapia was observed when supplied with algae as nutritional source in feed. Also, *Phorphidium valderianum* was successfully used as feed for aquaculture (based on its nutritional performance and non-toxic properties) [5]. The diatom *Thalassiosira pseudonana* is widely cultivated to feed variety of mollusks, including the Pacific oyster *Crassostrea gigas* and rock scallops [1]. The long chain PUFAs are highly susceptible to oxidation because of their unsaturated nature; the addition of natural antioxidants to the highly unsaturated fish oil may protect it from oxidation [92].

Shrimp Feed

The total global production of farmed shrimp reached more than 1.6 million tonnes in 2003, representing a value of nearly US$9,000 million (http://fisherymanagement.wikia.com/wiki/Shrimp_farm). Shrimp farming production reached 737,200 tonnes in 1998, an increase of 12% from 1997. This increased production was mainly reported in subtropical regions of America (28%, 457 hatcheries) and SE Asia (72%, 3,718 hatcheries) [59]. According to Shrimp Production Review by Food and Resource Economics Department, University of Florida USA, the shrimp farming was estimated to be about 4 million metric tonnes in 2013, with 10.3% projected annual growth rate (mainly attributed to SE Asian countries and China) (http://www.gaalliance.org/update/GOAL11/DiegoValderrama.pdf). However, disease outbreaks in recent years have affected farmed Atlantic salmon in Chile, oysters in Europe, and marine shrimp farming in several countries in Asia, South America and Africa, resulting in partial or sometimes total loss of production (http://www.fao.org/docrep/016/i2727e/i2727e.pdf). In 2010, aquaculture in China suffered production losses of 1.7 million tons caused by natural disasters, diseases and pollution. Disease outbreaks virtually wiped out marine shrimp farming production in Mozambique in 2011 (http://www.fao.org/docrep/016/i2727e/i2727e.pdf). Microalgae species *Hypnea cervicornis*and *Cryptonemia crenulata* particularly rich in protein were tested in shrimp diets. Amount of algae in fish meal resulted in significant increase in shrimp growth rates [5]. The requirement for DHA in marine fish and shrimp nutrition has been established via feeding diets containing DHA and AA; the findings revealed that shrimp rose on diets rich in DHA and AA showed significant improvement in immune parameters, such as total hemocyte count, phenoloxidase activity, superoxide dismutase activity, and bactericidal activity [93]. Recently, it has been reported that the addition of 0.5% Spirulina meal in a complete diet for shrimp (*L. vannamei* juveniles), with 14% of Peruvian fishmeal, has proved as a nutritionally efficient feeding attractant [94]. Through short-term feeding studies, two marine algal products (MAP) named MAP3 and MAP8 were found suitable as substitute of protein source in the feeds of three prominent farmed species [89]. The studies regarding the utilization of fish oil and algae as a source of PUFAs are going side by side but the trends are more inclined towards algal products. Menhaden fish oil can also be used as a source of dietary fatty acids which influence the composition of Pacific white shrimp muscle tissue by exhibiting significantly higher shrimp muscle DHA level [95]. As fish meal content, DHA is required for the production of ecdysone for molting, growth and egg production. The mechanism involves the biological membranes rich in di-22:6 (n-3) phosphoglycerides which is relatively constant in the face of changing environmental variables like temperature, pressure, and salinity [59].

CONCLUSION

Microalgae have multiple potential to produce high-valued products and there is an urgent need of awareness makes these biomolecules popular in the world to meet the increasing demands with respect to population. Most of these biomolecules are not produced in the animal/human body but termed as essential; therefore, it is highly recommended to make these biomolecules available for food and feed purposes. Although the production of biomolecules from algae requires some awareness regarding technology and market value at the industrial and farmer level, it has been proved that these biomolecules can be produced using algae as a potential source at small or large scale. The demanding market of these biomolecules is quite convincing to stimulate research interests leading to production programs by the related organizations. Algae culturing got attention because of its biofuel producing characteristics but the awareness regarding cultivation technologies may also lead to set trends for biomolecule production as a pharmaceutical or nutritional product. The production of biomolecules from algae and their subsequent utilization in animal and aquaculture feed may support the objectives to establish industry from microalgae at domestic or industrial scale. Some of the biomolecules can be produced as a byproduct while producing biofuel from algae as a main targeted product. Eighty-two percent of the world fish stocks are overexploited, depleted or endangered, while demand for fish protein is exploding. With the world population expected to reach 9 billion by 2050, demand for critical sources of protein continues to outstrip supply.

CONSENT

Written informed consent was obtained from the patient for the publication of this report and any accompanying images.

ACKNOWLEDGEMENTS

Authors, acknowledge the University Kebangsaan Malaysia for providing infrastructure and funding to facilitate the preparation of this review report.

COMPETING INTERESTS

The author's declare that they have no competing interests.

REFERENCES

1. Graham JE, Wilcox LW, Graham LE: *Algae*. City: Pearson Education; 2009.
2. Preisig HR, Andersen RA: Historical review of algal culturing techniques. In *Algal Culturing Techniques*. Edited by: Andersen RA. London: Elsevier Academic Press; 2005:1–12.
3. Del Campo JA, García-González M, Guerrero MG: Outdoor cultivation of microalgae for carotenoid production: current state and perspectives. *Appl Microbiol Biot*2007, 74: 1163–1174. 10.1007/s00253-007-0844-9
4. Spolaore P, Joannis-Cassan C, Duran E, Isambert A: Commercial applications of microalgae. *J Biosci Bioeng* 2006, 101: 87–96. 10.1263/jbb.101.87
5. Harun R, Singh M, Forde GM, Danquah MK: Bioprocess engineering of microalgae to produce a variety of consumer products. *Renew Sust Energ Rev* 2010, 14: 1037–1047. 10.1016/j.rser.2009.11.004
6. Naguib YMA: Antioxidant activities of astaxanthin and related carotenoids. *J Ag Food Chem* 2000, 48: 1150–1154. 10.1021/jf991106k
7. Kupcinskas L, Lafolie P, Lignell Å, Kiudelis G, Jonaitis L, Adamonis K, Andersen LP, Wadström T: Efficacy of the natural antioxidant astaxanthin in the treatment of functional dyspepsia in patients with or without Helicobacter pylori infection: A prospective, randomized, double blind, and placebo-controlled study.*Phytomedicine* 2008, 15: 391–399. 10.1016/j.phymed.2008.04.004
8. Nakagawa K, Kiko T, Miyazawa T, Burdeos GC, Kimura F, Satoh A, Miyazawa T:Antioxidant effect of astaxanthin on phospholipid peroxidation in human erythrocytes. *Bri J Nutr* 2011, 105: 1563–1571. 10.1017/S0007114510005398
9. Canales-Gómez E, Correa G, Viana MT: Effect of commercial carotene pigments (astaxanthin, cantaxanthin and β-carotene) in juvenile abalone Haliotis rufescens diets on the color of the shell or nacre. *Vet Mexico* 2010, 41: 191–200.
10. Del Campo JA, Rodríguez H, Moreno J, Vargs MA, Rivas J, Guerrero MG: Accumulation of astaxanthin and lutein in Chlorella zofingiensis (Chlorophyta). *Appl Microbiol Biot* 2004, 64: 848–854. 10.1007/s00253-003-1510-5
11. Zhao LY, Chen G, Zhao G, Hu X: Optimization of microwave-assisted extraction of astaxanthin from Haematococcus pluvialis by response surface methodology and antioxidant activities of the extracts. *Separ Sci Technol* 2009, 44: 243–262. 10.1080/01496390802282321
12. Rodríguez-Sáiz M, de la Fuente JL, Barredo JL: Xanthophyllomyces dendrorhous for the industrial production of astaxanthin. *Appl Microbiol Biot* 2010, 88: 645–658. 10.1007/s00253-010-2814-x

13. Yuan J-P, Peng J, Yin K, Wang J-H: Potential health-promoting effects of astaxanthin: a high-value carotenoid mostly from microalgae. *Mol Nutr Food Res* 2011, 55: 150–165. 10.1002/mnfr.201000414
14. Yang Y, Seo JM, Nguyen A, Pham TX, Park HJ, Park Y, Kim B, Bruno RS, Lee J:Astaxanthin-rich extract from the green alga Haematococcus pluvialis lowers plasma lipid concentrations and enhances antioxidant defense in apolipoprotein E knockout mice. *J Nutr* 2011, 141: 1611–1617. 10.3945/jn.111.142109
15. Tolasa S, Cakli S, Ostermeyer U: Determination of astaxanthin and canthaxanthin in salmonid. *Eur Food Res Technol* 2005, 221: 787–791. 10.1007/s00217-005-0071-5
16. Dominguez A, Ferreira M, Coutinho P, Fábregas J, Otero A: Delivery of astaxanthin from Haematococcus pluvialis to the aquaculture food chain. *Aquaculture* 2005,250: 424–430. 10.1016/j.aquaculture.2005.08.005
17. Guerin M, Huntley ME, Olaizola M: Haematococcus astaxanthin: applications for human health and nutrition. *Trends Biotechnol* 2003, 21: 210–216. 10.1016/S0167-7799(03)00078-7
18. Sheikhzadeh N, Tayefi-Nasrabadi H, Oushani AK, Enferadi MHN: Effects of Haematococcus pluvialis supplementation on antioxidant system and metabolism in rainbow trout (Oncorhynchus mykiss). *Fish Physiol Biochem* 2012,38: 413–419. 10.1007/s10695-011-9519-7
19. Lyons NM, O'Brien NM: Modulatory effects of an algal extract containing astaxanthin on UVA-irradiated cells in culture. *J Dermatol Sci* 2002, 30: 73–84. 10.1016/S0923-1811(02)00063-4
20. Cysewski GR, Lorenz RT: Industrial production of microalgal cell-mass and secondary products - species of high Potential: *Haematococcus* . In *Handbook of Microalgal Culture Biotechnology and Applied Phycology*. Edited by: Richmond A. Oxford: Blackwell Publishing Ltd; 2004:281–288.
21. Landrum JT, Bone RA: Lutein, zeaxanthin, and the macular pigment. *Arch Biochem Biophys* 2001, 385: 28–40. 10.1006/abbi.2000.2171
22. Shi X-M, Chen F, Yuan J-P, Chen H: Heterotrophic production of lutein by selected Chlorella strains. *J Appl Phycol* 1997, 9: 445–450. 10.1023/A:1007938215655
23. Zhang X-W, Shi X-M, Chen F: A kinetic model for lutein production by the green microalga Chlorella protothecoides in heterotrophic culture. *J Ind Microbiol Biot* 1999, 23: 503–507. 10.1038/sj.jim.2900760
24. Wu Z-Y, Shi C-L, Shi X-M: Modeling of lutein production by heterotrophic Chlorella in batch and fed-batch cultures. *World J Microb Biot* 2007, 23: 1233–1238. 10.1007/s11274-007-9354-2
25. Bhattacharyya S, Datta S, Mallick B, Dhar P, Ghosh S: Lutein content and in vitro antioxidant activity of different cultivars of Indian marigold flower (Tagetes patula L.) extracts. *J Agr Food Chem* 2010, 58: 8259–8264. 10.1021/jf101262e
26. Christaki E, Bonos E, Giannenas I, Florou-Paneri P: Functional properties of carotenoids originating from algae. *J Sci Food Agr* 2012, 93: 5–11.
27. Wei D, Chen F, Chen G, Zhang XW, Liu LJ, Zhang H: Enhanced production of lutein in heterotrophic Chlorella protothecoides by oxidative stress. *Sci China Ser C*2008, 51: 1088–1093. 10.1007/s11427-008-0145-2
28. Becker W: Microalgae in human and animal nutrition. In *Handbook of Microalgal Culture Biotechnology and Applied Phycology*. Edited by: Richmond A. Oxford: Blackwell Publishing Ltd; 2004:312–351.
29. Lers A, Biener Y, Zamir A: Photoinduction of massive β-carotene accumulation by the alga Dunaliella bardawil: Kinetics and dependence on gene activation. *Plant Physiol* 1990, 93: 389–395. 10.1104/pp.93.2.389
30. Ben-Amotz A: Industrial production of microalgal cell-mass and secondary products - major industrial species *Dunaliella* . In *Handbook of Microalgal Culture Biotechnology and Applied Phycology*. Edited by: Richmond A. UK: Blackwell Publishing Ltd; 2004:273–280.
31. Kar HK: Efficacy of beta-carotene topical application in melasma: an open clinical trial. *Indian J Dermatol Ve* 2002, 68: 320–322.
32. Neuman I, Nahum H, Ben-Amotz A: Prevention of exercise-induced asthma by a natural isomer mixture of β-carotene. *Ann Allerg Asthma Im* 1999, 82: 549–553. 10.1016/S1081-1206(10)63165-1
33. Hr T, Tikellis G, Robman LD, McCarty CA, McNeil JJ: Vitamin E supplementation and macular degeneration: randomized controlled trial. *BMJ* 2002, 325: 11–14. 10.1136/bmj.325.7354.11
34. Hall NF: Prevention of age related, macular degeneration. *BMJ* 2002, 325: 1–2. 10.1136/bmj.325.7354.1
35. Humphrey AM: Chlorophyll. *Food Chem* 1980, 5: 57–67. 10.1016/0308-8146(80)90064-3
36. Humphrey AM: Chlorophyll as a color and functional ingredient. *J Food Sci* 2004,69: C422-C425. 10.1111/j.1750-3841.2004.tb18019.x
37. Ferruzzi MG, Blakeslee J: Digestion, absorption, and cancer preventative activity of dietary chlorophyll derivatives. *Nutr Res* 2007, 27: 1–12. 10.1016/j.nutres.2006.12.003
38. Gandul-Rojas B, Gallardo-Guerrero L, Mínguez-Mosquera IM: Influence of the chlorophyll pigment structure on its transfer from an oily food matrix to intestinal epithelium cells. *J Agr Food Chem* 2009, 57: 5306–5314. 10.1021/jf900426h
39. Scotter MJ, Castle L, Roberts D: Method development and HPLC analysis of retail foods and beverages for copper chlorophyll (E141[i]) and chlorophyllin (E141[ii]) food colouring materials. *Food Addit Contam* 2005, 22: 1163–1175. 10.1080/02652030500306885
40. Yoshida A, Yokono O, Oda T: Therapeutic effect of chlorophyll-a in the treatment of patients with chronic pancreatitis. *Gastroenterol Jpn* 1980, 15: 49–61.
41. Sekar S, Chandramohan M: Phycobiliproteins as a commodity: trends in applied research, patents and commercialization. *J Appl Phycol* 2008, 20: 113–136.

10.1007/s10811-007-9188-1

42. Boussiba S, Richmond AE: Isolation and characterization of phycocyanins from blue green algae Spirulina platensis. *Arch Microbiol* 1979, 120: 155–159. 10.1007/BF00409102

43. Benedetti S, Benvenuti F, Pagliarani S, Francogli S, Scoglio S, Canestrani F: Antioxidant properties of a novel phycocyanin extract from the blue-green alga Aphanizomenon flos-aquae. *Life Sci* 2004, 75: 2353–2362. 10.1016/j.lfs.2004.06.004

44. Hu Q: Industrial production of microalgal cell-mass and secondary products - major industrial species *Arthrospira* (*Spirulina*) *platensis* . In *Handbook of Microalgal Culture Biotechnology and Applied Phycology*. Edited by: Richmond A. Oxford: Blackwell Publishing Ltd; 2004:264–272.

45. Shapiro HM, Glazer AN, Christenson L, Williams JM, Strom TB: Immunofluorescence measurement in a flow cytometer using low-power helium-neon laser excitation. *Cytometry* 1983, 4: 276–279. 10.1002/cyto.990040314

46. Patel A, Mishra S, Ghosh PK: Antioxidant potential of C-phycocyanin isolated from cyanobacterial species Lyngbya, Phormidium and Spirulina spp. *Indian J Biochem Bio* 2006, 43: 25–31.

47. Chen T, Wong YS: In vitro antioxidant and antiproliferative activities of selenium-containing phycocyanin from selenium-enriched Spirulina platensis. *J Agr Food Chem* 2008, 56: 4352–4358. 10.1021/jf073399k

48. Givens DI, Gibbs RA: Very long chain n-3 polyunsaturated fatty acids in the food chain in the UK and the potential of animal-derived foods to increase intake.*Nutrition Bull* 2006, 31: 104–110. 10.1111/j.1467-3010.2006.00554.x

49. Frankel EN, Satué-Gracia T, Meyer AS, German BJ: Oxidative stability of fish and algae oils containing long-chain polyunsaturated fatty acids in bulk and in oil-in-water emulsions. *J Agri Food Chem* 2002, 50: 2094–2099. 10.1021/jf0111458

50. Bradbury J: Docosahexaenoic Acid (DHA): an ancient nutrient for the modern human brain. *Nutrients* 2011, 3: 529–554. 10.3390/nu3050529

51. Cottin SC, Sanders TA, Hall WL: The differential effects of EPA and DHA on cardiovascular risk factors. *P Nutr Soc* 2011, 70: 215–231. 10.1017/S0029665111000061

52. Tanigai T, Ueki S, Kihara J, Kamada R, Yamauchi Y, Sokal A, Takeda M, Ito W, Kayaba H, Adachi T, Ohta K, Chihara J: Docosahexaenoic Acid exerts anti-inflammatory action on human eosinophils through peroxisome proliferator-activated receptor-independent mechanisms. *Int Arch Allergy Imm* 2012, 158: 375–386. 10.1159/000332965

53. Thiombiano-Coulibaly N, Rocquelin G, Eymard-Duvernay S, Kiffer-Nunes J, Tapsoba S, Traoré SA: Seasonal and environmental effects on breast milk fatty acids in Burkina Faso and the need to improve the omega 3 PUFA content. *Acta Paediatr*2003, 92: 1388–1393. 10.1111/j.1651-2227.2003.tb00820.x

54. Da Silva TL, Mendes A, Mendes RL, Calado V, Alves SS, Vasconcelos JMT, Reis A: Effect of n-dodecane on Crypthecodinium cohnii fermentations and DHA production. *J Ind Microbiol Biot* 2006, 33: 408–416. 10.1007/s10295-006-0081-8

55. Jiang Y, Chen F, Liang S-Z: Production potential of docosahexaenoic acid by the heterotrophic marine dinoflagellate Crypthecodinium cohnii. *Process Biochem*1999, 34: 633–637. 10.1016/S0032-9592(98)00134-4

56. Milledge JJ: Commercial application of microalgae other than as biofuels: a brief review. *Rev Environ Sci Biotechnol* 2011, 10: 31–41. 10.1007/s11157-010-9214-7

57. Zittelli GC, Rodolfi L, Tredici MR: Industrial production of microalgal cell-mass and secondary products - species of high potential: mass cultivation of*Nannochloropsis* in Closed Systems. In *Handbook of Microalgal Culture Biotechnology and Applied Phycology*. Edited by: Richmond A. Oxford: Blackwell Publishing Ltd; 2004:298–303.

58. Zamora O, Richmond A: Microalgae production for aquaculture. In *Handbook of Microalgal Culture Biotechnology and Applied Phycology*. Edited by: Richmond A. Oxford: Blackwell Publishing Ltd; 2004:365–379.

59. Hemaiswarya S, Raja R, Kumar RR, Ganesan V, Anbazhagan C: Microalgae: a sustainable feed source for aquaculture. *World J Microb Biot* 2011, 27: 1737–1746. 10.1007/s11274-010-0632-z

60. Zhang CW, Richmond A: Sustainable, high-yielding outdoor mass cultures of Chaetoceros muelleri var. subsalsum and Isochrysis galbana in vertical plate reactors. *Mar Biotechnol* 2003, 5: 302–310. 10.1007/s10126-002-0075-0

61. Wen Z-Y, Chen F: Heterotrophic production of eicosapentaenoic acid by microalgae. *Biotechnol Adv* 2003, 21: 273–294. 10.1016/S0734-9750(03)00051-X

62. Arad S, Richmond A: Industrial production of microalgal cell-mass and secondary products - Species of high potential: *Porphyridium* sp. In *Handbook of Microalgal Culture Biotechnology and Applied Phycology*. Edited by: Richmond A. Oxford: Blackwell Publishing Ltd; 2004:289–297.

63. Johnson MM, Swan DD, Surette ME, Stegner J, Chilton T, Fonteh AN, Chilton FH:Dietary supplementation with g-linolenic acid alters fatty acid content and eicosanoid production in healthy humans. *J Nutr* 1997, 127: 1435–1444.

64. Fan Y-Y, Chapkin RS: Importance of dietary γ-linolenic acid in human health and nutrition. *J Nutr* 1998, 128: 1411–1414.

65. Yamaguchi K: Recent advances in microalgal bioscience in Japan, with special reference to utilization of biomass and metabolites: a review. *J Appl Phycol* 1997,8: 487–502.

66. Riggi SJ, Di Luzio NR: Identification of a reticuloendothelial stimulating agent in zymosan. *Am J Physiol* 1961, 200: 297–300

67. Sahoo PK, Mukherjee SC: Effect of dietary beta-1,3

glucan on immune responses and disease resistance of healthy and aflatoxin B1-induced immunocompromised rohu (Labeo rohita Hamilton). *Fish Shellfish Immun* 2001,11: 683–695. 10.1006/fsim.2001.0345

68. Iwamoto H: Industrial production of microalgal cell-mass and secondary products - major industrial species *Chlorella* . In *Handbook of Microalgal Culture Biotechnology and Applied Phycology*. Edited by: Richmond A. Oxford: Blackwell Publishing Ltd; 2004:255–263.

69. Vetvicka V, Yvin J-C: *Int Immunopharmacol*. 2004, 4: 721–730. 10.1016/j.intimp.2004.02.007

70. Blaylock RL: Yeast β-1,3-glucan and its use against anthrax infection and in the treatment of cancer. *J Am Nutraceutical Assoc* 2002, 5: 3–4.

71. Vetvicka V, Terayama K, Mandevill R, Brousseau P, Kournikakis B, Ostroff G: Pilot study: orally-administered yeast β1,3-glucan prophylactically protects against anthrax infection and cancer in mice. *J Am Nutraceutical Assoc* 2002, 5: 1–5.

72. Amparyup P, Sutthangkul J, Charoensapsri W, Tassanakajon A: Pattern recognition protein binds to lipopolysaccharide and beta-1,3-glucan and activates shrimp prophenoloxidase system. *J Biol Chem* 2012, 287: 10060–10069. 10.1074/jbc.M111.294744

73. Gustafson KR, Cardellina JH 2nd, Fuller RW, Weislow OS, Kiser RF, Snader KM, Patterson GM, Boyd MR: AIDS-antiviral sulfolipids from cyanobacteria (blue-green algae).*J Natl Cancer I* 1989, 81: 1254–1258. 10.1093/jnci/81.16.1254

74. Archer SD, McDonald KA, Jackman AP: Effect of light irradiance on the production of sulfolipids from Anabaena 7120 in a fed-batch photobioreactor. *Appl Biochem Biotechnol* 1997, 67: 139–152. 10.1007/BF02787848

75. Carmeli S, Moore RE, Patterson GML: Tolytoxin and new scytophycins from three species of Scytonema. *J Nat Prod* 1990, 53: 1533–1542. 10.1021/np50072a021

76. Patterson GML, Carmeli S: Biological effects of tolytoxin (6-hydroxy-7-O-methyl-scytophycin b), a potent bioactive metabolite from cyanobacteria. *Arch Microbiol* 1992, 157: 406–410. 10.1007/BF00249096

77. Han B, Gross H, Goeger DE, Moobery SL, Gerwick WH: Aurilides B and C, cancer cell toxins from a Papua New Guinea collection of the marine cyanobacterium Lyngbya majuscula. *J Nat Prod* 2006, 69: 572–575. 10.1021/np0503911

78. Gross H, Goeger DE, Hills P, Moobery SL, Ballantine DL, Murray TF, Valeriote FA, Gerwick WH: Lophocladines, bioactive alkaloids from the red alga Lophocladia sp. *J Nat Prod* 2006, 69: 640–644. 10.1021/np050519e

79. Singh J, Gu S: Commercialization potential of microalgae for biofuels production. *Renew Sust Energ Rev* 2010, 14: 2596–2610. 10.1016/j.rser.2010.06.014

80. Kulshreshtha A, Zacharia AJ, Jarouliya U, Bhadauriya P, Prasad GBKS, Bisen PS:Spirulina in health care management. *Curr Pharm Biotechno* 2008, 9: 400–405. 10.2174/138920108785915111

81. Costa JA, de Morais MG, Dalcanton F, Reichert CC, Durante AJ: Simultaneous cultivation of Spirulina platensis and the toxigenic cyanobacteria Microcystis aeruginosa. *Z Naturforsch C* 2006, 61: 105–110.

82. Holman B, Kashani A, Malau-Aduli AEO: Growth and body conformation responses of genetically divergent Australian sheep to Spirulina (Arthrospira platensis) supplementation. *Am J Experiment Agric* 2012, 2: 160–173.

83. Stamey JA, Shepherd DM, de Veth MJ, Corl BA: Use of algae or algal oil rich in n-3 fatty acids as a feed supplement for dairy cattle. *J Dairy Sci* 2012, 95: 5269–5275. 10.3168/jds.2012-5412

84. Boeckaert C, Vlaeminck B, Dijkstra J, Issa-Zacharia A, Van Nespen T, Van Straalen W, Fievez V: Effect of dietary starch or micro algae supplementation on rumen fermentation and milk fatty acid composition of dairy cows. *J Dairy Sci* 2008,91: 4714–4727. 10.3168/jds.2008-1178

85. Subhadra B, Grinson G: Algal biorefinery-based industry: an approach to address fuel and food insecurity for a carbon-smart world. *J Sci Food Agr* 2011, 91: 2–13. 10.1002/jsfa.4207

86. Qin S, Jiang P, Tseng C: Transforming kelp into a marine bioreactor. *Trends Biotechnol* 2005, 23: 264–268. 10.1016/j.tibtech.2005.03.010

87. Becker EW: Micro-algae as a source of protein. *Biotechnol Adv* 2007, 25: 207–210. 10.1016/j.biotechadv.2006.11.002

88. Ginzberg A, Cohen M, Sod-Moriah UA, Shany S, Rosenshtrauch A, Arad SM: Chickens fed with biomass of the red microalga Porphyridium sp. have reduced blood cholesterol level and modified fatty acid composition in egg yolk. *J Appl Phycol*2000, 12: 325–330. 10.1023/A:1008102622276

89. Kiron V, Phromkunthong W, Huntley M, Archibald I, De Scheemaker G: Marine microalgae from biorefinery as a potential feed protein source for Atlantic salmon, common carp and whiteleg shrimp. *Aquacul Nutr* 2012, 18: 521–531. 10.1111/j.1365-2095.2011.00923.x

90. Pulz O, Gross W: Valuable products from biotechnology of microalgae. *Appl Microbiol Biot* 2004, 65: 635–648. 10.1007/s00253-004-1647-x

91. Anemaet IG, Bekker M, Hellingwerf KJ: Algal photosynthesis as the primary driver for a sustainable development in energy, feed, and food production. *Mar Biotechnol* 2010, 12: 619–629. 10.1007/s10126-010-9311-1

92. Keithly JI, Kott RW, Berardinelli JG, Moreaux S, Hatfield PG: Thermogenesis, blood metabolites and hormones, and growth of lambs born to ewes supplemented with algae-derived docosahexaenoic acid. *J Anim Sci* 2011, 89: 4305–4313. 10.2527/jas.2010-3391

93. Nonwachai T, Purivirojkul W, Limsuwan C, Chuchird N, Velasco M, Dhar AK: Growth, nonspecific immune characteristics, and survival upon challenge with Vibrio harveyi in Pacific white shrimp (Litopenaeus vannamei) raised on diets containing algal meal. *Fish Shellfish Immun* 2010, 29: 298–304. 10.1016/j.fsi.2010.04.009

94. Silva-Neto JF, Nunes AJP, Sabry-Neto H, Sá MVC: Spirulina meal has acted as a strong feeding attractant for Litopenaeus vannamei at a very low dietary inclusion level. *Aquac Res* 2012, 43: 430–437. 10.1111/j.1365-2109.2011.02846.x

95. Perez-Velazquez M, González-Félix ML, Navarro-García G, Valenzuela-Escalante E:Nutritional value of various ray fish liver oils to the pacific white shrimp Litopenaeus vannamei. *Lipids* 2008, 43: 1009–1016. 10.1007/s11745-008-3231-0

CHAPTER 4

IMPORTANCE OF NUTRIENTS, RESOURCE ALLOCATION, AND ECOLOGY IN MICROALGAL CULTIVATION FOR LIPID ACCUMULATION

Matthew W. Fields, Adam Hise, Egan J. Lohman, Tisza Bell, Rob D. Gardner, Luisa Corredor, Karen Moll, Brent M. Peyton, Gregory W. Characklis, Robin Gerlach

Keywords: Biofuel, Recycle, Algal biofilm, Biofuel ecology

INTRODUCTION

In modern societies, petroleum-based products and fuels have strongly influenced human culture and infrastructure. For example, energy, food, and chemicals make up approximately 70 % of commerce on the planet (www.eia.gov), and petroleum/hydrocarbons directly and indirectly impact these commodities. Petroleum/hydrocarbon markets have become increasingly unpredictable and cause destabilized commodity prices (e.g., fuel, food). In addition, the environmental impacts from increased carbon dioxide (CO_2) without balanced CO_2 sequestration has contributed to increases in atmospheric CO_2 levels. The amount of carbon released in 1 year from the consumption of fossil fuels is more than 400-fold the amount of carbon that can be fixed via net global primary productivity (Dukes 2003). In order to offset the massive influx of CO_2 into the atmosphere, the utilization of renewable biofuels (e.g., ethanol, butanol, H_2, CH_4, and biodiesel) is needed.

Bacillariophyta (diatoms) and Chlorophyta (green algae) are eukaryotic photoautotrophs that can utilize inorganic carbon (e.g., CO_2) as a carbon source and sunlight as an energy source, and many microalgae can store carbon and energy in the form of neutral lipids [e.g., triacylglycerides (TAGs)]. Moreover, different diatoms and algae can produce and accumulate different precursors (e.g., carbohydrates, fatty acids, and pigments) that are value added products. In addition to accumulating useful compounds for biofuels and chemical feed stocks, the use of autotrophic microorganisms can further contribute to reduced CO_2 emissions through utilization of atmospheric CO_2. For these reasons, eukaryotic photoautotrophs have been studied in the context of lipid accumulation for over 50 years and were a focus of the US Department of Energy's Aquatic Species Program in the 1980s and 1990s (Sheehan 1998). However, low petroleum prices eventually eroded monetary support for alternative (and renewable) energy sources until increasing petroleum prices over the last two decades reinvigorated interest in alternatives.

The advent and increased use of fracking technologies has opened up new petroleum and hydrocarbon reservoirs,

Citation: Matthew W. Fields, Adam Hise, Egan J. Lohman, Tisza Bell, Rob D. Gardner, Luisa Corredor, Karen Moll, Brent M. Peyton, Gregory W. Characklis, Robin Gerlach, Sources and resources: importance of nutrients, resource allocation, and ecology in microalgal cultivation for lipid accumulation, DOI: 10.1007/s00253-014-5694-7

and almost \$190 × 10^9 was spent in the USA in 2012 to drill and "frac" for conventional hydrocarbons (www.eia.gov). However, the process of fracking increases the production rate and not the ultimate supply of hydrocarbons, and peak hydrocarbon production is predicted to occur around 2030 (www.eia.gov). Regardless of current market conditions and availability of conventional sources, alternatives are needed to circumvent future economic and environmental impacts from continued exploration and harvesting of conventional hydrocarbons.

Conservative estimates predict (assuming a lipid content of 25–30 % (*w*/*w*) in microalgae) that an area equivalent to 3 % of the arable cropland in the USA would be required to grow sufficient microalgae to replace 50 % of the transportation fuel needs in the USA (Chisti 2007; Georgianna and Mayfield 2012). Although the interest in algal biofuels has been reinvigorated (Courchesne et al. 2009; Greenwell et al. 2010; Razghefard 2013), significant fundamental and applied research is still needed to fully maximize algal biomass and biochemical production for biofuels and other products.

The accumulation of lipids is of substantial interest because these compounds are energy-rich biodiesel precursors (Dismukes et al. 2008; Hu et al. 2008). Much of the reported research has focused on increasing algal lipid accumulation upon exposing cultures to a range of environmental stresses prior to harvest (Hu et al. 2008; Valenzuela et al. 2012, 2013; Mus et al.2013; Lohman et al. 2013; references therein). Temperature variations, pH, salinity, light, and osmotic and chemical stress inducements have also been investigated with varying success (Sharma et al. 2012). While a stress event can increase lipid accumulation, it can also limit biomass production, but the stress scenario provides a tractable method to study and understand lipid accumulation at the laboratory scale (Valenzuela et al. 2013). Because of the inherent connection between carbon (C), nitrogen (N), and phosphorus (P) in biological systems, macronutrient deprivation has been proven to significantly enhance lipid accumulation in different diatom and algae species. While nitrogen limitation is the most commonly studied stress in green algae and diatoms; the effect of silica limitation is regularly studied in diatoms (Valenzuela et al. 2012; Lohman et al. 2013; Chu et al. 2013; Schnurr et al.2013). Light and temperature are also known stressors that can impact lipid accumulation (Hu et al. 2008), and particular wavelengths have been shown to impact the rate and amount of accumulated lipid in *Chlorella* (Atta et al. 2013). Keeping in mind that a vast majority of living pools of C, N, and P resides in the microbial realm (Whitman et al. 1998), much work is needed to understand the link between C, N, and P in controlling resource allocation both with respect to natural and man-made systems. In this context, a 50 % replacement of transportation fuel by renewable biological sources would impose a vast nutrient demand (Pate et al. 2011). However, microalgal biomass/product production can be coupled to wastewater resources (e.g., water, N, and P), and wastewater from agricultural, industrial, and municipal activity may provide a cost-effective source of nutrients. Agricultural and municipal wastewater can be high in N and P (Aslan and Kapdan 2006; Hoffmann 2002; Mallick 2002; Pittman et al. 2011), and thus, there is a great potential for the integration of wastewater treatment and algal biofuel/biomass production (Fig. 1). However, an improved understanding of the relationship between the effects of N, P, and micronutrient availability on cellular resource allocation (cell growth versus lipid storage) in microalgae is needed. This mini-review will briefly discuss the current literature on the use of nutrient deprivation and other conditions to control and optimize microalgal culture growth in the context of cell and lipid accumulation.

Figure 1: The biological recycling of carbon, nitrogen, and phosphorus to harvest fuel and food linked to sunlight to reduce net consumption of N and P and net production of C

Nutrient-dependent Lipid Accumulation

Under optimal growth conditions, (i.e., adequate supply of nutrients including C, N, P, and sunlight), algal biomass productivity can exceed 30 g dry weight per square meter per day (Gordon and Polle 2007); however, the lipid content of the biomass is typically very low (<5 %*w*/*w*) and is species-dependent (Gordon and Polle 2007). The low-lipid content is due to lipid biosynthesis being a metabolic process that is typically stimulated by stress inducement. Essentially, biomass synthesis and lipid biosynthesis compete for photosynthetic assimilation of inorganic carbon, and a fundamental metabolic switch is required to shift from biomass production to energy storage metabolism (Schuhmann et al. 2012; Valenzuela et al. 2012). As denoted by Odum (1985), stress is a syndrome that consists of inputs and outputs, and the input is the stressor that is contrasted to the stress, or the output. Lipids (the output) are typically believed to provide a storage function within the cell that enables the organism to endure adverse environmental conditions, i.e., the stressor. The output can be viewed as the cessation of cell production and the accumulation of lipids in response to the input of unbalanced resources (e.g., N, P, and/or sunlight). It is likely that there are tradeoffs in terms of biomass versus lipid accumulation depending on the different levels of perturbation (Fig. 2).

Recent research has provided evidence that lipids may also act as a reservoir for specific fatty acids such as polyunsaturated fatty acids (PUFAs) (Ratha et al. 2013). PUFAs play a key role in the structural components of cell membranes, and as antioxidants (PUFAs can counteract free radical formation during photosynthesis). As such, PUFA-rich TAGs might donate specific compounds necessary to rapidly reorganize membranes through adaptive metabolic responses to sudden changes in environmental conditions (Khozin-Goldberg and Cohen 2006). However, a recent study showed that PUFA content in lipids can negatively impact biodiesel quality based upon lipids from *Chlorella pyrenoidosa* (Shekh et al. 2013), and this result suggests that lipid composition, and not just amounts, should be considered. In either case, lipid is an energy-rich storage compound that can be chemically transesterified to produce fatty acid methyl esters (FAME), the biological equivalent to diesel fuel (a.k.a., biodiesel). However, to maximize lipid biosynthesis, the producing organism is typically induced through environmental stress conditions (Hu et al. 2008). In addition, most studies have been based upon axenic cultures with limited understanding of potential bacterial "contamination," and thus, lipid accumulation may be different at different scales of biological resolution (discussed below).

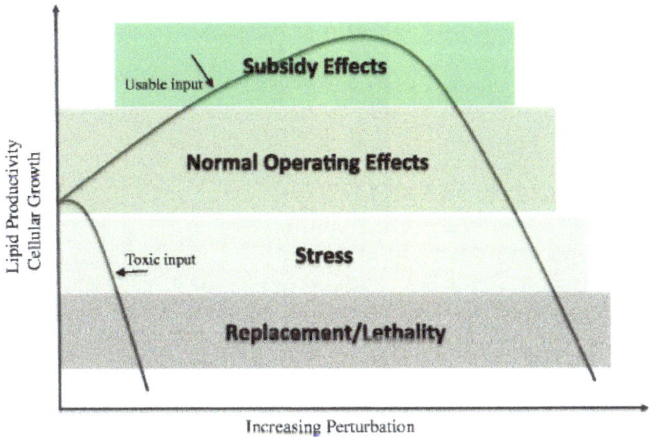

Figure 2: Hypothetical performance curve for an increasingly perturbed (i.e., stressed) microalgal system being used to produce photoautotrophic biomass and/or lipids. Adapted from Odum et al. (1979)

Significant work has been done to identify and optimize stress-inducement strategies that enhance lipid accumulation in microalgal species. Nutrient deprivation, specifically nitrogen depletion, is the most prevalent technique employed (Hu et al. 2008). This may be due to two factors: (1) Lack of requisite nutrients such as nitrogen limits the capacity to synthesize proteins necessary for biomass production (e.g., cellular division). In order to compensate, the organism must take advantage of alternative metabolic pathways for inorganic carbon fixation, such as fatty acid synthesis and hence, store those de novo fatty acids as TAG (Msanne et al. 2012). (2) Photosynthesis and the electron transport chain in eukaryotic microalgae produce ATP and NADPH as energy "storage" and electron carrier metabolites, respectively (Halsey et al. 2013). These metabolites are consumed during biomass production resulting in ADP and $NADP^+$, which in turn, are regenerated via photosystems. Under normal growth conditions, this cycle maintains a balanced ratio of the reduced and oxidized forms of these metabolites; however, when biomass production is impaired due to a lack of requisite nutrients, the pool of $NADP^+$and ADP can become depleted. This can lead to a potentially dangerous situation for the cell because photosynthesis is mainly controlled by light availability, and cannot be shut off completely. Fatty acid synthesis consumes NADPH and ATP; therefore, increased fatty acid synthesis replenishes the pool of required electron acceptors in the form of $NADP^+$, and de novo fatty acids are most frequently stored as lipid (Brown et al. 2009). Here, we will review the most successful strategies involving nutrient stress to induce lipid accumulation in commonly studied microalgal species.

Nitrogen and Phosphorus

Nutrient availability is critical for cell division and intracellular metabolite cycling, and once nutrients such as N or P become depleted or limited in the medium, invariably, a steady decline in cellular reproduction rate ensues. Once this occurs, the activated metabolic pathways responsible for biomass production are down-regulated and cells instead divert and deposit much of the available C into lipid (Wang et al. 2009; Valenzuela et al. 2013). There have been numerous studies to compare different N sources in the context of maximal biomass or lipid accumulation, and the results are different dependent upon the organism. Breuer et al. (2012) accumulated previous literature on 56 eukaryotic, photoautotrophic genera studied in the context of lipid accumulation (Table 1). The authors chose *Chlorella vulgaris*, *Chlorella zofingiensis*, *Nannochloris* UTEX 1999, *Neochloris oleoabundans*, *Scenedesmus obliquus*,*Dunaliella tertiolecta*, *Isochrysis galbana*, *Phaeodactylum tricornutum*, and *Prophyridium cruentum* to conduct normalized growth and lipid accumulation studies with nitrate as the N source (Breuer et al. 2012). Under N deprivation, *Chlorella vulgaris*, *Chlorella zofingiensis*,*Neochloris oleoabundans*, and *Scenedesmus obliquus* accumulated over 35 % dry weight as TAG, and *Scenedesmus obliquus* and *Chlorella zofingiensis* had the highest TAG productivity (240–320 mg L^{-1} day^{-1}) among the nine compared strains.

Table 1: Genera of 56 eukaryotic, photoautotrophs previously studied and reported for the accumulation of lipids. Modified from Breuer et al. (2012)

Amphora	Ankitrodesmus	Biddulphia	Botryococcus
Bracteacoccus	Chaetoceros	Chlamydomonas	Chlorella
Chlorococcum	Chroomonas	Cryphecodinium	Cryptomonas
Cylindrotheca	Dictyosphaerium	Dunaliella	Ellipsoidion
Emuliania	Enteromorpha	Euglena	Fragilaria
Glossomastrix	Gymnodinium	Haematococcus	Hantzchi
Hemiselmis	Isochrysis	Monallantus	Monodus
Nannochloris	Nannochloropsis	Navicula	Neochloris
Nephroselmis	Nitzschia	Ochromonas	Parietochloris
Pavlova	Phaeodactylum	Pheomonas	Polytoma
Porphyridium	Protosiphon	Prototheca	Rhodomonas

Rhodosorus	Scenedesmus	Scrippsiella	Selenastrum
Skeletonema	Stichococcus	Tetraselmis	Thalassiosira
Ulothirx	Volvox		

When the model Chlorophyte *Chlamydomonas reinhardtii* was cultivated under N limitation, an increase in lipid was also observed. Interestingly, fully saturated C_{16} fatty acids were the most abundantly synthesized compounds, whereas polyunsaturated C_{18} fatty acids remained relatively unchanged in this organism under the tested conditions (Lohman et al. 2013). While nitrate supported increased biomass compared to ammonium in *Monoraphidium* sp. SB2 (Wu et al. 2013), *Chlorococcum ellipsoideum* exhibited elevated lipid levels with urea compared to nitrate (Li et al. 2013). A different *Scenedesmus* strain (sp. R-16) was shown to have the highest lipid accumulation with nitrate compared to urea, peptone, or yeast extract (Ren et al. 2013). To date, nitrate is a commonly studied N source used to understand nutrient deprivation to induce lipid accumulation; however, different N sources have different effects dependent upon the organism. This is most likely a consequence of typical habitat for the organism as well as long-term life history that is common for the respective species. As the need for nutrient recycling becomes more evident, different types and mixtures of nutrients (e.g., human, agriculture, industrial) must continue to be evaluated. For example, two recent studies investigated the ability of *Chlamydomonas polypyrenoideum* and *Chlorella pyrenoidosa* to grow and accumulate lipids during cultivation on dairy wastewater (Kothari et al. 2012, 2013), and we recently grew a green alga isolated from storage ponds of coal-bed water that produced lipids under nutrient deprivation (Fields, Nagy, and Barnhart, unpublished results). Nitrogen deprivation was shown to induce lipid accumulation in the wastewater isolates, *Scenedesmus*sp. 131 and *Monoraphidium* sp. 92 with ammonium, nitrate, or urea (Eustance et al. 2013) or nitrate depletion in *Skeletonema marinoi* (Bertozzini et al. 2013). Interestingly, *Ettlia oleoabundans* initiated lipid accumulation in response to increased temperature before nitrate was completely depleted (Yang et al. 2013). These results suggest that different combinations of potential stressors could impact lipid accumulation in different ways.

In addition to N, P starvation to induce lipid accumulation in microalgae has been studied as a sole stress or in combination with N limitation. In general, greater lipid accumulation due to N deprivation has been observed compared to P deprivation as reported for various *Chlorella*species (Feng et al. 2012; Liang et al. 2012). When the marine diatom *Phaeodactylum tricornutum* was grown under N and P limitation, an increase in lipid accumulation was noticed in all limiting conditions (Valenzuela et al. 2012; Burrows et al. 2012). However, cultures of*Phaeodactylum tricornutum* that were limited exclusively in N showed a more significant increase in TAG than cultures that were limited solely in P. The combined limitation of both N and P resulted in the highest lipid concentrations in *Phaeodactylum tricornutum* (Valenzuela et al. 2012, 2013). Given the commonly accepted N/P ratio of 16:1 in microalgal biomass (Redfield1958), the *Phaeodactylum tricornutum* work demonstrated that the external N/P ratio was 27 and the cellular N/P ratio was between 8:1 and 9:1 when lipid accumulation was observed (Valenzuela et al. 2012).

Both N and P deprivation result in cell cycle cessation, but the relative lipid accumulation response is different, and this observation is most likely a consequence of cellular resource allocation (e.g., protein/chlorophyll vs. nucleotides). Based upon results in *Phaeodactylum tricornutum*, we observed a fivefold greater increase in specific fluorescence of Nile Red, a commonly used indicator of lipid accumulation (Gardner et al. 2011), when cells were depleted of nitrate compared to cells depleted of phosphate. In addition, resupplementation of N or P promoted cellular growth, cessation of lipid accumulation, and increased lipid consumption in*Phaeodactylum tricornutum* (Valenzuela et al. 2013).

Carbon

It is important to keep in mind that when comparing different nutrient-deprived states, carbon, above all else, is absolutely required for lipid biosynthesis (Palmqvist et al. 1988; Spalding2008). Without carbon, independent of nutrient deprivation, biomass or lipid biosynthesis is impossible. Therefore, the most successful reports of lipid induction techniques in microalgal lipid production typically involve elevated concentrations of inorganic carbon in tandem with N and/or P limitation (Gardner et al. 2011, 2012; Sharma et al. 2012). These strategies often employ a CO_2 sparge to increase dissolved CO_2 above atmospheric concentrations, or addition of soluble inorganic carbon during inoculation or just prior to nutrient depletion (Gardner et al.2011, 2012). It should be kept in mind that the addition of soluble inorganic carbon (e.g., bicarbonate) can also affect pH and osmolarity. The addition of large amounts of dissolved inorganic carbon via a CO_2 gas sparge can contribute significantly to the production cost in an algal biorefinery (e.g., Liu et al. 2013), and alternative methods to gaseous CO_2-based carbon supply should be considered in conjunction with pH control. Gardner et al. (2011, 2012) demonstrated that the dosage of small amounts of bicarbonate, solely or in combination with a CO_2 sparge, can achieve similar algal growth and lipid production yields compared to continuous CO_2 sparging. The use of bicarbonate addition, versus CO_2 sparging, could result in significantly lower equipment costs. In either case, elevated concentrations of C, combined with N or other nutrient deprivation, has been shown to induce lipid accumulation in virtually every microalgal species tested. However, an improved understanding of cellular and population responses to not only the respective concentrations but the ratios of macronutrients (e.g., C, N, and P) will improve resource utilization and promote efficient, cost-effective processes.

Silicon Limitation

Reports on silicon limitation have revealed that both marine and freshwater diatoms will accumulate lipid under Si-limiting conditions (Sharma et al. 2012), and diatoms possess immense potential as contributors to biodiesel production. When faced with Si-limitations, most diatoms appear to direct carbon storage towards lipid (Roessler 1988), albeit the response is dependent on the degree of Si content in the cell wall. Diatoms incorporate biologically available Si as monomeric or dimeric silicic acid into silicious cell walls (frustules) and require approximately 7 % of the energy expenditure required for polysaccharide cell wall formation characteristic of green algae (Hildebrand et al. 2012; Kroger and Poulsen 2008; Raven 1983). Diatoms produce comparatively less cellular starch, such that fixed carbon has increased potential to be allocated to lipid accumulation (Burrows et al. 2012; Gardner et al. 2011; Roessler 1988; Smith et al. 2012). In fact, diatom cells can accumulate enough TAG to cause the frustules to break under silica deplete conditions (Hildebrand et al. 2012), potentially reducing the need for energy intensive procedures associated with lipid extraction in green algae.

Numerous studies have shown increased lipid accumulation when diatoms are cultured in silica deplete media (Lombardi and Wangersky 1991; McGinnis et al. 1997; Obata et al. 2013; Yu et al. 2009). However, the majority of these studies were performed on marine diatoms (e.g.,*Cylindrotheca spp*., *Thalassiosira pseudonana*, and *Phaeodactylum tricornutum*) grown in media containing comparatively lower silica concentrations (Taguchi et al. 1987; Roessler 1988; Yu et al. 2009). The results of Moll et al. (2014) indicate that increasing the silica concentration will increase cell numbers, which is vital for improving algal biodiesel productivity in terms of increased biomass. Therefore, while research on marine diatoms for use in biofuel applications may be advantageous for use in large-scale raceway ponds due to the ability to tolerate saline environments, the actual use may be limited until conditions are optimized for diatom cell growth and lipid accumulation.

While silica limitation is known to increase lipid accumulation, combined with other physiological stresses, lipid accumulation may be enhanced. A recent study investigated the effect of coincident silica and nitrate limitation and HCO_3^- addition to promote lipid accumulation in a freshwater diatom. Moll et al. (2014) observed that combined silica and nitrate limitation, as well as sodium bicarbonate addition increased lipid accumulation compared to individual stressors with or without HCO_3^-. One hypothesis for this observation is the effect on the cell cycle. Olson et al. (1986) and Vaulot et al. (1987) revealed that for *Thalassiosira weisflogii* and *Hymenomonas carterae*, nitrate and silica limitation resulted in halting the cell cycle at G_1 and the G_1/S and G_2/M boundaries, respectively (Darley and Volcani1969). It is possible that the two combined nutrient limitations at different periods within the cell cycle may contribute to cellular stress and ultimately lead to enhanced lipid accumulation in diatoms.

Iron Limitation

As mentioned above, N, P, and C are the most important macronutrients, but Fe is the most versatile and important trace element for biochemical catalysis. Approximately 30 to 40 % of the world's oceans are iron limited, and studies have investigated "iron fertilization" experiments whereby iron is added to high-nutrient low chlorophyll (HNLC) areas to induce phytoplankton growth and CO_2 fixation (Buesseler et al. 2004). Iron-limited conditions are thought to alter cell physiology by reducing cell volume, chlorophyll content, and photosynthetic activity, and thus appear to impact cellular accumulation more than lipid accumulation per se. Specifically in *Phaeodactylum tricornutum*, the following enzymes were down-regulated during iron-starvation: β-carbonic anhydrase, phosphoribulo kinase (PRK), two RuBisCO enzymes and a HCO_3^- transporter, likely resulting in decreased carbon fixation and cellular growth (Allen et al. 2008). The results suggest that iron limitation greatly impacts cell growth and accumulation, and that approximately 10 μmol Fe/mol C is needed by marine algae (Morel et al. 1991). Iron limitation has also been linked to increased rates of silicification, thus increasing cell density and cell sinking. According to Allen et al. (2008), cells grown under Fe-limited conditions fixed carbon 14 times slower compared to cells grown in iron-replete conditions. Since iron limitation can result in detrimental physiological effects, it is pertinent to determine the potential for these processes to be useful for commercial scale lipid accumulation.

Biofilm Growth

One of the most significant limitations to the economical use of algae is the high cost of harvesting and concentrating the biomass (Johnson and Wen 2010; Christenson and Sims2012; Ozkan et al. 2012; Schnurr et al. 2013; Bernstein et al. 2014). To date, research has been focused on microalgae in suspended phase for lipid production, and few studies have focused on the biofilm growth state. However, the biofilm growth state provides some advantages over suspended growth systems in terms of biomass accumulation and maintenance that would be beneficial for biomass harvesting and concentrating prior to processing. Algal suspensions are often between 0.02 and 0.06 % total suspended solids (TSS), and significant energy is required to harvest and concentrate the cells to 5 to 25 % TSS. Biofilms can range from 6 to 16 % TSS (Schnurr et al. 2013), and could potentially minimize biomass-processing costs (Johnson and Wen 2010; Christenson and Sims 2012; Ozkan et al. 2012). In general, the available algal biofilm studies are based upon wastewater treatment, biofilm structure and development, and aquaculture applications (Johnson and Wen 2010; Christenson and Sims 2012; Patil and Anil2005; Irving and Allen 2011; Avendaño-Herrera and Riquelme 2007). There is a small amount of research on biofilm systems for the production of biomass and lipids in eukaryotic photoautotrophs (Schnurr et al. 2013; Bernstein et al. 2014), but very little in relation to the influence of environmental stresses.

Recently, Schnurr et al. (2013) reported biofilm growth under nutrient starvation to stimulate lipid accumulation. A semi-continuous flat plate parallel horizontal photobioreactor system (PBR) was designed to control the bulk medium nitrogen and silicon concentrations until nutrient depletion and biofilm onset. Wastewater was used to seed biofilm growth and was later replaced by synthetic medium and pure cultures of *Nitzschia palea* and *Scenedesmus obliquus*. Well-attached, thick algal biofilms were observed in all experiments, until N and Si levels decreased to below detection limits, resulting in detachment from the substratum. In contrast to suspended algae, the algal biofilms did not accumulate more neutral lipids when exposed to nutrient-deficient conditions in these studies. Similar results were reported by Bernstein et al. (2014) who observed little lipid accumulation in mixed culture wastewater biofilms on the field scale or in laboratory-scale algal biofilm reactors seeded with a *Botryococcus* sp. (strain WC2B). Based upon these results, there appears to be fundamental differences in the way suspended cultures and biofilm cultures respond to nutrient deprivation. The exact reasons for differences between suspended and biofilm cells are unknown, but may be a consequence of altered nutrient cycling in biofilms due to altered carbon flow for cellular turnover and compound accumulation. A result of "community growth" (i.e., biofilms) may be to accumulate excess C and reducing equivalent as cells and exopolymer rather than internal storage molecules (e.g., lipid). Future work is needed to discern the differences between the physiological states of biofilm and free-living cells in multiple species.

It is possible that benthic microorganisms would prove more useful for biofilm growth modes, and we have recently grown a benthic diatom in biofilm reactors that could accumulate lipids (Fields, Whitney, and Valenzuela, unpublished results). These results suggest that the two growth modes can elicit different behaviors, and numerous research approaches and questions need to be explored to better understand the feasibility and cellular responses of microalgal biofilms for biomass and lipid accumulation.

Ecological Effects

The literature offers many examples of increased lipid production in numerous algal species cultivated as monocultures in closed photobioreactor (PBR) or open raceway systems under varying nutrient limitations. However, as demonstrated by mathematical models and field experiments, phytoplankton biodiversity can be correlated to increased productivity (Tilman1981; Downing and Leibold 2002; Striebel et al. 2009). Furthermore, in natural freshwater systems, productivity, measured as biomass, was highest when there was abundant nutrient availability (Interlandi and Kilham 2001). These observations underlie the challenge of needing high biomass loads to maximize overall lipid production. Obviously, productivity can fall under the guise of several metrics ranging from biomass, cell number, chlorophyll/pigments, and more recently, lipid accumulation. Despite the success of increased lipid content in nutrient-deprived monocultures, recent studies indicate that comparable lipid production can also be achieved in nutrient-rich systems with a diverse community.

A study by Stockenreiter et al. (2012) demonstrated increased lipid production in naturally occurring algal communities compared to that of single species cultivated in PBRs (Stockenreiter et al. 2012). For freshwater systems, P is the key nutrient responsible for eutrophication and can greatly alter productivity when limited (Schindler 1977; Smith and Bennett 1999; Schindler et al. 2008; Stockenreiter et al. 2012). Based upon observations with PBRs with deplete nutrient availability, one would expect to see substantially higher lipid content in the oligotrophic communities than the eutrophic. However, Stockenreiter et al. (2012) showed a linear increase in total algal lipid content in correlation with species richness of the examined communities and that lipid content in communities did not differ significantly from 22 laboratory monocultures (1.4×10^6 versus 3.3×10^6 pg ml^{-1}). Although in no way conclusive, results such as these suggest comparable lipid values in nutrient replete and deplete systems and indicate the need to further investigate the relationship between nutrient type and abundance in the context of lipid production in mixed communities (Stockenreiter et al. 2013).

A diverse community is also more resistant to invasion from other species that could outcompete the desired algal species (Tilman 1977, 1981). Higher nutrient availability may also aid in algal cultivation by making algae less susceptible to viral infection. Coupled with community diversity, the relative health of each species is an important component. "Healthy" algae (i.e., cells not under nutrient stress) can be more resistant to viral infection that leads to cell lysis (Murray 1995). Rhodes and Martin (2010) developed a theoretical model that implicated high-nutrient availability in significantly reduced viral infection, and such scenarios will be important to consider in microalgal cultivation processes.

In contrast, despite the success of increased lipid accumulation in PBR monocultures under nutrient-limiting conditions, economic assessments indicate that PBRs operating on a large scale may not be commercially viable (Smith 2009). However, some have argued for hybrid systems that utilize a combination of both closed and open systems (Singh and Dhar 2011), or modified PBRs such as solid-state reactors (Naumann et al. 2013). In addition, if ponds are not well mixed, biomass loss due to dark respiration may impact performance for some microalgae (Huesemann et al. 2013). The ecology of open and closed systems will have different parameters and inputs that need to be considered in order to control and optimize ecosystem function (e.g., biomass, lipids, high value compounds). Therefore, life cycle analyses should help direct research to identify complementarity between water footprint, nutrient sources, regional light availability, process design, and targeted lipid-producing organisms.

Integrating Life Cycle Analysis

Algal biofuels have the potential to provide a substantial fraction of US transportation fuel while imposing a relatively small (arable) land footprint (Sander and Murthy 2010) and providing opportunities for reducing water and nutrient consumption relative to first generation biofuels (Clarens et al. 2011). The degree to which biology and engineering can contribute to these goals will, however, be a function of the entire life cycle (Zaimes and Khanna 2013). A circumscribed version of that life cycle consistent with this review, one involving only the production cycle (distinct from the usage cycle), includes microbial growth, dewatering/drying, extraction/conversion energy/input recovery stages, with each stage involving a number of choices (Fig. 3).

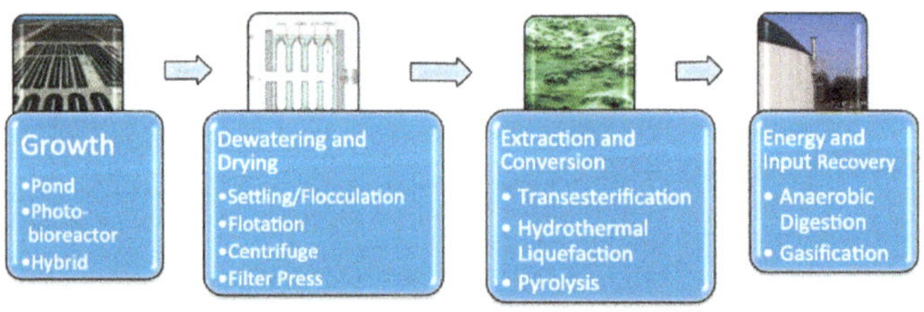

Figure 3: Primary stages and (alternative processes) in the microalgae to fuel production process

With respect to benefits, life cycle analysis (LCA) has promoted system optimization by highlighting processing alternatives that produce a net increase in system performance, while also avoiding environmental "burden shifting" that can be obscured when viewing the production system less holistically (Klöpffer 1997). In addition, the growth stage will be significantly affected by choices made in the other three production stages, with the extraction/conversion technology having particular importance (Brennan and Owende 2010). While each has respective strengths and weaknesses, two of the most critical distinctions from a life cycle perspective are (a) the degree of pre-conversion drying required (Lardon et al. 2009) and (b) whether the conversion process involves all of the algal biomass or only the lipid fraction (Kirrolia et al. 2013).

The dependence of transesterification processes on algal lipid content can impose extra costs in the growth stage (Brennan and Owende 2010), and lipid accumulation procedures typically come at the cost of algal productivity (Davis et al. 2011). As noted by Quinn et al. (2013) and Chowdhury et al. (2012), increasing lipid content can result in an increase in processing greenhouse gas (GHG) emissions, because less residual biomass is used in a potential energy recovery stage. Thus, the grid energy requirement increases proportionally with the lipid fraction. Wet extraction transesterification processes, while significantly reducing the drying energy input, typically involve solvent-based extractions that lead to concerns over solvent disposal (Torres et al. 2013). In addition, solvent recycling can be challenging and energy intensive due to the high volumes and accompanying wet slurry (Ríos et al. 2013). Simultaneous extraction and transesterification processes (i.e., "reactive extraction") offer the potential for increased oil yields and lower process costs (Rawat et al. 2013), but the effectiveness of these processes at an industrial scale is still untested (Nagarajan et al. 2013).

Hydrothermal liquefaction (HTL) processes, despite greater capital expense, also reduce drying/dewatering requirements through the utilization of a wet feedstock (López Barreiro et al.2013), while converting up to 60 % of the total biomass into a useable fuel (Liu et al. 2013). As a result, HTL can increase fuel yields relative to transesterification (Frank et al. 2013), and this technology may reduce the importance of advanced culturing methods to enhance algal lipid accumulation for biofuel production (Elliott et al. 2013). However, thermochemical conversion methods such as HTL make nutrient recycling less efficient, as the nutrient-rich byproducts are poorly suited for direct recycling into the growth process or anaerobic digestion (López Barreiro et al 2013). In addition, N loss during the conversion process results in a substantially increased nutrient requirement in the growth stage (Liu et al. 2013).

When attempting to assess the economic competitiveness of algal biofuels, an important source of uncertainty in existing studies is the extrapolation of lab and pilot-scale data to industrial scale production (Collet et al. 2013). The use of harmonization procedures (Sun et al. 2011) to reduce variability between multiple studies has allowed for more direct comparison of similar production frameworks given different assumptions. By harmonizing results from several sources, the base cost of production converged to \$11.57 gal^{-1} with a standard deviation of \$1.17 gal^{-1}, this from a pre-harmonization value of \$19.10 gal^{-1} and a standard deviation of \$6.22 gal^{-1} (Slade and Bauen 2013). Given energy content, algae are predicted to require a production cost of \$2.83 gal^{-1} to be competitive with petroleum-based fuels (based on March 2014 oil prices) (Chisti 2007). However, many technological advances must occur to reach this point, and one issue that has received significant attention is the choice of microbial growth technology.

A wide range of different, and often contradicting, results have arisen from economic comparisons of microbial growth via open raceway ponds (ORPs) and photobioreactors (PBRs). The scale-up of PBRs is challenged by proper gas exchange and pH maintenance, but has mostly been discouraged by concerns over higher capital expense (Rawat et al. 2013). The latter has often led to pessimism regarding the use of PBRs for large-scale algae production, but recent studies have suggested that the higher productivity resulting from control of growth parameters leads to lower overall costs than those observed with ORPs (Norsker et al. 2011). Economies of scale arising from increasing production will also continue to reduce costs, and Acién et al. (2012) observed that increasing production by a factor of 50 decreased production cost per gallon by a factor of 5. Other studies have found that systems making use of a hybridized approach, involving both PBRs (for inoculation) and ORPs (for large-scale production and lipid accumulation) (Brennan and Owende 2010; Quinn et al. 2013), have the potential to greatly improve the cost competitiveness of algal biofuels. In all cases, however, the costs of inputs weigh heavily in the determination of overall production costs. Consequently, there is considerable interest in the impacts of co-location scenarios that involve low cost (or no cost) opportunities to acquire CO_2 (e.g., via flue gas from power generation) as well as water and nutrients (e.g., municipal or agricultural wastewater). Access to these low-cost inputs can combine to decrease production costs by more than 50 % (Fernandez et al. 2012); however, physiological issues related to growth on these nutrient sources (Kirrolia et al. 2013) and co-location challenges (Liu et al. 2013) continue to require further research.

Life cycle analysis has been, and continues to be, successfully utilized to identify optimal algal biofuel production pathways. Ongoing refinement and application of this analytical technique can lead to advances that will guide future research toward a better understanding of the implications of many important choices, and thereby promote the development of more cost-effective and environmentally benign biofuel production processes. LCA has successfully identified synergies and tradeoffs between the growth stage and other parts of the production process, and results suggest that parallel research efforts involving both experimental research and life-cycle modeling can be effective (Collet et al. 2013).

CONCLUSION

With the reinvigorated interest in alternative fuels, microalgae provide one option that will likely contribute to an overall plan for biomass, biochemical, and biofuel production in a more sustainable and efficient manner. Given the typical ratio of C/N/P in microalgal biomass ($C_{106}/N_{16}/P_1$), much of the research has focused on N and P (P to a lesser extent) and these two elements are linked in different ways to C through resource allocation at the cellular, population, and community levels. In addition, the supply of C either as CO_2 or bicarbonate at critical times in the growth cycle can significantly improve lipid and biomass productivity. Micronutrients also play a role in cellular responses and activity, and Si and Fe need to be further studied with respect to C/N/P ratios and the allocation of C into desired compounds (e.g., lipids). Diatoms have potential for important contributions to lipid and biomass production but are less studied than the green algae. Many of the nutrient-deprived states have been studied with monocultures (or nearly axenic) as suspended cultures, and regardless of the systems used (e.g., closed reactors vs. open ponds), communities will assemble with different characteristics of stability, resiliency, and productivity. In addition, biofilms will likely develop, and may even be desired for the traits of accumulated biomass that can provide advantages for harvesting.

Moreover, while not directly covered in this mini-review, other resources/conditions will affect the cultivation of microalgae and include water, climate (e.g., light and temperature), land, and location (i.e., geography). Water will be essential for any biological process, and the water recycle will be crucial as many parts of the globe become increasingly stressed for potable water. Light is obviously an important parameter for phototrophs, and is inherently related to temperature as the need for light energy and heat-regulation scale at different proportions. Land is an essential commodity whether bioreactors or ponds are used and should not compete with agricultural needs. The location of growth and processing facilities are crucial aspects to be considered via LCA both for economic implications as well as the biology/ecology (e.g., biogeography) that can differ from region to region. Therefore, targeted science and engineering research is needed to better inform life cycle analyses and process design to maximize productivity, efficiency, and cost ratios.

ACKNOWLEDGMENTS

The authors would like to thank all members of the MSU Algal Biofuels Group for their helpful discussions. Support was provided by the US Department of Energy-Advancements in Sustainable Algal Production (ASAP) program under contract DE-EE0005993 and the National Science Foundation Sustainable Energy Pathways (SEP) Program under NSF CHE-1230632. We also would like to acknowledge student support from the Molecular Biosciences Program at Montana State University and the NSF IGERT Program in Geobiological Systems (DGE 0654336).

REFERENCES

1. Acién FG, Fernández JM, Magán JJ, Molina E (2012) Production cost of a real microalgae production plant and strategies to reduce it. Biotechnol Adv 30:1344–1353
2. Allen AE, LaRoche J, Maheswari U, Lommer M, Schauer N, Lopez PJ, Finazzi G, Fernie AR, Bowler C (2008) Whole-cell response of the pennate diatom *Phaeodactylum tricornutum* to iron starvation. Proc Natl Acad Sci U S A 105:10438–10443 Central
3. Aslan S, Kapdan IK (2006) Batch kinetics of nitrogen and phosphorus removal from synthetic wastewater by algae. Ecol Eng 28:64–70
4. Atta M, Idris A, Bukhari A, Wahidin S (2013) Intensity of blue LED light: a potential stimulus for biomass and lipid content in fresh water microalgae *Chlorella vulgaris*. Bioresour Technol 148:373–378
5. Avendaño-Herrera RE, Riquelme CE (2007) Production of a diatom-bacteria biofilm in a photobioreactor for aquaculture applications. Aquac Eng 36:97–104
6. Bernstein HC, Kesaano M, Moll K, Smith T, Gerlach R, Carlson RP, Miller CD, Peyton BM, Cooksey KE, Gardner RD, Sims RC (2014) Direct measurement and characterization of active photosynthesis zones inside wastewater remediating and biofuel producing microalgal biofilms. Bioresource Technol (in press)
7. Bertozzini E, Galluzzi L, Ricci F, Penna A, Magnani M (2013) Neutral lipid content and biomass production in *Skeletonema marinoi* culture in response to nitrate limitation. Appl Biochem Biotechnol 170:1624–1636
8. Brennan L, Owende P (2010) Biofuels from microalgae—a review of technologies for production, processing, and extractions of biofuels and co-products. Renew Sust Energ Rev 14:557–577
9. Breuer G, Lamers PP, Martens DE, Draaisma RB, Wijffels RH (2012) The impact of nitrogen starvation on the dynamics of triacylgycerol accumulation in nine microalgae strains. Bioresour Technol 124:217–226
10. Brown AP, Slabas AR, Rafferty JB (2009) Fatty acid biosynthesis in plants—metabolic pathways, structure and organization. In: Govindjee, Wada H, Murata N (eds) Lipids in photosynthesis: essential and regulatory functions, 30th edn., pp 11–34
11. Buesseler KO, Andrews JE, Pike SM, Charette MA (2004) The effects of iron fertilization on carbon sequestration in the southern ocean. Science 304:414–417
12. Burrows EH, Bennette NB, Carrieri D, Dixon JL, Brinker A, Frada M, Baldassano SN, Falkowski PG, Dismukes GC (2012) Dynamics of lipid biosynthesis and redistribution in the marine diatom *Phaeodactylum tricornutum* under nitrate deprivation. BioEnergy Res 5:876–885
13. Chisti Y (2007) Biodiesel from microalgae. Biotechnol Adv 25(3):294–306
14. Chowdhury R, Viamajala S, Gerlach R (2012) Reduction of environmental and energy footprint of microalgal biodiesel production through material and energy integration. Bioresour Technol 108:102–111
15. Christenson LB, Sims RC (2012) Rotating algal biofilm reactor and spool harvester for wastewater treatment with biofuels by-products. Biotechnol Bioeng 109:1674–1684
16. Chu F, Chu P, Cai P, Li W, Lam PKS, Zeng RJ (2013) Phosphorus plays an important role in enhancing biodiesel productivity of *Chlorella vulgaris* under nitrogen deficiency. Bioresour Technol 134:341–346
17. Clarens AF, Resurreccion EP, White MA, Colosi LM (2011) Environmental life cycle comparison of algae to other bioenergy feedstocks. Environ Sci Technol 44:1813–1819
18. Collet P, Spinelli D, Lardon L, Hélias A, Steyer JP, Bernard O (2013) Life-cycle assessment of microalgal-based biofuels. Biofuels Algae 287–312
19. Courchesne NMD, Pariesien A, Wang B, Lan CQ (2009) Enhancement of lipid production using biochemical,

genetic and transcription factor engineering approaches. J Bacteriol 141:31–41

20. Darley WM, Volcani BE (1969) Role of silicon in diatom metabolism: a silicon requirement for deoxyribonucleic acid synthesis in the diatom *Cylindrotheca fusiformis* Reimann and Lewin. Exp Cell Res 58:334–342
21. Davis R, Aden A, Pienkos PT (2011) Techno-economic analysis of autotrophic microalgae for fuel production. Appl Energy 88:3524–3531
22. Dismukes GC, Carrieri D, Bennettee N, Ananyez GM, Posewitz MC (2008) Aquatic phototrophs: efficient alternatives to land-based crops for biofuels. Curr Opin Biotechnol 19:235–240
23. Downing AL, Leibold MA (2002) Ecosystme consequences of species richness and composition in pond food webs. Nature 416:837–841
24. Dukes JS (2003) Burning buried sunshine: human consumption of ancient solar energy. Clim Chang 61:31–44
25. Elliott DC, Neuenschwander GC, Hart TR (2013) Hydroprocessing biooil and products separation for coke production. ACS Sustain Chem Eng 1:389–392
26. Eustance E, Gardner RD, Mole KM, Menicucci J, Gerlach R, Peyotn BM (2013) Growth, nitrogen utilization, and biodiesel potential for two chlorophytes grown on ammonium, nitrate, or urea. J Appl Phycol 25:1663–1677
27. Feng P, Deng Z, Fan L, Hu Z (2012) Lipid accumulation and growth characteristics of *Chlorella zofingiensis* under different nitrate and phosphate concentrations. J Biosci Bioeng 114:405–410
28. Fernandez GA, González-López CV, Fernández Sevilla JM, Molina Grima E (2012) Conversion of CO_2 into biomass by microalgae: how realistic a contribution may it be to significant CO_2removal? Appl Microbiol Biotechnol 57:96–586
29. Frank ED, Elgowainy A, Han J, Wang ZC (2013) Life cycle comparison of hydrothermal liquefaction and lipid extraction pathways to renewable diesel from algae. Mitig Adapt Strateg Glob Chang 18:137–158
30. Gardner R, Peters P, Peyton BM, Cooksey KE (2011) Medium pH and nitrate concentration effects on accumulation of triacylglycerol in two members of the chlorophyta. J Appl Phycol 23:1005–1016
31. Gardner R, Cooksey KE, Mus F, Macur R, Moll K, Eustance E, Carlson RP, Gerlach R, Fields MW, Peyton BM (2012) Use of sodium bicarbonate to stimulate triacylglycerol accumulation in the chlorophyte *Scenedesmus* sp. and the diatom *Phaeodactylum tricornutum*. J Appl Phycol 24:1311–1320
32. Georgianna DR, Mayfield SP (2012) Exploiting diversity and synthetic biology for the production of algal biofuels. Nature 488:329–335
33. Gordon JM, Polle JEW (2007) Ultrahigh bioproductivity from algae. Appl Microbiol Biotechnol 76:969–975
34. Greenwell HC, Laurens LML, Shields RJ, Lovitt RW, Flynn KJ (2010) Placing microalgae on the biofuels priority list: a review of the technological challenges. J R Soc Interface 7:703–726
35. Halsey KH, O'Malley RT, Graff JR, Milligan AJ, Behrenfeld MJ (2013) A common partitioning strategy fro photosynthetic products in evolutionary distinct phytoplankton species. New Phytol 198:1030–1038
36. Hildebrand M, Davis AK, Smith SR, Traller JC, Abbriano R (2012) The place of diatoms in the biofuels industry. Biofuels 3:221–240
37. Hoffmann JP (2002) Wastewater treatment with suspended and nonsuspended algae. J Phycol 34:757–763
38. Hu Q, Sommerfeld M, Jarvis E, Ghirardi M, Posewitz M, Seibert M, Valenzuela AD (2008) Microalgal triacylglycerols as feedstocks for biofuel production: perspectives and advances. Plant J 54:621–639
39. Huesemann MH, Van Wagenen J, Miller T, Chavis A, Hobbs S, Crowe B (2013) A screening model to predict microalgae biomass growth in photobioreactors and raceway ponds. Biotechnol Bioeng 110:1583–1594
40. Interlandi SJ, Kilham SS (2001) Limiting resources and the regulation of diversity in phytoplankton communities. Ecology 82:1270–1282
41. Irving TE, Allen DG (2011) Species and material considerations in the formation and development of microalgal biofilms. Appl Microbiol Biotechnol 92:283–294
42. Johnson MB, Wen Z (2010) Development of an attached microalgal growth system for biofuel production. Appl Microbiol Biotechnol 85:525–534
43. Khozin-Goldberg I, Cohen Z (2006) The effect of phosphate starvation on the lipid and fatty acid composition of the fresh water eustigmatophyte *Monodus subterraneus*. Phytochemistry 67:696–701
44. Kirrolia A, Bishnoi NR, Singh R (2013) Microalgae as a boon for sustainable energy production and its future research and development aspects. Renew Sustain Energy Rev 20:642–656
45. Klöpffer W (1997) Life cycle assessment. Environ Sci Pollution Res 4:223–228
46. Kothari R, Pathak VV, Kumar V, Singh DP (2012) Experimental study for growth potential of unicellular alga *Chlorella pyrenoidosa* on diary wastewater: an integrate approach for treatment and biofuel production. Bioresour Technol 116:466–470
47. Kothari R, Prasad R, Kumar V, Singh DP (2013) Production of biodiesel from microalgae*Chlamydomonas polypyrenoideum* grown on dairy industry wastewater. Bioresour Technol 144:499–503
48. Kroger N, Poulsen N (2008) Diatoms, from cell wall biogenesis to nanotechnology. Ann Rev Genet 42:83–107
49. Lardon L, Helias A, Sialve B, Steyer JP, Bernard O (2009) Life-cycle assessment of biodiesel production from microalgae. Environ Sci Technol 43:6475–6481
50. Li T, Wan LL, Li AF, Zhang CW (2013) Responses in growth, lipid accumulation, and fatty acid composition of four oleaginous microalgae to different nitrogen sources and concentrations. Chin J Oceanol Limnol

31:1306–1314

51. Liang K, Zhang Q, Gu M, Cong W (2012) Effect of phosphorus on lipid accumulation in freshwater microalga *Chlorella* sp. J Appl Phycol 25:311–318
52. Liu X, Saydeh B, Ernki P, Colosi LM, Greg MB, Rhoses J, Clarens AF (2013) Pilot-scale data provide enhanced estimates of the life cycle energy and emissions profile of algae biofuels produced via hydrothermal liquefaction. Bioresour Technol 148:163–171
53. Lohman EJ, Gardner RD, Halverson L, Peyton BM, Gerlach R (2013) Carbon partitioning in lipids synthesized by *Chlamydomonas reinhardtii* when cultured under three unique inorganic carbon regimes. Bioresource Technol (in press)
54. Lombardi A, Wangersky P (1991) Influence of phosphorus and silicon on lipid class production by the marine diatom *Chaetoceros gracilis* grown in turbidostat cage cultures. Mar Ecol Prog Oldendorf 77:39–47
55. López Barreiro DL, Prins W, Ronsse F, Brilman W (2013) Hydrothermal liquefaction (HTL) of microalgae for biofuel production: state of the art review and future prospects. Biomass Bioenergy 53:113–127
56. Mallick N (2002) Biotechnological potential of immobilized algae for wastewater N, P, and metal removal: a review. BioMetals 15:377–390
57. McGinnis K, Dempster T, Sommerfeld M (1997) Characterization of the growth and lipid content of the diatom *Chaetoceros muelleri*. J Appl Phycol 9:19–24
58. Moll KM, Gardner RD, Eustance EO, Gerlach R, Peyton BM (2014) Combining multiple nutrient stresses and bicarbonate addition to promote lipid accumulation in diatoms. Algal Res (accepted with revisions)
59. Morel FMM, Rueter JG, Price NM (1991) Iron nutrition of phytoplankton and its possible importance in the ecology of ocean regions with high nutrient and low biomass. Oceanography 4:56–61
60. Msanne J, Xu D, Konda AR, Casas-Mollano JA, Awada T, Cahoon EB, Cerutti H (2012) Metabolic and gene expression changes triggered by nitrogen deprivation in the photoautotrophically grown microalgae *Chlamydomonas reinhardtii* and *Coccomyxa* sp. C-169. Phytochemistry 75:50–59
61. Murray AG (1995) Phytoplankton exudation—exploitation of the microbial loop as a defense against algal viruses. J Plankton Res 17:1079–1094
62. Mus F, Toussaint J, Cooksey KE, Fields MW, Gerlach R, Peyton BM, Carlson RP (2013) Physiological and molecular analysis of carbon source supplementation and pH stress-induced lipid accumulation in the marine diatom *Phaeodactylum tricornutum*. Appl Microbiol Biotechnol 97:3625–3642
63. Nagarajan S, Chou SK, Cao SY, Wu C, Zhou Z (2013) An updated comprehensive techno-economic analysis of algae biodiesel. Bioresour Technol 145:150–156
64. Naumann T, Cebi Z, Podola B, Melkonian M (2013) Growing microalgae as aquaculture feeds on twin-layers: a novel solid-state photobioreactor. J Appl Phycol 25:1413–1420
65. Norsker NH, Barbosa MJ, Vermuë MH, Wijffels RH (2011) Microalgal production—a close look at the economics. Biotechnol Adv 29:24–27
66. Obata T, Fernie AR, Nunes-Nesi A (2013) The central carbon and energy metabolism of marine diatoms. Metabolites 3:325–346
67. Odum EP (1985) Trends expected in stressed ecosystems. Bioscience 35:419–422
68. Odum EP, Finn JT, Franz EH (1979) Perturbation theory and the subsidy-stress gradient. Bioscience 29:349–352
69. Olson RJ, Vaulot D, Chisholm SW (1986) Effects of environmental stresses on the cell cycle of two marine phytoplankton species. Plant Physiol 80:918–925 Central
70. Ozkan A, Kinney K, Katz L, Berberoglu H (2012) Reduction of water and energy requirement of algae cultivation using an algae biofilm photobioreactor. Bioresour Technol 114:542–548
71. Palmqvist K, Sjöberg S, Samuelsson G (1988) Induction of inorganic carbon accumulation in the unicellular green algae *Scenedesmus obliquus* and *Chlamydomonas reinhardtii*. Plant Physiol 87:437–442
72. Pate R, Klise G, Wu B (2011) Resource demand implications for US algae biofuels production scale-up. Appl Energy 88:3377–3388
73. Patil JS, Anil AC (2005) Biofilm diatom community structure: influence of temporal and substratum variability. Biofouling 21:189–206
74. Pittman JK, Dean AP, Osundeko O (2011) The potential of sustainable algal biofuel production using wastewater resources. Bioresour Technol 102:17–25
75. Quinn JC, Catton KB, Johnson S, Bradley TH (2013) Microalgae to biofuels lifecycle assessment—multiple pathway evaluation. Bioenergy Res 6:591–600
76. Ratha SK, Babu S, Renuka N, Prasanna R, Prasad RBN, Saxena AK (2013) Exploring nutritional modes of cultivation for enhancing lipid accumulation in microalgae. J Basic Microbiol 53:440–450
77. Raven JA (1983) The transport and function of silicon in plants. Biol Rev 58:179–207
78. Rawat I, Kumar RR, Mutanda T, Bux F (2013) Biodiesel from microalgae: a critical evaluation from laboratory to large-scale production. Appl Energy 103:444–467
79. Razghefard R (2013) Algal biofuels. Photosynth Res 117:207–219
80. Redfield A (1958) The biological control of chemical factors in the environment. Am Sci 46:205–211
81. Ren HY, Liu BF, Ma C, Zhao L, Ren NQ (2013) A new lipid-rich microalgae *Scenedesmus* strain R-16 isolated using Nile Red staining: effects of carbon and nitrogen sources and initial pH on the biomass and lipid production. Biotechnol Biofuels 6:143
82. Rhodes CJ, Martin AP (2010) The influence of viral infection on a plankton ecosystem undergoing nutrient enrichment. J Theo Biol 265:225–237
83. Ríos SD, Torres CM, Torras C, Salvado J, Maeo-Sanz JM, Jimenez L (2013) Microalgae-based biodiesel: economic analysis of downstream process realistic

scenarios. Bioresour Technol 136:617–625

84. Roessler PG (1988) Effects if silicon deficiency in lipid composition and metabolism in the diatom *Cyclotella cryptica*. J Phycol 24:394–400
85. Sander K, Murthy GS (2010) Life cycle analysis of algae biodiesel. Int J Life Cycle Assess 15(7):704–714
86. Schindler DW (1977) Evolution of phosphorus limitation in lakes. Science 195:260–262
87. Schindler DW, Wolfe AP, Vinebrooke R, Crowe A, Blais JM, Miskimmin B, Freed R, Perren B (2008) The cultural eutrophication of Lac la Biche, Alberta, Canada: a paleoecological study. Can J Fish Aquat Sci 65:2211–2223
88. Schnurr PJ, Espie GS, Allen DG (2013) Algae biofilm growth and the potential to stimulate lipid accumulation through nutrient starvation. Bioresour Technol 136:337–344
89. Schuhmann H, Lim DKY, Schenk PM (2012) Perspectives on metabolic engineering for increased lipid contents in microalgae. Biofuels 3:71–86
90. Sharma KK, Schuhmann H, Schenk PM (2012) High lipid induction in microalgae for biodiesel production. Energies 5:1532–1553
91. Sheehan J, 1998. A look back at the US Department of Energy's aquatic species program (NREL/TP-580-24190)
92. Shekh AY, Shrivastava P, Krishnamurthi K, Mudliar SN, Devi SS, Kanade GS, Lokhande SK, Chakrabarti T (2013) Stress-induced lipids are unsuitable as a direct biodiesel feedstock: a case study with *Chlorella pyrenoidose*. Bioresour Technol 138:382–386
93. Singh NK, Dhar DW (2011) Microalgae as second generation biofuel. Agron Sustain Dev 31:605–629
94. Slade R, Bauen A (2013) Micro-algae cultivation for biofuels: cost, energy balance, environmental impacts and future prospects. Biomass Bioenergy 53:29–38
95. Smith SR (2009) Organic contaminants in sewage sludge and their significance for agricultural recycling. Phil Trans Royal Soc A Math Phy Eng Sci 367:4005–4041
96. Smith VH, Bennett SJ (1999) Nitrogen:phosporus supply ratios and phytoplankton community structure in lakes. Arch Hydrobiol 146:37–53
97. Smith SR, Abbriano RM, Hildebrand M (2012) Comparative analysis of diatom genomes reveals substantial differences in the organization of carbon partitioning pathways. Algal Res 1:2–16
98. Spalding MH (2008) Microalgal carbon-dioxide-concentrating mechanisms: *Chlamydomonas*inorganic carbon transporters. J Exp Bot 59:1463–1473
99. Stockenreiter M, Graber AK, Haupt F, Stibor H (2012) The effect of species diversity on lipid production by microalgal communities. J Appl Phycol 24:45–54
100. Stockenreiter M, Haupt F, Graber AK, Seppala J, Spilling K, Tamminen T, Stibor H (2013) Functional group richness: implications of biodiversity for light use and lipid yield in microalgae. J Phycol 49:838–847
101. Striebel M, Behl S, Stibor H (2009) The coupling of biodiversity and productivity in phytoplankton communities: consequences for biomass stoichiometry. Ecology 90:2025–2031
102. Sun A, Davis R, Starbuck M, Ben-Amotz A, Pate R, Pienkos PT (2011) Comparative cost analysis of algal oil production for biofuels. Energy 36:5169–5179
103. Taguchi S, Hirata JA, Laws EA (1987) Silicate deficiency and lipid synthesis of marine diatoms. J Phycol 23:260–267
104. Tilman D (1977) Resource competition between planktonic algae—experimental and theoretical approach. Ecology 58:338–348
105. Tilman D (1981) Tests of resource competition theory using 4 species of Lake Michigan algae. Ecology 62:802–815
106. Torres CM, Rios SD, Torras C, Salvado J, Mateo-Sanz JM, Jimenez L (2013) Microalgae-based biodiesel: a multicriteria analysis of the production process using realistic scenarios. Bioresour Technol 147:7–16
107. Valenzuela J, Mazurie A, Carlson RP, Gerlach R, Cooksey KE, Peyton BM, Fields MW (2012) Potential role of multiple carbon fixation pathways during lipid accumulation in*Phaeodactylum tricornutum*. Biotechnol Biofuels 5:1–17
108. Valenzuela J, Carlson RP, Gerlach R, Cooksey KE, Peyton BM, Bothner B, Fields MW (2013) Nutrient re-supplementation arrests bio-oil accumulation in *Phaeodactylum tricornutum*. Appl Microbiol Biotechnol 97:7049–7059
109. Vaulot D, Olson R, Merkel S, Chisholm S (1987) Cell-cycle response to nutrient starvation in two phytoplankton species, *Thalassiosira weissflogii* and *Hymenomonas carterae*. Mar Biol 95:625–630
110. Wang ZT, Ullrich N, Joo S, Waffenschmidt S, Goodenough U (2009) Algal lipid bodies: stress induction, purification, and biochemical characterization in wild-type and starchless*Chlamydomonas reinhardtii*. Eukaryotic Cell 8:1856–1868
111. Whitman WB, Coleman DC, Wiebe WJ (1998) Prokaryotes: the unseen majority. Proc Natl Acad Sci U S A 95:6578–6583
112. Wu LF, Chen PC, Lee CM (2013) The effects of nitrogen sources and temperature on cell growth and lipid accumulation of microalgae. Int Biodeterior Biodegrad 85:506–510
113. Yang Y, Mininberg C, Tarbet A, Weathers P (2013) At high temperature lipid production in*Ettlia oleoabundans* occurs before nitrate depletion. Appl Microbiol Biotechnol 97:2263–2273
114. Yu E, Zendejas F, Lane P, Gaucher S, Simmons B, Lane T (2009) Triacylglycerol accumulation and profiling in the model diatoms *Thalassiosira pseudonana* and *Phaeodactylum tricornutum*(*Baccilariophyceae*) during starvation. J Appl Phycol 21:669–681
115. Zaimes GG, Khanna V (2013) Environmental sustainability of emerging algal biofuels: a comparative life cycle evaluation of algal biodiesel and renewable diesel. Environ Prog Sustain Energy 32:926–936

CHAPTER 5 MICROALGAE AS A SUSTAINABLE BIOFUEL FEEDSTOCK AND ANIMAL FEED

Krystal K Lum, Jonggun Kim and Xin Gen Lei

Keywords: Algae, Biofuel, Feed, Health, Nutrition

INTRODUCTION

The projected rise of global population from the current 7 billion to over 9 billion within the next several decades[1, 2] garners urgent needs for renewable energy and alternative foods. As global supplies of petroleum continue to decline, renewable nature fuels including biofuel are being explored. Biofuels are defined as the energy derived from raw biological materials, and show promise in harnessing adequate energy and reducing greenhouse gas emissions associated with fossil fuels. As corn and soybean are staple food crops for humans as well as the two main conventional feedstuffs that provide energy and protein for food-producing animals, the projected global population rise, along with the expansion of animal production, presents a serious threat to nutrition security for both humans and animals. However, the increasing use of these feedstocks for biodiesel and bioethanol production has driven up their global prices. From 2007 to 2011, the worldwide production of bioethanol nearly doubled from 49.6 to 84.6 billion liters[3]. In parallel, the price of corn was increased from $163 to $291 per metric ton over the four years[4]. In 2011, the United States produced approximately 12.4 billion bushels of corn, and 38% of which was used to produce bioethanol or to generate other co-products[3]. Apparently, current allocations of corn and(or) soybean for the biofuel and animal productions are unsustainable. Alternatives to these ingredients are required to maintain a harmonious infrastructure among the fuel, food, and feed industries.

Marine microalgae bear attractive properties for biofuel production[5–8]. Subsequently, the fat-extracted microalgal biomass derived from the biofuel production may be a promising carbon-neutral animal feed supplement[9–11]. In fact, our laboratory has recently demonstrated that the de-fatted biomass of *Staurospira sp.* with 19% crude protein replaced 7.5% of corn and soybean meal without affecting the growth performance or health status of broiler chickens[12]. Notably, the dual application of microalgae as a new source of biofuel and animal feed will help alleviate the greenhouse effects associated with current energy and food production.

Citation: Krystal K Lum, Jonggun Kim and Xin Gen Lei, Dual potential of microalgae as a sustainable biofuel feedstock and animal feed, DOI: 10.1186/2049-1891-4-53

MICROALGAE FOR BIOFUEL PRODUCTION

Marine microalgae sequester carbon dioxide (CO_2) through photosynthesis, and may be used to produce biogas including methane and hydrogen via anaerobic processing[13–15]. While certain species of microalgae were recognized in the 1940s to yield high amounts of cellular lipids under selective growth pressures, it was not until the 1950's when algae were viewed as a potential energy source, and were tested for methane gas production via anaerobic digestion of their cell carbohydrates[16, 17]. The flexibility and(or) adaptivity of microalgal species to water and cultural conditions allows us to spare fresh water and arable land for crop production[18]. The land use efficiency of microalgae for biofuel production, grown with 30% oil content by weight, was 130 and 338 times greater than the conventional biodiesel feedstock soybean and corn, respectively[6].

While optimal growth conditions for microalgae are species-specific, photoautotrophic cultivation of these single cell species at large scales for biofuel and co-products depends on the technical and economic feasibility. At the present time, the photoautotrophic production of microalgae is marginally cost-effective only for generating value-added co-products or feed additives used in aquaculture[19, 20]. In such productions, microalgae are grown in the presence of light within constructions such as open raceway ponds. To extract the lipids, microalgae are first de-watered. The concentrated biomass is subsequently processed to optimize the solvent extraction through cell disruption, particle size reduction, and drying[21]. The remaining microalgae skeleton after lipid extraction is the so-called de-fatted microalgal biomass to be used as an animal feed. Without the presumed feed application, the commercial microalgae cultivation and processing for biofuel production[22, 23] remains largely cost-ineffective. Therefore, the feed application of the de-fatted biomass would not only create a new source of animal feed to mitigate the current competition with human food supply, but also help make the biofuel production of microalgae economically feasible.

MICROALGAE FOR ANIMAL NUTRITION

Feasibility of Microalgae as Animal Feed

Although the use of whole microalgae in animal diets has long been studied, dating as far back as in the 1950's, only recent literature has shown attempts to supplement lipid-extracted microalgae in animal diets[12, 24]. Initially, researchers sought out methods to culture algae in ponds, and the developments of which were quickly followed by studies on algae supplementation into animal diets as a protein source[25, 26]. Different sources of cultivated algae were effective in maintaining animal growth performance, and in some cases improving daily body weight gain. Sewage-grown *Chlorella* and *Scenedesmus* sp. gained attention as potential nutrient sources due to their high crude protein and carotenoid contents[27–30]. In 1952, Combs reported that supplementing 10% *Chlorella* sp. into a diet deficient in riboflavin and vitamin A improved feed efficiency and growth of chicks[27]. Later work showed no adverse response of growth by chicks to diets containing 20% sewage-grown, aluminum-free*Chlorella* and *Scenedesmus* sp., compared with those fed a corn-soybean meal based diet[28].

During the following decades, pond and tap water-grown algae were found to sustain fish growth in aquaculture, and 6-10% sewage-grown algae incorporated in a barley-based diet for growing-finishing pigs maintained their growth rate and feed conversion efficiency[31, 32]. Other algal variants, such as the blue-green algae *Spirulina* sp., were also investigated not only for their effects on overall growth performance, but also on organ health and reproductive characteristics of animals. Ross and Dominy supplemented diets for broiler chickens with up to 20% blue-green algae, and found the 3 wk-old broilers experienced depressed growth when algae inclusion levels were higher than 10%[33]. At the inclusion rate of 12%, broilers showed slower growth in comparison with those fed 0, 1.5, 3.0, or 6.0% algae. Toyomizu et al.[34] fed broilers with up to 8% *Spirulina* sp. and reported no differences in their body weight, liver weight, abdominal fat, or kidney weight at 16 d of age. When early-weaned piglets were fed a diet with *S. maxima* replacing up to 12% of the total protein from skim milk, apparent digestibilities of the diets were reduced, but the piglet growth remained insignificantly different from that of the control pigs[35]. Fevrier and Seve[36] supplemented *S. maxima* into the diets of sows, and showed that the algae-fed sows had lower net weight gains with more and heavier piglets at the first reproductive cycle. By the end of the second reproductive cycle, the growth and litter characteristics were similar between the algae-fed and control groups, although the algae-fed sows produced a higher culling rate of piglets. Another study showed that feeding weanling pigs with a mixture of *S. maxima* and *A. platensis*, or *Chlorella* sp. in replacing 33% of soybean meal did not affect their body weight gain, feed efficiency, or incidence of diarrhea compared with pigs fed a control diet. Apparent toxicity or gastro-intestinal lesion was absent in the algae-fed pigs[35].

In contrast, the de-fatted biomass of microalgal species, derived from the biofuel production research, has only recently been shown feasibility in replacing corn and soybean meal in diets for poultry, swine, and cattle. Studies supplementing the de-fatted biomass from *Staurospira*sp. to replace 7.5% corn and soybean meal in diets for weanling pigs did not affect their overall growth performance or plasma biochemical indicators[37, 38]. However, the pigs were incapable of tolerating a 15% replacement of corn and soybean meal[38]. Factors causing this intolerance might include amino acid imbalances, disruption of the acid:base balance, high ash content of the algal biomass, and(or) an overall reduction in the buffering capacity within the gastrointestinal tract of the weanling pigs. Austic et al.[12] replaced 7.5% of soybean meal with the de-fatted biomass of the *Staurosira* sp. in the diets of broilers, and showed decreased body weight gain and feed efficiency during the first three weeks of experiment. During the following three weeks, these differences were no longer seen. In the same study, broiler chicks fed a diet containing essential amino acids (Met, Lys, Ile, Thr, Trp, and Val) co-supplemented with 7.5% of the de-fatted biomass did not show growth performance differences from the control group. A diet with 10% lipid-extracted *Nannochloropsis oculata* meal was well tolerated by adolescent male rabbits, which showed similar final body weight, serum urea nitrogen, blood glucose, and organ histology in comparison with rabbits fed a control diet[39, 40]. Finishing wethers maintained similar growth performance and carcass characteristics (*longissimus* muscle area, dressing percentage, marbling score, hot carcass weight, and subcutaneous adipose depth) when fed up to 20% de-fatted algal biomass on a dry matter basis as a protein replacement, in comparison with wethers fed a control diet[24]. In a study that examined effects of supplementing dietary dry matter with de-fatted *Lithothamnium calcareum* meal in Holstein cows, up to 1% algal meal in the diet mediated venous acid–base balance after acidosis induction, but did not improve the tract digestibility or growth performance of the cows[41].

Nutritional Values of Microalgae

While the nutritional profiles of microalgae vary considerably with the species used, a large majority are characterized by protein, carbohydrate, and lipid contents that are comparable, if not superior, to conventional feedstuffs (Table 1). Dietary soybean meal typically contains up to 48% crude protein, and lays claim to a relatively well balanced amino acid profile. The diversity of microalgae makes certain species amenable to cultivation for diet-specific needs of humans and animals. A commonly cultivated algae species for human consumption: *S. maxima*, contains high levels of vitamin B_1, vitamin B_2, and β-carotene, and up to 71% crude protein with sufficient concentrations of all essential amino acids except for the sulfur-containing ones (Table 2)[5, 42]. Since protein is considered to be the most expensive nutrient in animal feed[43], developing natural alternatives to soybean meal may be cost competitive.

Table 1: Nutrient composition of conventional feedstuffs and various algae (% dry matter)[a]

Source	Crude protein	Carbohydrates	Lipids
Soybean	37	30	20
Corn	10	85	4
Wheat	14	84	2
Anabaena cylindrical	43-56	25-30	4-7
Arthrospira maxima	60-71	13-16	6-7
Chlorella vulgaris	51-58	12-17	14-22
Spirogyra sp.	6-20	33-64	11-21
Synechococcus sp.	73	15	11

[a]Adapted from[11, 44–46].

Table 2: Amino acid profile of conventional protein sources and various algae (g/100 protein)[a]

Source	Ala	Arg	Asp	Cys	Glu	Gly	His	Ile	Leu	Lys	Met	Phe	Pro	Ser	**Thr**	**Try**	**Tyr**	**Val**
Egg	-	6.2	11.0	2.3	12.6	4.2	2.4	6.6	8.8	5.3	3.2	5.8	4.2	6.9	5.0	1.7	4.2	7.2
Soybean	5.0	7.4	1.3	1.9	19.0	4.5	2.6	5.3	7.7	6.4	1.3	5.0	5.3	5.8	4.0	1.4	3.7	5.3
Dried Whole Milk	0.9	0.9	2.0	0.2	5.5	0.5	0.7	1.6	2.6	2.1	0.6	1.3	2.5	1.4	1.2	0.4	1.2	1.7
C. vulgaris	9.4	6.9	9.3	-	13.7	6.3	2.0	3.2	9.5	6.4	1.3	5.5	5.0	5.8	5.3	-	2.8	7.0
D. bardawil	7.3	7.3	10.4	1.2	12.7	5.5	1.8	4.2	11.0	7.0	2.3	5.8	3.3	4.6	5.4	0.7	3.7	5.8
S. platensis	9.5	7.3	11.8	0.9	10.3	5.7	2.2	6.7	9.8	4.8	2.5	5.3	4.2	5.1	6.2	0.3	5.3	7.1
Aphanizomenon flos-aquae	4.7	3.8	4.7	0.2	7.8	2.9	0.9	2.9	5.2	3.2	0.7	2.5	2.9	2.9	3.3	0.7	-	3.2

[a]Based on reference[11].

Among all dietary amino acids, lysine and methionine are the first and second limiting amino acids. Many microalgal species contain relatively high amounts of lysine, but as previously described, are somewhat deficient in the sulfur-containing amino acids: cysteine and methionine[5]. To maximize amino acid utilization by animals, diets are typically formulated by mixing different feedstuffs to balance amino acid profiles and(or) by supplementing synthetic amino acids to meet their nutrient requirements. Austic et al.[12] reported that the decreased growth performance of broilers fed the de-fatted *Staurospira* sp. biomass in the first three weeks was prevented by the supplementation of essential amino acids. In a laying hen study, a 7.5% replacement of corn and soybean meal with the de-fatted microalgal biomass did not negatively affect hen production or health parameters, yet significantly increased the redness and decreased both the lightness and yellowness of the egg yolks[47]. However, inclusions of the biomass at 10% reduced feed intake, plasma uric acid, and egg albumen weight. Up to a third of soybean meal was successfully replaced in the diets of weanling pigs by algae biomass from the cyanobacteria *S. maxima*, *Arthrospira platensis*, and *Chlorella* sp.[35]. In another study conducted on laying hens, an inclusion of 10% *Porphyridium* sp. red algal biomass did not affect their body weight, egg production rate, or egg weight, but lowered egg yolk cholesterol levels by 24%[48].

MICROALGAE FOR VALUE-ADDED ANIMAL PRODUCTS

The research on microalgal biomass supplementation in food-producing animals has opened a new gateway to improve human health. A study conducted in the last decade showed not only 10% enhancement of growth but also potential in producing iodine-rich pork for human consumption when pigs were fed algae containing naturally high iodine[49]. Likewise, microalgae may be cultivated for their advantageous fatty acid profile (Table 3), with notable enrichment in omega-3 (n-3) polyunsaturated fatty acids (PUFA) including arachidonic acid, docohexaenoic acid (DHA), eicosapentaenoic acid (EPA), and γ–linoleic acid[21]. The enriched concentrations of these omega-3 fatty acids by a variety of microalgal species represent a largely untapped natural resource with multiple health implications for both humans and animals. Among many fat sources, fish products represent the major source of n-3 fatty acids[50]. However, marine fish species are incapable of synthesizing n-3 fatty acids by themselves; they may obtain n-3 fatty acids by consuming microalgae or other algae-consuming fish[51]. Most notably, dietary intakes of n-3 fatty acids are associated with decreased risks of chronic diseases [52]. However, the average fish consumption in the U.S. has not changed since 1983, at approximately 6.5 kg per year[53]. To increase the public consumption of n-3 fatty acids, studies have been conducted with full-fatted or de-fatted microalgae to generate n-3 fatty acid-fortified animal products such as milk, meats, and eggs. Apparently, the profile of PUFA in the de-fatted microalgal biomass is affected by the lipid extraction procedure. While a practical solvent extraction can remove 95% of lipids from oilseeds, it is ineffective to extract microalgal lipid due to the relatively high moisture content and cellular elasticity [21]. Grima et al.[54] reported that the lipid extraction yields of dried algal biomass range from 50 to 93% depending on the polarity of solvent systems. In fact, a higher proportion of long chain polyunsaturated fatty acids in the de-fatted biomass may bolster its appeal as an n-3 fatty acids source for developing value-added animal products.

Table 3: Generalized fatty acid profiles (%) of oil extracts from *Spirulina maxima*(SP),*Chlorella vulgaris*(Cv),*Scenedesmus obliquus*(Sc),*Dunaliella tertiolecta*(Dt),*Nannochloropsis*sp, (Nanno), and*Neochloris oleabundans*(Neo)[a]

Fatty acid	Sp	Cv	Sc	Dt	Nanno	Neo
14:0	0.34	3.07	1.48	0.47	7.16	0.43
16:0	40.16	25.07	21.78	17.70	23.35	19.35
16:1	9.19	5.25	5.95	0.88	26.87	1.85
16:2	N.D.	N.D.	3.96	3.03	0.39	1.74
16:3	0.42	1.27	0.68	1.24	0.48	0.96
16:4	0.16	4.06	0.43	10.56	N.D.	7.24
18:0	1.18	0.63	0.45	N.D.	0.45	0.98
18:1	5.43	12.64	17.93	4.87	13.20	20.29
18:2	17.89	7.19	21.74	12.37	1.21	12.99
18:3	18.32	19.05	3.76	30.19	N.D.	17.43
18:4	0.08	N.D.	0.21	N.D.	N.D.	2.10
20:0	0.06	0.09	N.D.	N.D.	N.D.	N.D.
20:1	N.D.	0.93	N.D.	N.D.	N.D.	N.D.
20:2	0.48	N.D.	N.D.	N.D.	N.D.	N.D.
20:3	N.D.	0.83	N.D.	N.D.	N.D.	N.D.
20:4	N.D.	0.23	N.D.	N.D.	2.74	N.D.
20:5	N.D.	0.46	N.D.	N.D.	14.31	N.D.
SFA	41.74	28.86	23.71	18.17	30.96	20.76
MUFA	14.62	18.82	23.88	5.75	40.07	22.14
PUFA	37.35	33.09	30.78	57.39	19.13	42.46

[a]Based on reference[55].

Microalgal biomass or oil may be supplied in the feed of ruminants to manipulate their milk fatty acid composition. The content of n-3 fatty acids in milk, especially DHA, has been shown to increase with the inclusion of algal biomass or oil, without affecting the milk yield[56–58]. However, a study with similar dietary algae incorporation reported a decrease in milk fat content[59]. Many studies in ruminants have focused on producing n-3 fatty acid-fortified milk. In such an attempt, Stamey et al.[51] supplied either algal biomass or oil to dairy cows. The overall milk fat and yield were not affected by either supplement, but the milk DHA (C22:6) content was increased by both supplementations. Glover et al.[57] reported that feeding microalgae and fresh forage decreased total milk fat content, but elevated its DHA concentration. A parallel study was conducted with sheep to determine if algae and (or) their co-supplementation with sunflower oil in the diet could enhance the nutrient profile of milk[59]. While the milk yield was unaffected by the dietary treatments, the milk fat content was decreased and the milk DHA concentration was increased as dietary algae concentration rose. Franklin et al.[58] determined that feeding dairy cows with either algae protected against ruminal biohydrogenation or unprotected algae elevated the milk DHA concentration without affecting the milk fat content. Collectively, the supplementation of microalgae to the diets of ruminants increased the concentration of n-3 PUFA in milk, with a mixed effect on the milk fat content.

Likewise, fatty acid compositions in meats and eggs are also affected by dietary supplementations of microalgae. In lambs and horses, dietary microalgae increased the n-3 fatty acid content in meat and blood, respectively[60, 61]. In pigs, dietary microalgae increased DHA concentrations in the loin and subcutaneous fat[62]. However, this increment did not manifest as a dose-dependent response[63]. To enrich eggs with n-3 fatty acids, flaxseed is usually supplemented in diets of laying hens[64]. Alternatively, dietary algae have been shown to increase n-3 and decrease n-6 fatty acids in eggs[64, 65]. Supplementing flaxseed at 10% or higher usually decreases egg production and the body weight of hens. Moreover, such high flaxseed supplementations may induce fatty acid oxidation, resulting in off-odor eggs[66]. Herber and Van Elswyk[65] supplemented up to 4.8% of golden marine algae into the diet of hens, and reported increased levels of DHA with a concomitant reduction of n-6 fatty acids. The inclusion of algae in the diet affected neither the egg production nor egg quality in the 56 wk-old hens. When up to 4.3% of fermented *Schizochytrum* sp. was included in the diets of hens, their egg production and feed conversion ratio were elevated[67]. The 0.86% and 4.3% supplementation of the microalgae elevated the DHA content to 134 and 220 mg/egg, respectively.

SPECIAL CONSIDERATIONS OF MICROALGAE

Despite their high nutritional values and health implications, caution must be taken to avoid possible toxicity of the microalgal biomass[8]. Among over 200,000 existing algal species, approximately 35,000 species are characterized[68] without apparent toxicological property[5]. However, certain species contain biogenic toxins including purines and non-biogenic toxins such as heavy metals. Some algal species are known to rapidly accumulate heavy metals at concentrations higher than their surroundings[5], while others generate pathological metabolites that cause neuro-degenerative disorders[69]. Such species require establishment and evaluation prior to their commercialization as feed supplements. Currently, no official regulations for heavy metals in microalgal products exist[5].

In spite of the relatively high crude protein contents, many microalgae species may still show limited biological values for the proteinaceous biomass due to the presence of non-protein nitrogen that consists of nucleic acids, nitrogen-containing cell walls, and amines. As single cell proteins, microalgae are photosynthetic unicellular eukaryotes containing nucleic acids that may represent approximately 10% of the crude protein[5]. Ruminal bacteria are effective in utilizing non-protein nitrogen to synthesize protein[70]. Simple-stomached animals lack ruminal microbes and thus are incapable of utilizing the relatively high levels of non-protein nitrogen in the microalgal biomass. Because the individual amino acid content and digestibility are important factors for formulating diets to support the maximum growth of animals without excreting excess nitrogen, it is necessary to optimize dietary amino acid profile and digestibility when the microalgal biomass is used to replace conventional feedstuffs.

Notably, the microalgal cell wall is largely indigestible by simple-stomached animals, yet contains immunostimulating properties that may provide added value to the biomass[71, 72]. Approaches to improve the absorption of microalgal nutrients include employing enzymatic treatments to enhance cell wall digestibility[71], as well as developing genetically-altered microalgae that lack cell walls[73]. Microalgal solubility is high at alkaline conditions[74]. Comprising approximately 10% cellular dry matter[71], the cell wall of *Chlorella* sp. contains high levels of β-1,3-glucan, a bio-active immunostimulating free-radical scavenger[75]. Other species, such as those of cyanobacterial origin, have well-recognized anti-tumor, anti-bacterial, and anti-viral properties[76]. Microalgal cell wall-derived sulfated polysaccharides and carrageenans show strong T and B cell mitogenic effects[77]. Animals fed the whole *Spirulina*sp. samples accumulated greater degrees of cytotoxic lymphocytes that were critical to innate immunity, and had elevated phagocytic activity and antibody production[78].

CONCLUSIONS

The use of microalgal biomass in animal feed will not only improve human and animal food security, but also facilitate cost-effective biofuel production and reduces greenhouse gas production of agriculture[6, 79–81]. Recent estimates indicate that 30% of the global algal production is used by the animal feed industry[5], amounting to a fast-growing $300 million in retail value[82]. Pragmatic species-specific large-scale algae refinery techniques must be devised to reduce the cost of the biomass production. It is necessary to determine limiting factors of the microalgal biomass that hinder its digestion and utilization by animals. Novel technology should be explored to improve microalgal nutrient utilization, and their long term nutritional and metabolic effects should be assessed. In addition, the production of DHA from microalgal biomass alone is a rapidly growing market. The last estimate is that approximately 300 tons is produced to generate $1.5 billion market value each year[72, 82]. However, the tremendous potential of using the microalgal biomass in producing DHA/EPA enriched eggs, meats, and milk for improving human health is yet to be fully explored.

ACKNOWLEDGEMENT

Research on this subject in the authors' laboratory has been supported in part by a USDA/DOE Biomass R&D Initiative grant.

REFERENCES

1. United States Census Bureau: International Database: World population information.http://www.census.gov/population/international/data/idb/informationGateway.php.
2. Population Reference Bureau: 2012 World Population Data Sheet.http://www.prb.org/pdf12/2012-population-data-sheet_eng.pdf.

3. Accelerating Industry Innovation: 2012 Ethanol Industry Outlook.http://ethanolrfa.3cdn.net/d4ad995ffb7ae8fbfe_1vm62ypzd.pdf.
4. Monthly commodity futures price chart.http://futures.tradingcharts.com/chart/CN/M.
5. Becker W: Microalgae in human and animal nutrition. Handbook of Microalgal Culture. Edited by: Richmond A. 2004, Oxford, UK: Blackwell Publishing Ltd, 312-351.
6. Chisti Y: Biodiesel from microalgae. Biotechnol Adv. 2007, 25: 294-306. 10.1016/j.biotechadv.2007.02.001.
7. Alzate ME, Muñoz R, Rogalla F, Fdz-Polanco F, SI P -E: Biochemical methane potential of microalgae biomass after lipid extraction. Chem Eng J. in press
8. Kovač DJ, Simeunović JB, Babić OB, Mišan AČ, Milovanović IL: Algae in food and feed. Food Feed Res. 2013, 40: 21-32.
9. Brune DE, Lundquist TJ, Benemann JR: Microalgal biomass for greenhouse gas reductions: potential for replacement of fossil fuels and animal feeds. J Environ Eng. 2009, 135 (11): 1136-10.1061/(ASCE)EE.1943-7870.0000100.
10. Shields RJ, Lupatsch I: Algae for aquaculture and animal feeds. J Anim Sci. 2012, 21: 23-37.
11. Becker EW: Microalgae: Biotechnology and Microbiology. 1994, Cambridge: Cambridge University Press
12. Austic RE, Mustafa A, Jung B, Gatrell S, Lei XG: Potential and limitation of a new defatted diatom microalgal biomass in replacing soybean meal and corn in diets for broiler chickens. J Agric Food Chem. 2013, 61: 7341-7348. 10.1021/jf401957z.
13. Mostafa S: Microalgal biotechnology: prospects and applications. Plant Science. Edited by: Nabin Kumar D. INTech,http://www.intechopen.com/books/plant-science/microalgal-biotechnology-prospects-and-applications.
14. Li Y, Horsman M, Wu N, Lan CQ, Dubois-Calero N: Biofuels from microalgae. Biotechnol Prog. 2008, 24 (4): 815-820.
15. Hughes AD, Kelly MS, Black KD, Stanley MS: Biogas from microalgae: is it time to revisit the idea?. Biotechnol Biofuel. 2012, 5: 86-10.1186/1754-6834-5-86.
16. Meier RL: Biological cycles in the transformation of solar energy into useful fuels. J Anim Sci. 1955, 14: 179-183.
17. Oswald WJ, Golueke CG: Biological transformation of solar energy. Adv Appl Microbiol. 1960, 2: 223-262.
18. Mata TM, Martins AA, Caetano NS: Microalgae for biodiesel production and other applications: a review. Renew Sust Energ Rev. 2010, 14: 217-232. 10.1016/j.rser.2009.07.020.
19. Borowitzka M: Microalgae for aquaculture: opportunities and constraints. J Appl Physiol. 1997, 9: 393-401.
20. Slade R, Bauen A: Micro-algae cultivation for biofuels: Cost, energy balance, environmental impacts and future prospects. Biomass Bioenerg. 2013, 53: 29-38.
21. U.S. Department of Energy: National algal biofuels technology roadmap: 9–10 December 2008. 2010, Maryland,http://www1.eere.energy.gov/biomass/pdfs/algal_biofuels_roadmap.pdf.
22. Chisti Y: Constraints to commercialization of algal fuels. J Biotechnol. 2013, 167: 201-214. 10.1016/j.jbiotec.2013.07.020.
23. Saharan BS, Sharma D, Sahu R, Sahin O, Warren A: Towards algal biofuel production: a concept of green bio energy development. Innovat Rom Food Biotechnol. 2013, 12: 1-21.
24. Dib MD, Engle TE, Han H, Roman-Muniz IN, Archibeque SL: Effects of algal meal supplementation to finishing wethers on performance and carcass characteristics [abstract]. J Anim Sci. 2012, 90 (Suppl 3): 148.
25. Gotaas HB, Oswald WJ, Ludwig HF: Photosynthetic reclamation of organic wastes. Sci Monthly. 1954, 79: 368-378.
26. Oswald WJ, Gotaas HB: Photosynthesis in sewage treatment. Trans Am Soc Civ Eng. 1957, 122: 73-105.
27. Combs GF: Algae (chlorella) as a source of nutrients for the chick. Science. 1952, 116: 453-454. 10.1126/science.116.3017.453.
28. Grau CR, Klein NW: Sewage-grown algae as a feedstuff for chicks. Poult Sci. 1957, 36: 1046-1051. 10.3382/ps.0361046.
29. Hundley JM, Ing RB: Algae as sources of lysine and threonine in supplementing wheat and bread diets. Science. 1956, 124: 536-537. 10.1126/science.124.3221.536.
30. Fink H, Herold E: Biological value of the protein of unicellular algae. On the protein value of unicellular algae and of young green leaves of higher plants/A contribution on alimentary liver necrosis of the rat and the treatment of protein deficiency in the tropics. Naturwissenschaften. 1955, 42: 516-517.
31. Ahmad MR: Observations on the effect of feeding *Labeo rohita*(Ham.) with microcystis aeruginosa *Mycrocystis aeroginosa*. Hydrobiologia. 1967, 29: 388-388. 10.1007/BF00189903.
32. Hintz HF, Heitman H: Sewage-grown algae as a protein supplement for swine. Anim Prod. 1967, 9: 135-140. 10.1017/S0003356100038393.
33. Ross E, Dominy W: The nutritional value of dehydrated, blue-green algae (spirulina plantensis) for poultry. Poult Sci. 1990, 69: 794-800. 10.3382/ps.0690794.
34. Toyomizu M, Sato K, Taroda H, Kato T, Akiba Y: Effects of dietary spirulina on meat colour in muscle of broiler chickens. Br Poult Sci. 2001, 42: 197-202. 10.1080/00071660120048447.
35. Yap TN, Wu JF, Pond WG, Krook L: Feasibility of feeding spirulina-maxima, arthrospira-platensis or chlorella sp to pigs weaned to a dry diet at 4 to 8 days of age. Nutr Rep Int. 1982, 25: 543-552.
36. Fevrier C, Seve B: Incorporation of a spiruline (spirulina maxima) in swine food. Ann Nutr Aliment. 1975, 29: 625-650.

37. Isaacs R, Roneker KR, Huntley M, Lei XG: A partial replacement of soybean meal by whole or defatted algal meal in diet for weanling pigs does not affect their plasma biochemical indicators [abstract]. J Anim Sci. 2011, 89 (Suppl 1): 723.

38. Lum KK, Roneker KR, Lei XG: Effects of various replacements of corn and soy by defatted microalgal meal on growth performance and biochemical status of weanling pigs [abatract]. J Anim Sci. 2012, 90 (Suppl 3): 701.

39. Howe BA, Roman-Muniz IN, Willson BD, Archibeque SL: Evaluation of algae meal from Nannochloropsis oculata as a protein source for non-ruminant animals [abstract]. J Anim Sci. 2010, 88 (Suppl 2): 101-102.

40. Howe BA, Roman-Muniz IN, Willson BD, Archibeque SL: Nannochloropsis oculata meal did not alter nutrient usage and had no adverse health effects when fed to rabbits as a protein source [abstract]. J Anim Sci. 2011, 89 (Suppl. 2): 724.

41. Lopex NM, Pereira RAN, Pereira MN: Intake, milk yield, and blood acid–base balance of cows in response to marine algae meal [abstract]. J Anim Sci. 2013, 91 (Suppl. 2): 29.

42. Clement G, Giddey C, Menzi R: Amino acid composition and nutritive valve of the alga Spirulina maxima. J Sci Food Agric. 1967, 18: 497-501. 10.1002/jsfa.2740181101.

43. Rezaei R, Wang W, Wu Z, Dai Z, Wang J, Wu G: Biochemical and physiological bases for utilization of dietary amino acids by young Pigs. J Anim Sci Biotechnol. 2013, 4: 7-10.1186/2049-1891-4-7.

44. Aaronson S, Dubinsky Z: Mass production of microalgae. Experientia. 1982, 38: 36-40. 10.1007/BF01944523.

45. Fabregas J, Herrero C: Marine microalgae as a potential source of single cell protein (SCP). Appl Microbiol Biotechnol. 1985, 23: 110-113. 10.1007/BF00938962.

46. Miles R, Chapman R: The benefits of fish meal in aquaculture diets. 2009,https://edis.ifas.ufl.edu/pdffiles/FA/FA12200.pdf.

47. Leng XJ, Hsu HN, Austic RE, Lei XG: Defatted algae biomass may replace one-third of soybean meal in diets for laying hens [abstract]. J Anim Sci. 2012, 90 (Suppl 3). 701.

48. Ginzberg A, Cohen M, Sod-Moriah U, Shany S, Rosenshtrauch A, Arad S: Chickens fed with biomass of the red microalga Porphyridium sp have reduced blood cholesterol level and modified fatty acid composition in egg yolk. J Appl Phycol. 2000, 12: 325-330. 10.1023/A:1008102622276.

49. He ML, Hollwich W, Rambeck WA: Supplementation of algae to the diet of pigs: a new possibility to improve the iodine content in the meat. J Anim Physiol Anim Nutr (Berl). 2002, 86: 97-104. 10.1046/j.1439-0396.2002.00363.x.

50. Daviglus ML, Stamler J, Orencia AJ, Dyer AR, Liu K, Greenland P, Walsh MK, Morris D, Shekelle RB: Fish consumption and the 30-year risk of fatal myocardial infarction. N Engl J Med. 1997, 336: 1046-1053. 10.1056/NEJM199704103361502.

51. Stamey JA, Shepherd DM, de Veth MJ, Corl BA: Use of algae or algal oil rich in n-3 fatty acids as a feed supplement for dairy cattle. J Dairy Sci. 2012, 95: 5269-5275. 10.3168/jds.2012-5412.

52. Albert CM, Hennekens CH, O'Donnell CJ, Ajani UA, Carey VJ, Willett WC, Ruskin JN, Manson JE: Fish consumption and risk of sudden cardiac death. JAMA. 1998, 279: 23-28. 10.1001/jama.279.1.23.

53. United States Department of Agriculture: Food Availability (Per Capita) Data System.http://www.ers.usda.gov/data-products/food-availability-(per-capita)-data-system.aspx#.Ua6jU0DrzVU.

54. Grima EM, González MJI, Giménez AG: Solvent extraction for microalgae lipids. Algae for biofuels and energy. Edited by: Borowitzka MA, Moheimani NR. 2013, New York, London: Springer, 187-205.

55. Gouveia L, Oliveira AC: Microalgae as a raw material for biofuels production. J Ind Microbiol Biotechnol. 2009, 36: 269-274. 10.1007/s10295-008-0495-6.

56. Papadopoulos G, Goulas C, Apostolaki E, Abril R: Effects of dietary supplements of algae, containing polyunsaturated fatty acids, on milk yield and the composition of milk products in dairy ewes. J Dairy Res. 2002, 69: 357-365.

57. Glover KE, Budge S, Rose M, Rupasinghe HP, Maclaren L, Green-Johnson J, Fredeen AH: Effect of feeding fresh forage and marine algae on the fatty acid composition and oxidation of milk and butter. J Dairy Sci. 2012, 95: 2797-2809. 10.3168/jds.2011-4736

58. Franklin ST, Martin KR, Baer RJ, Schingoethe DJ, Hippen AR: Dietary marine algae (Schizochytrium sp.) increases concentrations of conjugated linoleic, docosahexaenoic and transvaccenic acids in milk of dairy cows. J Nutr. 1999, 129: 2048-2054.

59. Toral PG, Hervas G, Gomez-Cortes P, Frutos P, Juarez M, de la Fuente MA: Milk fatty acid profile and dairy sheep performance in response to diet supplementation with sunflower oil plus incremental levels of marine algae. J Dairy Sci. 2010, 93: 1655-1667. 10.3168/jds.2009-2769.

60. Cooper SL, Sinclair LA, Wilkinson RG, Hallett KG, Enser M, Wood JD: Manipulation of the n-3 polyunsaturated fatty acid content of muscle and adipose tissue in lambs. J Anim Sci. 2004, 82: 1461-1470.

61. Hess TM, Rexford JK, Hansen DK, Harris M, Schauermann N, Ross T, Engle TE, Allen KG, Mulligan CM: Effects of two different dietary sources of long chain omega-3, highly unsaturated fatty acids on incorporation into the plasma, red blood cell, and skeletal muscle in horses. J Anim Sci. 2012, 90: 3023-3031. 10.2527/jas.2011-4412.

62. Sardi L, Martelli G, Lambertini L, Parisini P, Mordenti A: Effects of a dietary supplement of DHA-rich marine algae on Italian heavy pig production parameters. Livest Sci. 2006, 103: 95-103. 10.1016/j.livsci.2006.01.009.

63. Marriott NG, Garrett JE, Sims MD, Abril JR: Performance characteristics and fatty acid composition of pigs fed a diet with docosahexaenoic acid. J Muscle Foods. 2002,

13: 265-277. 10.1111/j.1745-4573.2002.tb00335.x.

64. Sim JS, Qi G: Designing poultry products using flaxseed. Flaxseed in Human Nutrition. Edited by: Thompson LU, Cunnane LU. 1995, Champaign, IL: American Oil Chemists' Society Press, 315-333.

65. Herber SM, Van Elswyk ME: Dietary marine algae promotes efficient deposition of n-3 fatty acids for the production of enriched shell eggs. Poult Sci. 1996, 75: 1501-1507. 10.3382/ps.0751501

66. Leeson S, Caston L, MacLaurin T: Organoleptic evaluation of eggs produced by laying hens fed diets containing graded levels of flaxseed and vitamin E. Poult Sci. 1998, 77: 1436-1440.

67. Abril JR, Barclay WR, Abril PG: Safe use of microalgae (DHA GOLD) in laying hen feed for the production of DHA-enriched eggs. Egg nutrition and biotechnology. Edited by: Sim JS, Nakai S, Guenter W. 1999, New York: CABI Publishing, 197-202.

68. Munir N, Sharif N, Shagufta N, Saleem F, Manzoor F: Harvesting and processing of microalgae biomass fractions for biodiesel production (a review). Sci Tech and Dev. 2013, 32: 235-243.

69. Jin DQ, Lim CS, Sung JY, Choi HG, Ha I, Han JS: Ulva conglobata, a marine algae, has neuroprotective and anti-inflammatory effects in murine hippocampal and microglial cells. Neurosci Lett. 2006, 402: 154-158. 10.1016/j.neulet.2006.03.068.

70. McNaught ML, Owen EC: The utilization of non-protein nitrogen in the bovine rumen. 8. The nutritive value of the proteins of preparations of dried rumen bacteria, rumen protozoa and brewer's yeast for rats. Biochem J. 1954, 56: 151-156.

71. Becker EW: Micro-algae as a source of protein. Biotechnol Adv. 2007, 25: 207-210. 10.1016/j.biotechadv.2006.11.002.

72. Spolaore P, Joannis-Cassan C, Duran E, Isambert A: Commercial applications of microalgae. J Biosci Bioeng. 2006, 101: 87-96. 10.1263/jbb.101.87.

73. Davies DR: Single cell protein and the exploitation of mutant algae lacking cell walls. Nature. 1971, 233: 143-144. 10.1038/233143a0.

74. Pulz O, Gross W: Valuable products from biotechnology of microalgae. Appl Microbiol Biotechnol. 2004, 65: 635-648. 10.1007/s00253-004-1647-x.

75. Iwamoto H: Industrial production of microalgal cell-mass and secondary products - major industrial species: *Chlorella*. Handbook of microalgal culture: biotechnology and applied phycology. Edited by: Richmond A. 2004, New Jersey: Willey-Blackwell Publishing, 255-263.

76. Singh S, Kate BN, Banerjee UC: Bioactive compounds form cyanobacteria and microalgae: an overview. Crit Rev Biotechnol. 2005, 25: 73-95. 10.1080/07388550500248498.

77. Schaeffer DJ, Grylov VS: Anti-HIV activity of extracts and compounds form algae and cyanobacteria. Ectoxicol Environ Saf. 2000, 45: 208-227. 10.1006/eesa.1999.1862.

78. Luescher-Mattli M: Algae, a possible source for new drugs in the treatment of HIV and other viral diseases. Curr Med Chem Anti Infect Agents. 2003, 7: 219-225.

79. Vijffels RA, Barbosa M: An outlook of microalgal biofuels. Science. 2010, 329: 796-799. 10.1126/science.1189003.

80. Darzins A, Pienkos P, Edye L: Current status and potential for algal biofuels production.http://www.fao.org/uploads/media/1008_IEA_Bioenergy_-_Current_status_and_potential_for_algal_biofuels_production.pdf.

81. Singh J, Gu S: Commercialization potential of microalgae for biofuels production. Renew Sust Energ Rev. 2010, 14: 2596-2610. 10.1016/j.rser.2010.06.014.

82. Guil-Guerrero JL, Navarro-Juarez R, Lopez-Martinez JC, Campra-Madrid P, Rebolloso-Fuentes MM: Functional properties of the biomass of three microalgal species. J Food Eng. 2004, 65: 511-517. 10.1016/j.jfoodeng.2004.02.014.

CHAPTER 6 MICROALGAL BIOFACTORIES: A PROMISING APPROACH TOWARDS SUSTAINABLE OMEGA-3 FATTY ACID PRODUCTION

T. Catalina Adarme-Vega, David K Y Lim, Matthew Timmins, Felicitas Vernen, Yan Li and Peer M Schenk

Keywords: Docosahexaenoic acid DHA Eicosapentaenoic acid EPA Microalgae Omega-3 fatty acids Polyunsaturated fatty acids

INTRODUCTION

Omega-3 (ω-3) fatty acids are polyunsaturated fatty acids (PUFAs) and essential components for the growth of higher eukaryotes [1]. Nutritionally, eicosapentaenoic acid (EPA, 20:5) and docosahexaenoic acid (DHA, 22:6) are the most important fatty acids belonging to this group of bioactive compounds. These long chain omega-3 fatty acids provide significant health benefits to the human population, particularly in reducing cardiac diseases such as arrhythmia, stroke and high blood pressure [2, 3]. Additionally, they have been seen to offer beneficial effects to depression, rheumatoid arthritis and asthma [4–6]. Currently, the principal source of EPA and DHA for human consumption is marine fatty fish such as salmon, mullet and mackerel [7, 8]. However, global catches have been in decline since the late 1980s and the number of overfished stocks has been increasing exponentially since the 1950s [9, 10]. Furthermore, the presence of chemical contaminants (e.g. mercury) in fish oil can be harmful to consumers [11, 12]. In addition, fish oil is not suitable for vegetarians and the odour makes it unattractive. There is a variety of alternative EPA and DHA sources such as bacteria, fungi, plants and microalgae that are currently being explored for commercial production. Fungi require an organic carbon source and typically long growth periods [13], plants need arable land, have longer growth times and have no enzymatic activity for producing long chain PUFAs EPA and DHA, unless genetically modified [14]. Microalgae are the initial EPA and DHA producers in the marine food chain and can naturally grow fast under a variety of autotrophic, mixotrophic and heterotrophic culture conditions with high long chain ω-3 fatty acid production potential [15]. Autotrophic and mixotrophic microalgae fix atmospheric carbon dioxide during photosynthesis, can potentially grow on non-arable land and have short harvesting times [16, 17]. A comparison shows that microalgae can reach much higher EPA and DHA contents and productivities compared with other possible sources (Table 1). In particular heterotrophic microalgae are well established as an alternative source of DHA and are added to infant milk formula or other food [18]. Other

Citation: T Catalina Adarme-Vega, David K Y Lim, Matthew Timmins, Felicitas Vernen, Yan Li, Peer M Schenk, Microalgal biofactories: a promising approach towards sustainable omega-3 fatty acid production, DOI: 10.1186/1475-2859-11-96

microalgal products are used as food additives, animal feed (including aquaculture), vitamins, pigments, pharmaceutical compounds, cosmetics and potentially as a biofuel source [17, 19, 20]. The development of an efficient large-scale cultivation system for the commercial production of EPA and DHA would address a major global need. Here, we review the potential of autotrophic eukaryotic microalgae as biofactories for large-scale production of omega-3 fatty acids.

Table 1: Comparison of EPA and DHA fatty acid contents as percentage from total lipids in examples of bacteria, fungi, fish, transgenic plants and microalgae

Organism	% EPA and/or DHA production	Reference
Bacteria		
Shewanella putrefaciens	40.0 EPA	[21]
Alteromonas putrefaciens	24.0 EPA	[22]
Pneumatophorus japonicus	36.3 EPA	[23]
Photobacterium	4.6 EPA	[24]
Fungi		
Thraustochytrium aureum	62.9 EPA + DHA	[1]
Mortierella	20.0 EPA	[25]
Mortierella	13.0 EPA	[26]
Pythium	12.0 EPA	[27]
Pythium irregulare	8.2 EPA	[28]
Fish		
Merluccius productus	34.99 EPA + DHA	[29]
Theragra chalcogramma	41.35 EPA + DHA	[29]
Hypomesus pretiosus	33.61 EPA + DHA	[29]
Sebastes pinniger	29.8 EPA + DHA	[29]
Oncorhynchus gorbusha	27.5 EPA + DHA	[29]
Mallotus villosus	17.8 EPA + DHA	[29]
Sardinops sagax	44.08 EPA + DHA	[29]
Clupea harenguspallasi	17.32 EPA + DHA	[29]
Plant (transgenic)		
Soybean	20.0 EPA	[30]
Brassica carinata	25.0 EPA	[31]
Nicotiana benthamiana	26.0 EPA	[32]
Microalgae		
Nannochloropsis sp.	26.7 EPA + DHA	[33]
Nannochloropsis oceanica	23.4 EPA	[34]
Nannochloropsis salina	~28 EPA	[35]
Pinguiococcus pyrenoidosus	22.03 EPA + DHA	[36]
Thraustochytrium sp.	45.1 EPA + DHA	[37]
Chlorella minutissima	39.9 EPA	[38]
Dunaliella salina	21.4 EPA	[39]
Pavlova viridis	36.0 EPA + DHA	[40]
Pavlova lutheri	27.7 EPA + DHA	[41]
Pavlova lutheri	41.5 EPA + DHA	[42]
Isocrysis galbana	~28.0 EPA + DHA	[43]

MICROALGAE IN AQUATIC FOOD CHAINS: THE INITIAL OMEGA-3 PRODUCERS

Microalgae are by far the most abundant primary producers that can be found in most aquatic systems, photosynthetically converting light energy and carbon dioxide (CO_2) into biomass such as carbohydrates [44], proteins [45] and lipids [46].

Under high nutrient supply (eutrophic conditions), algae blooms commonly occur as microalgal cell density drastically increases [47]. During microalgal blooms the limitation of nutrients or light halters the increase of biomass. If nutrients, but not light, are limiting, this leads to the accumulation of photosynthetic bioproducts such as lipids and carbohydrates. These serve as storage products in order to survive the stressful growth limiting conditions, after which a large number of cells die [47, 48]. Algal biomass is subsequently degraded by microorganisms, consuming large amounts of oxygen. As a result an anaerobic zone in the water is formed (Figure 1). In extreme cases, this can lead to anaerobiosis of the entire water body, causing the death of plants and animals in the waterway; interestingly this process is also believed to have been the key factor for large-scale oceanic anoxic events that led to fossil mineral oil deposition [17].

Figure 1: Algal blooms in eutrophic aquatic systems use up nutrients and compete for light. If nutrients become limiting first, microalgae may accumulate large amounts of lipids and/or carbohydrates as a survival strategy. The decay of organic matter by bacteria uses up oxygen causing localized anaerobiosis zones. These zones (here shown as grey areas) are present in all aquatic systems but occur at much deeper levels under mesotrophic or oligotrophic conditions. Photosynthetic microalgae require polar polyunsaturated lipids in particular for membrane where fluidity is critical, while most storage lipid occurs in the form of lipid bodies containing triacylglycerides. These typically vary in their composition and typically contain a mixture of saturated and unsaturated fatty acids for storage.

Importantly, microalgae are also the primary producers of EPA and DHA that are eventually accumulated through the various trophic levels. Changes in microalgal lipid content are carried on up the food chain (Figure 2), impacting the growth and dietary make-up of zooplankton, crustacean larvae, mollusc and some fish [49]. This subsequently affects the accumulation of EPA and DHA fatty acids in higher organisms and humans. Consequently, lipid profiles in microalgae play a vital role in maintaining the integrity of the world's aquatic food webs.

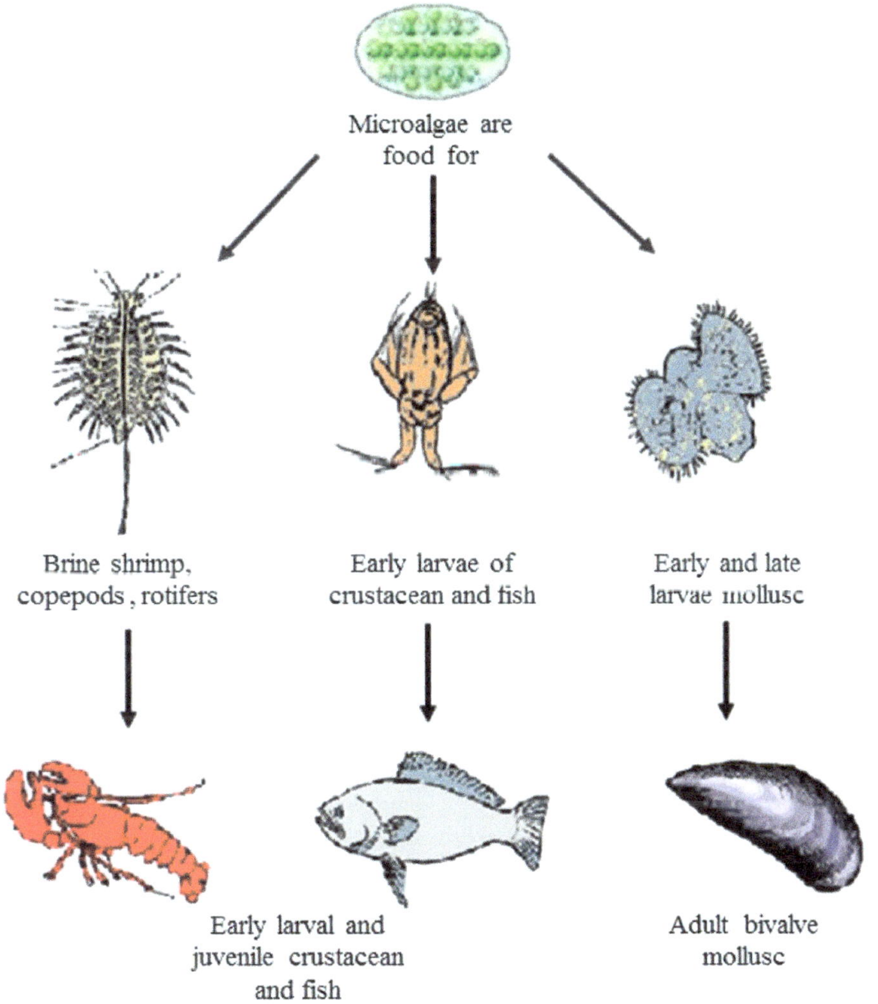

Figure 2: Microalgae are the primary food source of essentially all marine and freshwater food chains.

THE NUTRITIONAL IMPORTANCE OF MICROALGAE AND EPA CONTENT IN AQUACULTURE

Microalgae are essential to the aquaculture industry which has grown substantially over the last 10 years [50, 51]. The successful cultivation of oysters, scallops and mussels is dependent on the ω-3 fatty acids from microalgal feedstock. The polyunsaturated omega-3 fatty acids EPA and DHA derived from microalgae (e.g. *Isochrysis, Tetraselmis,Chaetoceros, Thalassiosira,Nannochloropsis*) are also known to be essential for healthy development of various bivalve larvae [52, 53]. Prior research on the scallop *Pecten maximus* has shown a direct relationship between the fatty acid profile of female gonads and the fatty acid composition of the eggs [54]. The increase of EPA and DHA from an algal diet significantly increased the concentration of fatty acids in the digestive gland (78%) of scallops as well as the female (57%) and male gonads (51%). It appears that dietary lipids are stored in the digestive gland and are later transferred to the developing female gonad. These dietary lipids are then incorporated into the eggs and can significantly improve their quality. This in turn improves the hatching rate of eggs and hatching rates have been linked to high contents of EPA and DHA [53]. Aside from bivalve culture, microalgae are also used as food additives to improve the flesh color of salmon [55], as well as inducing a range of other biological activities such as survival and resistance [19].

The selection of suitable microalgae species for aquaculture is very important. Firstly, a candidate species must be adaptable to mass culture with high growth rates and lipid content [34, 56]. Furthermore, it must tolerate moderate fluctuations of temperature, light and nutrients [57, 58]. A microalgae species used for aquaculture must also have the appropriate size for ingestion (e.g. from 1 to 15 μm for filter feeders; 10 to 100 μm for grazers) and be readily digestible [56]. Finally, they must possess a suitably high lipid composition with long chain polyunsaturated fatty acids and be free of toxins for target culture species [34, 56]. Selection of the suitable microalgal diet is of paramount importance to aquaculture hatchery and nursery success [58]. At present, the most widely cultured species for aquaculture hatcheries and nurseries include *Chaetoceros calcitrans,Isochrysis galbana,Pavlova lutheri,Pseudoisochrysis paradoxa,Tetraselmis suecica* and *Skeletonema costatum.* Other genera include*SpirulinaDunaliellaChlorellaThalassiosiraIsochrysis* and *Nannochloropsis*[49].

HEALTH BENEFITS OF MICROALGAL OMEGA-3 FATTY ACIDS

Omega-3 fatty acids represent an important structural component of human cell membranes, particularly neuronal cells [59]. The consumption of EPA and DHA supplements has been shown to prevent cardiovascular, nervous system and inflammatory conditions [60]. With regards to cardiovascular health, regular consumption of ω-3 fatty acids can help reduce the risk of hypertension, thrombosis, myocardial infarction and cardiac arrhythmias [61]. This occurs because ω-3 fatty acids increase the high-density lipoprotein/low-density lipoprotein (HDL/LDL) ratio and decrease the total cholesterol/HDL ratio [61]. In addition to cardiovascular benefits, omega-3 fatty acids have also demonstrated positive effects on brain function and the nervous system [62]. In pregnant women, the adequate intake of EPA and DHA is crucial for healthy development of the fetal brain [63]. In infants, arachidonic acid (ARA), an omega-6 fatty acid, and DHA are also required for normal growth and functional development [64]. Interestingly, increased consumption of DHA may also diminish the severity of depression [65]. Immuno-modulatory effects have been observed when ω-3 fatty acids were used in the treatment of inflammatory conditions such as rheumatoid arthritis, Crohn's disease, ulcerative colitis, psoriasis, asthma, lupus and cystic fibrosis [66, 67]. Children ingesting fish oil more than once a week had a lower probability of suffering from asthma [68]. Increasing the levels of DHA and EPA in patients with rheumatoid arthritis and ulcerative colitis has also been found to reduce pain and improve conditions, although the modes of operation are unclear at this point [69, 70].

There is currently a large demand for microalgae in the nutraceutical and pharmaceutical industry due to their health-promoting effects. Microalgal-derived PUFA, such as ARA and DHA are added as fortifications to infant formulae—an industry that is worth $10 billion per annum alone. To date, microalgal extracts can be found in many face and skin care products, e.g. anti-aging cream, refreshing or regenerative care products, sun cream, emollient and anti-irritant in peelers [19]. Dermochlorella is actually extracted from *Chlorella vulgaris,* which can stimulate collagen synthesis in skin supporting tissue regeneration and wrinkle reduction [71]. Protulines is a protein-rich extract from *Arthrospira* (*Spirulina*), which helps combat early skin aging, exerting a tightening effect and preventing wrinkle formation [72].

OMEGA-3 FATTY ACID PRODUCTION IN MICROALGAE

Microalgae produce a variety of compounds to help in the adaptation and survival of different environmental conditions. Many marine microalgal strains have oil contents of between 10–50%, (w/w) and produce a high percentage of total lipids (up to 30–70% of dry weight) [1]. The accumulation of fatty acids is closely linked to microalgal growth stages, functioning as an energy stockpile during unfavourable conditions or cell division. Omega-3 is accumulated due to its high energy content, as well as the good flow properties crucial for cellular functions [73,74]. To date, the ω-3 fatty acid content of numerous microalgae strains have been studied. Strains from the genera *Phaeodactylum, Nannochloropsis,Thraustochytrium* and*Schizochytrium* have demonstrated high accumulation of EPA and/or DHA. *Phaeodactylum tricornutum*[38] and *Nannochloropsis* sp. [75] demonstrated an EPA content of up to 39% of total fatty acids, while strains such as *Thraustochytrium*[76] and *Schizochytrium limacinum*[77] contained a DHA percentage of between 30–40% of total fatty acids when grown heterotrophically. High biomass and commercially acceptable EPA and DHA productivities are achieved with microalgae grown in media with optimized carbon and nitrogen concentrations and controlled pH and temperature conditions [78]. High oil production, including DHA from *Schizochytrium* (50% w/w), can be obtained as a result of high growth rate by controlling of nutrients such as glucose, nitrogen, sodium and some other environmental factors, such as oxygen concentrations as well as temperature and pH, achieving high cell densities and DHA productivities [1].

INDUCTION OF OMEGA-3 PRODUCTION IN AUTOTROPHIC MICROALGAE

An increase in microalgal lipid content can be induced by a sudden change of growth conditions. The accumulation of starch and/or lipids reserves is considered a survival mechanism in response to growth-limiting stresses [17], such as UV radiation [79], temperature [80] and shock or nutrient deprivation [81, 82], as long as light conditions are present that still allow efficient photosynthesis. For example, during nutritional deprivation (e.g. nitrogen) and under the provision of light, cellular division of many marine or brackish microalgae is put on hold and cells begin to accumulate lipids [83], leading to a 2–3 fold increase in lipid content. Both total lipid and omega-3 fatty acid production can be adjusted by varying growth conditions. The diatom *Phaeodactylum tricornutum* can be induced to increase its lipid level from 81.2 mg/g of culture dry weight to 168.5 mg/g dry weight [38]. Similarly,*Nannochloropsi* s sp. [84] and *Dunaliella* sp. [85] can achieve a total lipid content of up to 47% and 60% of dry ash weight by modifying the light intensity, temperature and salinity levels. Lipid abundance has also been shown to increase due to anaerobic sulphur deprivation [86] or the addition of extra nutrients [87].

Omega-3 fatty acid biosynthesis can be stimulated by a number of environmental stresses, such as low temperature, change of salinity or UV radiation. For example, *Pavlova lutheri* increased its relative EPA content from 20.3 to 30.3 M % when the culture temperature was reduced to 15°C [88]. Similarly, *Phaeodactylum tricornutum* had a higher EPA content when the temperature was shifted from 25°C to 10°C for 12 h [89]. An increase in PUFAs is expected as these fatty acids have good flow properties and would be predominately used in the cell membrane to maintain fluidity during low temperatures. Salinity may also regulate PUFA biosynthesis, although not in a consistent manner. For example, *Crythecodinium cohniiATCC 30556* increased its total DHA content up to 56.9% of total fatty acids when cultured in 9 g/L NaCl. Other treatments that cause the generation of reactive oxygen species and lipid peroxidation also result in higher PUFA contents. For example, *Phaeodactylum tricornutum*increased its EPA content up to 19.84% when stressed with UV light [90]. Some of the increased PUFAs are used to repair membrane damage but as PUFAs contain many double bonds, these also act as an antioxidant by scavenging free radicals.

METABOLIC ENGINEERING OF MICROALGAE FOR HIGHER OMEGA-3 CONTENTS

Apart for external stresses, metabolic engineering is another promising approach to increase the production of fatty acids in microalgae (for a recent review see Schuhmann et al. [91]). Genes encoding key enzymes involved in the fatty acid biosynthesis have been identified in*Ostreococcus tauri*[92], *Thalassiosirapseudonana*[93–95], *Phaeodactylum tricornutum*[96,97] and in particular the model organism *Chlamydomonas reinhardtii*[98]. At present, the mechanisms involved in the fatty acid biosynthetic pathways in microalgae have not been extensively studied and most information has been gathered from studies on plant metabolism. Briefly, *de novo* fatty acid synthesis occurs in the chloroplast and involves the carboxylation and condensation of acetyl-CoA to malonyl-CoA, with further elongation reactions occurring with malonyl ACP as substrate to create long chain fatty acids. Long chain fatty acids are transferred to glycerol-3-phosphate to form triacylglycerol (TAG) via the metabolic intermediate phosphatidic acid in the endoplasmic reticulum [99]. Synthesis of ω-3 fatty acids occurs via the elongation and desaturation of long chain fatty acids (Figure 3).

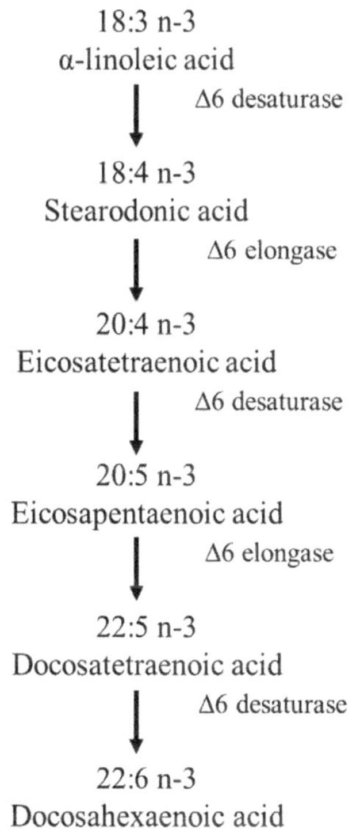

Figure 3: Conventional **Δ6** pathway for biosynthesis of EPA and DHA polyunsaturated fatty acids [100].

Work has been performed to create recombinant sources of ω-3 fatty acids in a variety of systems with some success [101, 102]. Canola (*Brassica napus*) seeds have been produced which overexpress the *B. napus* Δ15 desaturase, as well as the Δ6 and Δ12 desaturases from the commercially grown fungus *Mortierella alpina* to synthesize the ω-3 fatty acid stearidonic acid (SDA) [14]. It may be possible in the future to produce ω-3 fatty acids in microalgae in much larger quantities by regulating the expression of similar enzymes. A promising cisgenic approach for microalgae maybe to increase EPA or DHA production by overexpressing at least some of their native elongases and desaturases. It may be necessary to use promoters inducible by external stimuli rather than constitutive promoters that may interfere with normal cell function and growth. Another, yet unexplored option may lie in the inhibition of PUFA degradation. β-oxidation of fatty acids occurs in the peroxisomes but before PUFAs can be metabolized, saturases are required to fill in the double bonds. Mutations in one or several saturases may result in less efficient β-oxidation of PUFA and a higher percentage of these fatty acids. However, at present the mechanism behind the selection and storage of fatty acids for triacylglycerol production remains unclear.

EXTRACTION AND PURIFICATION OF OMEGA-3 FATTY ACIDS FROM MICROALGAL BIOMASS

Figure 4 summarizes an integrated system for the large-scale production of microalgal bio-products. A microalgae strain is cultivated to increase cell density using photobioreactors, open ponds, race ways or hybrid systems. Algal cells are separated from culture media by filtration, flocculation or centrifugation, followed by drying to improve extraction [1]. Lipid extraction is then commonly performed using a non-water miscible organic solvent. A typical extraction protocol in small scale is often based on the method of Bligh and Dyer [103], which uses a solvent mixtures made of methanol/chloroform for the cell disruption and lipid extraction. Larger scale extraction is typically carried out with hexane as a solvent. Subsequently, unsaturated fatty acids are separated from the total lipids by fractional (molecular) distillation or winterization, whereby oil temperature is reduced to precipitate the more saturated lipids. Further processing to improve the quality, shelf-life and quantity of PUFA oil can include filtration, bleaching, deodorization, polishing and antioxidant addition [1, 104] (Table 2).

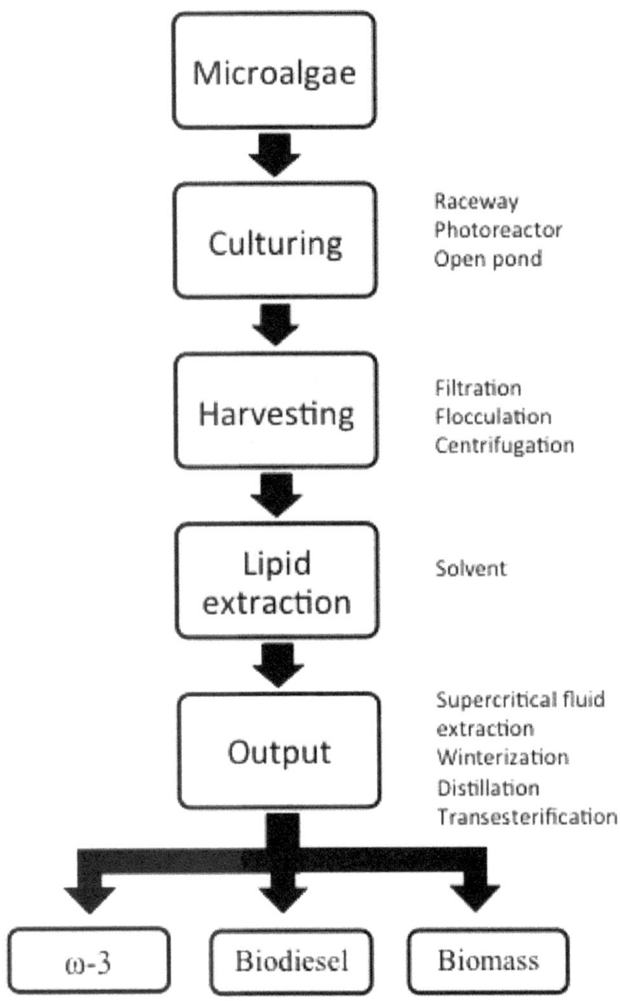

Figure 4: Examples of a bioprocess production chain in a microalgal biorefinery. Apart from omega-3 fatty acids (ω-3), the product portfolio includes biodiesel and protein-rich animal feed from the remaining biomass.

Table 2: Summary of PUFA enrichment processes

Method	Procedure
Molecular distillation (Fractional distillation)	Purification of fatty acid esters in a vacuum system based on the different boiling points of different fatty acids [105].
Molecular sieves	Separation via membrane permeability and selectivity [106].
PUFA transformations	Esterification of PUFA and free fatty acids to produce esters (ethyl-, glyceryl-, sugar-, other). Inter-esterification to enrich lowly unsaturated fatty acids with PUFA [107].
Super Critical Fluid Extraction	Optimization of lipid solubility and fractionation in supercritical CO_2[108].
Urea complexation	Solubilization of fatty acids, adding urea and ethanol to saturation point exposing it to heat. Recovery of product by filtration [109].
Winterization	Temperature reduction to render more saturated fats insoluble [110].

Efforts have been made to use lipases, hydrolysis and esterification processes to selectively enrich PUFAs. The main application of lipases on PUFAs is the generation of non-natural esters of these products for use as pharmaceutical products or other synthetic bioactive compounds or their precursors [1]. The effectiveness of harvesting and extraction techniques depends on the microalgal strain's physical characteristics (e.g. cell size and cell wall properties) and the use of the end product. In aquaculture, microalgae are used as a fresh product or as dry pellets which preserve the nutritional content of microalgae [57, 58, 111]. In this case, microalgal biomass is first de-watered either by filtration, dissolved air flotation, flocculation or sedimentation and then dried to form pellets or directly administrated to livestock [111]. When produced for the pharmaceutical industry, further extraction and purification processes are required. Currently, methods such as supercritical fluid extraction, winterization and fractional (molecular) distillation are used for the extraction and purification of PUFA from microalgae [112, 113] (Table 2)

OMEGA-3 FATTY ACID PRODUCTION: A BIOREFINERY APPROACH

The natural capacity of microalgae to produce multiple products, (e.g. oils, proteins and carbohydrates) has encouraged the development of a biorefinery concept for processing. Akin to the petrochemical industry, where crude oil is processed to yield petroleum and a range of other chemicals, microalgae can be processed to produce a range of bioproducts. Different industries are able to use different algal products. For instance, the pharmaceutical and nutraceutical industries use high value bioactive products such as ω-3 fatty acids and carotenoids; the transport industry can use fatty acids from TAG for biodiesel, the chemical industry can use products such as glycerine, while the majority of the biomass can be used by agriculture and aquaculture as animal feed [114, 115]. Additional processes that address nutrient recycling and carbon sequestration can be used by anaerobic digestion of wet biomass and pyrolysis for the production of biochar. Undoubtedly, the biggest interest in microalgal use is for biodiesel production. It potentially represents a more sustainable alternative to fossil fuels as microalgal production facilities do not need to compete for arable land or freshwater. Furthermore, in comparison to land plants, 10–400 times more energy per acre can potentially be produced from microalgae. Although there has been considerable interest and research over the past years into microalgal biofuel production [83], no commercial enterprise has successfully established itself as a supplier of autotrophically derived algal biofuels for any duration. Nevertheless, decreasing fossil fuel reserves and increasing fuel costs continue to drive research targeted towards economically viable production of microalgal biodiesel, with the level of improvement necessary now appearing attainable [15, 17]. There is confidence among companies producing microalgae that the production of a high value product, such as omega-3 from microalgae, will further assist in the establishment of the microalgae industry. Several companies have (at least temporarily) shifted their focus from algal biodiesel production, to high value products such as omega-3 and protein-rich biomass as animal feed (e.g. Aurora Algae, MBD, Cellana).

CONCLUSIONS

Global fish stocks are declining and cannot provide a sustainable source of omega-3 fatty acids. Heterotrophic microalgae have been used for the production of omega-3 fatty acids, in particular DHA. However, as the primary producers of PUFAs, the use of autotrophic microalgae for large-scale production of omega-3 fatty acids has recently attracted a lot of interest. Autotrophic microalgae do not require an organic carbon source and hence may avoid the problems faced for heterotrophic cultures that can easily get contaminated with other microorganisms. In a biorefinery concept, omega-3 fatty acids can be separated from microalgal lipids which would be widely used for biodiesel production, while biomass can find uses as valuable protein-rich animal feed which could free up arable land for food production. If carried out at a large scale this would address three major areas of importance: human health, transportable energy and food security.

Over the past decade, algae biotechnology has grown steadily into a global industry with increasing numbers of entrepreneurs attempting to utilize its biochemical diversity for a wide array of applications. At present, achieving economically viable production of microalgal lipids is still a major challenge, but strong potential stems from the fact that these microbial cell factories have not been domesticated and are not as well studied compared to agricultural crops [102]. Indeed, of approximately 40,000 algal species, only a few thousand strains are kept in collections, a few hundred are investigated for chemical content and approximately half a dozen are cultivated in industrial quantities. Therefore, continued isolation and screening of microalgae is required, as well as more in depth studies into algal physiology, biochemistry and genetics. Meanwhile the processes for algae cultivation, harvesting and oil extraction need to be further improved in efficiency and costs. As omega-3 fatty acids are one of the most valuable products from microalgae, they are likely to be the "game-changer" towards large-scale economical microalgae cultivation that will catalyze the production of other important algal bioproducts.

ACKNOWLEDGEMENTS

This work was supported by the Australian Research Council and Queensland Sea Scallops Pty Ltd. The funding bodies had no influence in the writing of the manuscript and in the decision to submit the manuscript for publication.

REFERENCES

1. Ward OP, Singh A: Omega-3/6 fatty acids: alternative sources of production. Process Biochem. 2005, 40 (12): 3627-3652. 10.1016/j.procbio.2005.02.020.
2. Romieu I, Tellez-Rojo MM, Lazo M, Manzano-Patino A, Cortez-Lugo M, Julien P, Belanger MC, Hernandez-Avila M, Holguin F: Omega-3 fatty acid prevents heart rate variability reductions associated with particulate matter. Am J Respir Crit Care Med. 2005, 172 (12): 1534-1540. 10.1164/rccm.200503-372OC
3. Von Schacky C: Omega-3 fatty acids: antiarrhythmic, proarrhythmic or both?. Curr Opin Clin Nutr Metab Care. 2008, 11 (2): 94-99. 10.1097/MCO.0b013e3282f44bdf
4. Von Schacky C, Harris WS: Cardiovascular benefits of omega-3 fatty acids. Cardiovasc Res. 2007, 73 (2): 310-315. 10.1016/j.cardiores.2006.08.019
5. Balk E, Chung M, Lichtenstein A, Chew P, Kupelnick B, Lawrence A, DeVine D, Lau J: Effects of omega-3 fatty acids on cardiovascular risk factors and intermediate markers of cardiovascular disease. Evid Rep Technol Assess (Summ). 2004, 1-6. 93,
6. Adams PB, Lawson S, Sanigorski A, Sinclair AJ: Arachidonic acid to eicosapentaenoic acid ratio in blood correlates positively with clinical symptoms of depression. Lipids. 1996, 31 (1): 157-161. 10.1007/BF02637069.
7. Gunstone FD: Fatty acid and lipid chemistry. 1996, London: Black Academic and Professional:,
8. Whitehead S: FAO species catalogue. Clupeoid fishes of the world. 1985, Rome: UNITED NATIONS,
9. AGDAFF: Australian government department of agriculture, fisheries and forestry. Fishery status reports 2007. 2007,
10. Worm B, Barbier EB, Beaumont N, Duffy JE, Folke C, Halpern BS, Jackson JBC, Lotze HK, Micheli F, Palumbi SR, et al: Impacts of biodiversity loss on ocean ecosystem services. Science. 2006, 314 (5800): 787-790. 10.1126/science.1132294
11. Mahaffey KR, Clickner RP, Jeffries RA: Methylmercury and omega-3 fatty acids: co-occurrence of dietary sources with emphasis on fish and shellfish. Environ Res. 2008. 10. (1): 20-29.
12. Bourdon J, Bazinet T, Arnason T, Kimpe L, Blais J, White P: Polychlorinated biphenyls (PCBs) contamination and aryl hydrocarbon receptor (AhR) agonist activity of omega-3 polyunsaturated fatty acid supplements: implications for daily intake of dioxins and PCBs. Food Chem Toxicol. 2010,
13. Barclay WR, Meager KM, Abril JR: Heterotrophic production of long-chain omega-3-fatty-acids utilizing algae and algae-like microorganisms. J Appl Phycol. 1994, 6 (2): 123-129. 10.1007/BF02186066.

14. Ursin VM: Modification of plant lipids for human health: development of functional land-based omega-3 fatty acids. J Nutr. 2003, 133 (12): 4271-4274.

15. Li Y, Qin JG, Moore RB, Ball AS: Perspectives of marine phytoplankton as a source of nutrition and bioenergy. Marine phytoplankton. 2009, 14-New York: Nova Science Pub Inc,

16. Rubio-Rodríguez N, Beltrán S, Jaime I, de Diego SM, Sanz MT, Carballido JR: Production of omega-3 polyunsaturated fatty acid concentrates: a review. Innovat Food Sci Emerg Tech. 2010, 11 (1): 1-12. 10.1016/j.ifset.2009.10.006.

17. Schenk PM, Thomas-Hall SR, Stephens E, Marx UC, Mussgnug JH, Posten C, Kruse O, Hankamer B: Second generation biofuels: high-efficiency microalgae for biodiesel production. Bioenergy Res. 2008, 1 (1): 20-43. 10.1007/s12155-008-9008-8.

18. Van Tol EAF, Willemsen LEM, Koetsier MA, Beermann C, Stahl B: Improvement of intestinal barrier integrity. EP patent 1, 815, 755. 2009,

19. Spolaore P, Joannis-Cassan C, Duran E, Isambert A: Commercial applications of microalgae. J Biosci Bioeng. 2006. 10. (2): 87-96.

20. Yamaguchi K: Recent advances in microalgal bioscience in Japan, with special reference to utilization of biomass and metabolites: a review. J Appl Phycol. 1996, 8 (6): 487-502. 10.1007/BF02186327.

21. Yazawa K: Production of eicosapentaenoic acid from marine bacteria. Lipids. 1996, 31 (Suppl): S297-300.

22. Yazawa K, Araki K, Okazaki N, Watanabe K, Ishikawa C, Inoue A, Numao N, Kondo K: Production of eicosapentaenoic acid by marine bacteria. J Biochem (Tokyo). 1988, 103 (1): 5-7.

23. Yazawa K, Araki K, Watanabe K, Ishikawa C, Inoue A, Kondo K, Watabe S, Hashimoto K: Eicosapentaenoic acid productivity of the bacteria isolated from fish intestines. Nippon Suisan Gakkaishi. 1988, 54 (10): 1835-1838. 10.2331/suisan.54.1835.

24. Ryan J, Farr H, Visnovsky S, Vyssotski M, Visnovsky G: A rapid method for the isolation of eicosapentaenoic acid-producing marine bacteria. J Microbiol Methods. 2010, 82 (1): 49-53. 10.1016/j.mimet.2010.04.001

25. Jareonkitmongkol S, Shimizu S, Yamada H: Production of an eicosapentaenoic acid-containing oil by a Δ12 desaturase-defective mutant of mortierella alpina 1S-4. J Am Oil Chem Soc. 1993, 70 (2): 119-123. 10.1007/BF02542612.

26. Jermsuntiea W, Aki T, Kawamoto S, Ono K: Metabolism and synthesis of lipids in the polyunsaturated fatty acid-producing fungus mortierella alliacea. J Oleo Sci. 2011, 60 (1): 11- 10.5650/jos.60.11

27. Athalye SK, Garcia RA, Wen Z: Use of biodiesel-derived crude glycerol for producing eicosapentaenoic acid (EPA) by the fungus pythium irregulare. J Agric Food Chem. 2009, 57 (7): 2739-2744. 10.1021/jf803922w

28. Liang Y, Zhao X, Strait M, Wen Z: Use of dry-milling derived thin stillage for producing eicosapentaenoic acid (EPA) by the fungus<i>Pythium irregulare</i>. Bioresour Technol. 2012, 1: 1-

29. Huynh MD, Kitts DD: Evaluating nutritional quality of pacific fish species from fatty acid signatures. Food Chem. 2009, 114 (3): 912-918. 10.1016/j.foodchem.2008.10.038.

30. Kinney AJ, Cahoon EB, Damude HG, Hitz WD, Kolar CW, Liu Z: Production of very long chain polyunsaturated fatty acids in oilseed plants. Patent WO. 2004, 71467: A2-

31. Cheng B, Wu G, Vrinten P, Falk K, Bauer J, Qiu X: Towards the production of high levels of eicosapentaenoic acid in transgenic plants: the effects of different host species, genes and promoters. Transgenic Res. 2010, 19 (2): 221-229. 10.1007/s11248-009-9302-z

32. Petrie JR, Shrestha P, Mansour MP, Nichols PD, Liu Q, Singh SP: Metabolic engineering of omega-3 long-chain polyunsaturated fatty acids in plants using an acyl-CoA [delta] 6-desaturase with [omega] 3-preference from the marine microalga micromonas pusilla. Metab Eng. 2010, 12 (3): 233-240. 10.1016/j.ymben.2009.12.001

33. Hu H, Gao K: Optimization of growth and fatty acid composition of a unicellular marine picoplankton, Nannochloropsis sp., with enriched carbon sources. Biotechnol Lett. 2003, 25 (5): 421-425. 10.1023/A:1022489108980

34. Patil V, Källqvist T, Olsen E, Vogt G, Gislerød HR: Fatty acid composition of 12 microalgae for possible use in aquaculture feed. Aquac Int. 2007, 15 (1): 1-9. 10.1007/s10499-006-9060-3.

35. Van Wagenen J, Miller TW, Hobbs S, Hook P, Crowe B, Huesemann M: Effects of light and temperature on fatty acid production in nannochloropsis salina. Energies. 2012, 5 (3): 731-740. 10.3390/en5030731.

36. Sang M, Wang M, Liu J, Zhang C, Li A: Effects of temperature, salinity, light intensity, and pH on the eicosapentaenoic acid production of pinguiococcus pyrenoidosus. J Ocean Univ China (English Edition). 2012, 1-6.

37. Scott SD, Armenta RE, Berryman KT, Norman AW: Use of raw glycerol to produce oil rich in polyunsaturated fatty acids by a thraustochytrid. Enzyme Microb Technol. 2011, 48 (3): 267-272. 10.1016/j.enzmictec.2010.11.008

38. Yongmanitchai W, Ward OP: Growth of and omega-3 fatty acid production by phaeodactylum tricornutum under different culture conditions. Appl Environ Microbiol. 1991, 57 (2): 419-425.

39. Bhosale RA, Rajabhoj M, Chaugule B: Dunaliella salina Teod. as a prominent source of eicosapentaenoic acid. Int J Algae. 2010, 12 (2): 185-189. 10.1615/InterJAlgae.v12.i2.70.

40. Hu C, Li M, Li J, Zhu Q, Liu Z: Variation of lipid and fatty acid compositions of the marine microalga pavlova viridis (Prymnesiophyceae) under laboratory and outdoor culture conditions. World J Microbiol Biotechnol. 2008, 24 (7): 1209-1214. 10.1007/s11274-007-9595-0.

41. Carvalho AP, Malcata FX: Optimization of ω-3 fatty acid production by microalgae: crossover effects of CO 2 and light intensity under batch and continuous cultivation

modes. Mar Biotechnol. 2005, 7 (4): 381-388. 10.1007/s10126-004-4047-4

42. Guihéneuf F, Mimouni V, Ulmann L, Tremblin G: Combined effects of irradiance level and carbon source on fatty acid and lipid class composition in the microalga pavlova lutheri commonly used in mariculture. J Exp Mar Biol Ecol. 2009, 369 (2): 136-143. 10.1016/j.jembe.2008.11.009.
43. Yago T, Arakawa H, Morinaga T, Yoshie-Stark Y, Yoshioka M: Effect of wavelength of intermittent light on the growth and fatty acid profile of the haptophyte isochrysis galbana. Glob Chang: Mank-Mar Environ Interact. 2011, 43-45.
44. Park J-H, Yoon J-J, Park H-D, Kim YJ, Lim DJ, Kim S-H: Feasibility of biohydrogen production from gelidium amansii. Int J Hydrogen Energ. 2011, 36 (21): 13997-14003. 10.1016/j.ijhydene.2011.04.003.
45. Becker EW: Micro-algae as a source of protein. Biotechnol Adv. 2007, 25 (2): 207-210. 10.1016/j.biotechadv.2006.11.002
46. Harwood JL, Guschina IA: The versatility of algae and their lipid metabolism. Biochimie. 2009, 91 (6): 679-684. 10.1016/j.biochi.2008.11.004
47. Sellner KG, Doucette GJ, Kirkpatrick GJ: Harmful algal blooms: causes, impacts and detection. J Ind Microbiol Biotechnol. 2003, 30 (7): 383-406. 10.1007/s10295-003-0074-9
48. Anderson D, Glibert P, Burkholder J: Harmful algal blooms and eutrophication: nutrient sources, composition, and consequences. Estuar Coast. 2002, 25 (4): 704-726. 10.1007/BF02804901.
49. Brown MR: Nutritional value and use of microalgae in aquaculture. Avances en Nutrición Acuícola VI Memorias del VI Simposium Internacional de Nutrición Acuícola. 2002, 3: 281-292.
50. Foster C: Aquaculture production to double by 2015. The fish site. 2008,
51. Australian bureau of agricultural and resource economics. Australian fisheries production falls.http://www.thefishsite.com/fishnews/10374/australian-fisheries-production-falls,
52. Caers M, Coutteau P, Sorgeloos P, Gajardo G: Impact of algal diets and emulsions on the fatty acid composition and content of selected tissues of adult broodstock of the chilean scallop argopecten pupuratus (Lamarck, 1819). Aquaculture. 2003, 217 (1–4): 437-452.
53. Soudant P, Marty Y, Moal J, Samain J: Fatty acids and egg quality in great scallop. Aquac Int. 1996, 4 (3): 191-200. 10.1007/BF00117381.
54. Utting SD, Millican PF: The role of diet in hatchery conditioning of pecten maximus L.: a review. Aquaculture. 1998, 165 (3–4): 167-178.
55. Torrissen OJ: Pigmentation of salmonids—a comparison of astaxanthin and canthaxanthin as pigment sources for rainbow trout. Aquaculture. 1986, 53 (3–4): 271-278.
56. Brown M: Nutritional value of microalgae for aquaculture. Avances en nutrición acuícola VI memorias del VI simposium internacional de nutrición. Edited by: Cruz-Suárez LE, Ricque-Marie D, Tapia-Salazar M, Gaxiola-Cortés MG, Simoes NE. 2002, Cancún, Quintana Roo, México,
57. Borowitzka M: Microalgae for aquaculture: opportunities and constraints. J Appl Phycol. 1997, 9 (5): 393-401. 10.1023/A:1007921728300.
58. Benemann J: Microalgae aquaculture feeds. J Appl Phycol. 1992, 4 (3): 233-245. 10.1007/BF02161209.
59. Brunner E: Oily fish and omega 3 fat supplements. BMJ. 2006, 332 (7544): 739-740. 10.1136/bmj.38798.680185.47
60. Sijtsma L, Swaaf ME: Biotechnological production and applications of the n-3 polyunsaturated fatty acid docosahexaenoic acid. Appl Microbiol Biotechnol. 2004, 64 (2): 146-153. 10.1007/s00253-003-1525-y
61. Horrocks LA, Yeo YK: Health benefits of docosahexaenoic acid (DHA). Pharmacol Res. 1999, 40 (3): 211-225. 10.1006/phrs.1999.0495
62. Simopoulos AP, Bazán NG, Karger S: Omega-3 fatty acids, the brain and retina. 2009, Karger,
63. Damude HG, Kinney AJ: Enhancing plant seed oils for human nutrition. Plant Physiol. 2008, 147 (3): 962-968. 10.1104/pp.108.121681
64. Dyerberg J, Leaf A, GA C: ISSFAL board statement: recommendations for the essential fatty previous termacidnext term requirement for infant formulas. J Am Coll Nutr. 1995, 14: 2-
65. Hibbeln JR, Salem N: Dietary polyunsaturated fatty acids and depression: when cholesterol does not satisfy. Am J Clin Nutr. 1995, 62 (1): 1-9.
66. Simopoulos AP: Omega-3 fatty acids in health and disease and in growth and development. Am J Clin Nutr. 1991, 54 (3): 438-463.
67. Calder PC: Sir david cuthbertson medal lecture. Immunomodulatory and anti-inflammatory effects of n-3 polyunsaturated fatty acids. Proc Nutr Soc. 1996, 55 (2): 737-774. 10.1079/PNS19960069
68. Hodge L, Salome CM, Peat JK, Haby MM, Xuan W, Woolcock AJ: Consumption of oily fish and childhood asthma risk. Med J Aust. 1996, 164 (3): 137-140.
69. Stenson WF, Cort D, Rodgers J, Burakoff R, DeSchryver-Kecskemeti K, Gramlich TL, Beeken W: Dietary supplementation with fish oil in ulcerative colitis. Ann Intern Med. 1992, 116 (8): 609-614.
70. Simopoulos AP: Omega-3 fatty acids in inflammation and autoimmune diseases. J Am Coll Nutr. 2002, 21 (6): 495-505.
71. CODIF Recherche & Nature: DERMOCHLORELLA D - DERMOCHLORELLA DP skin restructuring. St Malo cedex. 2006, France: CODIF Recherche & Nature,
72. Exymol SAM: Protulines: spirulina extract. Exymol SAM. 2012, Monaco,
73. Tiez L, Zeiger E: Plant physiology. 2010, Publishers, Sunderland: Sinauer Associates Inc, 5,
74. Cohen Z, Khozin-Goldberg I, Adlerstein D, Bigogno C: The role of triacylglycerol as a reservoir of polyunsaturated fatty acids for the rapid production of

chloroplastic lipids in certain microalgae. Biochem Soc Trans. 2000, 28 (6): 740-744. 10.1042/BST0280740

75. Sukenik A: Ecophysiological considerations in the optimization of eicosapentaenoic acid production by Nannochloropsis sp. (Eustigmatophyceae). Bioresour Technol. 1991, 35 (3): 263-269. 10.1016/0960-8524(91)90123-2.

76. Burja AM, Radianingtyas H, Windust A, Barrow CJ: Isolation and characterization of polyunsaturated fatty acid producing thraustochytrium species: screening of strains and optimization of omega-3 production. Appl Microbiol Biotechnol. 2006, 72 (6): 1161-1169. 10.1007/s00253-006-0419-1

77. Zhu L, Zhang X, Ji L, Song X, Kuang C: Changes of lipid content and fatty acid composition of schizochytrium limacinum in response to different temperatures and salinities. Process Biochem. 2007, 42 (2): 210-214. 10.1016/j.procbio.2006.08.002.

78. Griffiths MJ, Harrison STL: Lipid productivity as a key characteristic for choosing algal species for biodiesel production. J Appl Phycol. 2009, 21 (5): 493-507. 10.1007/s10811-008-9392-7.

79. Singh SC, Sinha RP, Hader D: Role of lipids and fatty acids in stress tolerance in cyanobacteria. Acta protozoologica. 2002, 41 (4): 297-308.

80. de Castro AS, Garcia VMT: Growth and biochemical composition of the diatom chaetoceros cf. wighamii brightwell under different temperature, salinity and carbon dioxide levels. I. Protein, carbohydrates and lipids. Aquaculture. 2005, 246 (4): 405-412. 10.1016/j.aquaculture.2005.02.051.

81. Otero A, García D, Fábregas J: Factors controlling eicosapentaenoic acid production in semicontinuous cultures of marine microalgae. J Appl Phycol. 1997, 9 (5): 465-469. 10.1023/A:1007930804367.

82. Wen ZY, Chen F: Optimization of nitrogen sources for heterotrophic production of eicosapentaenoic acid by the diatom nitzschia laevis. Enzyme Microb Technol. 2001, 29 (6–7): 341-347.

83. Sheehan J, Dunahay T, Benemann J, Roessler P: A look back at the US department of energy›s aquatic species program: biodiesel from algae, close out report TP-580-24190, CO: national renewable energy laboratory. 1998,

84. Pal D, Khozin-Goldberg I, Cohen Z, Boussiba S: The effect of light, salinity, and nitrogen availability on lipid production by nannochloropsis sp. Appl Microbiol Biotechnol. 2011, 1-13.

85. Takagi M: Effect of salt concentration on intracellular accumulation of lipids and triacylglyceride in marine microalgae dunaliella cells. J Biosci Bioeng. 2006. 10. (3): 223-226.

86. Timmins M, Zhou W, Lim L, Thomas-Hall SR, Doebbe A, Kruse O, Hankamer B, Marx UC, Smith SM, Schenk PM: The metabolome of chlamydomonas reinhardtii following induction of anaerobic H2 production by sulphur deprivation. J Biol Chem. 2009,

87. Hu H, Gao K: Response of growth and fatty acid compositions of Nannochloropsis sp. to environmental factors under elevated CO 2 concentration. Biotechnol Lett. 2006, 28 (13): 987-992. 10.1007/s10529-006-9026-6

88. Tatsuzawa H, Takizawa E: Changes in lipid and fatty acid composition of pavlova lutheri. Phytochemistry. 1995, 40 (2): 397-400. 10.1016/0031-9422(95)00327-4.

89. Jiang H, Gao K: Effects of lowering temperature during culture on the production of polyunsaturated fatty acids in the marine diatom phaeodactylum tricornutum (bacillariophyceae) 1. J Phycol. 2004, 40 (4): 651-654. 10.1111/j.1529-8817.2004.03112.x.

90. Liang Y, Beardall J, Heraud P: Effect of UV radiation on growth, chlorophyll fluorescence and fatty acid composition of phaeodactylum tricornutum and chaetoceros muelleri (bacillariophyceae). Phycologia. 2006, 45 (6): 605-615. 10.2216/04-61.1.

91. Schuhmann H, Lim DKY, Schenk PM: Perspectives on metabolic engineering for increased lipid contents in microalgae. Biofuels. 2012, 3 (1): 71-86. 10.4155/bfs.11.147.

92. Wagner M, Hoppe K, Czabany T, Heilmann M, Daum G, Feussner I, Fulda M: Identification and characterization of an acyl-CoA: diacylglycerol acyltransferase 2 (DGAT2) gene from the microalga<i>O. tauri</i>. Plant Physiol Biochem. 2010, 48 (6): 407-416. 10.1016/j.plaphy.2010.03.008

93. Xu J, Zheng Z, Zou J: A membrane-bound glycerol-3-phosphate acyltransferase from thalassiosira pseudonana regulates acyl composition of glycerolipids. Botany. 2009, 87 (6): 544-551. 10.1139/B08-145.

94. Tonon T, Sayanova O, Michaelson LV, Qing R, Harvey D, Larson TR, Li Y, Napier JA, Graham IA: Fatty acid desaturases from the microalga thalassiosira pseudonana. FEBS J. 2005, 272 (13): 3401-3412. 10.1111/j.1742-4658.2005.04755.x

95. Tonon T, Qing R, Harvey D, Li Y, Larson TR, Graham IA: Identification of a long-chain polyunsaturated fatty acid acyl-coenzyme A synthetase from the diatom thalassiosira pseudonana. Plant Physiol. 2005, 138 (1): 402-408. 10.1104/pp.104.054528

96. Domergue F, Lerchl J, Zähringer U, Heinz E: Cloning and functional characterization of phaeodactylum tricornutum front–end desaturases involved in eicosapentaenoic acid biosynthesis. Eur J Biochem. 2002, 269 (16): 4105-4113. 10.1046/j.1432-1033.2002.03104.x

97. Domergue F, Spiekermann P, Lerchl J, Beckmann C, Kilian O, Kroth PG, Boland W, Zähringer U, Heinz E: New insight into phaeodactylum tricornutum fatty acid metabolism. Cloning and functional characterization of plastidial and microsomal Δ12-fatty acid desaturases. Plant Physiol. 2003, 131 (4): 1648-1660. 10.1104/pp.102.018317

98. Chi X, Zhang X, Guan X, Ding L, Li Y, Wang M, Lin H, Qin S: Fatty acid biosynthesis in eukaryotic photosynthetic microalgae: identification of a microsomal delta 12 desaturase in chlamydomonas reinhardtii. J Microbiol. 2008, 46 (2): 189-201. 10.1007/s12275-007-0223-3

99. Hu Q, Sommerfeld M, Jarvis E, Ghirardi M, Posewitz M, Seibert M, Darzins A: Microalgal triacylglycerols

as feedstocks for biofuel production: perspectives and advances. Plant J. 2008, 54 (4): 621-639. 10.1111/j.1365-313X.2008.03492.x

100. Pereira SL, Leonard AE, Mukerji P: Recent advances in the study of fatty acid desaturases from animals and lower eukaryotes. Prostaglandin Leukot Essent Fat Acid. 2003, 68 (2): 97-106. 10.1016/S0952-3278(02)00259-4.

101. Amiri Jami M, Griffiths M: Recombinant production of omega 3 fatty acids in escherichia coli using a gene cluster isolated from shewanella baltica MAC1. J Appl Microbiol. 2010,

102. Li YT, Li MT, Fu CH, Zhou PP, Liu JM, Yu LJ: Improvement of arachidonic acid and eicosapentaenoic acid production by increasing the copy number of the genes encoding fatty acid desaturase and elongase into pichia pastoris. Biotechnol Lett. 2009, 31 (7). 10.1-1017.

103. Bligh E, Dyer W: A rapid method of total lipid extraction and purification. Can J Biochem Physiol. 1959, 37 (8): 911-917. 10.1139/o59-099

104. Engström K, Saldeen AS, Yang B, Mehta JL, Saldeen T: Effect of fish oils containing different amounts of EPA, DHA, and antioxidants on plasma and brain fatty acids and brain nitric oxide synthase activity in rats. Ups J Med Sci. 2009, 114 (4): 206-213. 10.3109/03009730903268958

105. Hickman KCD: Vacuum distillation apparatus. Google patents. 1939,

106. Barrer RM, Ruiz JLLÓP: Glycerolysis of methyl esters of fatty acids using molecular sieves. J Appl Chem Biotechnol. 1973, 23 (3): 189-194.

107. Schlenk H, Gellerman JL: Esterification of fatty acids with diazomethane on a small scale. Anal Chem. 1960, 32 (11): 1412-1414. 10.1021/ac60167a011.

108. Francis AW: Ternary systems of liquid carbon dioxide. J Phys Chem. 1954, 58 (12). 10.9-1114.

109. Bengen F: German patent application OZ 12438. March. 1940, 18: 135-139.

110. Eckey EW: Directed interesterification in glycerides. Ind Eng Chem. 1948, 40 (7): 1183-1190. 10.1021/ie50463a005.

111. Reitan KI, Rainuzzo JR, Øie G, Olsen Y: A review of the nutritional effects of algae in marine fish larvae. Aquaculture. 1997, 155 (1–4): 207-221.

112. Andrich G, Nesti U, Venturi F, Zinnai A, Fiorentini R: Supercritical fluid extraction of bioactive lipids from the microalga Nannochloropsis sp. Eur J Lipid Sci Tech. 2005. 10. (6): 381-386.

113. Herrero M, Cifuentes A, Ibañez E: Sub- and supercritical fluid extraction of functional ingredients from different natural sources: plants, food-by-products, algae and microalgae: a review. Food Chem. 2006, 98 (1): 136-148. 10.1016/j.foodchem.2005.05.058.

114. Subhadra B: Algal biorefinery based industry: an approach to address fuel and food insecurity for a carbon smart world. J Sci Food Agric. 2011, 91 (1): 2-13. 10.1002/jsfa.4207

115. Subhadra BG: Sustainability of algal biofuel production using integrated renewable energy park (IREP) and algal biorefinery approach. Energ Pol. 2010, 38 (10): 5892-5901. 10.1016/j.enpol.2010.05.043

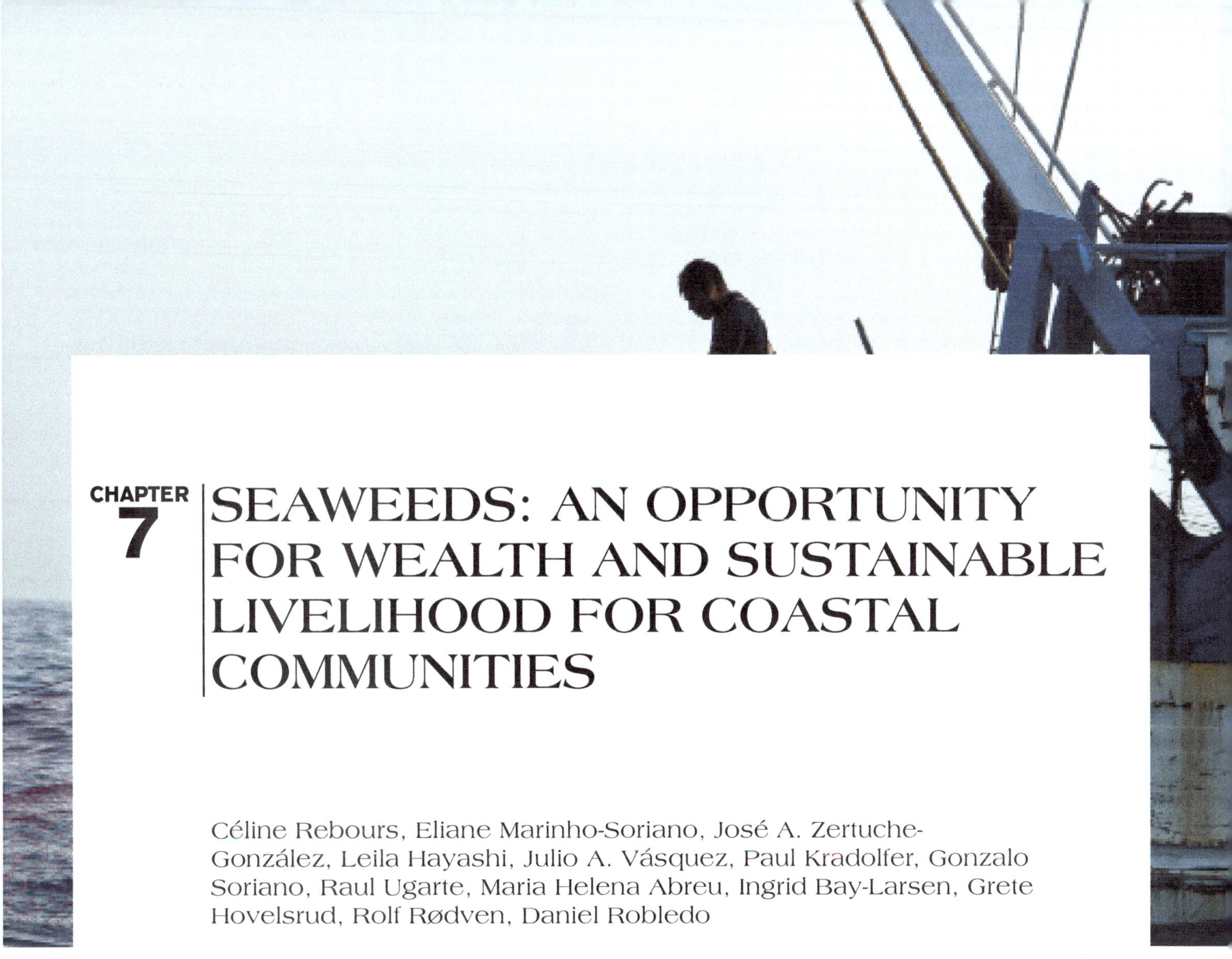

CHAPTER 7

SEAWEEDS: AN OPPORTUNITY FOR WEALTH AND SUSTAINABLE LIVELIHOOD FOR COASTAL COMMUNITIES

Céline Rebours, Eliane Marinho-Soriano, José A. Zertuche-González, Leila Hayashi, Julio A. Vásquez, Paul Kradolfer, Gonzalo Soriano, Raul Ugarte, Maria Helena Abreu, Ingrid Bay-Larsen, Grete Hovelsrud, Rolf Rødven, Daniel Robledo

Keywords: Coastal management Latin America Marine resources Seaweed industry

INTRODUCTION

The worldwide seaweed industry provides a wide variety of products for direct or indirect human uses that have an estimated total value of US$10 billion per year (Bixler and Porse 2011; FAO 2013). Sea vegetables for human consumption constitute about 83 % of production (Craigie 2011), while the remainder is used as fertilizers and animal feed additives, medical applications (Zimmermann et al. 2005; Ehrhart et al. 2013), and biotechnological applications (McHugh 2003). Worldwide, macroalgal production increases 5.7 % every year and more than 18 million tons of macroalgae were produced from global capture and aquaculture in 2011 (FAO2014). In 2011, 96 % of the global total production of macroalgae came from aquaculture, with Asian countries dominating seaweed culture production (99.05 % by quantity and 99.36 % by value, FAO 2014). Five genera (e.g., *Saccharina*, *Undaria*, *Porphyra*, *Eucheuma*/*Kappaphycus*, and *Gracilaria*) represented around 98 % of the world's cultivated seaweed production (Suo and Wang 1992; Pereira and Yarish 2008). *Saccharina japonica* was the most cultivated algae in the world until 2010 when the production of *Eucheuma*/*Kappaphycus* reached over 5.5 million tons for a value over US$1.3 billion (Suo and Wang 1992; McHugh 2003; FAO 2014).*Saccharina* and *Eucheuma*/*Kappaphycus* are mostly produced as raw materials for the food and food polymer industries. Aquaculture of seaweed is scarce outside of Asia, which triggered a worldwide search for hitherto

Citation: Céline Rebours, Eliane Marinho-Soriano, José A. Zertuche-González, Leila Hayashi, Julio A. Vásquez, Paul Kradolfer, Gonzalo Soriano, Raul Ugarte, Maria Helena Abreu, Ingrid Bay-Larsen, Grete Hovelsrud, Rolf Rødven, Daniel Robledo, Seaweeds: an opportunity for wealth and sustainable livelihood for coastal communities, DOI: 10.1007/s10811-014-0304-8

unexploited natural seaweed resources. In 2011, 786,466 t of seaweeds was commercially harvested in 28 countries, ranging from cold to tropical coastlines in both hemispheres, with over 55 % of the biomass harvested in Latin America and almost 32 % in Europe (FAO 2014). The top producers are Chile and Norway respectively accounting for 51.3 and 19.2 % of the global catches of natural seaweed (FAO 2014).

The Pacific coast of South America is naturally rich in marine resources (Thiel et al. 2007; Ortiz2008; Taylor et al. 2008; Vásquez et al. 2012; Vega et al. 2013). In Latin America, macroalgae are an important group for species richness of all regions, varying from 4.9 to 8.7 % of total marine species biodiversity (O'Dor et al. 2010). The highest biodiversity of macroalgal species is reported for the Brazilian region (10.6 species per 100 km of coast), followed by the Humboldt Current system (7.3 species per 100 km of coast), the Tropical West Atlantic (7.1 species per 100 km of coast), and the Tropical East Pacific (6.0 species per 100 km of coast), with the lowest diversity attributed to the Patagonian Shelf (4.7 species per 100 km of coast; Miloslavich et al.2011). The trends reported so far, however, may not truly reflect real patterns, as sampling has not been equal throughout the continent, and taxonomic capacity is very uneven from one country to another, as is the case in the Caribbean (Robledo and Townsend 2006; Miloslavich et al. 2010). In this context, there are still some efforts to be made on basic research to describe and evaluate some of the macroalgal resources in the region (Critchley et al. 2006).

Macroalgae or seaweeds are one of major components of primary biomass production in coastal maritime ecosystems and play an essential ecological role as habitat and substrata for invertebrates, fish, mammals, and birds (Vásquez 1992; Graham et al. 2007). Drastic reduction of any macroalgal community directly influences marine biodiversity, as well as reproduction, recruitment, and growth rates of marine fauna (Vásquez and Santelices 1984; Vásquez 1993). Furthermore, macroalgae may also protect coastlines against erosion (Dayton et al. 1984) and contribute significantly to the marine carbon cycle (Thiel et al. 2007; Ugarte et al. 2010; Vásquez et al. 2013). It is possible that overexploitation of natural seaweed resources could lead to significant ecological, economic, and social consequences at local, regional, and even global scales (Graham et al. 2007; Rebours and Karlsen 2007). In this regard, Alvarez and Vodden (2009) examined the relationships between human actors and *Chondracanthus chamissoi*harvesting in the community of Pisco in Peru. By using local ecological knowledge (LEK), these authors found that a disorderly extraction of *C. chamissoi* generates reductions in the resource that could be exacerbated by climatic change ("El Niño" and "La Niña" phenomena) and the rupture of ecological cycles due to pollution of the marine space and decline of marine species.

Accurate data about seaweed aquaculture or the exploitation of the natural resource stock are difficult to obtain through the international channels in most of the Latin American countries (except in Chile). The necessary information was extracted from database queries at the Harvest data from Instituto de Fomento Pesquero (IFOP) in Chile; Secretaría de Agricultura, Ganadería, Desarrollo Rural, Pesca y Alimentación (SAGARPA) in Mexico; numerous Dirección Regional de Produce (DIREPRO) in Peru; and the FAO Fisheries and Aquaculture Information and Statistics Service to discuss the trends for some Latin American countries, especially for those harvesting seaweeds. The potential for seaweed aquaculture, particularly for red seaweeds, has been addressed in Hayashi et al. (2013). In Latin America, exploitation of seaweed happened or has happened in Argentina, Brazil, Chile, Mexico, Peru, Uruguay, and Venezuela (Critchley et al.2006; FAO 2013). The main seaweed exploitation has been reported in Argentina, Brazil, Chile, Mexico, and Peru. In these countries, traditional harvesting and/or aquaculture is presently active or has been attempted and has generated an estimated biomass volume around 554,585 t in 2009 (equivalent to US$115 million in value), with an estimated decline in total volume to 444,686 t in 2011 (equivalent to US$22 million; Table 1). The main objective of this study is to present a short overview of the management and regulations in place for the exploitation and harvesting of seaweeds in the five above-mentioned Latin American countries. Some of the environmental, social, economic, and political obstacles to sustainable development for seaweed utilization in Latin America will be identified through multidisciplinary analyses, combining natural, ecological, and social sciences. The European Environmental Agency (EEA) also stressed the need for an analytical approach for coastal areas and to place this in an ecosystem-based management approach combining integrated spatial planning and management. Included in this is the need for a consolidated knowledge base and widespread information sharing to support informed policy development and management actions (EEA 2013). Examples for long-term management from Europe (Norway, Portugal) and North America (Canada) are then shortly presented, and successful practices in these Northern regions are then identified and presented as possible as "best practices" model for long-term sustainable seaweed exploitation in Latin America. Particular emphasis is given to identify the possible actions in particular to overcome social, economic, and political obstacles for using Latin American seaweeds as an economic resource for wealth and sustainable livelihoods for coastal communities.

Table 1: Harvesting of seaweeds in Argentina, Brazil, Chile, Mexico, and Peru and commercial aquaculture of seaweeds attempted in only four of these countries. Harvest and aquaculture volumes of seaweed biomass in tons

	2000	2001	2002	2003	2004	2005	2006	2007	2008	2009	2010	2011
Seaweed harvest[a]												
Argentina	3											
Brazil												
Chile	280,844	299,759	315,660	349,128	410,850	425,343	334,674	337,206	406,223	456,184	380,742	417,965

México	33,555	46,927	30,124	28,996	27,418	5,277	5,751	5,093	4,900	5,814	6,009	5,721
Peru	1,323	5,517	6,176	7,864	7,418	5,000	3,118	9,213	12,107	3,874	3,836	5,801
Seaweed aquaculture[b]												
Argentina	3											
Brazil									320[c]	520[c]	730[c]	730[c]
Chile	33,471	65,538	71,648	40,079	20,273	15,493	33,586	23,668	21,687	88,193	12,179	14,469
México												
Peru	11	12										

IFOP Instituto de Fomento Pesquero (www.ifop.cl), *SAGARPA* Secretaría de Agricultura, Ganadería, Desarrollo Rural, Pesca y Alimentación (http://www.sagarpa.gob.mx),*DIREPRO* Dirección Regional de Produce (Regional Direction from Production Ministry)

[a]Harvest data from Chile (IFOP), Mexico (SAGARPA), Peru (DIREPRO)

[b]Aquaculture data from FAO (18/12/2013)

[c]FAO estimates

STATE-OF-THE ART ON THE MANAGEMENT AND REGULATION IN LATIN AMERICA SEAWEED INDUSTRY

In Latin America, several species of red and brown seaweed are harvested as a raw material for phycocolloid extraction. Chile was one of the first agar producers in the world and supplies 10 % of the global biomass for alginates (Bixler and Porse 2011; Vásquez et al. 2012). In Chile, coastal communities have harvested primarily *Gracilaria* and *Lessonia* species since the 1980s, while in more recent years, *Macrocystis* species has been cultivated, first to feed abalone and most recently for bioenergy production (Buschmann et al. 2014). In Chile, several other benthic fisheries like sea urchins and molluscs depend on existence of seaweed beds. The exploitation of natural seaweed resources is therefore regulated with aims to protect the target species as well as the associated biodiversity. Several management strategies have been implemented successfully during the last 10 years, considering comanagement between fishermen unions and the state, using strategies with biological and ecological bases like morphological constraints, quotas by fishing areas, biological bans, rotation of harvesting, and also using experimental areas for harvesting and collection (Vásquez 2008; Vásquez et al. 2012). Lately, Vega et al. (2013) suggested an ecosystem approach and community indicators to manage wild brown seaweed stocks along the country.

In Argentina, different commercial species can be found such as *Undaria pinnatifida*,*Gracilaria gracilis*, *Gigartina skottsbergii*, *Lessonia vadosa*, *Macrocystis pyrifera*, *Porphyra columbina*, *Ulva lactuca*, and *Codium fragile*. In 1958, *G gracilis* and *G. skottsbergii* were already harvested for the agar and carrageenan industries, respectively. Since 1980, *L. vadosa*and *M. pyrifera* are exported to the USA and China to supply the alginate industry, and since 1999, the uses of the Argentinian seaweeds have expended to new markets, such as for human consumption, nutraceuticals, and cosmetic including the fucoidan industries. All seaweed is harvested in Patagonia, mostly in the provinces of Chubut (Fig. 1a, b) and Santa Cruz. Local farmers directly sell the seaweeds to the processing companies or companies with concessions which directly employ their own workers for harvesting during the year and contracted divers in the summer (Fig. 1 c–i). Harvest is regulated since 1970 by local government through special licenses for 3, 10, or 30 years. The duration of the license is decided based on the investment of the companies. Today, there is only one company (Soriano SA) that has worked steadily through the years producing agar and carrageenan. The National Center of Patagonia (CENPAT) guarantees that the harvesting methods are performed in a sustainable way. Regulations for the management of brown seaweeds and marine concessions are particularly well developed, and the supply in brown seaweed to the alginate industry is well managed and organized (Zaixso et al. 2006; Soriano SA unpublished report).

Figure 1: Argentina. Harvesting of seaweed in Patagonia. (**a**) Beach-casted seaweed is harvested. (**b**) Seabed of *U. pinnatifida* is harvested at low tide. (**c**) Transport in truck of the harvest biomass to the drying site. (**d**) Algae are harvested in nets. **e** Suspended net for drying the*U. pinnatifida*. (**f**) Selection of high-quality algae (processing of high-quality algae). (**g**) Preparation of high-quality *U. pinnatifida* for the Japanese market. (**h**) Storage of*Lessonia* sp. in bulk for the Chinese market. (**i**) Quality control on final product (photos: G Soriano)

In Mexico, the federal government manages all fisheries. However, under a law published in 2009, individual states can also manage sessile marine resources through an agreement to the federation. Presently, Mexico's seaweed biomass is sold to the phycocolloid industry, abalone farming, and seaweed extracts for agriculture (DOF 2012; Zertuche-González et al. 2013). None of the four commercial seaweed species harvested in Mexico (*M. pyrifera*, *Gelidium robustum*,*Chondracanthus canaliculatus*, *Gracilariopsis lemaneiformis*) are endangered, thanks to the application of proper harvesting methods (Hernández-Garibay et al. 2006; Robledo 2006; DOF2012). The highest reduction of the seaweed resource was observed along the Pacific Mexican coast during the years when El Niño occurred, but even then, the resources recovered successfully (Ladah et al. 1999; Casas-Valdez et al. 2003). In Mexico, seaweed harvesting may be classified into three broad categories of activity: by hand (at low tide), by diving, and by mechanized methods (specialized boats). Men and women participate both in harvesting and drying. Women and other family members are however mostly involved in the manual gathering, while men dominate the diving and mechanical harvesting, due to the hardship of the work. *C. canaliculatus* has been harvested for carrageenan, by hand, during low tide since 1966, and it is a sustainable harvest up to today (Fig. 2 a–d). The economic outcome from sustainable exploitation of natural resources does not yet necessarily assure a good livelihood to fishermen. Fishermen involved in seaweed harvesting by boat and diving (such as in the case of *G. robustum* and *M. pyrifera*) get better income than those harvesting by hand in low tide. *C. canaliculatus*, for instance, has been harvested for carrageenan, by hand, during low tide since 1966, and it is a sustainable harvest up to today (Fig. 2 a–d); however, harvesting and drying methods are artisanal and seasonal and therefore generate low income. Fishermen that harvest seaweeds from boats and diving gear complement their income by participating in other fisheries such as abalone or sea urchin. Although the seaweed fishery is economically or environmentally sustainable, the livelihood of the fishermen is not always attractive. It is therefore important to look at the fishery in the larger context, and the cultivation of seaweeds (as opposed to their harvesting) may actually offer a better alternative to coastal communities, as it can give them the opportunity to increase production and improve productivity and quality. In this regard, a socio-economic analysis of seaweed farming involving coastal communities was performed in the Gulf of Mexico coast using assumptions and data from the pilot cultivation of *Kappaphycus alvarezii* carried out in Dzilam, Yucatan Peninsula (Robledo et al. 2013). Based on this experience, various social and institutional factors in seaweed farming were discussed, which indicated a good potential for seaweed farming to become an integrated part of the local livelihood strategies for community development (Fig. 2e–g). One of the major conclusions of this study was that any integration of seaweed growing and industrialization in Mexico would require an interest from industry as well as local investors and government authorities in order to ensure successful implementation and development of the activity.

Figure 2: Mexico. **(a–d)** Harvesting of *Chondracanthus canaliculatus* in San Quintin, Baja California. *C. canaliculatus* is harvested by hand in low tide. In the harvest, men and women participate together. (Photos: J Zertuche). **(e–g)** *Kappaphycus alvarezii*aquaculture and harvesting in Dzilam de Bravo, Yucatan (photos: D Robledo)

Peru is just starting to establish regulations for management or exploitation of the seaweed resources including some harvesting restrictions associated with the brown seaweed populations (Vásquez et al. 2012). Southern Peru has approximately 1,000 km of exploitable coastline, with several major, natural populations of brown seaweed species, such as *Lessonia nigrescens*, *Lessonia trabeculata*, *Macrocystis integrifolia*, and *M. pyrifera* (Acleto 2006). The Peruvian production was previously based on the harvesting of *Gracilaria* species. Until 2001, drift brown seaweeds were collected along the shoreline and directly sold to one processing company. McHugh (2002) reported that *L. nigrescens* and *L. trabeculata* were available from natural beds in the south of Peru, although the biomass was strongly affected by El Niño events (Taylor et al. 2008). In the years when the natural marine resources are not negatively affected by El Niño events, about 3,000 t of dry *L. nigrescens* and dry *L. trabeculata* is exported to Asia (Peruvian seaweed exports 2012, PSW SA unpublished report). *M. pyrifera* is also widely available, and prior to 2001, it was exported only in small quantities. In 2012, 95.5 % of the Peruvian seaweed exports was done by the four main Peruvian-based companies (one Chilean and three Chinese). Today, there are four companies purchasing seaweed from the local communities: two Peruvian companies, one Chilean, and one Chinese-owned company. The occurrence of Chinese-owned companies in the Peruvian market has triggered an intensive competition for obtaining the seaweed biomass from the local fishers. As captured fisheries (including seaweed) decline because of overharvesting, the prices of target species often increase dramatically. Likewise, the most commonly harvested fish in Peru, anchoveta (Engraulis ringens), has had a dynamic history of overharvest and fluctuating production (Diana 2009). As a result, beach prices of dried seaweeds increased from US$60 to US$400 t^{-1}between 2006 and 2008 (Table 2, PSW SA unpublished report), creating in the short term an uncontrolled intensification of the harvesting and resulting in a depletion of the natural resources that could have long-term consequences for the Peruvian fishers and Peruvian seaweed industry. Thus, while anchoveta indeed is the fundamental fish species in the Peruvian ecosystem, there are other fisheries to be considered for management (including seaweed). There are trade-offs in managing fisheries, and ideally, such trade-offs should be known when setting fisheries policies (Christensen et al. 2014). This area is well known for experiencing large changes in the abundance and species composition of its main fish resources. Three decades ago, the Peruvian fishery also experienced an economic and social crisis due to an absence of adequate management actions in relation to overexploitation of the anchovy resource (Schreiber 2013). Particular favorable environmental conditions, good recruitment coupled with careful management, and a surveillance scheme have apparently contributed to the fast recovery of the anchovy resource at the end of the 1990s, provided by the political regulatory organism for fisheries resource PRODUCE (Ministry of Production) with the information given by their scientific and technical entity Marine Institute of Peru (IMARPE).

Table 2: Prices in US dollars per ton of dried seaweed paid to the Peruvian fishermen excluding the tax (IGV 19 %; PSW SA unpublished data)

Species	Early 2007	Early 2008	At present
Macrocystis pyrifera	74	170	400
Chondracanthus chamissoi	440	–	840

In the case of Brazil, historically, the seaweed industry is based on harvesting of seaweed from wild populations, a practice maintained until today. Industries based on *Kappaphycus* farming have also been established in the last 20 years but still need to address the issues relative to the acceptance of this new activity from the authorities (Hayashi et al. 2013). Despite the large number of economically commercial species, only few species of *Gracilaria* and *Hypnea* are exploited commercially in Brazil (Marinho-Soriano et al. 2006). Red seaweed exploitation began in the 1960s along the northeast coast and was undertaken without regulations. In 2006, the Brazilian Institute of Environment and Renewable Natural Resources (IBAMA) implemented a regulation on the harvest of natural resources along the coastline. However, the lack of taxation caused overexploitation of natural beds and most of activities that were then dependent on the harvesting of these species have nowadays stopped (Hayashi et al. 2013). Seaweeds are still a significant source of income and support the livelihoods of many people within coastal communities. Harvest is considered a secondary activity, undertaken without the participation of the family heads, who are generally fishermen (E Marinho-Soriano unpublished). On the other hand, for many women, harvesting is considered their main occupation, even if it is performed only part time, during low tide (Fig. 3a–c). The harvest is done on the shores or by diving from a small boat ("jangadas," Fig. 3d, e), when the seaweed is further out. This activity is mostly carried out by women and their children equipped with a net bag (Fig. 3b, c). After the harvest, the biomass is brought back to the shore and then sun-dried for 2 to 3 days on land (Fig. 3f). Finally, the dried seaweed is sold to an intermediate collector and will be used for agar or carrageenan production (E Marinho-Soriano unpublished).

Figure 3: Brazil. Harvesting and aquaculture in the community of Rio do Fogo, Rio Grande do Norte, Brazil. (**a–b**) Women and their children harvesting the natural bed of seaweeds at low tide. (**c–d**) Gathering of the seaweed in bags. (**d–e**) Harvesting of the seaweed by boat. (**f**) Sun-drying of the seaweeds. (**g**) Preparation of the seaweed seedling net by the women of the community. (**h**) Culture line of *Gracilaria birdie* (photos: E Marinho-Soriano)

To improve the situation of poor coastal fishing communities of Brazil, in 2001, the Brazilian government started an FAO Technical Cooperation Project (TCP/BRA/0065) with the objective of assisting the establishment of a sustainable seaweed farming sector that could benefit poor coastal communities in the northeast states (Rio Grande do Norte, Paraiba e Ceará). The scope of the project was testing farming techniques in pilot communities and verifying the technical and financial feasibility of the packages, promoting the associative work among the producers, monitoring the social impact of the introduction of these new techniques, and facilitating the establishment of an institutional framework for this new production sector (Freddi and Aguilar-Manjarrez 2003). Moreover, the project showed that community members easily learned techniques for seaweed cultivation, and red seaweed farming (*Gracilaria birdiae*) became a significant source of income, particularly for the poorer segments of the coastal fishing population (Fig. 3g, h). A fivefold increase in the prices of dry seaweed collected from natural beds or from the beaches was experienced by the seaweed-farming industries (Freddi and Aguilar-Manjarrez 2003). Nowadays, the kilogram of dried *Gracilaria* is sold for US$3.50; seaweeds can therefore bring to the coastal communities an additional income as described in the last data published by the Brazilian Institute of Geography and Statistics (IBGE 2014).

In Brazil, the size of the coastal zone and the large ecosystems and diversity produce a false idea of an inexhaustible exploitation potential, causing adoption of development policies that do not take into account the sustainable use of the resources (MMA/REVIZEE 2006). As a result, over 80 % of the resource is currently overexploited (Miloslavich et al. 2011). The importance of the fishing industry in Brazil is, however, incontestable. Marine fisheries contribute with 63

% of the national fish production and involve approximately 800,000 professionals creating 3.5 million jobs directly or indirectly on the sector (ACEB 2014). According to the IBGE (2014), approximately one quarter of the Brazilian population lives in the coastal area, accounting 50.7 million residents. The process of urbanization in which population and economic activities are concentrated particularly along the coastline has been one of the main processes responsible for the modification of habitat and communities, as well as increasing pressures on water resources (Marques et al. 2004). Only a small portion of the enormous Brazilian coastline is under some form of protection or management, and large areas are subject to anthropogenic pressures, encroachments, and overuse (Amaral and Jablonski 2005).

While Mexico, Chile, and Argentina have developed coastal management plans and appropriate regulations for harvesting of their natural resources, Peru and Brazil do not have well-established regulations or sustainable practices in regard to the exploitation of their natural seaweed resources to avoid depletion of their natural beds (Santelices and Doty 1989; Vásquez et al. 2012). Due to the significance of the seaweed forest for coastal ecosystems and its place in the maritime food chain, a widespread reduction in South American seaweed resources may have an impact on the overall productivity and stability of the coastal communities, with only moderate capacity to adopt to societal and environmental changes. In Peru, the Humboldt Current Large Marine Ecosystem (HCLME) Project therefore attempts to develop terms of reference and comprehensive plan of action (PIA), for the repopulation of natural seaweed beds along the Peruvian coast. In the Bay of Paracas ecosystem (Ica Region, Peru), through laboratory cultivation, transplanting, and harvesting, the PIA enables the recovery of the kelp populations, which were negatively affected by indiscriminate harvesting. The PIA also allows for the sustainability of the economic activity for the community of artisanal fishermen in the area in order to start a sustainable production that includes the transfer of appropriate technologies to them. The understanding of the synergy between ecology, economy, and sociology involved in the exploitation of selected seaweeds provides a sustainable platform for seaweed exploitation in Latin America (Clausen and York 2008).

EUROPEAN AND AMERICAN EXPERIENCES: WHAT ARE THE "LESSONS TO LEARN" ON REGULATIONS AND MANAGEMENT FOR A LONG-TERM SUSTAINABLE SEAWEED INDUSTRY IN LATIN AMERICA?

In Europe, the Norwegian seaweed industry is producing over 60 % of the biomass and is almost completely reliant on natural beds of *Ascophyllum nodosum* and *Laminaria hyperborea*. *L. hyperborea* accounts for about 90 % of the national harvest (Meland and Rebours 2012a), and the alginate industry is by far the most important sector. The remaining biomass is used directly for food, fodder, cosmetics, and the health sectors. Regulations for harvesting seaweed (Meland and Rebours 2012b) are applied to seabed algae such as *L. hyperborea*, and private owner rights only regulate the harvest of foreshore algae such as *A. nodosum*. However, environmental protection laws and other regulations can restrict areas for harvesting. Even a strong interest for seaweed aquaculture, particularly as part of integrated multitrophic aquaculture (Leonczek 2013), the development of any new industrial sector in Norway follows the principle of precaution and is for now highly restricted and controlled by the existing regulation (Alexander et al. 2014) as regulations are under development (Meland and Rebours 2012b).

Portugal seaweed activities are now reduced to the seasonal and low-volume harvest of agarophyte and carrageenophyte species, whereas in the 1980s it was one of the world's largest agar producer (Sousa-Pinto and Araújo 2006). Despite the existence of several economical valuable species, regulations only exist for residual (once important) activities such as Gelidium harvesting and beach cast seaweed activities. A revision of the existing regulation that allows the exploitation of new species is required not only in order to establish sustainable harvesting activities but also to prevent the uncontrolled harvest of economically important sensitive species (Araújo et al. 2009). On the other hand, the country's recent focus on aquaculture development opens the door for seaweed cultivation in monoculture or integrated multitrophic aquaculture (IMTA) systems (Abreu et al. 2011a).

Commercial exploitation of seaweeds in eastern Canada began in the early 1940s with the harvest of the carrageenophyte *Chondrus crispus*. From 1965 to 1980, Canada was the world's main supplier of raw material for the carrageenan industry, with a peak landing over 50,000 wet t in 1974. By the 1990s, the landings sharply declined to less than 10,000 wet t due to a combination of international competition and biological fluctuations in the abundance of the resource (Chopin and Ugarte 2006). Today, *A. nodosum* is the main economic seaweed resource in Canada, with over 40,000 t landed in 2010 (Ugarte and Sharp 2011). This seaweed is being used for among other applications as a biostimulant extract for crops and animal feed supplements, incorporating more than US$40 million to the local economy. *A. nodosum* has been managed since 1995 under an ecosystem approach, following strict regulations established in the Ocean Act (Ugarte and Sharp 2001). To solve problems mainly of fluctuating supply and avoid the risks of overexploitation, the aquaculture of *C. crispus* was developed in the 1970s. *C. crispus* production is the main commercial Canadian seaweed aquaculture, with the culture of*Porphyra* species and *Saccharina latissima* as organic extractive components of integrated multitrophic aquaculture system (Levine 2004; Chopin and Ugarte 2006; Chopin 2008).

Like other European and American countries, the Brazilian and Peruvian dependency on traditional coastal fisheries and eventual growth in aquaculture is followed by an urgent need for robust institutions for planning and managing the coastal zone with focus on developing an integrated and sustainable management plan for the seaweed resources (Prates 2007; Castello2010; Johnsen et al. 2014). The steep growth in the use of marine biological resources represents a fundamental change in the way humans derive benefits from the oceans. Marine aquaculture production is growing at 7 % per year and the sector of domestication of marine species is growing at about 3 % per year, while the number of natural marine products of commercial interest is growing at a rate of 4 % per year (Duarte et al. 2007). Large-scale integrated conservation and management plans are thus urgently required in order to address the sustainable development of various economic and societal sectors. In addition to developing and implementing management plans for sustainable use, these countries also face huge tasks connected to mapping, investigating, and monitoring their marine biological diversity. Information and "lessons learned" may be provided by the scientific communities and management plans in Latin American countries such as Chile or Mexico (Vásquez et al. 2006;2012; Ávila et al. 2013; Vega et al. 2013) and European directives such as those from Norway (Meland and Rebours 2012a) or the ecosystem-based management regulated by the Ocean Act in Canada (Ugarte and Sharp 2001). National regulations should be established on best practices for seaweed harvesting, management, and cultivation, and the knowledge should be passed on to the coastal communities, which could then make seaweed a sustainable opportunity for income.

Alternatively, the seaweed biomass could be obtained through cultivation (McHugh 2002). The seaweed aquaculture industry still requires technological and management improvements, institutional changes, and appropriate environmental and social frameworks (Valenti 2008; Oliveira 2009; Abreu et al. 2011b; Marroni and Asmus 2013). As seaweed aquaculture is at its infancy both in Europe and America, its development will have to overcome numerous challenges by introducing innovations at different levels, i.e., technology, biology, and policy. However, cultivation techniques are standardized and economically sustainable especially in Asia (Miura 1980) and cultivated macroalgae now supply 96 % of the global demand (FAO2014). The aquaculture of macroalgae is also strongly recommended by the EU regulation (EC No 710/2009) for the organic production of animals in aquaculture systems. The development of regulations for supporting the aquaculture of seaweed is a necessary action in recent years, and these regulations should be guided by best practices. The combined effect of rapidly increasing domestication and production with increasing demand for seaweed products is likely to promote innovation in seaweed biotechnology, and these efforts must be accompanied by substantial scientific effort (Mazarrasa et al. 2013). In addition, revision of the Common Agricultural Policy (CAP) intends to include management with High Nature Value in organic certification. Nordic countries such as Denmark, Sweden, and Norway are also aiming to increase organic products to 15 % of the home market. Several food and food supplement products directly prepared from algae are already certified in Europe and in Norway as organic food and given opportunities for producers to access to a niche market (Friis Pedersen et al.2013).

STEPS TOWARDS WEALTH AND SUSTAINABLE LIVELIHOODS FOR LATIN AMERICA COASTAL COMMUNITIES

The seascapes are increasingly managed for multiple functions and services in addition to provision of food, and this requires the integration of ecological and socio-economic research, policy innovation, and public education. The multiuse dilemma has driven many researchers, experts, and policy makers to try and address issues related to the sustainability of coastal development from disciplinary/sectorial perspectives addressing the interactions and functioning within the wider ecosystem, social, economic, and political contexts (Buchholz et al.2012). A review by Harley et al. (2012) addressed the significant gaps in understanding, which hamper ability to predict the outcomes of global change in seaweed-dominated systems. In particular, it addressed the lack of general or even basic understanding of (a) the importance of rates, timing, magnitude, and duration of environmental change; (b) nonadditive effects of multiple stressors; (c) population-level implications of variable environmental impacts among life history stages; (d) the scope for population- or species-level adaptation to environmental change; and (e) ecological responses at the level of communities and ecosystems, including tipping points and sudden phase shifts. In this regard, biological responses (i.e., ecophysiological) to key environmental drivers or combinations of drivers can be incorporated into demographic models to better describe and predict changes in population growth or decline. The expansion of seaweed cultivation particularly in tropical regions contributes significantly to carbon sequestration since the rapid turnover in seaweed culture, approximately 3 months per crop (in the tropics) with yields of over 2,500 wet t ha^{-1} (De Silva and Soto 2009; Vásquez et al. 2013). Nevertheless, some authors have pointed out that a significant proportion (estimates range up to 60 %) of the carbon they fix photosynthetically is released into the water, and a proportion of this released dissolved organic carbon (DOC) is highly labile, entering in the bacterial loop and rapidly remineralizing back to CO_2 (Hughes et al. 2012). Environmental impacts of seaweed farming in the tropics have been reviewed by Zemke-White and Smith (2006). Some authors have also pointed out other environmental impacts of algal farming both positive (i.e., increase on fish assemblages; Bergman et al. 2001) and/or negative (i.e., effect on the meiobenthos; Olafsson et al. 1995). All of these impacts should be considered when the environmental effects of seaweed aquaculture are taken into account. The concept of ecosystem-based management approaches based on an integrated approach of entire ecosystem including humans should also be considered to develop the coastal spatial planning and the best practices guidelines for the exploitation of seaweed (both harvesting and aquaculture) in order to avoid spatial and temporal mismatches of the governance (Crowder and Norse 2008).

The benefits for the well-being of coastal communities as increase of direct permanent employment in previously disadvantaged coastal communities, where unemployment is not only an economic issue but also a socio-political concern, have been exemplified in an IMTA farm of abalone and seaweed (Nobre et al. 2010). Sustainable management of coastal resources creates new economic activities based on the exploitation of a raw material and could participate in the local socio-economic development in coastal areas and communities in Latin America. Developing long-term management plans will also produce fundamental long-term results of interest for the international research community. Socio-economic benefits derived from seaweeds have been already exemplified in the Philippines where approximately 116,000 families consisting of one million individuals were farming more than 58,000 ha of seaweed, making seaweed farming the largest and most productive form of livelihood among the coastal population. In Zanzibar (Tanzania), more than 90 % of seaweed farmers are women. Like in Latin America, changing the life in the villages by women gaining independent economic power will contribute in reducing childhood malnutrition (as an indicator that the health of their mothers has improved), increasing the numbers of children attending schools regularly, and reversing the trend of rural depopulation by self-employment of the village youths (Msuya2006; Msuya et al. 2007).

Innovation should be promoted when trying to integrate seaweed harvesting or aquaculture as part of the wealth for coastal communities. In this regard, Castellacci (2010) pointed out that the technology dynamics of a country depend on three main factors: its innovative intensity, its human capital, and its technological infrastructures. Nowadays, Latin American countries show a much lower innovative strength than Norway or Canada, where the innovation gap has been quantified with a ratio of 16:1 for patents and 10:1 for scientific articles (Castellacci 2010). In order to get closer and eventually jump to the innovation development stage, developing economies such as in some Latin American countries should implement an appropriate combination of policies that takes into account the need to simultaneously develop R & D activities, traditional infrastructures, information and communication technologies, and advanced human skills. Human capital education explains differences in economic performance across countries; education is therefore a necessity to promote social inclusion and cohesion as well as employment. By focusing on marine resources involving low-cost technology requirements, such as the production of seaweed, developing countries are provided an opportunity to access an emerging market, propelled by a diversification of the demand for seaweed products from traditional uses to bioenergy, cosmetics, and biomedicine applications.

CONCLUSIONS

The European, Canadian, and Latin American seaweed industries rely on the harvesting of their natural resources—activity limited by the biomass availability and the potential for the harvested species to recover. Different scenarios associated to seaweed resource exploitation can be observed, and several countries are in need of long-term management plans for the sustainable exploitation of their natural seaweed (Hersoug and Revold 2012). As this activity increases, there is also an urgent need to develop and implement ecosystem-based management models and integrated coastal zone planning. Policy makers must develop regulations and directives that enable a sustainable exploitation of the natural resource, not only to preserve marine and coastal ecosystems but also to ensure social stability and economic income of local communities.

Both European and Latin American countries need to address capacity building and adaptive governance towards seaweed resources. Norway and Canada have well-established management regimes for the sustainable exploitation of their seaweed resources based on the development and application of a sound knowledge and cross-sectorial management and spatial plans. Portugal, on the contrary, needs to review the national and regional management plans concerning the exploitation of the seaweed resources. In Latin America, it is important to stress that Chile has succeeded in establishing a sustainable harvesting in the south of the country. The further expansion of the seaweed industry in this region thus depends on reliable access to raw material, the development of value-added products, and the transfer of expertise between developed and less developed regions. An alternative solution to produce seaweed biomass for this growing sector is offered by aquaculture, which will also require guidelines for evolving into a responsible and well-managed farming.

Finally, trained human resources are required in order to provide education to coastal communities, based on best practices for harvesting and cultivation, in order to establish profitable businesses which could provide socio-economic development leading to better living conditions to the coastal rural communities. The current weak participation of developing countries in the seaweed market could then be reversed by strengthening cooperation to transfer technological knowledge and investment to developing nations active in seaweed management and aquaculture, thereby promoting sustainable development based on their own natural resources, as encouraged by the Convention on Biological Diversity (http://www.cbd.int/doc/legal/cbd-en.pdf). In this context, South American, European, and North American experts should therefore collaborate to develop guidelines to plan an integrated coastal management for Latin American seaweed resources. It is also of great importance for the legitimacy of these guidelines that these originate from transnational and cross-sectorial cooperation, including political, cultural, commercial and industrial actors, NGOs, and research communities. As a first action of knowledge transfer, research and technology development (RTD) and small and medium enterprises (SME) representatives from Norway, Argentina, Brazil, Canada, Chile, Mexico, Peru, and Portugal have established a common working framework in order to support the development of the Latin American seaweed sector. These industry/academia networks encourage cooperation among the seaweed stakeholders; across the project areas; in all aspects related to seaweed production, research, ecosystem services, and management of artisanal and small-scale aquaculture; and traditional and alternative market and economics.

ACKNOWLEDGMENTS

The NorLARNET program (Norway), the County Council of Nordland (Norway), Bioforsk (Norway), and Ministry of Fisheries and Aquaculture (Brazil) financially supported this work. The LA Seaweed Consortium met in Brazil at the Latin American Congress of Algal Biotechnology (IV CLABA).

REFERENCES

1. Abreu MH, Pereira R, Yarish C, Buschmann A, Sousa-Pinto I (2011a) IMTA with *Gracilaria vermiculophylla*: productivity and nutrient removal performance of the seaweed in a land-based pilot scale system. Aquaculture 312:77–87
2. Abreu M, De Mattos P, Lima P, Padula A (2011b) Shrimp farming in coastal Brazil: reasons for market failure and sustainability challenges. Ocean Coast Manag 54:658
3. ACEB (2014) Anuário da Pesca e Aquicultura Brasileira 2014. 1st Brazilian Fishery and Aquaculture Yearbook 2014: http://www.mpa.gov.br/index.php/publicidade/publicacoes
4. Acleto CO (2006) Seaweed resources in Peru. In: Critchley AT, Ohno M, Largo DB (eds) CD-ROM World Seaweed Resources—an authoritative reference system.

Version:1.0 Margraf Publishers GmbH

5. Alexander K, Potts TP, Freeman S, Israel D, Johansen J, Kletou D, Meland M, Pecorino D, Rebours C, Shorten M, Angel DL (2014) The implications of aquaculture policy and regulation for the development of integrated multi-trophic aquaculture in Europe. Aquaculture (manuscript number: AQUA-D-14-00294, Submitted 17.02.2014)
6. Alvarez JC, Vodden K (2009) Local ecological knowledge and the impacts of global climatic change on the community of seaweed extractors in Pisco-Perú. PICMET - Proceedings, August 2–6, Portland, Oregon USA
7. Amaral ACZ, Jablonski S (2005) Conservation of marine and coastal biodiversity in Brazil. Conserv Biol 19:625–631
8. Araújo R, Vaselli S, Almeida M, Serrão EA, Sousa-Pinto I (2009) Disturbance on marginal populations: effects of human trampling on *Ascophyllum nodosum* assemblages at its southern distribution limit. Mar Ecol Prog Ser 378:81–92
9. Ávila M, Piel MI, Alcapan A (2013) Indoor and outdoor culture of *Callophyllis variegata* (Bory) Kützing (Gigartinales, Rhodophyta) in southern Chile. J Appl Phycol. doi:10.1007/s10811-013-0109-1
10. Bergman KC, Svensson S, Öhman MC (2001) Influences of algal farming on fish communities. Mar Pollut Bull 42:1379–1389PubMed
11. Bixler HJ, Porse H (2011) A decade of change in the seaweed hydrocolloids industry. J Appl Phycol 23:321–335
12. Buchholz CM, Krause G, Buck BH (2012) Seaweed and man. In: Wiencke C, Bischof K (eds) Seaweed biology. Springer, Berlin, pp 471–494
13. Buschmann AH, Prescott S, Potin P, Faugeron S, Vásquez JA, Camus C, Infante J, Hernández-González MC, Gutierrez A, Varela DA (2014) The status of kelp exploitation and marine agronomy, with emphasis on Macrocystis pyrifera, in Chile. Adv Bot Res 71
14. Casas-Valdez M, Serviere Zaragoza E, Lluch Belda D, Marcos R, Aguila Ramírez R (2003) Effect of climatic change on the harvest of the kelp *Macrocystis pyrifera* on the Mexican Pacific Coast. Bull Mar Sci 73:545–556
15. Castellacci F (2010) Closing the technology gap? MPRA Paper No. 27586, posted Online athttp://mpra.ub.uni-muenchen.de/27586/
16. Castello JP (2010) O futuro da pesca e da aquicultura marinha no Brasil: A pesca costeira. Cienc Cult 62(3):32–35
17. Chopin T (2008) Integrated multi-trophic aquaculture (IMTA) will also have its place when aquaculture moves to the open ocean. Fish Farmer 31(2):40–41
18. Chopin T, Ugarte R (2006) Seaweed resources of Eastern Canada. In: Critchley AT, Ohno M, Largo DB (eds) CD-ROM World Seaweed Resources—an authoritative reference system. Version:1.0 Margraf Publishers GmbH
19. Christensen V, de la Puente S, Sueiro JC, Steenbeek J, Majluf P (2014) Valuing seafood: the Peruvian fisheries sector. Marine Policy 44:302–311
20. Clausen R, York R (2008) Global biodiversity decline of marine and freshwater fish: a cross-national analysis of economic, demographic, and ecological influences. Soc Sci Res 37:1310–1320
21. Craigie JS (2011) Seaweed extract stimuli in plant science and agriculture. J Appl Phycol 23:321–335
22. Critchley AT, Ohno M, Largo DB (2006) World Seaweed Resources: an authoritative reference system. CD-ROM World Seaweed Resources—an authoritative reference system.Version:1.0 Margraf Publishers GmbH, Germany
23. Crowder L, Norse E (2008) Essential ecological insights for marine ecosystem-based management and marine spatial planning. Mar Policy 32:772–778
24. Dayton PK, Currie V, Gerrodette T, Keller B, Rosenthal R, Ven Tresca D (1984) Patch dynamics and stability of some California kelp communities. Ecol Monogr 54:253–289
25. De Silva SS, Soto D (2009) Climate change and aquaculture: potential impacts, adaptation and mitigation. In: Cochrane K, De Young C, Soto D, Bahri T (eds) Climate change implications for fisheries and aquaculture: overview of current scientific knowledge. FAO Fisheries and Aquaculture Technical Paper. No. 530. FAO, Rome, pp 151–212
26. Diana J S (2009) Aquaculture Production and Biodiversity Conservation. Bioscience 59(1):27–38
27. DOF (2012) Diario Oficial de la Federación: Plan de Manejo para la Pesquería de Macroalgas en Baja California, México. Secretaría de Agricultura, Ganadería, Desarrollo Rural, Pesca y Alimentación. http://dof.gob.mx/nota_detalle.php?codigo=5280840&fecha=30/11/2012
28. Duarte CM, Marbá N, Holmer M (2007) Rapid domestication of marine species. Science 316:382–383PubMed
29. EEA (2013) Balancing the future of Europe's coasts—knowledge base for integrated management. European Environmental Agency, Denmark, p 67, Report 12
30. Ehrhart F, Mettler E, Böse T, Weber MM, Vásquez JA et al (2013) Biocompatible coating of encapsulated cells using ionotropic gelation. PLoS ONE 8(9):e73498
31. FAO (2013) Fisheries and Aquaculture Information and Statistics Service – 18/12/2013
32. FAO (2014) Fisheries and Aquaculture Information and Statistics Service - 16/03/2014
33. Freddi A, Aguilar-Manjarrez J (2003) Small-scale seaweed farming in North East Brazil. TCP/BRA/0065 35 FAO Aquaculture Newsletter 32
34. Friis Pedersen S, Meland M, Rebours C (2013) Macroalgae for an increasing organic market. Bioforsk Fokus 8(2):337–338
35. Graham MH, Vásquez JA, Buschmann AH (2007) Global ecology of the giant kelp *Macrocystis*: from ecotypes to ecosystems. Oceanogr Mar Biol Annu Rev 45:39–88
36. Harley CDG, Anderson KM, Demes KW, Jorve JP, Kordas RL, Coyle TA, Graham MH (2012) Effects of climate change on global seaweed communities. J

Phycol 48:1064–1078

37. Hayashi L, Bulboa C, Kradolfer P, Soriano G, Robledo D (2013) Cultivation of red seaweeds: a Latin American perspective. J Appl Phycol. doi: 10.1007/s10811-013-0143-z
38. Hernández-Garibay E, Guardado-Puentes J, Bautista-Alcantar J, Reyes-Tisnado R (2006) Macroalgas del Oceáno Pacífico. In: Sustentabilidad y Pesca Responsable en Mexico. Ed J Cuellar. Instituto Nacional de la Pesca. SAGARPA: 235–244
39. Hersoug B, Revold J (2012) Assessment criteria for Norwegian assistance to aquaculture in developing countries. A draft report to Norad. The Norwegian College of Fishery Science, University of Tromsø, Tromsø, p 47
40. Hughes AD, Black KD, Campbell I, Davidson K, Kelly MS, Stanley MS (2012) Does seaweed offer a solution for bioenergy with biological carbon capture and storage? Greenh Gas Sci Technol 2:402–407
41. IBGE (2014) Instituto Brasileiro de Geografia e Estatística. Available: saladeimprensa.ibge.gov.br/en/noticias
42. Johnsen JP, Hersoug B, Solås A-M (2014) The creation of coastal space—how local ecological knowledge becomes relevant. Mar Stud 13:2. doi:10.1186/2212-9790-13-2
43. Ladah LB, Zertuche-González JA, Hernández-Carmona G (1999) Giant kelp (*Macrocystis pyrifera*, Phaeophyceae) recruitment near its southern limit in Baja California after mass disappearance during ENSO 1997–1998. J Phycol 35:1106–1112
44. Leonczek A (2013) Traditional and integrated aquaculture. Today's environmental challenges and solutions of tomorrow. The Bellona Foundation, report, 117pp
45. Levine IA (2004) *Porphyra* Cultivation in Maine, USA and New Brunswick, Canada Aquaculture Compendium 1–13
46. Marinho-Soriano E, Moreira WSC, Carneiro MAA (2006) Some aspects of the growth of*Gracilaria birdiae* (Gracilariales, Rhodophyta) in an estuary in northeast Brazil. Aquacult Int 14:327–336
47. Marques M, Costa MF, Mayorga MIO, Pinheiro PRC (2004) Water environments: anthropogenic pressures and ecosystem changes in the Atlantic drainage basins of Brazil. Ambio 33:68–77PubMed
48. Marroni EV, Asmus ML (2013) Historical antecedents and local governance in the process of public policies building for coastal zone of Brazil. Coast Ocean Manag 76:30–37
49. Mazarrasa I, Olsen YS, Mayol E, Marbà N, Duarte CM (2013) Rapid growth of seaweed biotechnology provides opportunities for developing nations. Nat Biotechnol 31:591–592PubMed
50. McHugh DJ (2002) Prospects for seaweed production in developing countries. FAO Fisheries Circular No. 968 FIIU/C968
51. McHugh DJ (2003) A guide to the seaweed industry. FAO Fisheries Technical paper, No 441: 105 p
52. Meland M, Rebours C (2012a) Short description of the Norwegian seaweed industry. Bioforsk Fokus 7(2):278–279
53. Meland M, Rebours C (2012b) Introduction to the management and regulation of Norwegian seaweed industry. Bioforsk Fokus 7(2):275–277
54. Miloslavich P, Díaz JM, Klein E, Alvarado JJ, Díaz C et al (2010) Marine biodiversity in the Caribbean: regional estimates and distribution patterns. PLoS ONE 5(8):e11916
55. Miloslavich P, Klein E, Díaz JM et al (2011) Marine biodiversity in the Atlantic and Pacific coasts of South America: knowledge and gaps. PLoS ONE 6(1):e14631
56. Miura A (1980) Seaweed cultivation: present practices and potentials. In: Carter A (ed), Living resources, University of Chicago, pp 57–68
57. MMA/REVIZEE (2006) Avaliação do potencial sustentável de recursos vivos na zona econômica exclusiva: relatório executivo/ MMA, Secretaria de Qualidade Ambiental. Brasília: Brazil
58. Msuya FE (2006) The impact of seaweed farming on the social and economic structure of seaweed farming communities in Zanzibar, Tanzania. In: Critchley AT, Ohno M, Largo DB (eds) CD-ROM World Seaweed Resources—an authoritative reference system. Version:1.0 Margraf Publishers GmbH Germany
59. Msuya FE, Shalli MS, Sullivan K, Crawford B, Tobey J, Mmochi AJ (2007) A comparative economic analysis of two seaweed farming methods in Tanzania. The sustainable coastal communities and ecosystems program. Coastal Resources Center, University of Rhode Island and the Western Indian Ocean Marine Science Association, pp 27
60. Nobre AM, Robertson-Andersson D, Neori A, Sankar K (2010) Ecological–economic assessment of aquaculture options: comparison between abalone monoculture and integrated multi-trophic aquaculture of abalone and seaweeds. Aquaculture 306:116–126
61. O'Dor R, Miloslavich P, Yarincik K (2010) Marine biodiversity and biogeography—regional comparisons of global issues, an introduction. PLoS ONE 5(8):e11871
62. Ólafsson E, Johnstone RW, Ndaro SGM (1995) Effects of intensive seaweed farming on the meiobenthos in a tropical lagoon. J Exp Mar Biol Ecol 191:101–117
63. Oliveira RC (2009) O panorama da aquacultura no Brasil: a prática com foco na sustentabilidade. Rev Intertox Toxicol Risco Ambient Soc 2(1):71–89
64. Ortiz M (2008) Mass balanced and dynamic simulations of trophic models of kelp ecosystems near the Mejillones Peninsula of northern Chile (SE Pacific): comparative network structure and assessment of harvest strategies. Ecol Model 216:31–46
65. Pereira R, Yarish C (2008) Mass production of marine macroalgae. Encyclopedia of Ecology, 2236–2247
66. Prates APL (2007). National Plan for Protected Areas Marine and Coastal Areas Context. In: MMA/SBF. 2007. Aquatic Protected Areas as Fisheries Management

tools. 18–24 p

67. Rebours C, Karlsen Å (2007) Seaweeds in the North: new scopes for coastal farming. Bioforsk Fokus 2(13):73–76
68. Robledo D (2006) Seaweed resources of Mexico. In: Critchley AT, Ohno M, Largo DB (eds) CD-ROM World Seaweed Resources—an authoritative reference system. Version:1.0 Margraf Publishers GmbH Germany
69. Robledo D, Townsend W (2006) Seaweed and mangroves: improving environmental practices in coastal communities to secure sustainable livelihoods. In: Breton Y, Brown D, Davy B, Haughton M, Ovares L (eds), Coastal resource management in the wider Caribbean: resilience, adaptation, and community diversity. Ian Randle Publishers and International Development Research Center
70. Robledo D, Gasca-Leyva E, Fraga J (2013) Social and economic dimensions of carrageenan seaweed farming in Mexico. In: Valderrama D, Cai J, Hishamunda N, Ridler N (eds) Social and economic dimensions of carrageenan seaweed farming. FAO, Rome, pp 185–204, Fisheries and Aquaculture Technical Paper No. 580
71. Santelices B, Doty MS (1989) A review of *Gracilaria* farming. Aquaculture 78:95–133
72. Schreiber MA (2013) Institutions for sustainable fisheries governance—the case of the commercial Peruvian anchovy fishery. PhD Thesis, Universität Bremen, 132 p
73. Sousa-Pinto I, Araújo R (2006) Seaweed resources of Portugal. In: Critchley AT, Ohno M, Largo DB (eds) CD-ROM World Seaweed Resources—an authoritative reference system. Version:1.0 Margraf Publishers GmbH, Germany
74. Suo R, Wang Q (1992) *Laminaria* culture in China. INFO Fish Int 1(92):40–42
75. Taylor MH, Wolff M, Mendo J, Yamashiro C (2008) Changes in trophic flow structure of Independence Bay (Peru) over an ENSO cycle. Prog Oceanogr 79:336–351
76. Thiel M, Macaya EC, Acuña E et al (2007) The Humboldt Current System of northern-central Chile: oceanographic processes, ecological interactions and socioeconomic feedback. Oceanogr Mar Biol Annu Rev 45:195–344
77. Ugarte R, Sharp G (2001) A new approach to seaweed management in Eastern Canada: the case of *Ascophyllum nodosum*. Cah Biol Mar 42:63–70
78. Ugarte R, Sharp G (2011) Management and production of the brown algae *Ascophyllum nodosum* in the Canadian maritimes. J Appl Phycol 24:409–416
79. Ugarte RA, Craigie JS, Critchley AT (2010) Fucoid flora of the rocky intertidal of the Canadian Maritimes: implications for the future with rapid climate change. In: Israel A, Einav R, Seckbach J (eds) Seaweeds and their roles in globally changing environments. Springer, Dordrecht, pp 69–90
80. Valenti WC (2008) Aqüicultura Brasileira é sustentável? Palestra apresentada durante o IV Seminário Internacional de Aqüicultura, Maricultura e Pesca, Aquafair 2008, Florianópolis, 13–15 de maio de 2008. pp 1–11 (www.avesui.com/anais)
81. Vásquez JA (1992) *Lessonia trabeculata*, a subtidal bottom kelp in northern Chile: a case of study for a structural and geographical comparison. In: Seeliger U (ed) Coastal plant communities of Latin America. Academic, San Diego, pp 77–89
82. Vásquez JA (1993) Natural mortality of giant kelp *Macrocystis pyrifera* affecting the fauna associated with its holdfasts. Pac Sci 47:180–184
83. Vásquez JA (2008) Production, use and fate of Chilean brown seaweeds: resources for a sustainable fishery. J Appl Phycol 20:457–467
84. Vásquez JA, Santelices B (1984) Comunidades de macroinvertebrados en discos de adhesión de*Lessonia nigrescens* en Chile central. Rev Chil Hist Nat 57:131–154
85. Vásquez JA, Vega JMA, Buschmann AH (2006) Long term studies on El Niño-La Niña in northern Chile: effects on the structure and organization of subtidal kelp assemblages. J Appl Phycol 18:505–519
86. Vásquez JA, Piaget N, Alonso Vega JM (2012) The *Lessonia nigrescens* fishery in northern Chile: "how you harvest is more important than how much you harvest". J Appl Phycol 24:417–426
87. Vásquez JA, Zuñiga S, Tala F, Piaget N, Rodriguez DC, Vega JMA (2013) Economic evaluation of kelp forest in northern Chile: values of good and service of the ecosystem. J Appl Phycol. doi:10.1007/s.10811-013-0173-6
88. Vega JMA, Broitman B, Vásquez JA (2013) Monitoring the sustainability of *Lessonia nigrescens*complex (Laminariales, Phaeophyta) in northern Chile under strong harvest pressure. J Appl Phycol. doi: 10.1007/s.10811-013-0167-4
89. Zaixso AB, Ciancia M, Cerezo AS (2006) Seaweed resources in Argentina. In: Critchley AT, Ohno M, Largo DB (eds) CD-ROM World Seaweed Resources—an authoritative reference system. Version:1.0 Margraf Publishers GmbH, Germany
90. Zemke-White WL, Smith JE (2006) Environmental impacts of seaweed farming in the tropics. In: Critchley AT, Ohno M, Largo DB (eds) CD-ROM World Seaweed Resources—an authoritative reference system. Version:1.0 Margraf Publishers GmbH, Germany
91. Zertuche-González JA, M Sánchez-Barredo, JM Guzmán-Calderón, Z Altamirano-Gómez (2013) *Eisenia arborea* J.E. Areschough as abalone diet on an IMTA farm in, Baja California, México. J Appl Phycol. doi: 10.1007/s10811-013-0138-9
92. Zimmermann H, Zimmermann D, Reuss R et al (2005) Towards a medically approved technology for alginate-based microcapsules allowing long-term immune isolated transplantation. J Mater Sci Mater Med 16:491–501

CHAPTER 8 BIOGAS FROM MACROALGAE

Adam D Hughes, Maeve S Kelly, Kenneth D Black and Michele S Stanley

Keywords: Biogas Methane Anaerobic digestion Seaweed Macroalgae Aquaculture

INTRODUCTION

Growing terrestrial crops for biofuel may make a negligible contribution to net greenhouse gas emissions [1, 2] and may cause other environmental impacts while reducing freshwater resources and food security [3]. Given these limitations there has been renewed / increased interest in aquatic and marine production for biofuels [4, 5]. This interest can be divided into two principal components: biofuels derived from macroalgae (seaweed) and biofuels derived from microalgae (single cell plants). Microalgal derived biofuels have received much attention as a source for biodiesel [6 8], however production costs are an order of magnitude too expensive [3]. Although there is currently enormous research investment into the bulk production of microalgae for biodiesel, photo bioreactors are unlikely to be economically competitive for bioenergy production, and culture in outdoor ponds is only suited to regions with a relatively high number of sunlight hours and even then may still be uncompetitive in the biofuels market [9].

Macroalgae as a source of bioenergy first received intensive scrutiny as part of the US Ocean Food and Energy Farm project as proposed by Wilcox [10], initiated in 1973 and lasting over a decade [11]. It resulted in the construction of ocean farms for cultivation of the giant kelp *Macrocystis*[12]; reviewed by Kelly and Dworjanyn, [13]. While farming this species of seaweed in this truly offshore environment presented many technical challenges, the biogasification of macroalgal biomass gave excellent results [10, 12, 14, 15]. This and subsequent research highlights some of the major advantages of macroalgae over other sources of biofuels (see Table 1).

Table 1: Environmental and societal risk associated with terrestrial biofuels (after Koh and Ghazoul,[16]) and macroalgae biofuels

Environmental and societal advantages of macroalgae production for biofuels	
Net GHG emissions from land-use change	The culture of macroalgae for biofuel would be entirely marine based and would not have the associated GHG emissions associated with land use change.

Citation: Adam D Hughes, Maeve S Kelly, Kenneth D Black, Michele S Stanley, Biogas from Macroalgae, DOI: 10.1186/1754-6834-5-86.

Threats to biodiversity	Macroalgae cultivation takes place in the water column above the seabed. Impacts of large scale macroalgae production on benthic biodiversity are currently unquantified. Likely impacts will include shading and competition for nutrients. However, most production will be in waters where the seabed is deeper than the photic zone, and where terrestrial nutrient run off creates hypernutrified water. It is likely that biodiversity would increase in the vicinity of macroalgae farms as a result of increased habitat structural complexity.
Impacts on food prices	Currently most macroalgae cultivation is for human consumption. Large scale production of macroalgae for biofuels is bound to distort this market. However the impacts on the supply of macroalgae to human food chain is likely to be small due to a clear market segregation and the far higher value of macroalgae as food compared to the price of energy.
Competition for water resources	Mass cultivation of macroalgae has a zero freshwater requirement and only modest amounts are required in anaerobic digestion

With microalgae much of the research interest has focused on their conversion to liquid biofuels such as ethanol [17–20]. However, in this review we focus on anaerobic digestion of cultivated macroalgae for the production of biogas. Since this original gasification / culture research was conducted there have been substantial advances in macroalgal cultivation and offshore engineering. However the concept of ocean farming for biogas production has received relatively little attention in the 21st century.

In coastal temperate regions this technology may have significant potential to meet local energy demand, particularly in areas where the expansion of terrestrial biofuel production is limited due to high percentages of net primary productivity (NPP) already being appropriated for human use, such as the north Atlantic Coast of North America and Europe, and the western seaboard of South America (Figure 1). In this review the wild harvest of seaweeds for biofuel is not considered a viable option either in terms of potential yield or its environmental impact. Although macroalgal harvest for high value products takes place in some countries, very careful management is required to prevent serious ecosystem damage [21–24] hence it would be impossible to justify harvest on the massive scale necessary [25–27] to make a significant energy contribution. Macroalgal forests, as with other biogenic structures in the marine environment such as corals and seagrasses, are considered to be biodiversity hot spots providing important habitat to a wide range of organisms including fish and birds [28]. It is worth noting that in Norway where approximately 170,000 tonnes pa *Laminaria hyperborea*are harvested, even a 4–5 year rotation is not always sufficient to allow recovery [29]. In addition, as wild stocks are generally dispersed around coastlines this would result in high costs of transport to processing plants.

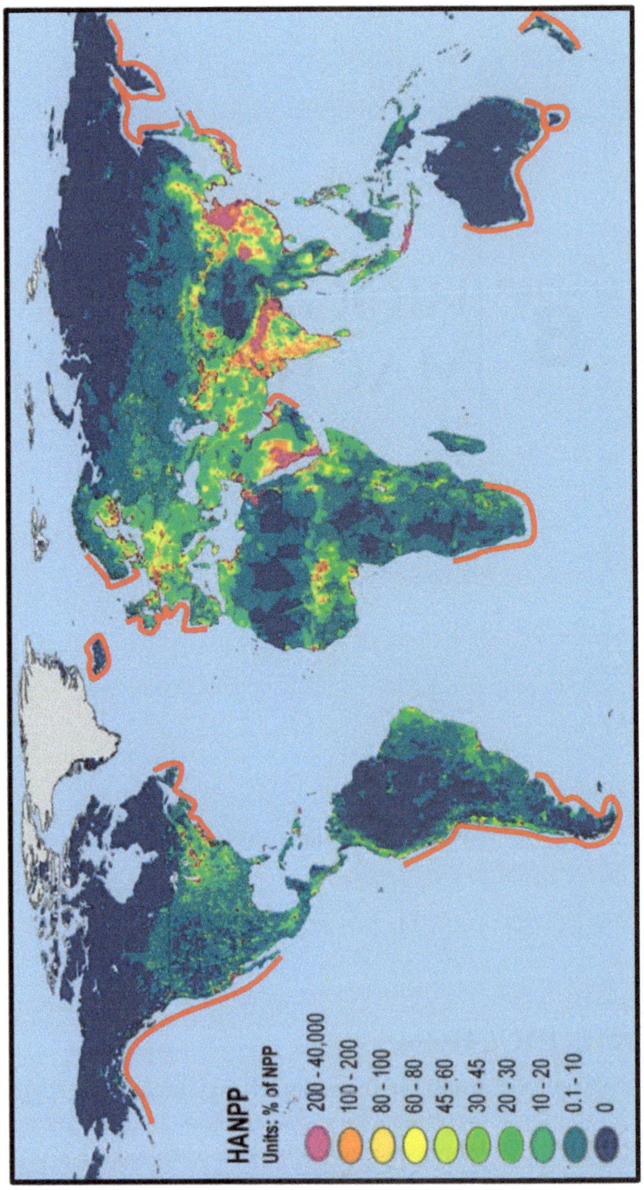

Figure 1: Natural distribution of shallow water macroalgae (red line) indicating coastal areas with the potential to culture macroalgae for biogas, and human appropriation of net primary production as a percentage of local net primary productivity (NPP) (NASA 2004 [30]). Redrawn from Santelices (2007) [31] and NASA Earth Observatory 2004.

Macroalgal Culture

Currently over 100 species of macroalgae are used for food, in medicine, or as fertiliser and in the processing of phycolloids and chemicals [32]. Although used for millennia, their domestication only began in the twentieth century as a fuller understanding of their life cycle was achieved [33]. Several species are now in culture on a large-scale in east Asia. China is the world's largest producer of cultivated seaweed, mostly grown on long-line systems where hatchery produced seedlings are transplanted to sea on ropes suspended vertically from a horizontal top-line. The large brown *L. japonica* known as haidai or 'sea-strap' and originally introduced to China from Japan is the world's most cultivated species by volume and value. It was the first seaweed to be subjected to the entire process of seeding, tending and planting out and to have the status of a marine plant crop [32]. Global production of *L. japonica* alone in 2010 was 5.14 million tonnes with a value of 3.01 billion USD [34]; it is grown primarily for food but also for iodine and alginates. Its fast growth and high productivity make this and several other species of brown macroalgae particularly suited to culture for energy crops. Estimates of macroalgal primary productivity rates, in terms of carbon capture during photosynthesis, are approximately 1600 g $Cm^{-2}y^{-1}$[35], comparing favourably to a global net primary productivity of crop land of 470 g $Cm^{-2}y^{-1}$[36].

Selective breeding of macroalgae began in China in the 1960s with *Laminaria* species. This has resulted in a number of varieties that show the enhancement of desirable characteristics over wild varieties. These characteristics include:

i. increased frond growth rate at higher temperatures, resulting in a longer frond and higher production (20-58% higher)
ii. a higher (8-40%) iodine content as compared with the natural population
iii. a lower water content

Since the early seventies these selectively bred strains have been widely adopted by the*Laminaria* cultivation industry in North China. There are thus good prospects for the development of strains having traits desirable for biofuel production, such as increased sugar content or altered seasonality of production cycles [37].

Macroalgal Production

In Europe, hatchery raised macroalgae have been cultured successfully on long-line systems, similar to those used for mussel production. Positioned adjacent to salmon cages in Scottish sea lochs [38], a 100 m horizontal long-line bearing vertical strings carrying seaweeds every 50 cm, indicated average yields of >50 kg (native *Saccharina latissima*) per horizontal meter of long-line. If this were extrapolated to consider 40 such 100 m longlines, then yields of 200 t wet weight ha^{-1} (approximately 20 t dry weight) would be obtainable. This is comparable to yields achieved in China without fertiliser (H. Liu pers.comm. citing China Fish Annals, 2003). However if macroalgal crops are to make a significant contribution to fuel supply then very large areas would have to be farmed. MacKay (2009) [39] makes it clear that biomass energy will need to be a country-scale activity to make a meaningful contribution to UK energy needs. This will require significant changes in societal attitudes to use of the marine environment and, in many countries, regulatory changes. Inshore areas are already under significant pressure so the culture of macroalgae at the scale required for biofuel production must be largely located on continental shelves. Globally there is a very large amount of continental shelf suitable for such a massive aquaculture expansion; presently aquaculture occupies only about 0.04% of continental shelf area [40]. However culturing seaweed in an European offshore environment will require the development of more mechanised technologies for outplanting and harvest than the labour intensive methods on which the large-scale culture in Asia currently depends. This in turn may lead to the development of more specialised vessels than the mussel/salmon-farm work boats currently employed. The growth rate and productivity of seaweeds, grown on a large and dense scale, and in a different nutrient regime (offshore) to that of the inshore waters (Scottish sea lochs) has yet to be verified.

Seaweed to Biogas: Anaerobic Digestion

Macroalgae can be converted to biofuels by various processes including thermal treatment [41] and fermentation [19, 42] but the most direct route to obtaining biofuel from macroalgae is via its anaerobic digestion (AD) to biogas (~ 60% methane). Methane can be used to produce heat and electricity or compressed for use as a transport fuel. Research conducted in the 1980's [43,44] still provides a bench mark for biogas yields for a number of macroalgal species, but since this time there have been developments in AD technology and an enormous increase in its use.

In comparison to terrestrial biomass crops, macroalgae contain little cellulose and no lignin and therefore undergo a more complete hydrolysis. Gas yield is related both to ash content (and its inverse relationship with volatile solids content) and the level of storage sugars; and, as seaweed biochemical composition varies with season, gas yield will vary [45, 46]. The C:N ratio is also an important part of optimising digester diet and strengthens the argument for the co-digestion of seaweeds with other more N rich substrates, for example waste food or agricultural slurries. Biogas yields are also dependent on a wide range of other variables such as inoculum, digester system configuration and feed stock composition.

Perhaps the most realistic estimate of the true industrial potential of methane production from macroalgae were obtained by Matusi (2006) [47] using a commercial scale 4 stage anaerobic digester for over 150 days, with a daily input between 0.2-1.0 tonnes of seaweed and a retention time of 15 to 25 days. This resulted in an average production of 22 m^3 of methane per tonne wet weight of brown seaweed (*Laminaria* sp). The potential energy yield from the AD of marine biomass compares favourably with that of terrestrial crops (Figure 2) on an energy per unit area basis. These figures are based on a production of 22 m^3 of methane per tonne of macroalgae wet weight, and a production of 200 tonnes of macroalgae ha^{-1}. However recent advances suggest there is still potential for further optimising biogas yields through co-digestion with a more nitrogenous substrate [48, 49] and manipulation of the microbial composition of the inoculum [50].

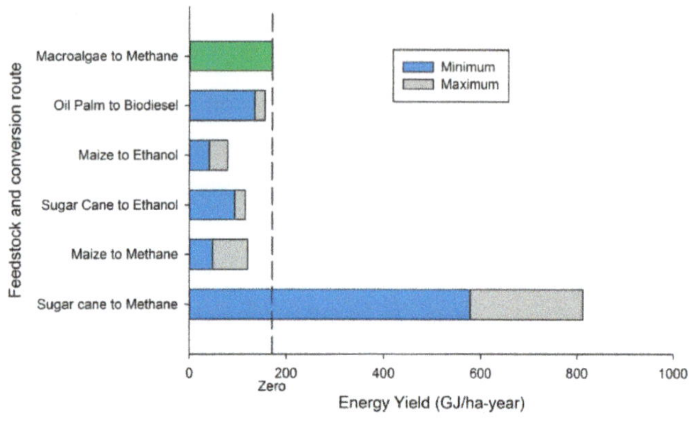

Figure 2: The energy production of biofuel crops ha^{-1}based on macroalgal production of 200 t ww ha^{-1}, terrestrial biofuel crop estimates are from Shilton and Guieysse (2010)[3].

Figure 3: Area needed to grow enough seaweed to meet domestic gas requirements for homes on the Isle of Mull, West coast of Scotland, based on a production of 200 wet tonnes ha −1.

Bioenergy Potential – A Question of Scale

If we use a realistic estimate of macroalgal production [26] (200 t ha $^{-1}$) and a conservative estimates of biogas yield after conversion (22 m^3 tonne wet weight (ww)) yielding 171 GJ ha $^{-1}$we can see that to make a significant contribution to bioenergy targets there will need to be macroalgal cultivation on a massive and unprecedented scale. For example if all of the brown algae currently produced in culture (6.8 million tonnes p.a. [34]) was converted to biogas using the parameters above it would yield approximately 5.7 PJ which is approx. 0.06% of the UK total energy demand for 2010 (9518 PJ [51]). To meet 1% of UK total energy demand would require an area of cultivation of approximately 5440 km^2. This is equivalent to half of the entire global area currently used for aquaculture production. However, if this is put in context of available space, this area accounts for only approximately 3% of the UK territorial waters (161200 km^2). By comparison with terrestrial biofuel production in the UK, to produce 1% of the UK's total energy demand using maize to methane would require a land area of 7700 km^2, equivalent to 18% of the UK's cropland (45000 km^2[52]). Although neither scenario seems attractive, such comparisons clearly illustrate the potential advantages of scale in moving UK biofuel production into the marine environment. At a regional level large-scale macroalgal culture for biofuels offers real potential for rural coastal communities. A good example would be for the Isle of Mull on the west coast of Scotland which has no domestic gas supply, the main fuels being heating oil and electricity. The Isle of Mull has 1278 households [53] and the average UK annual domestic gas consumption is 57.6 GJ [54], so to provide all the households on Mull with gas would require the methane from 430 ha of macroalgae cultivation, the extent of which can be seen on Figure 3. We envisage this would both increase local employment and improve rural fuel security. The cultivation of macroalgae for biofuels could be developed through a modular approach in a European/American context where the distances between culture, processing and AD facilities are minimised, similar to that proposed for Japan [55]. Seaweed farms would supply local, coastal processing facilities where high value products can be first extracted from the crop before transfer to an on-site or a shared AD facility. From here the biogas produced could be piped directly to augment the local natural gas supply.

Environmental Impact

Environmental impacts of large scale seaweed farms may arise from; changes to local hydrodynamics and resulting sedimentation patterns, benthic impacts from increased organic matter supply, changes to water column nutrient availability and from shading of the sea-floor (in shallow sites). Although we anticipate some types of interactions may well be positive [25] a measure of the extent and nature of interactions with fish, cetaceans and birds as well as other users of the marine environment for aquaculture, fisheries, energy generation and shipping is required.

During the growth cycle a portion of the macroalgae and the associated biota from the culture lines will be lost to the benthos either through erosion of the blade tips or shearing of cultured material creating an organically enriched zone [56]. In 'fed' aquaculture, that of fin-fish for example, where high energy feeds are supplied to the system, measurement of the extent of the zone of deposition is required [49] and has in turn led to the development of regulatory tools [57]. Although the macroalgal cultures are not 'fed' i.e. artificially supplied with additional nutrients or fertilised, the extent and effect of the zone of organic enrichment should be described. In enclosed water bodies, there may be competition for dissolved nutrients with phytoplankton but, in more open shelf systems, nutrient supply is likely to be sufficient provided that farms are spatially arranged for optimal nutrient exchange. In any event, nutrients taken up by macroalgal culture, on the scale required for biofuel production, would be far less than that produced by agricultural, urban sources and fin-fish aquaculture. If macroalgae is subjected to the AD process then a proportion of the nitrogen may be lost through denitrification depending on the conditions in the reactor. Digestates are typically higher in ammonia and lower in organic nitrogen than ingestates [58]. The digestate will most likely be used in fertilisers and so find its way back into the hydrological cycle.

There may also be a number of positive benefits; the macroalgal farms effectively acting as no-take zones for mobile gear fisheries and thus enhancing less destructive static gear fisheries within the cultivation zone and providing spill over benefits to adjacent waters [59]. In addition, providing the crop is not removed in its entirety at the end of the cycle it will provide a refuge and a substrate to enhance local biodiversity. The digestate after AD may be either a valuable by-product or an expensive waste. This will depend on a number of factors including its contaminant metal burden and whether the macroalgae has been mixed with other organic waste streams in the digestor. A study on the AD of lipid-extracted microalgal biomass [60] suggested that 80% of the nitrogen in the biomass was recoverable as ammonium/ammonia from the liquid supernatant fraction, and that the remaining nitrogen in the solid digestate fraction had a 40% bioavailability when applied to soil. A similarly detailed analysis of the fate of nitrogenous emissions following AD of macroalgal biomass is required. Overall the global effect of using macroalgal culture for biofuel is likely to be positive and an initial full life cycle analysis of biomethane production from offshore cultivation of macroalgae has shown a 69% reduction in fossil fuel utilisation when compared to natural gas, a 54% reduction in greenhouse gas emissions and an improvement in the marine eutrophication index [61].

Making it Pay

Costing the culture of large amounts of seaweed in a European context is currently highly uncertain as there are too many unconstrained parameters, such as scalability, location and the degree of mechanisation readily achievable. However, our analysis based on inshore production suggest that at 2011 wellhead value for natural gas (US $3.95 [62] per thousand cubic feet (equivalent to £0.09 m^3)) based on a production of 20 tonnes dry weight (dw) ha^{-1} the production costs for macroalgal biogas would have to be less than £400 ha^{-1} to be competitive with fossil fuels without additional subsidy. It is unlikely that in the short term such production costs could be achieved. However under the UK Renewable Heat Incentive 2011 [63] scheme injection of biomethane into the natural gas grid attracts a price of £0.068 kWh. This is equivalent to £3230 hectare which would make the cultivation of macroalgae for methane production highly competitive. In addition the identification and extraction of higher value products, prior to AD, is advisable, as is the quantification of how the prior extraction affects biogas yield. Added value could be achieved by processing part of the crop for human and animal foodstuffs, and food supplements, for its mineral content for animal feeds, as an organic slow release fertiliser, and potential bio-active compounds [64].

CONCLUSION

Our analysis of growth data from hatchery-raised macroalgal sporelings outplanted to conventional long-line systems in Scotland suggests there are no major biological obstacles to the culture process in a European context. A fuller understanding of the impacts and performance of native macroalgae grown in dense large-scale cultures can only be achieved through pilot scale trials. Technological advancement is required to mechanise the outplanting and harvest process. The biological gasification of macroalgae was well proven in the later decades of the 20th century and AD technology has sufficiently matured to offer a range of possibilities to further optimise methane yields. Compared to first generation biofuels, macroalgae have inherent advantages that make them environmentally sustainable. Given that fossil fuel prices are likely to increase and that macroalgal production costs will inevitably fall as production is expanded and intensified, it is prudent to develop the technology required to obtain significant quantities of biofuel from marine biomass in time to help meet Europe's energy needs and climate change targets.

ACKNOWLEDGEMENTS

MSK, KDB and MSS acknowledge funding for the BioMara project (http://www.biomara.ac.uk). The Biomara project is supported by the European Regional Development Fund through the INTERREG IVA Programme, Highlands and Islands Enterprise, Crown Estate, Northern Ireland Executive, Scottish Government and Irish Government. ADH received funding from the MASTS pooling initiative (The Marine Alliance for Science and Technology for Scotland), MASTS is funded by the Scottish Funding Council (grant reference HR09011) along with contributing institutions, and from the European Union's Seventh Framework Programme (FP7/2007-2013) project AT~SEA under grant agreement n° 280860. Their support is gratefully acknowledged.

Competing Interests

The authors declare that they have no competing interests.

REFERENCES

1. Scharlemann JPW, Laurance WF: Environmental science - How green are biofuels?Science 2008, 319: 43-44. 10.1126/science.1153103
2. Gibbs HK, Johnston M, Foley JA, Holloway T, Monfreda C, Ramankutty N, Zaks D:Carbon payback times for crop-based biofuel expansion in the tropics: the effects of changing yield and technology. Environ Res Lett 2008., 3: 10.1088/1748-9326/3/3/034001
3. Shilton A, Guieysse B: Sustainable sunlight to biogas is via marginal organics.Curr Opin Biotechnol 2010, 21: 287-291. 10.1016/j.copbio.2010.03.008
4. Singh J, Cu S: Commercialization potential of microalgae for biofuels production. Renew Sust Energ Rev 2010, 14: 2596-2610. 10.1016/j.rser.2010.06.014
5. Williams PJL, Laurens LML: Microalgae as biodiesel & biomass feedstocks: review & analysis of the biochemistry, energetics & economics. Energ Environ Sci 2010,3: 554-590. 10.1039/b924978h
6. Brennan L, Owende P: Biofuels from microalgae-a review of technologies for production, processing, and extractions of biofuels and co-products. Renew Sust Energ Rev 2010, 14: 557-577. 10.1016/j.rser.2009.10.009
7. Chisti Y: Biodiesel from microalgae. Biotechnol Adv 2007, 25: 294-306. 10.1016/j.biotechadv.2007.02.001
8. Mata TM, Martins AA, Caetano NS: Microalgae for biodiesel production and other applications: a review. Renew Sust Energ Rev 2010, 14: 217-232. 10.1016/j.rser.2009.07.020
9. Wilcox HA: The U.S. Navy's Ocean food and energy farm project. In Book the U.S. Navy's Ocean food and energy farm project Edited by: Monney NT. 83-104.

City; 1977:83–104

10. North WJ: Oceanic farming of Macrocystis, the problems and non-problems. Seaweed cultivation for renewable resources. In Seaweed cultivation for renewable resources. Volume 16. Edited by: Bird KT, Benson PH. Amsterdam: Elsevier; 1987:39-68.
11. Bird KT, Benson PH: Seaweed cultivation for renewable resources. Amsterdam; New York: Elsevier; 1987.
12. Leese TM: The conversion of ocean farm kelp to methane and other products. InClean fuels from biomass, sewage, urban refuse, agricultural wastes; Proceedings of the Symposium, Orlando, Fla., January 27-30, 1976. (A77-37652 17-44). Chicago: Institute of Gas Technology; 1976:253-266. Research sponsored by the American Gas Association and ERDA
13. Kelly M, Dworjanyn S: The potential of marine biomass for anaerobic biogas production. The potential of marine biomass for anaerobic biogas production: sThe Crown Estate; 2008:103.
14. Wise DL, Augenstein DC, Ryther JH: Methane fermentation of aquatic biomass.Resour Recover Conserv 1979, 4: 217-237. 1979, 4:217–237 10.1016/0304-3967(79)90002-7
15. Wargacki AJ, Leonard E, Win MN, Regitsky DD, Santos CNS, Kim PB, Cooper SR, Raisner RM, Herman A, Sivitz AB, et al.: An engineered microbial platform for direct biofuel production from brown macroalgae. Science 2012, 335: 308-313. 10.1126/science.1214547
16. Koh LP, Ghazoul J: Biofuels, biodiversity, and people: understanding the conflicts and finding opportunities. Biol Conserv 2008, 141: 2450-2460. 10.1016/j.biocon.2008.08.005
17. Brennan L, Owende P: Biofuels from microalgae—a review of technologies for production, processing, and extractions of biofuels and co-products. Renew Sustain Energy Rev 2010, 14: 557-577. 10.1016/j.rser.2009.10.009
18. Bastianoni S, Coppola F, Lezzi E, Colacevich A, Borghini F, Focardi S: Biofuel potential production from the Orbetello lagoon macroalgae: a comparison with sunflower feedstock. Biomass Bioenergy 2008, 32: 619-628. 10.1016/j.biombioe.2007.12.010
19. Goh CS, Lee KT: A visionary and conceptual macroalgae-based third-generation bioethanol (TGB) biorefinery in Sabah, Malaysia as an underlay for renewable and sustainable development. Renew Sust Energ Rev 2010, 14: 842-848. 10.1016/j.rser.2009.10.001
20. John RP, Anisha GS, Nampoothiri KM, Pandey A: Micro and macroalgal biomass: a renewable source for bioethanol. Bioresour Technol 2011, 102: 186-193. 10.1016/j.biortech.2010.06.139
21. Ortiz M: Dynamic and spatial models of kelp forest of Macrocystis integrifolia and lessonia trabeculata (SE pacific) for assessment harvest scenarios: short-term responses. Aquat Conserv Mar Freshwat Ecosyst 2010, 20: 494-506. 10.1002/aqc.1126
22. Springer YP, Hays CG, Carr MH, Mackey MR: Toward ecosystem-based management of marine macroalgae-the bull kelp, Nereocystis luetkeana. Oceanogr Mar Biol Annu Rev 2010, 48: 1-41.
23. Thompson SA, Knoll H, Blanchette CA, Nielsen KJ: Population consequences of biomass loss due to commercial collection of the wild seaweed Postelsia palmaeformis. Mar Ecol Prog Ser 2010, 413: 17-31.
24. Vasquez J: Production, use and fate of chilean brown seaweeds: resources for a sustainable fishery. J Appl Phycol 2008, 20: 457-467. 10.1007/s10811-007-9308-y
25. Hughes AD, Black KD, Campbell I, Heymans JJ, Orr KK, Stanley MS, Kelly MS:Comments on 'prospects for the use of macroalgae for fuel in Ireland and UK: an overview of marine management issues'. Marine Policy
26. Kraan S: Mass-cultivation of carbohydrate rich macroalgae, a possible solution for sustainable biofuel production. Mitig Adapt Strat Glob Chang 2010, 1-20.
27. MacKay D: Sustainable energy without the hot air. UIT Cambridge Ltd., Cambridge; 2009.
28. Lorentsen SH, Sjotun K, Gremillet D: Multi-trophic consequences of kelp harvest.Biol Conserv 2010, 143: 2054-2062. 10.1016/j.biocon.2010.05.013
29. Christie H, Fredriksen S, Rinde E: Regrowth of kelp and colonization of epiphyte and fauna community after kelp trawling at the coast of norway. Hydrobiologia1998, 375–376: 49-58.
30. Tseng CK: Some remarks on the kelp cultivation industry of china. Dev Aquac Fish Sci 1987, 16: 147-155.
31. Haug A, Jensen A: Seasonal variations in the chemical composition of Alaria esculenta. Laminaria saccharina:Laminaria hyperborea and Laminaria digitata from Northern Norway. Reports of the Norwegian Institute of Seaweed Research No. 4 1954.
32. Santelices B: The discovery of kelp forests in deep-water habitats of tropical regions. Proc Natl Acad Sci 2007, 104: 19163-19164. 10.1073/pnas.0708963104
33. Ortiz J, Romero N, Robert P, Araya J, Lopez-Hernandez J, Bozzo C, Navarrete E, Osorio A, Rios A: Dietary fiber, amino acid, fatty acid and tocopherol contents of the edible seaweeds Ulva lactuca and Durvillaea antarctica. Food Chem 2006, 99: 98-104. 10.1016/j.foodchem.2005.07.027
34. FAO: The state of world fisheries and aquaculture - 2010 (SOFIA). 2010.
35. Duarte CM, Middelburg JJ, Caraco N: Major role of marine vegetation on the oceanic carbon cycle. Biogeosciences 2005, 2: 1-8.
36. Field CB, Campbell JE, Lobell DB: Biomass energy: the scale of the potential resource. Trends Ecol Evol 2008, 23: 65-72. 10.1016/j.tree.2007.12.001
37. Wu CY, Pang SJ: World seaweed resources - an authoritative reference system.Book world seaweed resources - An authoritative reference system 2006. City
38. Sanderson CJ: Reducing the environmental impact of fish cage farming through the cultivation of seaweeds. SAMS PhD Thesis: University of the Highlands and Islands; 2006.
39. Rodger ANS: Sea-based integrated multi-trophic

aquaculture: investigation of a fish, bivalve and macroalgal co-culture system. SAMS PhD Thesis: University of the Highlands and Islands; 2010.

40. Duarte CM, Holmer M, Olsen Y, Soto D, Marba N, Guiu J, Black K, Karakassis I: Will the oceans help feed humanity? BioScience 2009, 59: 967-976. 10.1525/bio.2009.59.11.8

41. Zhou D, Zhang LA, Zhang SC, Fu HB, Chen JM: Hydrothermal liquefaction of macroalgae Enteromorpha prolifera to Bio-oil. Energy Fuel 2010, 24: 4054-4061. 10.1021/ef100151h

42. Adams JM, Gallagher JA, Donnison IS: Fermentation study on Saccharina latissima for bioethanol production considering variable pre-treatments. J Appl Phycol2009, 21: 569-574. 10.1007/s10811-008-9384-7

43. Chynoweth DP, Klass DL, Ghosh S: Anaerobic digestion of kelp. In Biomass conversion processes for energy and fuels. Edited by: Sofer SS, Zaborsky OR. New York: Plenum Press; 1981:315-318.

44. Chynoweth DP, Fannin KF, Srivastava VJ: Biological gasification of marine algae. InSeaweed cultivation for renewable resources. Developments in aquaculture and fisheries science. Edited by: Bird KT, Benson PH. Elsevier, Amsterdam; 1987:287-303.

45. Adams JMM, Toop TA, Donnison IS, Gallagher JA: Seasonal variation in Laminaria digitata and its impact on biochemical conversion routes to biofuels. Bioresour Technol 2011, 102: 9976-9984. 10.1016/j.biortech.2011.08.032

46. Matsui JT, Amano T, Koike Y, Saiganji A, Saito H: Methane fermentation of seaweed biomass. In American institute of chemical engineers. San Francisco; 2006.

47. Cecchi F, Pavan P, MataAlvarez J: Anaerobic co-digestion of sewage sludge: application to the macroalgae from the Venice lagoon. Resour Conserv Recycl1996, 17: 57-66. 10.1016/0921-3449(96)88182-1

48. Nkemka VN, Murto M: Evaluation of biogas production from seaweed in batch tests and in UASB reactors combined with the removal of heavy metals. J Environ Manage 2010, 91: 1573-1579. 10.1016/j.jenvman.2010.03.004

49. Lee S-M, Kim GH, Lee J-H: Bio-gas production by co-fermentation from the brown algae, Laminaria japonica. J Ind Eng Chem 2012,18(4):1512-1514. 10.1016/j.jiec.2012.02.014

50. MacLeay I, Harris K, Annut A: Digest of united kingdom energy statistics 2011. InBook digest of united kingdom energy statistics 2011. National Statistics publication; 2011.

51. DEFRA: Crop areas, yeilds and production,livestock populations and the size of the agricultural workforce:2011 UK final results. In Book crop areas, yeilds and production,livestock populations and the size of the agricultural workforce:2011 UK final results. Department for Environmental Food and rural Affairs, UK; 2012.

52. ABC: Argyll and bute housing strategy 2011–2016. In Book Argyll and Bute housing strategy 2011–2016. Argyll and Bute Council; 2011.

53. Typical domestic energy consumption figures.

54. Yokoyama S, Jonouchi K, Imou K: Energy production from marine biomass: fuel cell power generation driven by methane produced from seaweed. Proc World Acad Sci Eng Tech 2007, 22: 320-323.

55. Zhang JH, Hansen PK, Fang JG, Wang W, Jiang ZJ: Assessment of the local environmental impact of intensive marine shellfish and seaweed farming-application of the MOM system in the Sungo Bay. China. Aquaculture 2009,287(3-4):304-310.

56. Cromey CJ, Nickell TD, Treasurer J, Black KD, Inall M: Modelling the impact of cod (Gadus morhua L) farming in the marine environment-CODMOD. Aquaculture2009, 289: 42-53. 10.1016/j.aquaculture.2008.12.020

57. Adani F, Scievano A, Boccasile G: In Anaerobic digestion: opportunity for agriculture and environment. In Anaerobic digestion: opportunity for agriculture and environment. Milan; 2008:27-37.

58. Gell FR, Roberts CM: Benefits beyond boundaries: the fishery effects of marine reserves. Trends Ecol Evol 2003,18(9):448-455. 10.1016/S0169-5347(03)00189-7

59. Frank ED, Han J, Palou-Rivera I, Elgowainy A, Wang MQ: Methane and nitrous oxide emissions affect the life-cycle analysis of algal biofuels. Environ Res Lett 2012., 7:

60. Langlois J, Sassi J-F, Jard G, Steyer J-P, Delgenes J-P, Hélias A: Life cycle assessment of biomethane from offshore-cultivated seaweed. Biofuels, Bioprod Biorefin 2012. n/a-n/a

61. Natural Gas. http://205.254.135.7/dnav/ng/ng_pri_sum_dcu_nus_a.htm

62. Anon: Departmental note: support for renewable heat technologies in the domestic and non domestic sectors. In Book departmental note: support for renewable heat technologies in the domestic and non domestic sectors. UK Gov; 2012.

63. Cumashi A, Ushakova NA, Preobrazhenskaya ME, D'Incecco A, Piccoli A, Totani L, Tinari N, Morozevich GE, Berman AE, Bilan MI, et al.: A comparative study of the anti-inflammatory, anticoagulant, antiangiogenic, and antiadhesive activities of nine different fucoidans from brown seaweeds. Glycobiology 2007, 17: 541-552.

64. Imhof M, Bounoua L: Human consumption of primary productivity. In Book human consumption of primary productivity 2011. Earth Observatory NASA; 2004.

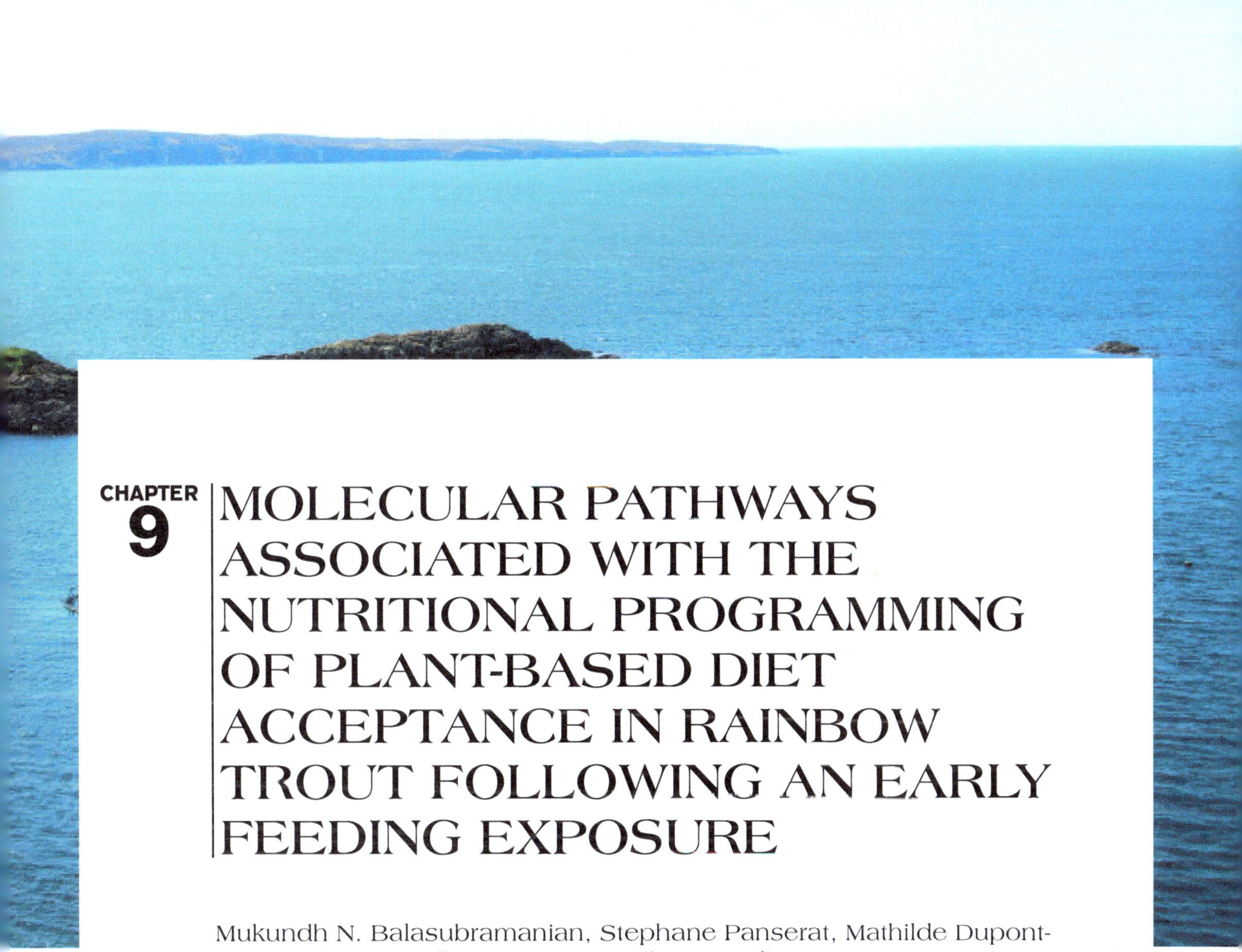

CHAPTER 9

MOLECULAR PATHWAYS ASSOCIATED WITH THE NUTRITIONAL PROGRAMMING OF PLANT-BASED DIET ACCEPTANCE IN RAINBOW TROUT FOLLOWING AN EARLY FEEDING EXPOSURE

Mukundh N. Balasubramanian, Stephane Panserat, Mathilde Dupont-Nivet, Edwige Quillet, Jerome Montfort, Aurelie Le Cam, Francoise Medale, Sadasivam J. Kaushik and Inge Geurden

BACKGROUND

Achieving independence on wild fish-derived feed inputs is a priority for the aquaculture industry to ensure its further sustainable growth [1]. In this respect, the reduced growth performance seen in farmed fish, including rainbow trout, when fed high levels of plant-based ingredients in lieu of fishery derived fish meal and fish oil is a problem [2, 3]. Despite meeting all known nutrient requirements [4], the poor feed intake and utilisation efficiency of terrestrial plant-based ingredients in fish, including salmon and trout, is not fully understood. In order to analyse the metabolic consequences of partial or complete substitution of fishmeal and fish oil with plant-based sources, studies based on nutrigenomic approaches have been undertaken, comparing hepatic transcriptome changes in trout fed fishmeal or fish oil-free diets [5], total replacement of fishmeal or fish oil with plant sources in trout [6] and European sea bass [7], fishmeal substitution with plant proteins in Atlantic salmon [8], and a proteomic study on the effect of plant-protein substitution on hepatic metabolism in trout [9]. Taken together, these studies showed that plant-feeding altered the expression of actors involved in a large variety of biological functions. Low acceptance, in terms of food intake and also nutrient utilisation, is often attributed to the presence of various plant secondary metabolites that can act as anti-nutritional factors (ANFs) [10, 11]. Examples of the lack of adaptation of rainbow trout to purified ANFs concern alkaloids found in legumes like pea and lupins [12, 13] and saponins found in soybean [14]. Strategies to overcome the poor acceptance of plant-based ingredients in trout have included the use of selective breeding techniques to select for fish with a greater ability to grow on plant-based feeds [15–17].

In mammals, it is widely accepted that nutritional intervention at critical periods of early development, such as during gestation of the foetus or lactation of neonates, can have life-long impact on the organism's physiology and metabolism, long after the original nutritional intervention, a phenomenon known as nutritional programming [18–21]. Nutritional programming is believed to confer adaptive advantages to the organism to better sustain itself later in life when the adult

Citation: Mukundh N. Balasubramanian, Stephane Panserat, Mathilde Dupont-Nivet, Edwige Quillet, Jerome Montfort, Aurelie Le Cam, Francoise Medale, Sadasivam J. Kaushik, Inge Geurden, Molecular pathways associated with the nutritional programming of plant-based diet acceptance in rainbow trout following an early feeding exposure, DOI: 10.1186/s12864-016-2804-1

nutritional environment corresponds with that encountered during early development. The metabolic and physiological changes induced by nutritional programming in mammals are often accompanied by changes in protein or mRNA levels or, in some cases, persistent epigenetic changes in DNA methylation or histone modifications at regulatory enhancer and promoter regions of candidate genes [22, 23]. Such persistent modifications can even lead to trans-generational metabolic changes as shown from the Dutch famine study on starvation-induced nutritional programming during World War II [24]. Some recent studies in fish also report long-term metabolic modifications, albeit at the molecular level, due to early nutritional intervention at the first-feeding stage, such as high carbohydrate intake [25, 26] or changed dietary fatty acid profiles [27, 28]. The results of the above studies validate the first-feeding stage as an effective nutritional programming window. Instead, only few long-term effects were noted in zebrafish following injection of glucose at the embryo stage, directly into the egg [29]. In addition to the nutritional programming of metabolic pathways, studies in mammalian models suggest the ability of early flavour experiences to drive adult flavour and feeding preferences in humans [30]. Similarly, studies on conditioned flavour aversion in rats indicate that the early exposure to either pleasant or noxious flavours may affect adult behaviour through the programming of sensory and cognitive systems [31–33].

We recently attempted to exploit the phenomenon of nutritional programming to improve the acceptance of plant-based feed in rainbow trout [34]. Two groups of trout swim-up fry transitioning to exogenous feeding were exposed to either a control fish meal and oil diet (M-diet) or a plant-based diet (V-diet) for three weeks. Both groups were maintained on the control diet M for 7 months. When the two groups of juvenile trout were then challenged for 25 days with the plant-based V diet, we observed a significantly higher feed intake, growth rate and feed utilization in the V- compared to M-fish (see Fig. 1 and [34]). These long-term positive effects due to the early plant-diet exposure on later plant-diet acceptance suggest that the fish have been nutritionally programmed [34].

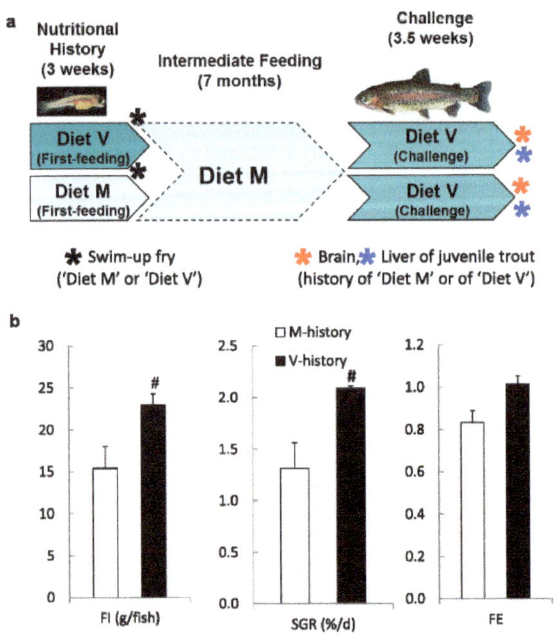

Figure1: Experimental design evaluating impact of early-feeding exposure of plant-based diet on its future acceptance. **a** Feeding protocol. Rainbow trout swim-up fry (50–150 mg) received for 3 weeks either the plant-based diet V or diet M (first-feeding period, details on diet are given in text) after which both groups received diet M (7-month intermediate growth phase). Both groups were then challenged for 25 days by feeding diet V during which voluntary feed intake, growth and feed utilisation efficiency were monitored (V-challenge. No difference in initial body weight was noted at the start of the V-challenge. The asterix indicate the sampling times for the present microarray study. **b** During the final V-challenge, fish of nutritional history V vs M displayed a higher ($P < 0.05$) feed intake (FI) and specific growth rate (SGR) and a tendency for improved efficiency of feed utilization (FE, $P = 0.06$), underlining the positive impact of the early diet V feeding (see [34] for extra details on the rearing conditions and V-challenge results). # denotes a significant difference due to nutritional history ($P < 0.05$)

To identify the molecular pathways mediating the nutritional programming phenotype observed in our study in trout [34], we utilised microarray expression analysis in order to monitor molecular changes related with the early plant-feeding in 3-week fry fed diet M or diet V and in the liver and brain of juvenile M-fish and V-fish challenged with the same plant-based V diet (Fig. 1a).

RESULTS AND DISCUSSION

Expression Profiling

The present study aims to identify the molecular mechanisms that govern the positive effect of prior plant-diet exposure on the acceptance (feed intake and utilisation efficiency) of the same plant-diet when given 7 months later (see [34]). We therefore performed transcriptomic analysis on swim-up fry collected at the end of early-feeding exposure fed either diet M (fish-based) or diet V (plant-based) as well as brain and liver of juveniles sampled at the end of the V-challenge when both groups received diet V (M-fish and V-fish). The two-way ANOVA performed on differentially expressed probes in whole fry and juvenile brain indicates that 1787 were persistently altered (3-week + Brain; see Additional file 1 and Fig. 2), whereas the 2-way ANOVA of differentially expressed probes in whole fry and juvenile liver showed 924 mRNA probes to be persistently altered by early-feeding exposure (3-week + Liver; see Additional file 1and Fig. 2). Supervised clustering (as shown in Additional file 2) gives a visual overview of the 1112 up-regulated mRNA probes and the 675 down-regulated mRNA probes in the brain, as well as the 573 up-regulated mRNA probes and the 351 down-regulated mRNA probes in the liver based on nutritional V- vs M-history.

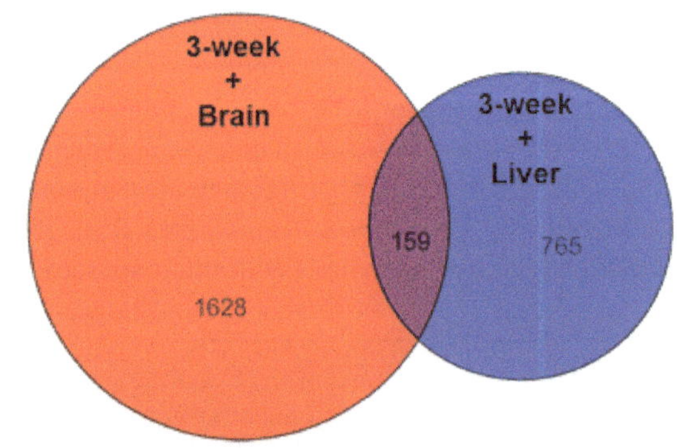

Figure 2: Venn diagram showing differentially expressed mRNA probes by microarray analysis. Summary of number

of mRNA probes that are differentially expressed following early nutritional exposure (3-week) in trout swim-up fry, and after V-challenge in the juvenile trout (Brain or Liver). The mRNA probes are all significantly different in expression (p [nutritional history] ≤ 0.05, and fold change ≥ 1.5) in both swim-up fry and juvenile trout (3-week + Brain; 3 week + Liver). The full lists of mRNA probes are outlined in Additional file 1

The gene ontology analysis, performed to ascertain the biological significance of the differentially expressed mRNA probes, revealed that the genes that are responsive to prior plant-feeding in the brain and liver belong to distinct and interconnected functional categories (Figs. 3 and 4). We acknowledge that the correlation between mRNA expression in whole trout fry (3-week) and that in the liver or brain in juvenile trout cannot be definite for all genes identified by the microarray analysis. However, by using DAVID to identify networks in which specific subset of the differentially expressed genes contribute to a biological process, followed by application of GeneMania for pathway analysis [35–37], we observed congruity in the relationship of co-expression and co-localization that exists between the genes (see Fig. 5 and Additional files 3 and 4). This strengthens our discussion on the implication of the differential mRNA expression at first-feeding and later in brain or liver of juvenile trout in relation with the observed positively programmed V-fish phenotype.

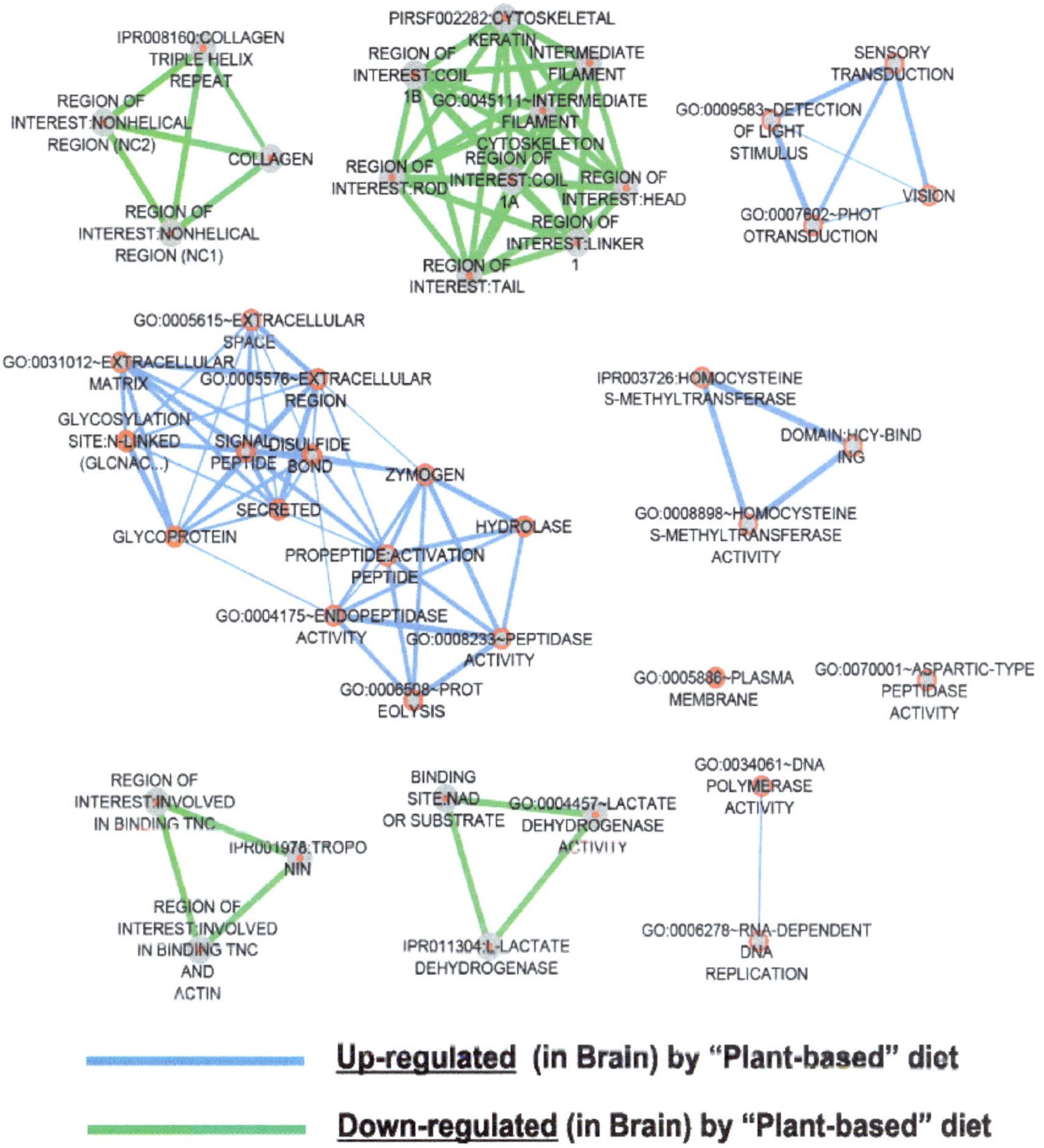

Figure 3: Summary of functional gene ontological analyses significantly enriched based on early nutritional history in the brain. The biological significance of the 1112 up-regulated mRNA probes and 675 down-regulated mRNA probes in the brain of trout based on nutritional history was ascertained by performing gene ontology (GO) analysis using the functional annotation cluster and chart tools from the DAVID (Database for Annotation, Visualization and Integrated Discovery) bioinformatics resource [137, 138]. The resulting data was used as input in the DAVID Enrichment Map plugin from the Bader Lab [139,140] in the Cytoscape network visualization tool (version 2.8.3) [141–143]. The nodes (circles) represent individual GO terms and the lines represent the relationship between the genes assigned to respective GO terms. Blue and green lines represent up-regulated and down-regulated mRNA probes in the brain respectively

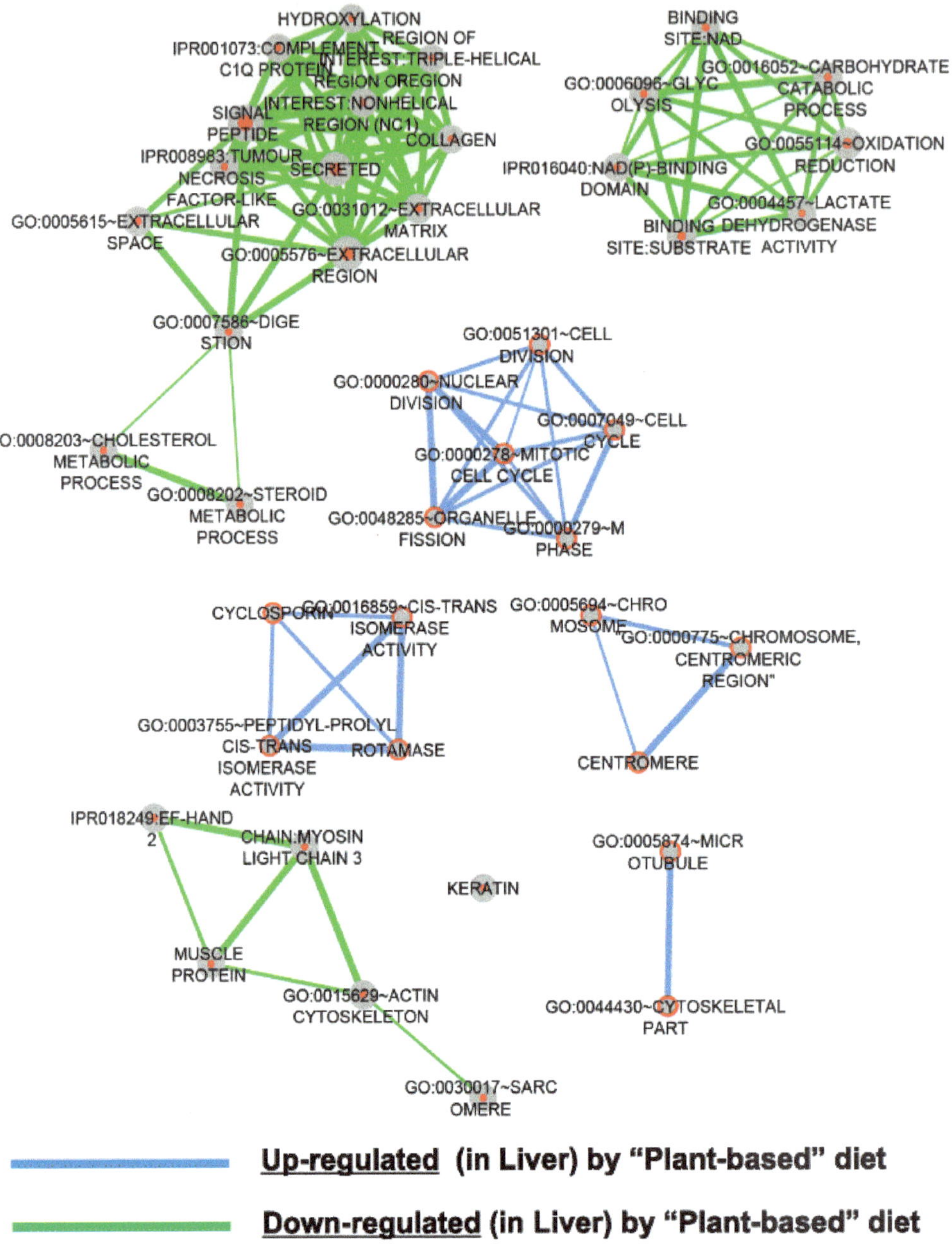

Figure 4: Summary of functional gene ontological analyses significantly enriched based on early nutritional history in the liver. The biological significance of the 573 up-regulated mRNA probes and 351 down-regulated mRNA probes in the liver of trout based on nutritional history was ascertained by performing gene ontology (GO) analysis using the functional annotation cluster and chart tools from the DAVID (Database for Annotation, Visualization and Integrated Discovery) bioinformatics resource [137, 138]. The resulting data was used as input in the DAVID Enrichment Map plugin from the Bader Lab [139,140] in the Cytoscape network visualization tool (version 2.8.3) [141–143]. The nodes (circles) represent individual GO terms and the lines represent the relationship between the genes assigned to respective GO terms. Blue and green lines represent up-regulated and down-regulated mRNA probes in the liver respectively

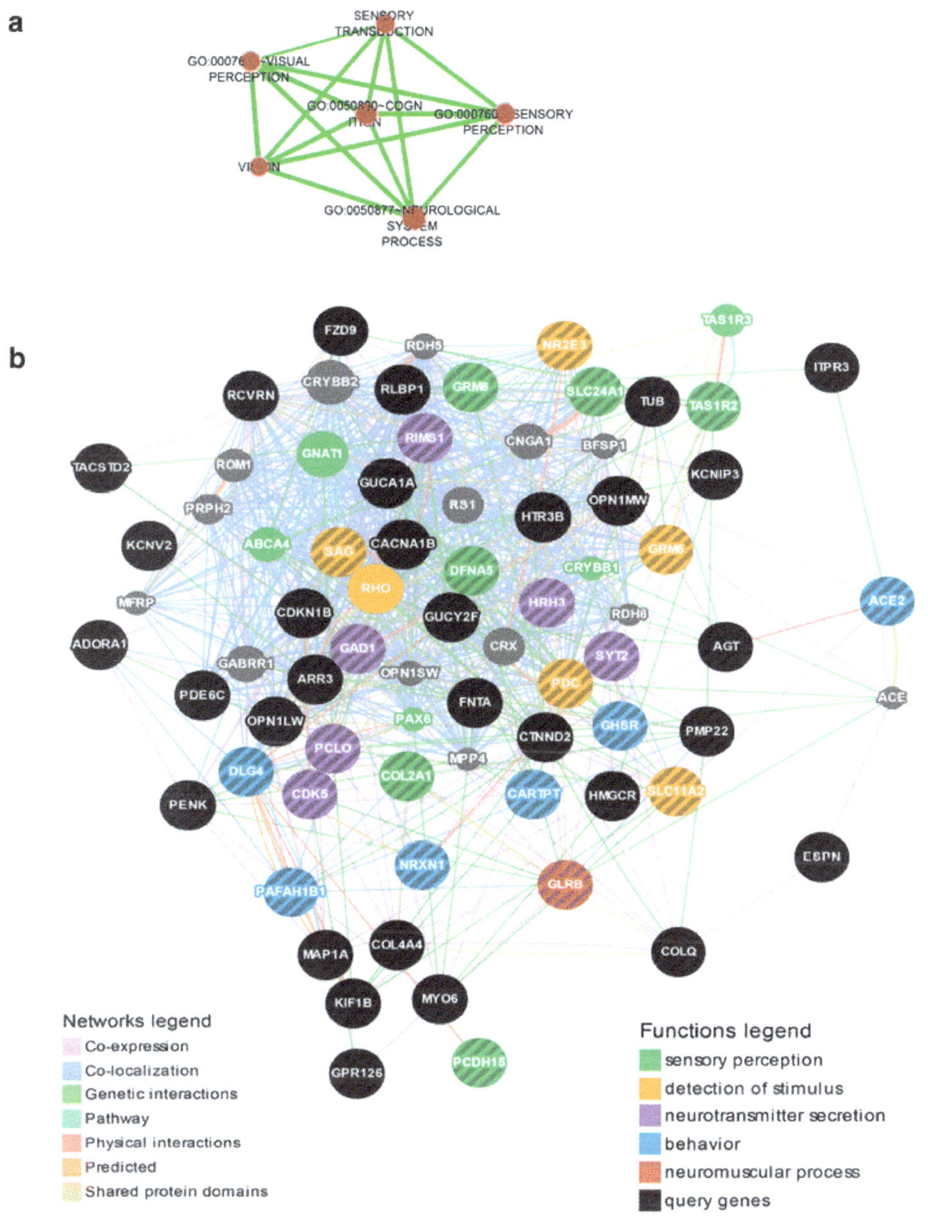

Figure 5: Pathways significantly enriched based on early nutritional history in the brain: Neurological system process. **a** An interconnected cluster of biological processes assigned through GO analysis using DAVID that involve cognition and sensory pathways contributing to neurological system process was identified and visualized as described in Fig. 3. **b** The mRNA probes (see list in Additional file 5) that were assigned to be part of the GO process-neurological system process (represented by nodes with black stripes) were used as input in the GeneMania pathway analysis tool [35–37] to generate networks. The functions legend (nodes) represents the sub-network of the mRNA probes and the network legend (lines) represents the relationship between the genes (seeMethods)

Genes responsive to Nutritional History in the Brain

The brain being the central organ controlling sensory and homeostatic processes is perhaps one of the more sensitive organs in reacting to early changes in quality of available feed. To our knowledge, this study is the first to probe transcriptomic changes in the brain of trout or other carnivorous-like fish in response to feeding plant-based ingredients. Functional annotation analysis of the differentially expressed mRNAs (Fig. 3) revealed persistent effects of the early plant-diet exposure on processes such as cognition (Fig. 5), sensory transduction (Fig. 5), methionine metabolism (Additional file 3) and on several genes encoding neuropeptides and their receptors that regulate central and peripheral feeding responses (Additional file 3).

Sensory Perception and Transduction

Early flavour experiences are important drivers of life-long flavour acceptance in mammals. Transmittance of flavours connected to the diet of the mother either through amniotic fluid to the foetus or via lactation to the neonatal offspring may help the offspring in future dietary choices [30]. In our study, one prominent cluster of genes identified in the juvenile trout brain to be affected by early diet exposure was assigned to the interconnected biological pathways affecting sensory perception, cognition and neurological system processes (Fig. 5 and Additional file 5).

Several genes in this cluster were associated with visual transduction processes (Fig. 5 and Additional file 5) such as photoreceptor components (Nr2e3, Pdc, Sag, arr3, Pcdh15), genes involved in phototransduction (Gucy2F, Pde6c, Kcnv2,

Slc24a1) and photosensitive proteins (Opsin, Pinopsin, Recoverin, Retinaldehyde-binding protein, Arrestin). These generally had a higher expression level in the V-fry and later in the brain of the V-fish when challenged again with the plant-based diet relative to M-fish. A functional sense of visual perception contributes to the ability of salmonids to consume various types of pelleted feed [38]. More recently, transcriptomic analysis of brain of hybrid mandarin fish showed differential expression of genes encoding for retinal photosensitivity and cognitive function depending on their feeding habits [39]. Possibly, the early feeding experience modified functional visual acuity affecting the later feeding behaviour in rainbow trout. Other interesting genes involved in sensory perception whose expression showed both short and long-term modification in the trout brain are olfactomedin (OLFM), espin (ESPN) and a member of the taste receptor family (Tas1r2) (Fig. 5and Additional file 5). Tas1r2 (also known as T1R2) constitutes one-half of the heterodimeric taste receptor involved in recognition of sweet taste [40]. In our study, brain Tas1r2 was upregulated by early exposure to plant-ingredients. Though there appears to be a loss of Tas1r2 in obligate carnivores such as the domestic cat unable to taste sweet substances [41], recent studies in rainbow trout detected extraoral Tas1r2 expression as documented in intestine [42] and in hypothalamus or hindbrain [43]. Tas1r2 expression in the latter study was found to be responsive to changes in glucose levels. Also in mammalian brain, Tas1r2 has been suggested to act as a glucosensor in energy-sensing hypothalamic neurons [44]. In the case of salmonids, gustation is believed to play a minor role in driving feeding preference when compared to olfaction [45]. Additionally, olfactory imprinting is suggested to be the primary mechanism used by adult salmon to return to their natal streams (homing) with free (L-) amino acids assigned as the olfactory cues directing this phenomenon [46–48]. Olfactomedin (Fig. 5 and Additional file 5) is a secreted glycoprotein involved in regulating maintenance or differentiation of chemosensory cilia in olfactory neurons, as well as axonal growth, branching and optic nerve extension [49]. ESPN, also responsive to nutritional programming in our study, is an intriguing candidate in terms of genes that encode structural proteins that mediate sensory perception. ESPN (Fig. 5 and Additional file 5) is a multifunctional actin-bundling protein expressed in brain tissues, retina and the inner ear, mediating sensory transduction in various mechano- and chemosensory cells, important for vertebrates to sense their environment [50]. The observations from studies elaborating on the perception of the umami flavour [51] together with evidences of visual stimuli activating gustatory responses [52] indicate the synergism of primary senses in the formation of flavour preferences. Thus, it appears that early plant-diet exposure in trout may mediate feed acceptance of the same diet at a later life stage by affecting pathways regulating the sensory perception of taste, odour and possibly vision.

Cognition and Synaptic Plasticity

In mammalian models, the exploration of flavour perception for evaluating the reward value of food and its behavioural outcomes, has revealed a remarkable integration of the primary cortex centres of taste, olfaction and vision to the higher learning, reward and decision centres in the brain (reviewed in [53]. Whenever an animal interacts with its surrounding environment, its experience drives learning and the formation of memory to modify its behaviour and better adapt its responses to ever-changing conditions. Synaptic plasticity, the ability to change the strength and stability of synapses dynamically in response to various stimuli, forms the fundamental basis of the complex neuronal functions of learning and memory formation based on prior experience [54]. A multitude of proteins mediating processes of cytoskeletal remodelling, cell adhesion, neurotransmitter production and secretion, receptor function and signal transduction, either pre- or post-synaptically, play an essential role in synaptic plasticity. As part of a larger and connected network of neurological system process, the early plant feeding induced short (following first-feeding) and long-term (during V-challenge) changes on the transcriptomic profile of genes involved in cognitive process (Fig. 5 and Additional file 5) like receptors involved in stimulus detection (SLC11A2, Grm6, Grm8), neurotransmitter secretion (RIMS1, GAD1, PCLO, CDK5, HRH3, Syt2), response to external stimuli (Ghsr, CARTPT, NRXN1, DFNA5, PAFAH1B1, Pafah1b1b, DLG4, Kcnip3) and synaptic transmission and memory formation (COL2A1, Glrb, Dlg4, CTNND2, ADORA1, Gpr126, Itpr3, FZD9).

The maintenance of sustained synaptic plasticity to encode the information and memory of any particular stimulus, including that related to given taste, known as long term potentiation (LTP), often relies on tight regulation of neurotransmitter levels [55]. Neurotransmitter release occurs at the presynaptic active zone (AZ), where pools of synaptic vesicles containing neurotransmitters are maintained. Piccolo (Pclo), a cytoskeletal matrix protein expressed in the brain of mammals [56] and of teleosts [57], is associated with the AZ and is involved in the functioning of glutamine- and γ-amino butyric acid (GABA)-responsive synapses [56]. Pclo knockdown in mice leads to impairment of spatial learning and a decrease in LTP [58], highlighting its role in establishing functional synapses for stimulus-driven memory formation. Like Pclo, regulating synaptic membrane exocytosis protein 1 (RIMS1) is another evolutionary conserved scaffolding protein of AZ that regulates neurotransmitter release [59]. RIMS1 controls the pool of synaptic vesicles and functions to maintain homeostatic synaptic plasticity that is essential for the efficiency of synaptic network activity [54]. A key role in the regulation of the pool of synaptic vesicles during homeostatic plasticity was also suggested recently for cyclin-dependent protein kinase 5 (CDK5) [60] as perturbing the levels of CDK5 by pharmacological inhibition of CDK5 or gene deletion in mice impairs the regulation of synaptic vesicle pools and disrupts homeostatic synaptic plasticity [54]. The persistent changes in the expression of genes that control neurotransmitter release through several distinct mechanisms (Pclo, RIMS1, CDK5) following the early plant-diet exposure suggest that neurotransmitter release may be one process involved in programming the enhanced feed acceptance of plant-based diet in juvenile trout confronted with this diet later in life. Besides neurotransmitter release, enzymes involved in neurotransmitter synthesis might contribute to synaptic adaptation. Dynamic changes in transcript levels of glutamate decarboxylase (GAD1), as observed in this study, control GABA synthesis and regulate the inhibitory GABAergic potentiation at interneurons, important for the functioning of activity-driven synaptic plasticity [61, 62]. Of interest, GAD1 expression has been found to be tightly controlled during gustatory habituation [63]. Furthermore, the activity-dependent epigenetic changes at the GAD1 proximal promoter in olfactory bulb interneurons [64] and the conserved function of GAD1 in teleost forebrain development [65] hint at GAD1 as a target associated with the programming of plant-based diet acceptance in the trout brain. This warrants further detailed exploration.

Early plant-diet exposure was found to exert long-term effects on several other classes of genes mediating synaptic plasticity, in addition to those regulating neurotransmitter levels. Platelet-activating factor acetyl hydrolase IB subunit alpha b (Pafah1b1b), also called Lis1, plays an important role in neuronal migration during development. Its mutation in mammals [66] and in zebrafish [67] causes lissencephaly, a severe brain malformation in which cognitive defects are associated with loss of complexity of the cerebral cortex. Given the role of Pafah1b1b in synaptic transmission in adult neurons [68] and its expression in the neurons of the early sensory system in the teleost embryo [69], the long-term changes in Pafah1b1b expression due to the early plant feeding in brain of trout displaying the nutritional programming phenotype is intriguing. Moreover, in rodents, changes in the lissencephaly-1 gene mRNA levels due to environmental stimulation have been suggested to modify synaptic plasticity and animal learning and memory performance [70]. Catenin delta 2 (CTNND2), a member of conserved p-120 cadherin family [71] is a constituent protein at adherens junctions between neurons where it plays a crucial role in maintenance of synapses and in ensuring stable memory storage [72]. There is also evidence of delta catenins in neuroplastic processes in the peripheral retinal [73] and olfactory pathways [74], which make it an interesting candidate gene for our study. The expression of genes encoding neurotransmitter receptors (Fig. 5 and Additional file 5) was also persistently modified by the early diet exposure including the adenosine receptor A1 (ADORA1). As adenosine, formed from the catabolism of ATP, can be directly released by neurons and glia, activation of ADORA1 modulates synaptic transmission. Adenosine and ADORA1 activation have been reported to be involved in LTP impairments caused by distinct types of prior experience (e.g. stress) susceptible to affect memory formation [75]. In the teleost nervous system, ADORA1 expression is highly affected by stimulation during zebrafish development [76] and has been implicated in processes of anxiety and arousal [77] and of photoreceptor function [78], rendering its association with the present nutritional programming phenotype and cognition noteworthy.

The interpretation of the functional implications regarding 'how does early diet experience act on the brain to store information and change behaviour' based on the current microarray analysis in the whole trout brain is challenging, and we are aware that we have possibly overlooked relevant region-specific neuronal processes. However, there is precedence for obtaining useful insights into the molecular mechanisms underlying behavioural responses from whole-brain transcriptomic studies in teleosts [79–83]. Moreover, as discussed earlier, the use of pathway analysis to ascertain co-expression and co-localization patterns for the genes within a network aids in proving greater biological robustness regarding the relevance of the data in relation with the observed phenotype. Certainly, histochemical analysis of selected genes in the fry and juvenile brain [84, 85] would be useful to ascertain the site of action influencing cognition, learning and memory formation regarding plant-diet acceptance in rainbow trout. Despite these concerns, the finding that the early diet exposure induced long-term changes in expression of genes involved in sensory perception, sensory transmission, cognition, memory formation and neuronal development is an exciting observation for further studies on the potential of nutritional programming of plant-based diet acceptance in trout or other teleosts.

Homocysteine and Methionine Metabolism

Several mRNA probes encoding genes involved in homocysteine-methionine metabolism were differentially expressed according to the nutritional history (Additional files 5 and 3). The expression of methionine synthase (MTR), up-regulated in brain of V- compared to M-fish, is involved in the recycling of methionine from homocysteine using 5-methyl tetrahydrofolate (THF) as the one-carbon substrate [86]. Similarly, BHMT and BHMT2 also convert homocysteine to methionine using betaine (N,N,N tri-methyl glycine) as substrate [86]. Thus, the expression of three different genes involved in conversion of homocysteine to methionine was enhanced by the prior plant-feeding. Homocysteine is considered as a risk factor for neurodegenerative diseases and deficiencies in MTR correlate with cognitive impairment [87]. In addition, age-dependent reduction in MTR mRNA in humans may be a source of risk for neurological disorders across the lifespan via their impact on methylation reactions, including epigenetic regulation of gene expression [88] as methionine is required for the generation of SAM (S-adenosylmethionine), the primary methyl donor [86]. Indeed, the relative mRNA levels of Mat1a, which catalyzes conversion of methionine to SAM, was affected by early feeding history, being higher in the brain of V-fish compared to M-fish (Additional files 5 and 3). Mat1a was identified as candidate gene in a study on evolutionary conserved longevity genes impacting human cognitive abilities in elderly cohorts [89]. Thus, the long-term effects of the early plant-diet exposure on the regulation of central homocysteine-methionine metabolism warrants further studies from the perspective of maintenance of cognitive function and regulation of epigenetic enzyme substrates in the brain of nutritionally programmed rainbow trout.

Appetite and Feeding

The functional annotation analysis identified a number of differentially expressed genes associated with the regulation of feeding behaviour (Additional files 5 and 3), of interest in view of the positive effect of the early plant-diet exposure on feed intake during the V-challenge. This set of mRNA probes include neuropeptide hormones (CCK, bombesin, somatolactin, POMCB CRF2, CARTPT), hormone receptors (LEPR, GHSR, HTR3B), transcription regulators (NR1H3, PPARG, Tub) and regulators of neuropeptide hormone levels (AANAT, MAOA). In terms of expression pattern, MAOA, CRF2 and CARTPT were up-regulated following early plant-diet exposure but down-regulated in the brain of V-fish during V-challenge. In contrast, CCK, bombesin, somatolactin, POMCB, LEPR, GHSR, HTR3B, NR1H3, PPARG, Tub and

AANAT, all remained upregulated by the prior plant-diet exposure (Additional files 5 and 3). Regarding neuropeptide transcripts, we hypothesized to find higher levels of anorectic peptides in brain of M-fish during the V-challenge, in line with the higher reluctance of these fish to consume the plant-based diet. However, gene expression of bombesin, pro-opiomelanocortin B (POMCB) or cholecystokinin (CCK), known as feeding-inhibitory in fish [90], was higher in brain of V- instead of M-fish. POMCB encodes the inactive precursor that is post-translationally processed to several biogenically active neuropeptides including adrenocorticotropic hormone (ACTH), melanotropin alpha 2 (α-MSH 2) and melanotropin beta 2 (β-MSH 2), collectively referred to as melanocortins. In Atlantic salmon brain, POMCB expression increases postprandially [91]. The cholestokinine gene CCK-T has been found before to be highly expressed in trout brain [92] and, as for POMC, CCK increases after feeding in Atlantic salmon, in line with its role in mediating satiety in fed condition [91]. In our study, the brain samples in the V-challenge were sampled 8 h postprandially. As such, the higher CCK and POMC mRNA levels in the brain of the V- relative to M-fish possibly signal enhanced satiety resulting from the higher intakes during the V-challenge. They would therefore reflect the downstream response rather than the actual basis of the programmed enhanced feed intake. On the other hand, some other peptides known to exert anorectic effects in teleosts [90] had lower expression in the V-fish brain, as initially hypothesized. This was the case for e.g. corticotropin releasing factor (CRF) and cocaine- and amphetamine-regulated transcript protein (CARTPT), which warrant a more targeted approach in order to confirm their possible role in the positive programming effect on plant-feed acceptance due to early plant feeding.

Besides neuropeptides, enzymes that mediate neuropeptide levels may also affect feeding behaviour [93]. An example of this is arylalkylamine N-acetyltransferase (AANAT), the rate limiting enzyme in melatonin synthesis, whose transcription and enzyme activity are crucial in the central regulation of circadian rhythmicity [94]. Some teleosts including rainbow trout have two AANAT genes [95]. In chum salmon *Oncorhynchus keta*, a close teleost homolog of rainbow trout, ANAAT 1 and −2 mRNA abundance has been found in various regions in the brain [95], which may relate with the various neural and endocrine roles attributed to melatonin in fish [96], including feeding. In rainbow trout, treatments with melatonin altered gene expression of brain neuropeptides involved in feeding [97], whereas feed deprivation altered melatonin production partly regulated by reduced AANAT activity [98]. Accordingly, lower AANAT-2 mRNA was seen in the brain of M-fish, possibly reflecting the lower acceptance of the plant-based diet. Given the propensity of AANAT mRNA to fluctuate with meal timing, it would be interesting in the future to measure via a time course, the rhythmicity of mRNA levels and enzymatic activity of AANAT in order to assess possible effects of nutritional programming on circadian rhythms in trout, as seen in the hypothalamus transcriptome of rat following in utero protein restriction [99].

Somatolactin (SL) is a member of the growth hormone/prolactin family and is highly expressed in rainbow trout pituitary and brain [100] where it plays a role in energy homeostasis. In gilthead seabream, plant-protein was found to lower plasma SL levels coincident with lower feed intake [101]. In our study, SL transcripts were significantly lower in brain of M-fish which consumed less than V-fish when challenged with the plant-based diet. Since increases in ration size were found to enhance plasma SL in fish [102], it remains however unclear whether the observed changes in SL mRNA mediate or result from changes in feed acceptance in our study. Transcript levels of several nutrient-sensitive receptors were also modified (Additional files 5and 3). Ghrelin receptor or growth hormone secretagogue receptor (GHSR) is an important gene of the somatotropic axis, playing an essential role in energy expenditure and food intake [103]. In mammals, GHSR expression is co-localized in neurons with CCK expression, and its high expression in brain reward circuits suggests a role for GHSR in the induction of feed intake resulting from central administration of ghrelin [104]. This may explain the higher GHSR expression in brain of juvenile V-fish showing a relatively superior feed intake when challenged again with the plant-diet. Leptin receptor (LEPR) is a membrane bound receptor that mediates the leptin-induced suppression of feed intake in specific populations of neurons located in the hypothalamus [105]. Juvenile salmon fed restrictively (40 % of the amount of feed given to the control group) displayed decreased growth along with significantly lower LEPR mRNA in the brain [106]. Along these lines, the lower LEPR expression in brain of M relative to V-fish possibly reflects the lower feed intake of that group during the V-challenge. Nuclear receptor subfamily 1 group H member 3 (Nr1h3) or Liver Xa receptor, is an oxysterol-activated nuclear receptor involved in the transcriptional control of whole body cholesterol homeostasis [107]. In mammalian brain, Nr1h3 plays a role in memory functions and the development of age-dependent neurodegenerative changes [108]. In zebrafish midbrain, endogenous Nr1h3 ligands (cholic acid, epoxycholesterol) have been shown to selectively regulate dopaminergic neurogenesis [109]. In our study, rainbow trout exposed at early life to the plant-diet have higher Nr1h3 expression in the brain (Additional files 5 and 3). It would be interesting to investigate if any diet-related or endogenous oxysterol ligands of Nr1h3 are involved in driving memory or cognitive processes related with the positive plant-diet acceptance phenotype observed in this study. 5-Hydroxytryptamine receptor 3B (HTR3B) is a serotonin receptor involved in neural processes related to cognition and emotion in mammalian brain [110] and in mediating a number of satiation signals [111]. Also in rainbow trout, the serotonergic system plays a role in the control of food intake, as shown by the altered feed intake after pharmacological stimulation of serotonin receptor subfamilies [112] or after pre-treatment with the serotonin precursor tryptophan which counteracts stress-induced anorexia [113]. Thus, higher HTR3B expression in brain of juvenile V-trout exposed to the plant-diet at early life may drive improved feed intake through neural functions of satiety and memory.

When we considered using an early exposure to nutritionally program the acceptance of a plant-based diet in rainbow trout [34], we deliberately targeted the critical first-feeding stage during which the reliance on endogenous yolk as nutrient source is diminishing and swim-up fry begin to consume exogenous feed [114]. This developmental first-feeding period in salmonids is accompanied by synchronized anatomical, physiological and behavioural changes, including plasticity of olfactory and taste responses [115, 116]. The present brain transcriptome reveals a very large number of genes affected by the early plant-feeding. These encode a large variety of proteins that regulate sensory perception, cognitive processes, epigenetic changes and neuropeptides mediating feed intake which align contextually with the observed phenotype of enhanced plant-diet acceptance at the juvenile stage, attributed to the initial plant-diet exposure.

Genes Responsive To Nutritional History in the Liver

The functional annotation analysis of differentially expressed mRNAs in liver (Fig. 4) reveals that the early plant-diet exposure has both short-term and long-lasting effects on intermediary metabolic processes (Additional file 4), zymogens mediating protein degradation (Additional file 4), protein folding and immunomodulatory activity of peptidyl-prolyl isomerases (Additional file 4) and on cytoskeletal proteins involved in stress response and cell cycle (Additional file 4).

Previous studies have utilized nutrigenomic approaches in order to identify changes in hepatic function due to feeding plant- versus fish-based ingredients [5–9]. These all compared the direct diet effects in naïve unconditioned fish fed the different feed for a relatively long period, though never from the first-feeding fry stage onwards. In contrast, our study compares changes in molecular response at first-feeding (short-term direct diet effect) with those induced by the early feeding at the juvenile stage (indirect long-term nutritional programming effect) when fed the same plant-based diet (V-challenge). Nevertheless, several of the candidate genes and functional pathways identified by the previous nutrigenomic studies were found relevant for discussing our data, in particular regarding responses to plant-feeding in the unconditioned M-fish.

Intermediary Metabolism

A large number of mRNA probes in the liver found to be affected by the early diet exposure, were members of intermediary metabolic processes. These could be further classified as mediating glucose, alpha-amino acid or steroid metabolic process (Additional files 6 and 4). Some of the mRNAs encoding mediators of glucose metabolism (e.g. Pgam1 and Dlat) were up-regulated by early plant exposure, whereas others (e.g. Ldha, LdhB, Gapdh, Pgam2, Aldoa) were permanently down-regulated by the plant-diet history. The higher Aldoa, Pgam2 and Gapdh expressions in the liver of unconditioned M-fish during V-challenge agree with the overall induction of glycolytic mediators by plant protein feeding in fish [5–9], as shown for higher aldolase expression in liver of European sea bass [7] or in the hepatic proteome in trout [9] fed plant- versus fish-based ingredients. In our study, the lower Aldoa expression in V- than in M-fish hence may indicate a programming effect due to the early plant-diet exposure giving a reversal in the trend of the hepatic glycolytic response to plant-based feed. Several probes of lactate dehydrogenase (LDHA and LDHB) were also affected by nutritional history, all having lower hepatic transcript levels in V- versus M-fish. The significance of the higher LDH expressions in the liver of M-fish is not clearly apparent in the absence of further data on plasma lactate and hepatic LDH activity. LDH, through regulating the cellular redox potential by modifying NAD^{+}/NADH levels, has also a role in oxidative stress response, which is another pathway identified to be affected in liver by the early plant-feeding (Additional files 6 and 4). Several oxidative stress-responsive genes, which include enzymes involved in hepatic carbohydrate and lipid intermediary metabolism, had higher mRNA levels in the unconditioned M-fish, consistent with previous observations of higher cellular oxidative stress in salmon fed plant proteins [8].

A group of mRNAs identified to regulate steroid metabolic processes (Fabp6, Tm7sf2, Apoa4, Fdps, Hmgcs1, Srd5a2, Cel, Cyp27a1, sult1st2, sult1st3) were found to be downregulated due to the prior plant-feeding (Additional files 6 and 4). In general, fish fed diets containing soybean meal show upregulated capacity of cholesterol and bile acid synthesis as illustrated by enhanced expression of FDPS, sterol transport genes (APOA1 and APOB100) [7] or genes involved in bile acid metabolism such as HMGCS1 [117], which was attributed to the compensatory response of liver to the lower cholesterol content in plant-based feed. This is unlikely to explain the differential response in our study since the trout during the V-challenge received the same plant-diet and no difference in plasma cholesterol was noted between M and V fish at the end of the V-challenge (6.4 and 6.3 mmol/L, respectively). Alternatively, a study with liver HepG2 cells reported direct effects of soy-extracts and isoflavones on SREBP-2-regulated genes involved in cholesterol biosynthesis and homeostasis such as HMG CoA synthase (HMGCS) mRNA [118]. As such, the higher expression of genes involved in cholesterol biosynthesis (HMGCS1, Tm7sf2 and FDPS) and transport (APOA4) in liver of the unconditioned M-fish during V-challenge possibly reflects the typical molecular response to soy products, seen in fish without an early-life plant-diet experience. Alternatively, as suggested above for glycolytic markers, the lower hepatic response of cholesterol biosynthetic and transport genes to plant-feeding in juvenile V-fish may reflect an adaptive metabolic process induced by the prior plant-feeding. Xenobiotic sterols such as soy phytoestrogens (e.g. genistein or daidzein) have been postulated as a reason for poor palatability of plant-based feed in fish [10] and may accumulate in the bile of rainbow trout [119]. M-fish also had higher hepatic expression of cytosolic sulfotransferases sult1st2 and sult1st3 than V-fish (Additional files 6 and 4). The respective mammalian homologs SULT1E1 and SULT1A1 are known to mediate the detoxification of genistein and daidzein [120] as also shown in HepG2 liver cells [121]. Similarly, the induction of genes mediating the xenobiotic metabolism of phytoestrogens in the juvenile M-fish may indicate an acute adaptive response to the V-challenge. The other way around, this may also be interpreted as reduced sensitivity of V-fish to the presence of anti-nutritional factors, which possibly contributes to the positive programming effect related to the prior plant-diet experience in the V-fish.

Zymogens

A group of mRNAs encoding zymogens (pro-enzymes) possessing peptidase activity (Additional files 6 and 4) was found to be differentially expressed following first exposure to plant-based diet (3-week) and after V-challenge (liver). While zymogens were expressed at either a lower (Mmmp2, CASP3, Cela2a, Prss1, Gzma, CTSK, MEP1A, PRSS7) or higher level (Psmb7, CPB1, Prss2, Ctrb1, Lce, Ctrc, Prss3, F5) in in the V-fish after V-challenge, most of the above mRNAs were down-regulated in the fry following the early plant-diet exposure. Changes in the expression of genes encoding proteolytic activity, including members of proteasome complex, have been documented before in studies on the replacement of fish meal by plant-ingredients in fish, albeit with contradictory trends. For example, feeding a plant-diet increased hepatic gene expression of several proteasome subunit members (PSMB) including PSMB7 in European sea bass [7] and also the hepatic protein levels of PSMB in rainbow trout [9]. In contrast, another study with rainbow trout fed plant-based feed showed reduced abundance of PSMB7 and of other transcripts encoding proteasome subunits [5]. This is in line with the lower expression of PSMB7 in the present fry when fed plant-ingredients and in liver of M-fish during the V-challenge (Additional file 6). Alternatively, the prior plant diet experience enhanced PSMB7 expression in liver of V-fish when confronted again to the plant-diet. These findings do not relate with the notion in trout that growth rate and efficiency of protein utilization are negatively correlated to hepatic proteasome activity [122]. The reason for the inconsistent hepatic responses of proteasome subunit to plant-based diets in fish is not apparent and a concurrent evaluation of effects on protein turnover and the activity state of proteases may be useful in future studies [123]. Additionally, several of the zymogen genes identified here may function in lysosomal autophagy-mediated protein degradation [124], which, by regulating levels of key metabolic enzymes, may play a role in the response of liver to nutritional challenges [125]. Several targets of autophagy in the hepatic lysosome, such as Aldoa, Gapdh, Ldha, Ldhb and Cyp27a1 are also modified in the trout liver by the prior plant-diet exposure (Additional files 6 and 4). Possibly, the expression of lysosomal proteases and enzymes mediating intermediary metabolism is co-regulated in trout by the early plant-diet exposure in order to adapt hepatic metabolism to the plant-diet when presented later, as might be the case in the nutritionally programmed V-fish.

Intracellular Protein Folding

V-fish have higher hepatic expression of several peptidyl prolyl cis-trans isomerases (PPIases) important for intracellular protein folding [126], including cyclophillins (PPIB, PPIG) and FK506 binding proteins (FKBP2, FKBP7, FKBP11) (Additional files 6 and 4). PPIB, involved in protein folding in the endoplasmic reticulum (ER), plays a significant role in protecting cells against ER stress [127]. The differential expression of hepatic PPIase mRNA seen in salmon when switched to a plant-based diet may be related with the presence of anti-nutritional factors, increasing reactive oxygen species [8]. Thus, an increase of the multiple genes encoding peptidyl-prolyl cis-trans isomerase activity, as seen for PPIB in liver of the nutritionally programmed V-fish when challenged again to the plant diet may serve to counteract oxidative and ER stress better than in the unconditioned M-fish.

Cytoskeleton

A large cluster of mRNAs encoding cytoskeletal proteins (Additional files 6 and 4) was differentially expressed following the first plant-diet exposure (3-week) and after V-challenge (liver). These groups of mRNAs represent members of the actin, intermediate filament and microtubule cytoskeletal components, as well as regulators of mitosis. The differential expression of genes mediating remodelling of cytoskeleton is consistent with previous observations in rainbow trout [6] where substitution of fish-based by plant-based ingredients altered the expression of actin, actin related protein 2/3 complex (Arpc4), keratin, tubulin alpha chain, kinesin members (KIF2C) and dynein 1 intermediate chain (Dync1i2); all of which were identified to be differentially expressed in the trout liver as a result of the early plant-diet exposure (Additional files 6 and 4). Several genes in this pathway also play a key role in cell cycle regulation (Additional files 6 and 4) and are all expressed at a lower level in M-fish during the V-challenge. Trying to unravel the postnatal consequences of nutrient manipulation during foetal life, a study with rat identified genes encoding cell cycle regulators and cytoskeletal remodelling proteins as key drivers of nutritional programming in rat [128]. Of interest, a number of candidate genes from that study, including Arpc4, Ube2c, actin and tubulin members, are also persistently altered by the early plant-diet exposure in our study (Additional file 6), which may indicate a role of these cytoskeletal components in the programmed V-fish phenotype, though the underlying mechanisms remain to be elucidated. Also, it cannot be excluded that these changes are the consequence of the faster growth in the V- versus M-fish during the V-challenge phase.

Confirmation of Selected Microarray Expressed Targets by Real-time PCR

Real time PCR analysis (Additional file 7) was performed on selected genes that represent the different pathways (see Additional files 1, 5 and 6) enriched by nutritional history in the brain or liver in order to confirm the significance of differential mRNA expression pattern observed in the microarray data. For all genes tested, despite some differences in the scale of differential expression, the expression pattern between microarray data and real time PCR data was congruent (Additional file 7).

Nutritional Programming of Plant-Based Diet Acceptance in Teleosts

The magnitude of fold-change difference in transcript levels was generally higher in the 3-week fry than in juvenile trout liver and brain. This is to be ascribed to the fact that the applied start-feeding stage is particularly sensitive to changes in the environment, and also in the setting of exogenous feed preferences in oviparous fish [114, 129], which translates into a more dynamic transcriptomic response than at the later stage, in the juvenile trout. Also, importantly, the molecular changes detected in the juveniles are to be ascribed to the early feeding experience (which occurred several months before the V-challenge) and hence do not represent a direct diet effect as in the early fry. As highlighted in a study dealing with molecular mechanisms involved in nutritional programming in rodents [128], the consistency in the response of key pathways between early developmental and adult stages is considered as a more relevant indicator of nutritional programming than consistency in the direction or magnitude of mRNA levels of individual genes. A follow-up proteomic analysis could be interesting, especially in the liver, to have a better overview of the functional changes in hepatic metabolic processes due to nutritional programming. In this respect, it is encouraging that our results on cholesterol and carbohydrate metabolism compare favourably with proteome data from a previous study in trout [9]. Additionally, candidate enzymatic studies, as in the case of protease regulation and xenobiotic metabolism, in both the 3-week fry and in the juvenile trout liver could help clarify the role of these enzymes in fish fed plant-ingredient rich feed.

Overall, the present molecular dataset, complementary to our previous data on the positive effect of prior plant-diet feeding on its later acceptance (feed intake and feed utilisation efficiency) in trout [34], fortifies the notion that the period of transition from reliance on yolk to exogenous feeding is a critical window for nutritional programming in rainbow trout, and possibly in other teleosts. Persistent molecular changes induced by dietary changes at the same ontogenic first-feeding stage have been observed in the field of hyperglucidic programming in several teleost species, such as trout [25, 26], zebrafish [130] and Siberian sturgeon [131]. Yet, the timing to evaluate the underlying molecular mechanisms deserves further consideration. We analysed the transcriptome changes in trout fry after a 3 week first-feeding exposure, and in liver and brain of juvenile trout at the end of the 25 day V-challenge period in which nutritionally programmed trout exhibit phenotypic enhancement of feed intake and growth. It might, however, be interesting to analyse parallel transcriptomic changes in juvenile trout just before the V-challenge, when both M-fish and V-fish are maintained on the control M-diet. This could be a pertinent sampling window to assess plausible epigenetic modifications of DNA methylation or histone modifications at genes that are nutritionally programmed by the early plant diet-exposure, but exhibit dormant transcriptional responses to the nutritional programming.

CONCLUSIONS

Our results elaborate the transcriptomic changes in trout fry at the first-feeding stage, and in the brain and liver of juvenile trout, accompanying the nutritional programming that enhances feed acceptance of a plant-based diet in rainbow trout. We observe that nutritional programming in the trout brain affects the pathways of sensory perception, synaptic transmission, cognitive processes and neuroendocrine peptides, suggesting the probable acquisition of flavour and feed preference due to early plant-diet exposure. In the liver, changes in pathways mediating intermediary metabolism, xenobiotic metabolism, proteolysis, and cytoskeletal regulation of cell cycle are affected by the prior plant-diet exposure, and some of these changes are congruous with those observed in previous nutrigenomic studies on plant-feeding in cultured fish species. Moreover, nutritional programming appears to reverse the trend in certain hepatic gene expression changes that were previously reported to be affected by plant-diet exposure. Overall, the work in this study highlights the first-feeding stage in trout and possibly other oviparous fish as a critical window for nutritional programming, and provides support for utilizing this strategy to achieve improvements in sustainability of feeding practices in aquaculture.

METHODS

Experimental Design and Animal Material

The present study aims to gain insight into mechanisms governing the positive effect of an early short term exposure of rainbow trout (*Oncorhynchus mykiss*) fry to a plant-based diet on its future acceptance and utilization, as demonstrated in [34]. Long-term changes in gene expression in brain and liver of juvenile rainbow trout fed diet V (during Challenge V) as a function of the first-feeding diet (diet M or diet V) were therefore analyzed. Details on diet preparation, fish rearing conditions and feeding protocol can be found in [34]. In summary, two experimental diets were prepared: the plant-based diet V (vegetable) contained a blend of protein-rich plant ingredients (wheat gluten, extruded peas, corn gluten meal, soybean meal and white lupin) and a plant oil blend (palm, rapeseed and linseed) as lipid source, whereas diet M (marine) had fishmeal as protein source and fish oil as lipid source. Both diets (Table 1) fulfilled the established nutrient requirements of rainbow trout [4]. The fish selected for the current study is the isogenic heterozygous family C1 (see [34]) of rainbow trout. The applied feeding protocol is based on the concept of nutritional programming in mammals, i.e. the evaluation of a long-term effect due to a specific early-life nutritional event, as applied in other studies with rainbow trout [25, 26]. *O. mykiss* swim-up fry (7 °C water temperature) at first-feeding were fed for 3 weeks either the plant-based diet V or the diet M, after which both groups received diet M during the 7-month intermediate growth phase (Fig. 1a). Both groups ($n = 3$ and 4 replicate tanks for fish of nutritional history V and M, respectively) were then challenged for 25 days by feeding diet V (Fig. 1a) during which voluntary feed intake, growth and feed utilisation efficiency were monitored (16.5 °C water temperature). No difference in initial body weight due to the early feeding experience was noted at the start of the V-challenge (33.6 g, see [34]). Remarkably, fish of nutritional history V compared to M displayed a higher feed intake and growth rate ($P < 0.05$) and a tendency for improved efficiency of feed utilization ($P = 0.06$) during the final V-challenge (Fig. 1), underlining the positive impact of the early diet V feeding history.

Table 1: Formulation and composition of diets

Ingredients (g 100 g^{-1} diet)	Diet M	Diet V
Fish oil	8.5	
Plant oil blend[a]	–	10.3
Fishmeal LT (CP 70 %)	63	–
White lupinseed meal (CP 40 %)	–	5.8
Corn gluten meal (CP 62 %)	–	17.4
Soybean meal (CP 46 %)	–	21.5
Wheat gluten (CP 80 %)	–	25.6
Whole wheat (CP 10 %)	25.4	5.1
Extruded dehulled peas (CP 24 %)	–	3.1
Soy-lecithin	–	2.0
L-Arginine	–	1.0
L-Lysine	–	1.5
CaHPO4.2H20 (18%P)	–	3.6

Mineral and vitamin premix[b]	3.0	3.0
Analysed composition		
Dry matter (DM, % diet)	93.3	92.4
Crude protein (% DM)	52.1	50.5
Crude fat (% DM)	17.9	17.0
Gross energy (kJ g^{-1} DM)	22.3	22.3

Reproduced from [34]. Formulation, approximate crude protein (CP) levels of ingredients and analysed composition of the experimental diets M (fishmeal and fish oil-based) and V (all fishmeal and fish oil replaced by plant protein and plant oil sources). [a]Consisting of (% blend): rapeseed oil (50), palm oil (30), linseed oil (20). [b]INRA UPAE, 78352 Jouy en Josas, France

RNA Isolation

As indicated on Fig. 1, samples were taken at the end of the 3-week early first-feeding exposure (whole fry, overnight fasted) and at the end of the V-challenge (8 h after the last meal) where liver and whole brain were carefully dissected, instantly snap-frozen in liquid nitrogen and stored at −80 °C prior to RNA extraction. The first sampling (fry) thus assesses the immediate short-term effect on the profile of gene expression, i.e. effect diet V vs. M, as opposed to the second sampling (juvenile tissue) used to evaluate potential long-term effects related to the early life exposure, i.e. effect nutritional history M vs. V. Total RNA was extracted from the trout swim-up fry, brain or liver by homogenization in 1–2 ml TRIzol (Invitrogen, Carlsbad, CA, USA) reagent following the manufacturer's instructions. Removal of genomic DNA from RNA samples was performed using RNeasy mini spin columns (Qiagen) through on-filter DNase digestion according to the manufacturers recommendations. Total RNA was quantified using spectrophotometry based on absorbance at 260 nm (Nanodrop ND1000, LabTech) and the integrity of the RNA was determined by electrophoresis (Agilent Bioanalyser 2100) and samples with RNA integrity number (RIN) ≥ 8 were used for microarray analysis.

Microarray Platform

Microarray experiments were performed on an Agilent-based microarray platform with 8 X 60 K probes per slide (Gene Expression Omnibus; GEO platform record: GPL19030). This platform is based on a 37 K high density rainbow trout oligonucleotide microarray resource designed by Yao and colleagues [132], as previously employed at Laboratoire de Physiologie et Génomique des Poissons (LPGP, Institut National de la Recherche Agronomique, Rennes, France) [133] and has been enriched with oligonucleotides designed utilizing recent NGS data from trout [134]. The microarray gene annotations were reanalysed by Sigenae (Institut National de la Recherche Agronomique, Toulouse, France). Microarray data sets have been submitted to the GEO-NCBI with the accession number GSE60010.

They may be accessed at http://www.ncbi.nlm.nih.gov/geo/query/acc.cgi?token=sfafokqmtlevbop&acc=GSE60010.

Microarray Hybridization, Data Acquisition and Analysis

Distinct RNA samples from trout swim-up fry exposed to diet M ($n=4$ fry) or diet V ($n=3$ fry) for 3 weeks, and whole brain ($n=4$ samples) and whole liver ($n=4$ samples) of juvenile trout with nutritional history of diet M or diet V after the 25 day V-challenge were used for labelling and hybridisation. For each sample, 150 ng of total RNA was amplified and labelled using Cy3-CTP according to the manufacturer's instructions (Agilent). Briefly, RNA was first reverse transcribed, using a polyDT T7 primer, Cy3 was then incorporated by a T7 polymerase mediated transcription and excess dye was removed using a RNeasy kit (Qiagen). The level of dye incorporation was evaluated using a spectrophotometer (Nanodrop ND1000, LabTech). Yield (>0.825 μg cRNA) and specific activity (>6pmol of Cy3 per μg of cRNA) of the Cy3-cRNA produced were checked with the Nanodrop ND1000 (LabTech). 600 ng of Cy3-cRNA was then fragmented in the appropriate buffer (Agilent) for 30 min at 60 °C and hybridized on a sub-array. Hybridisation was performed in a microarray hybridization oven (Agilent) for 17 h at 65 °C. The slides were then rinsed in gene expression wash buffers 1 and 2 (Agilent) and scanned with an Agilent scanner (Agilent DNA Microarray Scanner, Agilent Technologies, Massy, France) using the standard parameters for a gene expression 8x60K oligoarray (3 μm and 20 bits). Data were then obtained with the Agilent Feature Extraction software (10.7.1.1) according to the appropriate GE protocol (GE1_107_Sep09). Arrays were normalised using scale normalization with GeneSpring software (version 12.6.1). We performed a 2- way ANOVA on log 2-transformed normalized data, using a p-value with Benjamini-Hochberg correction [False Discovery Rate; FDR] ≤ 0.05 and fold-change ≥ 1.5, to determine the probes that are differentially expressed by early V vs M feeding in 3-week fry and in juvenile brain (further called '3-week + Brain') and, similarly, a 2-way ANOVA to determine the probes that are differentially expressed by early V vs M feeding in both 3-week fry and in juvenile liver (further called '3-week + Liver'). ANOVA were performed using Statistica 8.0 (StatSoft, OK, USA). For clustering analysis, data were log transformed, median-centred and an average linkage clustering was carried out using CLUSTER software [135]. The results were visualised using TREEVIEW [136].

Gene Ontology Analysis

Gene ontology (GO) analysis was performed using the functional annotation cluster and chart tools from the DAVID (Database for Annotation, Visualization and Integrated Discovery) bioinformatics resource [137, 138]. Briefly, the Uniprot accession ID's of the mRNA probes differentially expressed by nutritional history (Additional file 1, 3-week + Brain; 3-week + Liver) were used as input against a background Uniprot accession ID's of the entire 60 K array probe list and processed with the functional annotation cluster and chart tools from the DAVID bioinformatics resource. The resulting data was used as input in the DAVID Enrichment Map plugin from the Bader Lab [139, 140] in the Cytoscape network visualization tool (version 2.8.3) [141–143].

Pathway Analysis

Selected mRNA probes differentially expressed by nutritional history (3 week + Brain; 3-week + Liver) and those assigned to be part of a significantly affected GO process (Additional files 5 and6) were used as inputs in the GeneMania pathway analysis tool [35–37] using the inbuilt *Homo sapiens* pathway interaction database as background and processed according to the default settings in the GeneMania tool. The pathway maps generated by GeneMania are outlined in Fig. 5 and Additional files 3 and 4.

Real time PCR

Gene expression levels were determined by real-time RT-PCR using the same RNA samples ($n = 4$) that were also employed in the labelling and hybridisation for microarray analysis. Total RNA (1 µg) samples from trout swim-up fry exposed to diet M ($n = 4$ fry) or diet V ($n = 4$ fry) for 3 weeks, and whole brain ($n = 4$ samples) and whole liver ($n = 4$ samples) of juvenile trout with nutritional history of diet M or diet V after the 25 day V-challenge were used to reverse transcribed to cDNA with the Superscript III RT kit (Invitrogen, Carlsbad, CA, USA) using oligo dT primers. Real-time PCR was performed in the iCycler MyiQ Real-Time PCR detection system (Bio-Rad, Hercules, CA, USA). The genes chosen for the swim-up fry were keratin 13 (K13), Lissencephaly-1 homolog B (pafah1b1b), purpurin (rbp4l) and chymotrypsin B (Ctrb1). The genes chosen for the brain were Glucagon-2 (gcg2), purpurin (rbp4l), somatolactin (SL), Recoverin (Rcvrn) and cholecystokinin-Thr (CCK-T). The genes chosen for the liver were chymotrypsin B (Ctrb1), keratin 13 (K13) and cytosolic sulfotransferase 3 (ST1S3). Quantitative PCR analyses for gene expressions were performed on 10 µl of the RT reaction mixture using the iQ SYBR Green Supermix (Bio-Rad). The total volume of the PCR reaction was 25 µl containing 200 nM of each primer. The primers used are outlined in Additional file 8. Thermal cycling was initiated with the incubation at 95 °C for 90 s for hot-start iTaq DNA polymerase activation. Forty steps of PCR were performed, each one consisting of heating at 95 °C for 20 s for denaturing, and at 60 °C for 30 s for annealing and extension. Following the final cycle of the PCR, melting curves were systematically monitored (with a gradient of 0·5 °C per 10 s from 55 to 95 °C) to ensure that only one fragment was amplified. Samples without RT and samples without RNA were run for each reaction as negative controls. The mRNA expression levels were normalized using actin as the reference gene and the fold-change obtained by normalizing to the respective expression in fish fed diet M (M-diet or history of M-diet). The significance of mRNA expression between groups was determined using an unpaired t-test (Mann-Whitney test; $p \leq 0.05$).

ACKNOWLEDGEMENTS

We thank Marianne Cluzeaud for all technical help. Additionally, we thank all members of INRA, UR 1067 NUMEA for helpful discussion.

Funding

This work was part of a Postdoctoral fellowship programme to MNB funded by INRA, under the INRA-WUR Aquaculture platform. The study was partly funded by INRA (Department Phase, Crédit Incitatif), by the VEGEAQUA project (FUI funding, French government) and by the EU-funded ARRAINA project No. 288925: Advanced Research Initiatives for Nutrition & Aquaculture (EU Seventh Framework Programme).

Availability of Data and Material

Microarray may be accessed at GEO-NCBI with the accession number GSE60010 at:http://www.ncbi.nlm.nih.gov/geo/query/acc.cgi?token=sfafokqmtlevbop&acc=GSE60010.

Competing Interests

The authors declare that they have no competing interests.

Consent for Publication

Not applicable.

Ethics Approval and Consent to Participate

The experiments were conducted following the Guidelines of the National Legislation on Animal Care of the French Ministry of Research (Décret 2001–464, May 29, 2001) and in accordance with the boundaries of EU legal frameworks relating to the protection of animals used for scientific purposes (i.e. Directive 2010/63/EU). INRA NuMeA (UR1067) facilities are certified for animal services under the permit number A64.495.1 by the French veterinary services. No specific permit was needed for this study since the experimental diets fulfilled all known nutritional requirements for rainbow trout [4] and since tissue sampling took place post-mortem, according to the authorized ethical procedures.

REFERENCES

1. Naylor RL, Hardy RW, Bureau DP, Chiu A, Elliott M, Farrell AP, Forster I, Gatlin DM, Goldburg RJ, Hua K, et al. Feeding aquaculture in an era of finite resources. Proc Natl Acad Sci U S A. 2009;106(36):15103–10.
2. Gatlin DM, Barrows FT, Brown P, Dabrowski K, Gaylord TG, Hardy RW, Herman E, Hu G, Krogdahl Å, Nelson R, et al. Expanding the utilization of sustainable plant products in aquafeeds: a review. Aquacult Res. 2007;38(6):551–79.
3. Hua K, Bureau DP. Exploring the possibility of quantifying the effects of plant protein ingredients in fish feeds using meta-analysis and nutritional model simulation-based approaches. Aquaculture. 2012;356–357:284–301.
4. National Research Council (NRC). Nutrient requirements of fish and shrimp. Washington, D.C.: The National Academies Press; 2011. pp. 376.
5. Panserat S, Kolditz C, Richard N, Plagnes-Juan E, Piumi F, Esquerre D, Medale F, Corraze G, Kaushik S. Hepatic gene expression profiles in juvenile rainbow trout (Oncorhynchus mykiss) fed fishmeal or fish oil-free diets. Br J Nutr. 2008;100(5):953–67.
6. Panserat S, Hortopan GA, Plagnes-Juan E, Kolditz C, Lansard M, Skiba-Cassy S, Esquerré D, Geurden I,

Médale F, Kaushik S, et al. Differential gene expression after total replacement of dietary fish meal and fish oil by plant products in rainbow trout (Oncorhynchus mykiss) liver. Aquaculture. 2009;294(1–2):123–31.

7. Geay F, Ferraresso S, Zambonino-Infante JL, Bargelloni L, Quentel C, Vandeputte M, Kaushik S, Cahu CL, Mazurais D. Effects of the total replacement of fish-based diet with plant-based diet on the hepatic transcriptome of two European sea bass (Dicentrarchus labrax) half-sibfamilies showing different growth rates with the plant-based diet. BMC Genomics. 2011;12:522.

8. Tacchi L, Secombes CJ, Bickerdike R, Adler MA, Venegas C, Takle H, Martin SA. Transcriptomic and physiological responses to fishmeal substitution with plant proteins in formulated feed in farmed Atlantic salmon (Salmo salar). BMC Genomics. 2012;13:363.

9. Vilhelmsson OT, Martin SA, Medale F, Kaushik SJ, Houlihan DF. Dietary plant-protein substitution affects hepatic metabolism in rainbow trout (Oncorhynchus mykiss). Br J Nutr. 2004;92(1):71–80.

10. Francis G, Makkar HPS, Becker K. Antinutritional factors present in plant-derived alternate fish feed ingredients and their effects in fish. Aquaculture. 2001;199(3–4):197–227.

11. Collins SA, Mansfield GS, Desai AR, Van Kessel AG, Hill JE, Drew MD. Structural equation modeling of antinutrients in rainbow trout diets and their impact on feed intake and growth. Aquaculture. 2013;416–417:219–27.

12. Glencross B, Evans D, Rutherford N, Hawkins W, McCafferty P, Dods K, Jones B, Harris D, Morton L, Sweetingham M, et al. The influence of the dietary inclusion of the alkaloid gramine, on rainbow trout (Oncorhynchus mykiss) growth, feed utilisation and gastrointestinal histology. Aquaculture. 2006;253(1–4):512–22.

13. Serrano E, Storebakken T, Penn M, Øverland M, Hansen JØ, Mydland LT. Responses in rainbow trout (Oncorhynchus mykiss) to increasing dietary doses of lupinine, the main quinolizidine alkaloid found in yellow lupins (Lupinus luteus). Aquaculture. 2011;318(1–2):122–7.

14. Bureau DP, Harris AM, Young Cho C. The effects of purified alcohol extracts from soy products on feed intake and growth of chinook salmon (Oncorhynchus tshawytscha) and rainbow trout (Oncorhynchus mykiss). Aquaculture. 1998;161(1–4):27–43.

15. Le Boucher R, Dupont-Nivet M, Vandeputte M, Kerneis T, Goardon L, Labbe L, Chatain B, Bothaire MJ, Larroquet L, Medale F, et al. Selection for adaptation to dietary shifts: towards sustainable breeding of carnivorous fish. PLoS One. 2012;7(9):e44898.

16. Overturf K, Barrows FT, Hardy RW. Effect and interaction of rainbow trout strain (Oncorhynchus mykiss) and diet type on growth and nutrient retention. Aquacult Res. 2013;44(4):604–11.

17. Pierce LR, Palti Y, Silverstein JT, Barrows FT, Hallerman EM, Parsons JE. Family growth response to fishmeal and plant-based diets shows genotype × diet interaction in rainbow trout (Oncorhynchus mykiss). Aquaculture. 2008;278(1–4):37–42.

18. Gluckman PD, Hanson MA. Developmental and epigenetic pathways to obesity: an evolutionary-developmental perspective. Int J Obes (Lond). 2008;32 Suppl 7:S62–71.

19. Lucas A. Programming by early nutrition: an experimental approach. J Nutr. 1998;128(2 Suppl):401S–6.PubMed

20. Patel MS, Srinivasan M. Metabolic programming: causes and consequences. J Biol Chem. 2002;277(3):1629–32.

21. Waterland RA, Jirtle RL. Transposable elements: targets for early nutritional effects on epigenetic gene regulation. Mol Cell Biol. 2003;23(15):5293–300.

22. Lillycrop KA, Burdge GC. Epigenetic mechanisms linking early nutrition to long term health. Best Pract Res Clin Endocrinol Metab. 2012;26(5):667–76.

23. Petry CJ, Ozanne SE, Hales CN. Programming of intermediary metabolism. Mol Cell Endocrinol. 2001;185(1–2):81–91.

24. Roseboom TJ, Painter RC, van Abeelen AF, Veenendaal MV, de Rooij SR. Hungry in the womb: what are the consequences? Lessons from the Dutch famine. Maturitas. 2011;70(2):141–5.

25. Geurden I, Aramendi M, Zambonino-Infante J, Panserat S. Early feeding of carnivorous rainbow trout (Oncorhynchus mykiss) with a hyperglucidic diet during a short period: effect on dietary glucose utilization in juveniles. Am J Physiol Regul Integr Comp Physiol. 2007;292(6):R2275–83.

26. Geurden I, Mennigen J, Plagnes-Juan E, Veron V, Cerezo T, Mazurais D, Zambonino-Infante J, Gatesoupe J, Skiba-Cassy S, Panserat S. High or low dietary carbohydrate:protein ratios during first-feeding affect glucose metabolism and intestinal microbiota in juvenile rainbow trout. J Exp Biol. 2014;217(Pt 19):3396–406.

27. Vagner M, Robin JH, Zambonino-Infante JL, Tocher DR, Person-Le Ruyet J. Ontogenic effects of early feeding of sea bass (Dicentrarchus labrax) larvae with a range of dietary n-3 highly unsaturated fatty acid levels on the functioning of polyunsaturated fatty acid desaturation pathways. Br J Nutr. 2009;101(10):1452–62.

28. Vagner M, Zambonino Infante JL, Robin JH, Person-Le Ruyet J. Is it possible to influence European sea bass (Dicentrarchus labrax) juvenile metabolism by a nutritional conditioning during larval stage? Aquaculture. 2007;267(1–4):165–74.

29. Rocha F, Dias J, Engrola S, Gavaia P, Geurden I, Dinis MT, Panserat S. Glucose overload in yolk has little effect on the long-term modulation of carbohydrate metabolic genes in zebrafish (Danio rerio). J Exp Biol. 2014;217(Pt 7):1139–49.

30. Beauchamp GK, Mennella JA. Early flavor learning and its impact on later feeding behavior. J Pediatr Gastroenterol Nutr. 2009;48 Suppl 1:S25–30.

31. Graven SN, Browne JV. Sensory development in the fetus, neonate, and infant: introduction and overview. Newborn Infant Nurs Rev. 2008;8(4):169–72.

32. London RM, Snowdon CT, Smithana JM. Early

experience with sour and bitter solutions increases subsequent ingestion. Physiol Behav. 1979;22(6):1149–55.

33. Youngentob SL, Glendinning JI. Fetal ethanol exposure increases ethanol intake by making it smell and taste better. Proc Natl Acad Sci U S A. 2009;106(13):5359–64.

34. Geurden I, Borchert P, Balasubramanian MN, Schrama JW, Dupont-Nivet M, Quillet E, Kaushik SJ, Panserat S, Medale F. The positive impact of the early-feeding of a plant-based diet on its future acceptance and utilisation in rainbow trout. PLoS One. 2013;8(12):e83162.

35. Mostafavi S, Ray D, Warde-Farley D, Grouios C, Morris Q. GeneMANIA: a real-time multiple association network integration algorithm for predicting gene function. Genome Biol. 2008;9 Suppl 1:S4.

36. Montojo J, Zuberi K, Rodriguez H, Kazi F, Wright G, Donaldson SL, Morris Q, Bader GD. GeneMANIA Cytoscape plugin: fast gene function predictions on the desktop. Bioinformatics. 2010;26(22):2927–8.

37. Zuberi K, Franz M, Rodriguez H, Montojo J, Lopes CT, Bader GD, Morris Q. GeneMANIA prediction server 2013 update. Nucleic Acids Res. 2013;41(Web Server issue):W115–22.

38. Stradmeyer L, Metcalfe NB, Thorpe JE. Effect of food pellet shape and texture on the feeding response of juvenile Atlantic salmon. Aquaculture. 1988;73(1–4):217–28.

39. He S, Liang XF, Sun J, Li L, Yu Y, Huang W, et al. Insights into food preference in hybrid F1 of Siniperca chuatsi (female symbol) x Siniperca scherzeri (male symbol) mandarin fish through transcriptome analysis. BMC Genomics. 2013;14:601.

40. Treesukosol Y, Smith KR, Spector AC. The functional role of the T1R family of receptors in sweet taste and feeding. Physiol Behav. 2011;105(1):14–26.

41. Jiang P, Josue J, Li X, Glaser D, Li W, Brand JG, et al. Major taste loss in carnivorous mammals. Proc Natl Acad Sci U S A. 2012;109(13):4956–61.

42. Polakof S, Soengas JL. Evidence of sugar sensitive genes in the gut of a carnivorous fish species. Comp Biochem Physiol B Biochem Mol Biol. 2013;166(1):58–64.

43. Otero-Rodino C, Libran-Perez M, Velasco C, Lopez-Patino MA, Miguez JM, Soengas JL. Evidence for the presence of glucosensor mechanisms not dependent on glucokinase in hypothalamus and hindbrain of rainbow trout (Oncorhynchus mykiss). PLoS One. 2015;10(5):e0128603.

44. Ren X, Zhou L, Terwilliger R, Newton SS, de Araujo IE. Sweet taste signaling functions as a hypothalamic glucose sensor. Front Integr Neurosci. 2009;3:12.

45. Hara TJ. Feeding behaviour in some teleosts is triggered by single amino acids primarily through olfaction. J Fish Biol. 2006;68(3):810–25.

46. Scholz AT, Horrall RM, Cooper JC, Hasler AD. Imprinting to chemical cues: the basis for home stream selection in salmon. Science. 1976;192(4245):1247–9.

47. Shoji T, Yamamoto Y, Nishikawa D, Kurihara K, Ueda H. Amino acids in stream water are essential for salmon homing migration. Fish Physiol Biochem. 2003;28(1–4):249–51.

48. Yamamoto Y, Hino H, Ueda H. Olfactory imprinting of amino acids in lacustrine sockeye salmon. PLoS One. 2010;5(1):e8633.

49. Nakaya N, Sultana A, Lee HS, Tomarev SI. Olfactomedin 1 interacts with the Nogo A receptor complex to regulate axon growth. J Biol Chem. 2012;287(44):37171–84.

50. Sekerkova G, Freeman D, Mugnaini E, Bartles JR. Espin cytoskeletal proteins in the sensory cells of rodent taste buds. J Neurocytol. 2005;34(3–5):171–82.

51. McCabe C, Rolls ET. Umami: a delicious flavor formed by convergence of taste and olfactory pathways in the human brain. Eur J Neurosci. 2007;25(6):1855–64.

52. Simmons WK, Martin A, Barsalou LW. Pictures of appetizing foods activate gustatory cortices for taste and reward. Cereb Cortex. 2005;15(10):1602–8.

53. Rolls ET. Taste, olfactory and food texture reward processing in the brain and the control of appetite. Proc Nutr Soc. 2012;71(4):488–501.

54. Lazarevic V, Pothula S, Andres-Alonso M, Fejtova A. Molecular mechanisms driving homeostatic plasticity of neurotransmitter release. Front Cell Neurosci. 2013;7:244.

55. Gal-Ben-Ari S, Rosenblum K. Molecular mechanisms underlying memory consolidation of taste information in the cortex. Front Behav Neurosci. 2011;5:87.PubMed

56. Dick O, Hack I, Altrock WD, Garner CC, Gundelfinger ED, Brandstatter JH. Localization of the presynaptic cytomatrix protein Piccolo at ribbon and conventional synapses in the rat retina: comparison with Bassoon. J Comp Neurol. 2001;439(2):224–34.

57. Nonet ML. A window into domain amplification through Piccolo in teleost fish. G3 (Bethesda). 2012;2(11):1325–39.

58. Ibi D, Nitta A, Ishige K, Cen X, Ohtakara T, Nabeshima T, Ito Y. Piccolo knockdown-induced impairments of spatial learning and long-term potentiation in the hippocampal CA1 region. Neurochem Int. 2010;56(1):77–83.

59. Muller M, Liu KS, Sigrist SJ, Davis GW. RIM controls homeostatic plasticity through modulation of the readily-releasable vesicle pool. J Neurosci. 2012;32(47):16574–85.

60. Kim SH, Ryan TA. CDK5 serves as a major control point in neurotransmitter release. Neuron. 2010;67(5):797–809.

61. Lau CG, Murthy VN. Activity-dependent regulation of inhibition via GAD67. J Neurosci. 2012;32(25):8521–31.

62. Davis GW, Bezprozvanny I. Maintaining the stability of neural function: a homeostatic hypothesis. Annu Rev Physiol. 2001;63:847–69.

63. Paranjpe P, Rodrigues V, VijayRaghavan K, Ramaswami M. Gustatory habituation in Drosophila relies on rutabaga (adenylate cyclase)-dependent plasticity of GABAergic inhibitory neurons. Learn Mem. 2012;19(12):627–35.

64. Banerjee K, Akiba Y, Baker H, Cave JW. Epigenetic control of neurotransmitter expression in olfactory bulb interneurons. Int J Dev Neurosci. 2013;31(6):415–23.

65. MacDonald RB, Debiais-Thibaud M, Talbot JC, Ekker M. The relationship between dlx and gad1 expression indicates highly conserved genetic pathways in the zebrafish forebrain. Dev Dyn. 2010;239(8):2298–306.

66. Moon HM, Wynshaw-Boris A. Cytoskeleton in action: lissencephaly, a neuronal migration disorder. Wiley Interdiscip Rev Dev Biol. 2013;2(2):229–45.

67. Sun C, Xu M, Xing Z, Wu Z, Li Y, Li T, Zhao M. Expression and function on embryonic development of lissencephaly-1 genes in zebrafish. Acta Biochim Biophys Sin. 2009;41(8):677–88.

68. Hunt RF, Dinday MT, Hindle-Katel W, Baraban SC. LIS1 deficiency promotes dysfunctional synaptic integration of granule cells generated in the developing and adult dentate gyrus. J Neurosci. 2012;32(37):12862–75.

69. Drerup CM, Wiora HM, Morris JA. Characterization of the overlapping expression patterns of the zebrafish LIS1 orthologs. Gene Expr Patterns. 2010;10(1):75–85.

70. Rampon C, Jiang CH, Dong H, Tang YP, Lockhart DJ, Schultz PG, Tsien JZ, Hu Y. Effects of environmental enrichment on gene expression in the brain. Proc Natl Acad Sci U S A. 2000;97(23):12880–4.

71. Zhao ZM, Reynolds AB, Gaucher EA. The evolutionary history of the catenin gene family during metazoan evolution. BMC Evol Biol. 2011;11:198.

72. Sanhueza M, Lisman J. The CaMKII/NMDAR complex as a molecular memory. Mol Brain. 2013;6:10.

73. Paffenholz R, Kuhn C, Grund C, Stehr S, Franke WW. The arm-repeat protein NPRAP (neurojungin) is a constituent of the plaques of the outer limiting zone in the retina, defining a novel type of adhering junction. Exp Cell Res. 1999;250(2):452–64.

74. Steinke A, Meier-Stiegen S, Drenckhahn D, Asan E. Molecular composition of tight and adherens junctions in the rat olfactory epithelium and fila. Histochem Cell Biol. 2008;130(2):339–61.

75. Dias RB, Rombo DM, Ribeiro JA, Henley JM, Sebastiao AM. Adenosine: setting the stage for plasticity. Trends Neurosci. 2013;36(4):248–57.

76. Capiotti KM, Menezes FP, Nazario LR, Pohlmann JB, de Oliveira GM, Fazenda L, Bogo MR, Bonan CD, Da Silva RS. Early exposure to caffeine affects gene expression of adenosine receptors, DARPP-32 and BDNF without affecting sensibility and morphology of developing zebrafish (Danio rerio). Neurotoxicol Teratol. 2011;33(6):680–5.

77. Maximino C, Lima MG, Olivera KR, Picanco-Diniz DL, Herculano AM. Adenosine A1, but not A2, receptor blockade increases anxiety and arousal in Zebrafish. Basic Clin Pharmacol Toxicol. 2011;109(3):203–7.

78. Li H, Chuang AZ, O'Brien J. Regulation of photoreceptor gap junction phosphorylation by adenosine in zebrafish retina. Vis Neurosci. 2014;31(3):237–43.

79. Santos EM, Kille P, Workman VL, Paull GC, Tyler CR. Sexually dimorphic gene expression in the brains of mature zebrafish. Comp Biochem Physiol A Mol Integr Physiol. 2008;149(3):314–24.

80. Renn SC, Aubin-Horth N, Hofmann HA. Fish and chips: functional genomics of social plasticity in an African cichlid fish. J Exp Biol. 2008;211(Pt 18):3041–56.

81. Sanogo YO, Hankison S, Band M, Obregon A, Bell AM. Brain transcriptomic response of threespine sticklebacks to cues of a predator. Brain Behav Evol. 2011;77(4):270–85.

82. Drew RE, Settles ML, Churchill EJ, Williams SM, Balli S, Robison BD. Brain transcriptome variation among behaviorally distinct strains of zebrafish (Danio rerio). BMC Genomics. 2012;13:323.

83. Wong RY, Oxendine SE, Godwin J. Behavioral and neurogenomic transcriptome changes in wild-derived zebrafish with fluoxetine treatment. BMC Genomics. 2013;14:348.

84. Wong RY, Ramsey ME, Cummings ME. Localizing brain regions associated with female mate preference behavior in a swordtail. PLoS One. 2012;7(11):e50355.

85. Wada K, Howard JT, McConnell P, Whitney O, Lints T, Rivas MV, Horita H, Patterson MA, White SA, Scharff C, et al. A molecular neuroethological approach for identifying and characterizing a cascade of behaviorally regulated genes. Proc Natl Acad Sci U S A. 2006;103(41):15212–7.

86. Jimenez-Chillaron JC, Diaz R, Martinez D, Pentinat T, Ramon-Krauel M, Ribo S, Plosch T. The role of nutrition on epigenetic modifications and their implications on health. Biochimie. 2012;94(11):2242–63.

87. Jadavji NM, Bahous RH, Deng L, Malysheva O, Grand'maison M, Bedell BJ, Caudill MA, Rozen R. Mouse model for deficiency of methionine synthase reductase exhibits short-term memory impairment and disturbances in brain choline metabolism. Biochem J. 2014;461(2):205–12.

88. Muratore CR, Hodgson NW, Trivedi MS, Abdolmaleky HM, Persico AM, Lintas C, De la Monte S, Deth RC. Age-dependent decrease and alternative splicing of methionine synthase mRNA in human cerebral cortex and an accelerated decrease in autism. PLoS One. 2013;8(2):e56927.

89. Lopez LM, Harris SE, Luciano M, Liewald D, Davies G, Gow AJ, Tenesa A, Payton A, Ke X, Whalley LJ, et al. Evolutionary conserved longevity genes and human cognitive abilities in elderly cohorts. Eur J Hum Genet. 2012;20(3):341–7.

90. Volkoff H, Canosa LF, Unniappan S, Cerdá-Reverter JM, Bernier NJ, Kelly SP, Peter RE. Neuropeptides and the control of food intake in fish. Gen Comp Endocrinol. 2005;142(1–2):3–19.

91. Valen R, Jordal AE, Murashita K, Ronnestad I. Postprandial effects on appetite-related neuropeptide expression in the brain of Atlantic salmon, Salmo salar. Gen Comp Endocrinol. 2011;171(3):359–66.

92. Jensen H, Rourke IJ, Moller M, Jonson L, Johnsen AH. Identification and distribution of CCK-related peptides and mRNAs in the rainbow trout, Oncorhynchus mykiss.

Biochim Biophys Acta. 2001;1517(2):190–201.

93. Patton DF, Mistlberger RE. Circadian adaptations to meal timing: neuroendocrine mechanisms. Front Neurosci. 2013;7:185.

94. Besseau L, Vuilleumier R, Sauzet S, Boeuf G, Falcon J. Photoperiodic control of melatonin synthesis in fish pineal and retina. J Soc Biol. 2007;201(1):13–20.

95. Shi Q, Ando H, Coon SL, Sato S, Ban M, Urano A. Embryonic and post-embryonic expression of arylalkylamine N-acetyltransferase and melatonin receptor genes in the eye and brain of chum salmon (Oncorhynchus keta). Gen Comp Endocrinol. 2004;136(3):311–21.

96. Lima-Cabello E, Diaz-Casado ME, Guerrero JA, Otalora BB, Escames G, Lopez LC, Reiter RJ, Acuna-Castroviejo D. A review of the melatonin functions in zebrafish physiology. J Pineal Res. 2014;57(1):1–9.

97. Conde-Sieira M, Libran-Perez M, Lopez Patino MA, Soengas JL, Miguez JM. Melatonin treatment alters glucosensing capacity and mRNA expression levels of peptides related to food intake control in rainbow trout hypothalamus. Gen Comp Endocrinol. 2012;178(1):131–8.

98. Ceinos RM, Polakof S, Illamola AR, Soengas JL, Miguez JM. Food deprivation and refeeding effects on pineal indoles metabolism and melatonin synthesis in the rainbow trout Oncorhynchus mykiss. Gen Comp Endocrinol. 2008;156(2):410–7.

99. Orozco-Solis R, Matos RJ, Guzman-Quevedo O, Lopes de Souza S, Bihouee A, Houlgatte R, Manhaes de Castro R, Bolanos-Jimenez F. Nutritional programming in the rat is linked to long-lasting changes in nutrient sensing and energy homeostasis in the hypothalamus. PLoS One. 2010;5(10):e13537.

100. Yang BY, Chen TT. Identification of a new growth hormone family protein, somatolactin-like protein, in the rainbow trout (Oncorhyncus mykiss) pituitary gland. Endocrinology. 2003;144(3):850–7.

101. Vega-Rubin de Celis S, Rojas P, Gomez-Requeni P, Albalat A, Gutierrez J, Medale F, Kaushik SJ, Navarro I, Perez-Sanchez J. Nutritional assessment of somatolactin function in gilthead sea bream (Sparus aurata): concurrent changes in somatotropic axis and pancreatic hormones. Comp Biochem Physiol A Mol Integr Physiol. 2004;138(4):533–42.

102. Company R, Astola A, Pendon C, Valdivia MM, Perez-Sanchez J. Somatotropic regulation of fish growth and adiposity: growth hormone (GH) and somatolactin (SL) relationship. Comp Biochem Physiol Toxicol Pharmacol. 2001;130(4):435–45.

103. Ferrini F, Salio C, Lossi L, Merighi A. Ghrelin in central neurons. Curr Neuropharmacol. 2009;7(1):37–49.

104. Zigman JM, Jones JE, Lee CE, Saper CB, Elmquist JK. Expression of ghrelin receptor mRNA in the rat and the mouse brain. J Comp Neurol. 2006;494(3):528–48.

105. Sobrino Crespo C, Perianes Cachero A, Puebla Jimenez L, Barrios V, Arilla Ferreiro E. Peptides and food intake. Front Endocrinol. 2014;5:58.

106. Trombley S, Maugars G, Kling P, Bjornsson BT, Schmitz M. Effects of long-term restricted feeding on plasma leptin, hepatic leptin expression and leptin receptor expression in juvenile Atlantic salmon (Salmo salar L.). Gen Comp Endocrinol. 2012;175(1):92–9.

107. Jakobsson T, Treuter E, Gustafsson JA, Steffensen KR. Liver X receptor biology and pharmacology: new pathways, challenges and opportunities. Trends Pharmacol Sci. 2012;33(7):394–404.

108. Sodhi RK, Singh N. Liver X receptors: emerging therapeutic targets for Alzheimer's disease. Pharmacol Res. 2013;72:45–51.

109. Theofilopoulos S, Wang Y, Kitambi SS, Sacchetti P, Sousa KM, Bodin K, Kirk J, Salto C, Gustafsson M, Toledo EM, et al. Brain endogenous liver X receptor ligands selectively promote midbrain neurogenesis. Nat Chem Biol. 2013;9(2):126–33.

110. Niesler B. 5-HT(3) receptors: potential of individual isoforms for personalised therapy. Curr Opin Pharmacol. 2011;11(1):81–6.

111. Hayes MR, Covasa M. Dorsal hindbrain 5-HT3 receptors participate in control of meal size and mediate CCK-induced satiation. Brain Res. 2006;1103(1):99–107.

112. Perez Maceira JJ, Mancebo MJ, Aldegunde M. The involvement of 5-HT-like receptors in the regulation of food intake in rainbow trout (Oncorhynchus mykiss). Comp Biochem Physiol Toxicol Pharmacol. 2014;161:1–6.

113. Hoglund E, Sorensen C, Bakke MJ, Nilsson GE, Overli O. Attenuation of stress-induced anorexia in brown trout (Salmo trutta) by pre-treatment with dietary l-tryptophan. Br J Nutr. 2007;97(4):786–9.

114. Balon E. Types of feeding in the ontogeny of fishes and the life-history model. Environ Biol Fish. 1986;16(1–3):11–24.

115. Hara TJ, Zielinski B. Structural and functional development of the olfactory organ in teleosts. Trans Am Fish Soc. 1989;118(2):183–94.

116. Kasumyan AO, Døving KB. Taste preferences in fishes. Fish Fish. 2003;4(4):289–347.

117. Kortner TM, Gu J, Krogdahl A, Bakke AM. Transcriptional regulation of cholesterol and bile acid metabolism after dietary soyabean meal treatment in Atlantic salmon (Salmo salar L.). Br J Nutr. 2013;109(4):593–604.

118. Mullen E, Brown RM, Osborne TF, Shay NF. Soy isoflavones affect sterol regulatory element binding proteins (SREBPs) and SREBP-regulated genes in HepG2 cells. J Nutr. 2004;134(11):2942–7.PubMed

119. Kaushik SJ, Cravedi JP, Lalles JP, Sumpter J, Fauconneau B, Laroche M. Partial or total replacement of fish meal by soybean protein on growth, protein utilization, potential estrogenic or antigenic effects, cholesterolemia and flesh quality in rainbow trout, Oncorhynchus mykiss. Aquaculture. 1995;133(3–4):257–74.

120. Nishiyama T, Ogura K, Nakano H, Kaku T, Takahashi E, Ohkubo Y, Sekine K, Hiratsuka A, Kadota S, Watabe T. Sulfation of environmental estrogens by cytosolic human sulfotransferases. Drug Metab Pharmacokinet.

2002;17(3):221–8.

121. Chen Y, Huang C, Zhou T, Chen G. Genistein induction of human sulfotransferases in HepG2 and Caco-2 cells. Basic Clin Pharmacol Toxicol. 2008;103(6):553–9.

122. Dobly A, Martin SA, Blaney SC, Houlihan DF. Protein growth rate in rainbow trout (Oncorhynchus mykiss) is negatively correlated to liver 20S proteasome activity. Comp Biochem Physiol A Mol Integr Physiol. 2004;137(1):75–85.

123. Doucet A, Overall CM. Protease proteomics: revealing protease in vivo functions using systems biology approaches. Mol Aspects Med. 2008;29(5):339–58.

124. Kaushik S, Cuervo AM. Chaperone-mediated autophagy: a unique way to enter the lysosome world. Trends Cell Biol. 2012;22(8):407–17.

125. Schneider JL, Suh Y, Cuervo AM. Deficient chaperone-mediated autophagy in liver leads to metabolic dysregulation. Cell Metab. 2014;20(3):417–32.

126. Wang P, Heitman J. The cyclophilins. Genome Biol. 2005;6(7):226.

127. Kim J, Choi TG, Ding Y, Kim Y, Ha KS, Lee KH, Kang I, Ha J, Kaufman RJ, Lee J, et al. Overexpressed cyclophilin B suppresses apoptosis associated with ROS and Ca2+ homeostasis after ER stress. J Cell Sci. 2008;121(Pt 21):3636–48.

128. Swali A, McMullen S, Hayes H, Gambling L, McArdle HJ, Langley-Evans SC. Cell cycle regulation and cytoskeletal remodelling are critical processes in the nutritional programming of embryonic development. PLoS One. 2011;6(8):e23189.

129. Pittman K, Yúfera M, Pavlidis M, Geffen AJ, Koven W, Ribeiro L, Zambonino-Infante JL, Tandler A. Fantastically plastic: fish larvae equipped for a new world. Rev Aquac. 2013;5:S224–67.

130. Fang L, Liang XF, Zhou Y, Guo XZ, He Y, Yi TL, Liu LW, Yuan XC, Tao YX. Programming effects of high-carbohydrate feeding of larvae on adult glucose metabolism in zebrafish, Danio rerio. Br J Nutr. 2014;111(5):808–18.

131. Gong G, Xue M, Wang J, Wu X-f, Zheng Y-h, Han F, Liang X-f, Su X-o. The regulation of gluconeogenesis in the Siberian sturgeon (Acipenser baerii) affected later in life by a short-term high-glucose programming during early life. Aquaculture. 2015;436:127–36.

132. Salem M, Kenney PB, Rexroad CE, Yao J. Development of a 37 k high-density oligonucleotide microarray: a new tool for functional genome research in rainbow trout. J Fish Biol. 2008;72(9):2187–206.

133. Rescan PY, Montfort J, Fautrel A, Ralliere C, Lebret V. Gene expression profiling of the hyperplastic growth zones of the late trout embryo myotome using laser capture microdissection and microarray analysis. BMC Genomics. 2013;14:173.

134. Berthelot C, Brunet F, Chalopin D, Juanchich A, Bernard M, Noel B, Bento P, Da Silva C, Labadie K, Alberti A, et al. The rainbow trout genome provides novel insights into evolution after whole-genome duplication in vertebrates. Nat Commun. 2014;5:3657.

135. Eisen MB, Spellman PT, Brown PO, Botstein D. Cluster analysis and display of genome-wide expression patterns. Proc Natl Acad Sci U S A. 1998;95(25):14863–8.

136. Saldanha AJ. Java Treeview--extensible visualization of microarray data. Bioinformatics. 2004;20(17):3246–8.

137. da Huang W, Sherman BT, Lempicki RA. Bioinformatics enrichment tools: paths toward the comprehensive functional analysis of large gene lists. Nucleic Acids Res. 2009;37(1):1–13.

138. da Huang W, Sherman BT, Lempicki RA. Systematic and integrative analysis of large gene lists using DAVID bioinformatics resources. Nat Protoc. 2009;4(1):44–57.

139. Merico D, Isserlin R, Stueker O, Emili A, Bader GD. Enrichment map: a network-based method for gene-set enrichment visualization and interpretation. PLoS One. 2010;5(11):e13984.

140. Merico D, Isserlin R, Bader GD. Visualizing gene-set enrichment results using the Cytoscape plug-in enrichment map. Methods Mol Biol. 2011;781:257–77.

141. Cline MS, Smoot M, Cerami E, Kuchinsky A, Landys N, Workman C, Christmas R, Avila-Campilo I, Creech M, Gross B, et al. Integration of biological networks and gene expression data using Cytoscape. Nat Protoc. 2007;2(10):2366–82.

142. Smoot ME, Ono K, Ruscheinski J, Wang PL, Ideker T. Cytoscape 2.8: new features for data integration and network visualization. Bioinformatics. 2011;27(3):431–2.

143. Saito R, Smoot ME, Ono K, Ruscheinski J, Wang PL, Lotia S, Pico AR, Bader GD, Ideker T. A travel guide to Cytoscape plugins. Nat Methods. 2012;9(11):1069–76

CHAPTER 10

EVALUATION OF A CANOLA PROTEIN CONCENTRATE AS A REPLACEMENT FOR FISHMEAL AND POULTRY BY-PRODUCT MEAL IN A COMMERCIAL PRODUCTION DIET FOR ATLANTIC SALMON (SALMO SALAR)

Gary S Burr, William R Wolters, Frederic T Barrows and Alan W Donkin

BACKGROUND

Atlantic salmon (*Salmo salar*) is an important cultured carnivorous species that, in the past, has not tolerated high levels of most plant protein feed ingredients in their diet Gatlin et al. (2007). The inclusion of plant protein sources in aquafeeds is expanding due to the limited amount of fishmeal available for the production of animal feeds Glencross et al. (2005; Gatlin et al. 2007). Juvenile Senegalese sole (*Solea senegalensis*) fed diets that contain 5% fishmeal or less have been shown to have comparable growth to fish fed a 37% fishmeal diet when amino acids were supplemented Silva et al. (2009). For inclusion in the diets of rainbow trout (*Oncorhynchus mykiss*), many different plant protein sources have been examined, including plant protein meals and plant protein concentrates Lim and Webster (Lim et al. 2008). Canola protein concentrate (CPC), soybean meal and soy protein concentrates, pea protein concentrate, barley protein concentrate, rice protein concentrate, and wheat gluten meal have all been tested as fishmeal replacements with varying degrees of success Forster et al. (1999; Thiessen et al.2003; Barrows et al. 2007; Lim et al. 2008; Gaylord and Barrows 2009). Many of these same protein sources have been tested in Atlantic salmon (Lim et al. 2008). Atlantic salmon was shown to have slightly reduced, but not significantly lower, performance when fed diets that were fishmeal-free and contained wheat gluten, corn gluten, fish soluble, and crystalline amino acids (Espe et al. 2006). Atlantic salmon fed a diet that contained 12% fishmeal had slightly decreased growth but produced 2 kg of fish for every 1 kg of fishmeal consumed (Torstensen et al. 2008). Stickwater, the water-soluble fraction that is produced in making fishmeal, has been shown to stimulate the growth of Atlantic salmon when fed low fishmeal diets (10% of the diet) and high plant protein diets (62% of the diet) (Kousoulaki et al. 2009). Amino acid supplementation is often needed to maintain the growth performance of fish fed plant protein-based feeds (Espe et al. 2007; Lim et al. 2008; Gaylord and Barrows 2009). A major challenge is to increase the amount of plant or other non-animal protein in the diet of carnivorous fishes while maintaining acceptable growth, feed conversion, and production costs compared to traditional diets.

Citation: Gary S Burr, William R Wolters , Frederic T Barrows and Alan W Donkin, Evaluation of a canola protein concentrate as a replacement for fishmeal and poultry by-product meal in a commercial production diet for Atlantic salmon (Salmo salar), DOI: 10.1186/2008-6970-5-5

Recently, multiple studies have been reported on the digestibility of various plant feed ingredients by Atlantic salmon () (Storebakken et al. *Salmo salar*2000; Glencross et al. 2004a; Refstie et al. 2005; Aas et al. 2006; Aslaksen et al. 2007; Refstie et al. 2008). The protein digestibility was not significantly lower for the plant feed ingredients, except for extracted soybean meal, oat, canola, and sunflower. Canola meal contains a high level of carbohydrates and other antinutritional factors, such as phytic acid, tannins, sinapine, and goitrogenic compounds, lowering the nutrient availability of this feed ingredient to aquatic species (Kousoulaki et al. 2009). Further processing into a protein concentrate should remove more of the antinutritional factors and carbohydrates; however, significant amounts of phytic acid could remain (Burel and Kaushik 2008). In rainbow trout (*O. mykiss*), experimental diets that contained 20% fishmeal and 19.4% canola protein concentrate significantly lowered growth (Drew et al. 2007).

Both existing and new aquaculture operations are trying to increase the sustainability and reduce the environmental impacts by optimizing water use and reducing waste. Recirculating systems for cold water fish have undergone significant improvements within the last decade and can provide culture systems with reduced environmental impact. Such systems have greater control of the rearing environment, minimize water use, produce a concentrated and relatively small volume effluent, and can have improved biosecurity measures to control disease (Summerfelt et al. 2001).

The purpose of this study was to determine the effect of feeding multiple levels of canola protein concentrate in a commercial-type diet on the growth and nutrient efficiency of Atlantic salmon in a recirculating aquaculture system.

METHODS

The experimental protocols and methods used in the research at the NCWMAC were in compliance with Animal Welfare Act (9CFR) requirements and were approved by the location's Institutional Animal Care and Use Committee. A 38-week feeding trial was conducted at the National Cold Water Marine Aquaculture Center (Franklin, ME, USA) with healthy juvenile Atlantic salmon. A commercial diet (Signature Salmon 3.5 mm, Northeast Nutrition, Truro, Nova Scotia, Canada) was used and modified by the manufacturer to contain 10% or 20% canola protein concentrate (insoluble canola protein concentrate, CanPro, Arborfield Saskatchewan, Canada). The three diets were formulated to contain 45% to 46% protein and 21% to 23% lipid. Diets were formulated and prepared by the manufacturer to ensure practical relevance, and the proportional composition is shown in Table 1. The 10% CPC diet contained only 75% of the fish protein ingredients, and the 20% CPC diet contained only 57% of the fish protein ingredients compared to the commercial diet. Amino acids and lipids were balanced by the manufacturer. The feed was prepared before the trial and was stored in a cool dry room (temperature, 4°C to 12°C).

Table 1: Nutrient composition of canola protein concentrate and proportional ingredient composition of reference and experimental diets

	Canola protein concentrate	**0%**	**10%**	**20%**
		Canola protein	**Canola protein**	**Canola protein**
Ingredient[a] (g/kg)				
Canola protein concentrate		0	100	200
Plant protein ingredients (other than canola)		270.2	253.0	241.4
Animal protein ingredients		398.4	336.8	260.3
Fish protein ingredients		140.0	105.0	80.0
Fish oil and lecithin		165.2	177.9	189.7
Amino acids		4.4	5.5	6.8
Additives (nutritional)		18.3	18.3	18.3
Vitamin + trace mineral package		3.5	3.5	3.5
Analyzed levels (as fed basis)				
Dry matter (g/kg)	950.6	950.5	947.5	961.7
Protein (g/kg)	604.1	459.6	449.6	455.7
Fat (g/kg)	24.2	228.6	209.4	211.3
Ash (g/kg)	54.1	66.1	61.9	57.1
Calcium (g/kg)	6.0	15.3	13.8	10.6
Phosphorus (g/kg)	8.7	10.9	10.1	8.5
Sodium (g/kg)	6.2	3.0	3.1	3.1
Potassium (g/kg)	3.6	6.1	5.5	5.2
Magnesium (g/kg)	4.5	1.2	1.5	1.7
Manganese (ppm)	57.25	143.46	194.83	128.95
Copper (ppm)	11.52	16.03	11.97	14.36
Zinc (ppm)	79.07	113.08	104.99	131.19

[a]The basal diet contained poultry meal, wheat, fishmeal, fish oil, blood meal, crab meal, feather meal, CPSP special G fishmeal (fish hydrolysate from fresh whole fish or fresh offals; Sopropêche, Cedex, France), lecithin, betaine, XPC yeast culture (Diamond V, Cedar Rapids, Iowa, USA), choline chloride, amino acids, vitamin, and mineral premixes. The CPSP special G fishmeal was omitted from the CPC-containing diets.

Each experimental diet was randomly assigned to triplicate tanks (0.26 m^3 each) containing 20 Atlantic salmon (133.0 ± 7.8 g SD per fish, initial weight) at an initial stocking density of 10 kg/m^3. All tanks were connected to a common particulate (rotating drum filter) and a biological filter (fluidized sand) to maintain optimal water quality. The water in the system was brackish water (3%) and obtained from a ground source. Temperature (°C) and dissolved oxygen (mg/l) were continuously monitored with stationary probes. The temperature averaged to 10°C using well water and varied seasonally, ranging from 8°C in January to 13°C in July. Fish were subjected to a natural photoperiod (9 h light (L):15 h dark (D) initial to 15L: 9D final) maintained by artificial lighting controlled using a computer. Dissolved oxygen was maintained at or above saturation using an oxygen cone. The water quality of the system was monitored weekly and maintained within acceptable levels for Atlantic salmon.

The ArvoPro computer program (Arvo-tec, Huutokoski, Finland) was used to control the feed being dispensed using automatic feeders four times per day to the tanks, and the feed was adjusted daily with fish fed to apparent satiation. The computer program was developed from experimental growth models validated from commercial data for conditions and different genetic stocks (Ursin 1967; From and Rasmussen 1984; Ruohonen and Makinen 1992). The feed was slowly sinking. Fish were group weighed approximately every 4 weeks to confirm the projected weight (g) using the computer software. Throughout the trial, weight gain (g) was monitored. At the end of the trial, feed efficiency (weight gain/feed intake), protein efficiency ratio (g of weight gain/g of protein fed), and specific growth rate values were computed, and three fish were collected from each tank for subsequent measurement of whole-body composition according to established procedures (Webb et al. 2003).

RESULTS

Atlantic salmon fed the diet containing 20% canola protein concentrate showed the lowest weight gain compared to fish fed the diet containing 0% canola protein concentrate (p = 0.04; Figure 1; Table 2). These fish fed the 0% canola protein concentrate diet had the highest weight gain but were not statistically different from the fish fed the diet containing 10% canola protein concentrate (Figure 1; Table 2). Fish fed the fishmeal reference diet had the highest average weight throughout the study (Figure 1). Feed efficiency and protein efficiency ratio did not differ among the treatments (Table 2), indicating that the fish fed the reference diet consumed the most feed. Fish fed the higher fish protein diet had a significantly greater specific growth rate compared to fish fed the 20% canola concentrate diet (Table 2).

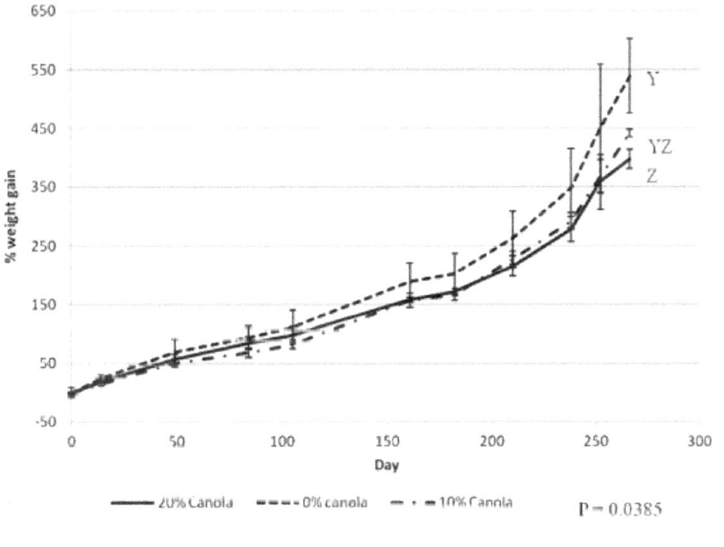

Figure 1: Percent weight gain over 38 weeks of Atlantic salmon fed the three diets. Values with common superscript letter do not differ significantly ($p > 0.05$).

Table 2: Feed efficiency (FE), protein efficiency ratio (PER), specific growth rate(SGR), and weight increase (%)

Diet	FE	PER	SGR	Weight increase (%)
0	0.80 ± 0.07	1.74 ± 0.29	0.73 ± 0.05 Y	539.6 ± 62.9 Y
10	0.78 ± 0.01	1.74 ± 0.05	0.67 ± 0.01 YZ	441.2 ± 6.8 YZ
20	0.68 ± 0.01	1.50 ± 0.04	0.64 ± 0.02 Z	397.3 ± 16.1 Z
ANOVA p	0.2170	0.2081	0.0268	0.0385

Values within columns that have a common uppercase letter do not differ significantly ($p > 0.05$). ANOVA, analysis of variance.

Proximate composition of the whole fish lipid, moisture, or ash concentrations did not vary among the treatments ($p = 0.05$; Table 3). Fish fed with the control diet had significantly lower crude protein concentrations compared to fish fed with the canola protein concentrate diets ($p = 0.04$; Table 3).

Table 3: Proximate composition of whole fish (g/kg of wet weight)

Diet	Protein	Lipid	Ash	Moisture
0	140.1 ± 15.9 b	107.5 ± 21.2	61.4 ± 36.3	716.0 ± 7.3
10	175.2 ±12.3 a	108.6 ± 15.0	57.9 ± 12.6	689.2 ± 2.8
20	167.7 ± 11.4 a	95.0 ± 3.6	67.8 ± 13.9	694.6 ± 3.0
ANOVA *p*	0.0401	0.5697	0.3961	0.1102

Values within columns that have a common lowercase letter do not differ significantly ($p > 0.05$).

DISCUSSION

Atlantic salmon fed the diet containing 20% canola protein concentrate had significantly lower weight gain and specific growth rate compared to fish fed diets containing higher levels of fish protein ingredients. However, feed efficiencies were similar, indicating that the increased weight gain of the fish on the fishmeal diet was the result of consuming more feed. There are several antinutritional factors associated with canola that could have influenced the consumption and resulting weight gain in this study, such as tannins, phytic acid, and glucosinolates (Kousoulaki et al. 2009). If antinutritional factors are indicated in future research, then each batch used in aquafeeds should be analyzed for antinutritional factors.

Several plant proteins have been evaluated in the diets of Atlantic salmon (Refstie et al. 2001; Refstie and Tiekstra 2003; Mundhein et al. 2004; Denstadli et al. 2007; Torstensen et al. 2008; Øverland et al. 2009). Potato protein concentrate (included up to 21% of the diet) and soy protein concentrate (included at 19.7% of the diet) did not inhibit the growth of Atlantic salmon compared to fishmeal-based diets (Refstie et al. 2001; Refstie and Tiekstra 2003). The growth of Atlantic salmon was inhibited by increasing plant proteins (full-fat soybean meal and corn gluten meal) up to 42% of the diet compared to a fishmeal-based diet (Kraugerud et al. 2007). Kraugerud et al. (2007) reported that a diet containing 46% soybean meal stunted the growth of Atlantic salmon (Torstensen et al. 2008). When included at 20% of the diet, canola protein concentrate also inhibited the growth of Atlantic salmon. These results indicate that a certain level of plant protein concentrates is tolerated in the diets of Atlantic salmon, but when the threshold is exceeded, either antinutritional factors or digestibility of the plant proteins hinders the intake or uptake of nutrients and limit growth. This would indicate that the digestibility of plant proteins need to be improved or the formulae require further modification to compensate. The antinutritional factors need to be mitigated or removed from the plant protein product prior to feeding (Denstadli et al. 2007).

Canola meal and canola protein concentrate have been investigated in a variety of finfish (Burel and Kaushik 2008). Protein apparent digestibility coefficients for rainbow trout fed rapeseed meal are similar to fishmeal (Mwachireya et al. 1999; Burel et al. 2000). However, apparent digestibility coefficients for energy and dry matter are significantly lower compared to a fishmeal reference diet (Mwachireya et al. 1999). There have been contradictory results obtained for rainbow trout fed a canola protein concentrate. Rainbow trout grew only slightly slower when fed a diet containing up to 42% canola protein concentrate compared to fish fed a fishmeal-based diet (Forster et al. 1999). Drew et al. (2007) also found that the growth of rainbow trout fed a canola protein concentrate was not affected when the levels were below 29%. However, Teskeredžić et al. (Teskeredzić et al. 1995) reported that rapeseed protein concentrate included in the diet above 26% did significantly lower the specific growth rate of rainbow trout. Hajen et al. (1993) reported that Chinook salmon (*Oncorhynchus tshawytscha*) fed a diet containing 30% rapeseed protein concentrate had the same growth rate as fish fed a fishmeal diet. The growth of the juvenile Chinook salmon fed diets that contained up to 23% canola meal was similar to a fishmeal reference diet (Higgs et al. 1983). Canola meal inclusion rates could be increased up to 30% when the diet was supplemented with 5 ppm 3, 5, 3'-triiodo-L-thyronine (Webster et al. 1997). In channel catfish, fish fed a diet containing 48% canola meal had significantly lower growth compared to fish fed a fishmeal diet or diets containing 12% or 36% canola meal (Webster et al. 1997). Sunshine bass (*Morone chrysops* × *Morone saxatilis*) fed a diet containing 20% canola meal did not have a significant difference in growth compared to fish fed a fishmeal diet (Webster et al. 2000). The processing method affected the digestibility of canola meals for red seabream, and protein digestibility was similar to soybean meal (Glencross et al. 2004a). The growth of red seabream fed diets that contained up to 60% canola meal was not adversely affected compared to a fishmeal reference diet (Glencross et al. 2004b). These results along with the results in the present study indicate that the source and processing of the canola meal/protein concentrate and species of fish affect growth performance. However, due to limited published studies, more research is needed to determine the upper limit of canola meal/protein concentrate that can be included in aquafeeds.

CONCLUSIONS

Overall, Atlantic salmon does not appear to be able to tolerate canola protein concentrate above 10% when included in their diet. Canola protein concentrate could be a candidate for a diet if included in the diet at low levels. This study did not address the antinutritional factors associated with canola meal/protein concentrate, but the viability of canola as a feed ingredient could be enhanced if the antinutritional factors could be removed or the effects could be mitigated.

ACKNOWLEDGEMENTS

We would also like to thank the staff of the National Cold Water Marine Aquaculture Center; Melissa Albert, Sharon Baron, Ryan Hastey, Davin O'Connell, and Amber Rankin for their technical assistance.

Disclaimer

The mention of trade names or commercial products in this publication is solely for the purpose of providing specific information and does not imply recommendation or endorsement by the US Department of Agriculture.

REFERENCES

1. Aas TS, Grisdale-Helland B, Terjesen BF, Helland SJ: Improved growth and nutrient utilization in Atlantic salmon (Salmo salar) fed diets containing a bacterial protein meal. Aquaculture 2006, 259: 365–376. 10.1016/j.aquaculture.2006.05.032
2. Aslaksen MA, Kraugerud OF, Penn M, Svihus B, Denstadli V, Jøgensen HY, Hillestad M, Krogdahl Å, Storebakken T: Screening of nutrient digestibilities and intestinal pathologies in Atlantic salmon, Salmo salar , fed diets with legumes, oilseeds, or cereals. Aquaculture 2007, 272: 541–555. 10.1016/j.aquaculture.2007.07.222
3. Barrows FT, Gaylord TG, Stone DAJ, Smith CE: Effect of protein source and nutrient density on growth efficiency, histology, and plasma amino acid concentration of rainbow trout (Oncorhynchus mykiss). Aquacult Res 2007, 38: 1747–1758.
4. Burel C, Kaushik SJ: Use of rapeseed/canola in diets of aquaculture species. InAlternative protein sources in aquaculture diets. Edited by: Lim CE, Webster CD, Lee C-S. New York: Haworth; 2008:343.
5. Burel C, Boujard T, Tulli F, Kaushik SJ: Digestibility of extruded peas, extruded lupin and rapeseed meal in rainbow trout (Oncorhynchus mykiss) and turbot (Psetta maxima). Aquaculture 2000, 188: 285–298. 10.1016/S0044-8486(00)00337-9
6. Denstadli V, Storebakken T, Svihus B, Skrede A: A comparison of online phytase pretreatment of vegetable feed ingredients and phytase coating in diets for Atlantic salmon (Salmo salar L.) reared in cold water. Aquaculture 2007, 269: 414–426. 10.1016/j.aquaculture.2007.02.033
7. Drew MD, Ogunkoya AE, Janz DM, Van Kessel AG: Dietary influence of replacing fish meal and oil with canola protein concentrate and vegetable oils on growth performance, fatty acid composition and organochlorine residues in rainbow trout (Oncorhynchus mykiss). Aquaculture 2007, 267: 260–268. 10.1016/j.aquaculture.2007.01.002
8. Espe M, Lemme A, Petri A, El-Mowafi A: Can Atlantic salmon (Salmo salar) grow on diets devoid of fish meal? Aquaculture 2006, 255: 255–262. 10.1016/j.aquaculture.2005.12.030
9. Espe M, Lemme A, Petri A, El-Mowafi A: Assessment of lysine requirement for maximal protein accretion in Atlantic salmon using plant protein diets. Aquaculture 2007, 263: 168–178. 10.1016/j.aquaculture.2006.10.018
10. Forster I, Higgs DA, Dosanjh BS, Rowshandeli M, Parr J: Potential for dietary phytase to improve the nutritive value of canola protein concentrate and decrease phosphorus output in rainbow trout (Oncorhynchus mykiss) held in 11°C fresh water. Aquaculture 1999, 179: 109–125. 10.1016/S0044-8486(99)00156-8
11. From J, Rasmussen G: A growth model, gastric evacuation, and body composition in rainbow trout, Salmo gairdneri Richardson, 1836. Dana 1984, 3: 61–139.
12. Gatlin DM III, Barrows FT, Brown P, Dabrowski K, Gaylord TG, Hardy RW, Herman E, Hu G, Krogdahl Å, Nelson R, Overturf K, Rust M, Sealey W, Skonberg D, Souza EJ, Stone D, Wilson R, Wurtele E: Expanding the utilization of sustainable plant products in aquafeeds: a review. Aquacult Res 2007, 38: 551–579. 10.1111/j.1365-2109.2007.01704.x
13. Gaylord TG, Barrows FT: Multiple amino acid supplementation to reduce dietary protein in plant-based rainbow trout, Oncorhynchus mykiss, feeds. Aquaculture2009, 287: 180–184. 10.1016/j.aquaculture.2008.10.037
14. Glencross B, Hawkins W, Curnow J: Nutritional assessment of Australian canola meals. I. Evaluation of canola oil extraction method and meal processing conditions on the digestible value of canola meals fed to the red seabream (Pagrus auratus, Paulin). Aquacult Res 2004, 35: 15–24. 10.1111/j.1365-2109.2004.00974.x
15. Glencross B, Hawkins W, Curnow J: Nutritional assessment of Australian canola meals. II. Evaluation of canola oil extraction method and meal processing conditions on the digestible value of canola meals fed to the red seabream (Pagrus auratus, Paulin). Aquacult Res 2004, 35: 25–34. 10.1111/j.1365-2109.2004.00975.x
16. Glencross B, Evans D, Dods K, Mcafferty P, Hawkins W, Maas R, Sipsas S: Evaluation of the digestible value of lupin and soybean protein concentrates and isolates when fed to rainbow trout, Oncorhynchus mykiss , using either stripping or settlement faecal collection methods. Aquaculture 2005, 245: 211–220. 10.1016/j.aquaculture.2004.11.033
17. Hajen WE, Beames RM, Higgs DA, Dosanjh BS: Digestibility of various feedstuffs by post-juvenile Chinook salmon (Oncorhynchus tshawytscha) in sea water.Aquaculture 1993, 113: 321–332.
18. Higgs DA, Fagerlund UHM, Mcbride JR, Plotnikoff MD, Dosanjh BS, Markert JR, Davidson J:Protein quality of Altex canola meal for juvenile Chinook salmon (Oncorhynchus tshawytscha) considering dietary protein and 3,5,3›-triiodo-L-thyronine content.Aquaculture 1983, 34: 213–238. 10.1016/0044-8486(83)90204-1
19. Kousoulaki K, Albrektsen S, Langmyhr E, Olsen HJ, Campbell P, Aksnes A: The water soluble fraction in fish meal (stickwater) stimulates growth in Atlantic salmon (Salmo salar L.) given high plant protein

diets. Aquaculture 2009, 289: 74–83. 10.1016/j.aquaculture.2008.12.034

20. Kraugerud OF, Penn M, Storebakken T, Refstie S, Krogdahl Å, Svihus B: Nutrient digestibilities and gut function in Atlantic salmon (Salmo salar) fed diets with cellulose or non-starch polysaccharides from soy. Aquaculture 2007, 273: 96–107. 10.1016/j.aquaculture.2007.09.013
21. Lim C, Webster CD, Lee C-S (Eds): Alternative protein sources in aquaculture diets. New York: Haworth; 2008. 571 pp 571 pp
22. Mundhein H, Aksnes A, Hope B: Growth, feed efficiency and digestibility in salmon (Salmo salar L.) fed different dietary proportions of vegetable protein sources in combination with two fish meals qualities. Aquaculture 2004, 237: 315–331. 10.1016/j.aquaculture.2004.03.011
23. Mwachireya SA, Beames RM, Higgs DA, Dosanjh BS: Digestibility of canola protein products derived from physical, enzymatic and chemical processing of commercial canola meal in rainbow trout Oncorhynchus mykiss (Walbum) held in fresh water. Aquacult Nutr 1999, 5: 73–82. 10.1046/j.1365-2095.1999.00089.x
24. Øverland M, Sørensen M, Storebakken T, Penn M, Krogdahl Å, Skrede A: Pea protein concentrate substituting fish meal or soybean meal in diets for Atlantic salmon (Salmo salar)—effect on growth performance, nutrient digestibility, carcass composition, gut health, and physical feed quality. Aquaculture 2009, 288: 305–311. 10.1016/j.aquaculture.2008.12.012
25. Refstie S, Tiekstra HAJ: Potato protein concentrate with low content of solanidine glycoalkaloids in diets for Atlantic salmon (Salmo salar). Aquaculture 2003, 216:283–298. 10.1016/S0044-8486(02)00434-9
26. Refstie S, Storebakken T, Baeverfjord G, Roem AJ: Long-term protein and lipid growth of Atlantic salmon (Salmo salar) fed diets with partial replacement of fish meal by soy protein products at medium or high lipid level. Aquaculture 2001, 193: 91–106. 10.1016/S0044-8486(00)00473-7
27. Refstie S, Sahlström S, Bråthen E, Baeverfjord G, Krogedal P: Lactic acid fermentation eliminates indigestible carbohydrates and antinutritional factors in soybean meal for Atlantic salmon (Salmo salar). Aquaculture 2005, 246: 331–345. 10.1016/j.aquaculture.2005.01.001
28. Refstie S, Glencross B, Landsverk T, Sørensen M, Lilleeng E, Hawkins W, Krogdahl Å:Digestive function and intestinal integrity in Atlantic salmon (Salmo salar) fed kernel meals and protein concentrates made from yellow or narrow-leafed lupins.Aquaculture 2008, 261: 1382–1395.
29. Ruohonen K, Makinen T: Validation of rainbow trout (Oncorhynchus mykiss) growth model under Finnish circumstances. Aquaculture 1992, 105: 353–362. 10.1016/0044-8486(92)90099-7
30. Silva JMG, Espe M, Conceição LEC, Dias J, Valente LMP: Senegalese sole juveniles (Solea senegalensis Kaup, 1858) grow equally well on diets devoid of fish meal provided the dietary amino acids are balanced. Aquaculture 2009, 296: 309–317. 10.1016/j.aquaculture.2009.08.031
31. Storebakken T, Shearer KD, Baeverfjord G, Nielsen BG, Åsgård T, Scott T, De Laporte A:Digestibility of macronutrients, energy and amino acids, absorption of elements and absence of intestinal enteritis in Atlantic salmon, Salmo salar , fed diets with wheat gluten. Aquaculture 2000, 184: 115–132. 10.1016/S0044-8486(99)00316-6
32. Summerfelt ST, Bebak-Williams J, Tsukuda S: Controlled Systems (2001) Water reuse and recirculation. In Fish hatchery management. 2nd edition. Edited by: Wedermeyer G. Bethesda: American Fisheries Society; 2001:285–395.
33. Teskeredzić Z, Higgs DA, Dosanjh BS, McBride JR, Hardy RW, Beames RM, Jones JD, Sirnell M, Vaarae T, Bridges RB: Assessment of undephytinized and dephytinized rapeseed protein concentrate as sources of dietary protein for juvenile rainbow trout (Oncorhynchus mykiss). Aquaculture 1995, 131: 261–277. 10.1016/0044-8486(94)00334-K
34. Thiessen DL, Campbell GL, Adelizi PD: Digestibility and growth performance of juvenile rainbow trout (Oncorhynchus mykiss) fed with pea and canola products.Aquacult Nutr 2003, 9: 67–75. 10.1046/j.1365-2095.2003.00203.x
35. Torstensen BE, Espe M, Sanden M, Stubhaug I, Waagbø R, Hemre G-I, Fontanillas R, Nordgarden U, Harvøy EM, Olsvik P, Berntssen MHG: Novel production of Atlantic salmon (Salmo salar) protein based on combined replacement of fish meal and fish oil with plant meal and vegetable oil blends. Aquaculture 2008, 285: 193–200. 10.1016/j.aquaculture.2008.08.025
36. Ursin E: A mathematical model of some aspects of fish growth, respiration, and mortality. J Fisheries Res Board Canada 1967, 24: 2355–2453. 14. 10.1139/f67-190
37. Webb KA Jr, Gatlin DM III: Effects of dietary protein level and form on production characteristics and ammonia excretion of red drum Sciaenops ocellatus .Aquaculture 2003, 225: 17–26. 10.1016/S0044-8486(03)00274-6
38. Webster CD, Tiu LG, Tidwell JH, Grizzle JM: Growth and body composition of channel catfish (Ictalurus punctatus) fed diets containing percentages of canola meal.Aquaculture 1997, 150: 103–112. 10.1016/S0044-8486(96)01471-8
39. Webster CD, Thompson KR, Morgan AM, Grisby EJ, Gannam AL: Use of hempseed meal, poultry by-product meal and canola meal in practical diets without fish meal for sunshine bass (Morone chrysops × M. saxatilis). Aquaculture 2000, 188: 299–309. 10.1016/S0044-8486(00)00338-0

CHAPTER 11 GLUTAMATE PROMOTES NUCLEOTIDE SYNTHESIS IN THE GUT AND IMPROVES AVAILABILITY OF SOYBEAN MEAL FEED IN RAINBOW TROUT

Chika Yoshida, Mayumi Mackawa, Makoto Bannai and Takeshi Yamamoto

Keywords: Fishmeal alternatives Feed ingredient Amino acids

BACKGROUND

Glutamate (Glu) is a nonessential amino acid which universally exists in living organism. It plays various roles directly or through conversions to other molecules in cell metabolisms and physiology. Glu serves as an energy source for enterocytes (Burrin and Stoll 2009; Blachier et al.2009), an excitatory neurotransmitter in the enteric nervous system (Kirchgessner 2001; Cartmell and Schoepp 2000; Zhang et al. 2013) and as a precursor for other biologically active molecules such as glutathione (Newsholme et al. 2003; Reeds et al. 2000). Supplementation of Glu into an enteral diet increases glutathione levels in jejunal mucosa of burned rat (Hasebe et al. 1999). The carbon skeleton of Glu are metabolized mainly into CO_2, lactate, or alanine and the nitrogen in Glu was utilized for the synthesis of other amino acids such as glutamine (Gln), proline, and arginine (Blachier et al. 2009; Newsholme et al. 2003; Reeds et al. 2000; Nakamura et al. 2013). Gln, in turn, has various functions in cellular metabolism such as a precursor for purine and pyrimidine nucleotide, NAD^+, and aminosugar or as an energy fuel (Blachier et al.2009; Newsholme et al. 2003). Although small intestine uses both Glu and Gln as energy fuel, the supply source of each molecule is different. While both dietary and arterial Gln are recruited into intestinal cells. On the other hand, almost all Glu utilized in the gut comes from the lumen (Windmueller and Spaeth 1980; Wu 1998). Glu, as a part of Glutathione synthesized at intestinal mucosa also derived largely from enteral Glu rather than arterial Glu (Reeds et al.2000). In piglet guts, it is shown that 95 % of the dietary Glu presented to the intestinal mucosa is metabolized in the intestinal cell, and that Glu is the largest resource of energy (Reeds et al.2000). Although Glu is nonessential amino acid which can be synthesized in the animal body, these studies indicate that utilization of dietary Glu in the gut has an important and particular role in the context of gut and systemic metabolism.

Citation: Chika Yoshida, Mayumi Maekawa, Makoto Bannai and Takeshi Yamamoto, Glutamate promotes nucleotide synthesis in the gut and improves availability of soybean meal feed in rainbow trout, DOI: 10.1186/s40064-016-2634-2

Motivated by the studies on these various functions in cell metabolism, several approaches have been made to utilize Glu as a solution for physiological problems. The most investigated application is for post-weaning piglet, since the intestinal tract of weaning period is not fully developed, and its digestive and absorption capability is poor (Berseth 1996). Besides, it is vulnerable to mechanical damage and infectious disease (Lallès et al. 2007). Thus, weaning piglet often goes into growth failure and is susceptible to diarrhea (Pluske et al. 1997). Supplementation of Glu has been shown to improve morphological abnormality in the small intestine, increase weight gain and feed efficiency, and reduce the occurrence of diarrhea (Rezaei et al. 2013; Wu et al. 2012). Glu increases gene expression of proliferating cell nuclear antigen (PCNA) in jejunum, indicating that it up-regulates cell proliferation (Wu et al. 2012). It is also shown that mixture of Glu and Gln has positive effect on weight gain and increases intestinal villus height in fish: tilapia (Da Silva et al. 2010). However, the primary effect of Glu on the gut health is poorly investigated so far.

In the aquaculture field, reduction of fishmeal in the feed has become inevitable issue because of the growth of aquaculture production and the rising market of fishmeal. Soybean meal (SBM) is one of the promising protein source for fish feed in the future. However, SBM is reported to have some anti-nutritional factors and suppress fish growth (Chikwati et al. 2012; Iwashita et al. 2008; Yamamoto et al. 2010). Among the negative impacts of SBM, tissue degeneration in the distal intestine is one of the most prominent abnormalities in salmonid fish (Suzuki and Yamamoto 2004; Buttle et al. 2001; Krogdahl et al. 2003; Burrells et al. 1999). Since cell turnover should be enhanced in degenerated tissue (Sanden et al. 2005), the energy consumption and biosynthesis of nucleic acids are estimated to be up-regulated. Thus, supplementation of Glu, which is main energy source in the intestine and provides nucleotide precursor thorough the conversion to Gln, to the SBM feed is expected to help restoration from the intestinal damage and contribute to growth promotion.

The aim of this study is to demonstrate the acute effect of oral Glu administration on cell proliferation in the gut (experiment 1) and then examine the chronic effect of Glu supplementation in the SBM feed on growth performance (experiment 2). In the experiment 1, we investigated the effect of single oral Glu administration on nucleotide-biosynthesis related gene expression and quantified cell proliferation by histological analysis. In the experiment 2, we investigated the growth performance, in addition to alteration of basal gene expression and cell proliferation in the gut after 8 weeks of Glu-supplemented feed.

METHODS

Fish Husbandry

Juvenile rainbow trout *Oncorhynchus mykiss* were purchased at Shiga Prefectual Samegai Trout Farm (Maibara, Shiga, Japan). The fish were fed a commercial trout feed (Nippon Formula Feed, Kanagawa, Japan) supplemented with 5 % Pollock oil twice a day to apparent satiation for 1 month. Then, fish were anesthetized in 100 mg/L ethyl 3-aminobenzoate methanosulfonate (Tricaine, Sigma Aldrich, St. Louis, USA) and weighed individually after 48 h-fasting. The fish were sorted into a uniform size (25.3, 9.53 g mean body weight (BW) in experiment 1, 2 respectively) and kept for 10 days for naturalization.

In experiment 1, the fish were anesthetized by Tricaine and 50 or 500 mg kg/BW mL Glu (suspended in 4 % carboxymethyl cellulose) were administered per os. Five fish from each group were additionally administered 50 mg kg/BW of bromo-deoxy-uridine (BrdU, Becton, Dickinson and Company, California, USA) intraperitoneally. Before or 30, 120, and 360 min after Glu administration, fish without BrdU injection were anesthetized again by Tricaine (6 fish for each time) and blood was collected using an ethylenediaminetetraacetic acid (EDTA) coated syringe for determination of plasma amino acids concentrations. Then, fish were sacrificed. The proximal and distal intestines were dissected out and immersed into RNA later (Thermo Fisher Scientific, Massachusetts, USA) to preserve RNA. The fish administered BrdU were sacrificed 24 h after the administration. Proximal and distal intestines were dissected and were immersed in 10 % phosphate buffered formarin (pH 7.4, Wako Pure Chemical Industries, Osaka, Japan) to fix the tissues.

In experiment 2, 52–53 fish per group were kept in 60 L volume tanks with water supply of 3 L/min. The water temperature was 15.6 ± 0.3 °C throughout the experiment. The fish were acclimatized to the experimental conditions and SBM-based diet for 10 days before the experiment. Then the experimental diet was fed by hand to apparent satiation twice a day, 6 days a week for 8 weeks. The fish were fasted for 48 h before the termination of the experiment and 8 fish from each tank were administered 50 mg kg/BW of BrdU intraperitoneally under anesthesia 24 h before the termination of the experiment. At the time of terminal BW measurement, 8 fish (without BrdU injection) from each tank were taken and blood was collected. Then the fish were sacrificed and the liver and gallbladder were dissected out and weighed. Subsequently, proximal and distal intestines were collected. The fish which were injected BrdU were sacrificed at the same time and intestines were collected.

Experimental Diet

In experiment 2, four iso-nitrogenous diets were prepared (Table 1). Chilean mackerel meal was used as the main protein source in the positive control diet (diet FM). In the three non-fish meal diets, SBM at a level of 49.7 %, together with corn gluten meal at a level of 17.11 % were included as the main protein sources (SBM, Glu 1 %, Glu 2 %). Diet Glu 1 %, Glu 2 % were supplemented crystalline Glu 1 %, 2 % wet weight respectively. Gelatin was used to balance the proportion of the ingredients among non-fish meal diets. To the non-fish meal diets, methionine and lysine were supplemented to simulate the digestible contents of diet FM (Yamamoto et al. 2002). Level of fish oil and soybean meal oil were adjusted to provide a diet with lipids from fish and those from vegetables being at ratio of 2:1. The ingredients were mixed well and made into dry pellets using a laboratory pellet mill (California Pellet Mill, San Francisco, CA, USA). The pellets were dried for 4 h at 60 °C and stored at −20 °C until use.

Table 1: Composition and nutritional contents of the experimental diet for rainbow trout in experiment 2

Ingredients (g/kg feed)	FM	SBM	Glu 1 %	Glu 2 %
Fish meal	480	0	0	0
Soybean meal	40	497	497	497
Corn gluten meal	20	171.1	171.1	171.1
Wheat flour	180	60	60	60
Gelatin	0	20	10	0
Fish oil	55.4	100.1	100.2	100.4
Soybean oil	28.3	20.4	20.8	21.1
α-Starch	60	28	28	28
Methionine	0	4.2	4.3	4.4
Lysine-HCl	0	12.2	12.7	13.3
Glutamate	0	0	10	20
Betaine	0	5	5	5
Vitamin mix	5	5	5	5
Choline chloride	2.5	2.5	2.5	2.5
Mineral mix	30	50	50	50
Cellulose	98.8	24.5	23.4	22.2
Total	1000	1000	1000	1000
Analytical contents (wet basis, %)				
Aspartic acid + asparagine	3.61	3.45	3.39	3.35
Threonine	1.67	1.37	1.32	1.32
Serine	1.68	1.93	1.85	1.83
Glutamic acid + glutamine	5.90	7.17	8.10	8.94
Glycine	2.34	1.84	1.59	1.34
Alanine	2.46	2.19	2.08	2.04
Valine	1.91	1.61	1.62	1.61
Isoleucine	1.63	1.50	1.53	1.51
Leucine	3.07	3.72	3.67	3.68
Tyrosine	1.31	1.49	1.47	1.49
Phenylalanine	1.66	1.95	1.95	1.92
Histidine	1.53	0.90	0.87	0.86
Lysine	2.86	2.65	2.68	2.67
Arginine	2.39	2.25	2.16	2.10
Proline	1.87	2.55	2.38	2.23
Cystine	0.40	0.54	0.53	0.55
Methionine	1.04	1.03	1.04	1.05
Tryptophan	0.477	0.378	0.385	0.381
Crude protein (%)	41.7	39.0	39.0	38.3
Moisture (%)	6.5	8.8	7.8	8.6

Plasma Amino Acid Analysis and Histological Analysis

For plasma amino acid analysis, the collected blood was centrifuged to obtain plasma (4 °C, 20 min, 2000*g*) and kept at −80 °C until analysis. To prepare sample for analysis, the plasma was mixed with 2 volumes of 5 % (w/w) trichloroacetic acid, and centrifuged immediately (4 °C, 20 min, 8000*g*) to remove precipitated protein. All samples were kept at 4 °C during all steps to minimize chemical reactions of thiol metabolites. The amino acid concentrations were measured by an automatic amino acid analyzer (L-8800; Hitachi, Tokyo, Japan). Briefly, amino acids, separated by cation exchange chromatography, were detected spectrophotometrically after post column reaction with ninhydrin reagent.

For histological analysis, intestine samples were fixed in 10 % phosphate-buffered formalin (pH 7.4). The fixed tissues were dehydrated in ethanol, equilibrated in xylene and embedded in paraffin according to standard histological techniques. For morphological observation, 4 μm of sections were cut out and stained with hematoxylin and eosine. Microvillus thickness was measured using calculating software (Sensiv Measure; Mitani Corporation, Tokyo, Japan) under microscope as the two-point distance of villus root. The microvillus thickness for each fish was calculated as the average of 5 microvilli per fish. For immunohistochemical detection of BrdU, 4 μm sections were cut out and stained using BrdU in situ detection kit (Becton, Dickinson and Company) following the manufacturer's instructions. The number of BrdU-positive cells per microvillus for each fish was counted and calculated as the average of BrdU positive cells in 5 microvilli per fish.

Quantitative Real-time PCR

Total RNA was extracted from the homogenized intestine using an RNeasy kit (Qiagen, California, USA) following the manufacturer's instructions. Equal amounts (2 ng) of RNA were reverse-transcribed using Omniscript RT kit (Qiagen) as per the manufacturer's instructions. Primers for quantitative RT-PCR were designed in reference to the sequence in Genbank (https://www.ncbi.nlm.nih.gov/genbank/) using the primer design software Primer3 (http://frodo.wi.mit.edu/). As for the genes with no reference on Genbank, expressed sequence tags (EST) similar to the known base sequence of teleost were searched from EST database by BLAST search (offered by The Computational Biology and Functional Genomics Laboratory of Dana-Farber Cancer institute and Harvard School of public Health (http://compbio.dfci.harvard.edu/tgi/cgi-bin/tgi/Blast/index.cgi). Quantitative RT-PCR was performed on ABI Prism® 7900 Sequence Detection System (Life Technologies, California, USA), with SYBR select PCR master mix (Life Technologies, California, USA) following the manufacturer's instructions. The PCR protocol consisted of one 10-min denaturation cycle at 95 °C followed by 40 cycles of denaturation at 95 °C for 15 s and finally annealing/extension at 60 °C for 1 min. Standard curves for each gene were obtained. All quantitative RT-PCR data were expressed as relative mRNA levels after normalizing to 60S acidic ribosomal protein P0 (arbp). The primer pair sequences and the reference base sequence for BLAST search are listed in Tables 2 and 3, respectively.

Table 2: Primer sequences used for qPCR analysis

Gene symbol	Gene name	Sequence 5′–3′	Product size (bp)	Gene ID[a]
arbp	60S acidic ribosomal protein P0	F: GCTGTAAAAGCGATCCTTCG	135	TC209405
		R: ATTGTCTGCACCCACAATGA		
cps2	Carbamoyl-phosphate synthetase II	F: ACATCCAGTACGCCCTCAAC	181	NCBI Ref. Seq.: AF014386.1
		R: GTACTTGGCACCGATGGACT		
		R: AAGTCCGCCTGTCTGAGTGT		
pfas	Phosphoribosylformylglycinamidine synthase	F: TGACGATACGCTGAGTCTGG	163	TC176935
		R: GGCCTTGATCCCTACACTGA		
ppat	Phosphoribosyl pyrophosphate amidotransferase	F: ACATCCTACTCGCTGCTGGT	195	TC216500
		R: GTACTTGGCACCGATGGACT		

[a]Gene IDs with TC are EST of DFCI R.trout Gene Index

Table 3: Reference sequences used for EST search

Gene ID (gene symbol)	Reference sequence (upper: protein, lower: mRNA)	Identity
TC176935 (pfas)	Phosphoribosylformylglycinamidine synthase [*Danio rerio*] (NP_001038667.1)	267/360 (74 %)
	Danio rerio phosphoribosylformylglycinamidine synthase (pfas), mRNA (NM_001045202.1)	746/1036 (72 %)
TC216500 (ppat)	Amidophosphoribosyltransferase [*Salmo salar*](NP_001133427.1)	424/438 (97 %)
	Salmo salar Amidophosphoribosyltransferase (pur1), mRNA	1446/1532 (94 %)

Statistics

The data are presented as the mean ± SEM. Data were analyzed by one-way ANOVA to determine main effects of treatment. In case normality test failed, one-way ANOVA on ranks was used. In experiment 1, Dunnett's post hoc test was used to identify statistical differences between initial group and treatment groups. In experiment 2, Tukey post hoc test was used for all multiple comparison procedure. Probability level of <0.05 was considered as significant.

RESULTS

Experiment 1

Plasma Amino Acid Concentration

The free amino acid concentrations in plasma are shown in Table 4. Plasma Glu concentrations significantly increased at 30, 120, and 360 min after 500 mg/kg Glu administration and they returned to basal level 24 h after administration. Plasma Glu concentrations in the group administered 50 mg/kg were not significantly different from the basal level. Plasma concentrations of neither Gln, alanine (Ala) nor aspartate (Asp), the amino acids which are directly generated from Glu in intestine, were significantly changed to the basal level throughout the experimental period in both 50 and 500 mg/kg administered fish (Table 4).

Table 4: Plasma amino acid concentrations by oral Glu administration (nmol/mL)

	0 h	**30 min**		**120 min**	
		50 mg/kg	**500 mg/kg**	**50 mg/kg**	**500 mg/kg**
Thr	170.4 ± 47.3	176.2 ± 37.9	177.0 ± 25.9	163.2 ± 32.6	223.2 ± 57.7
Ser	85.6 ± 23.9	117.9 ± 41.9	147.2 ± 43.3	96.8 ± 28.9	123.8 ± 18.8
Gly	290.5 ± 109.0	321.0 ± 161.9	410.6 ± 59.2	294.3 ± 104.4	390.3 ± 104.5
Ala	331.8 ± 82.0	428.2 ± 112.9	531.8 ± 82.8	358.6 ± 76.7	534.6 ± 97.4
Gln	221.9 ± 73.9	241.6 ± 80.1	288.7 ± 31.1	193.1 ± 55.7	282.2 ± 60.4
Glu	45.6 ± 5.1	92.3 ± 43.9	115.0 ± 38.0*	62.7 ± 22.9	406.6 ± 354.7*
Asn	73.8 ± 12.5	75.0 ± 21.6	93.9 ± 14.6	75.9 ± 17.1	77.3 ± 23.4
Asp	17.9 ± 6.1	17.1 ± 9.1	16.9 ± 8.0	14.2 ± 5.3	29.8 ± 10.9
Arg	88.3 ± 10.3	94.5 ± 30.9	92.6 ± 7.7	87.1 ± 3.9	95.4 ± 27.4
Lys	333.0 ± 84.5	452.5 ± 385.5	244.0 ± 28.7	282.8 ± 88.7	265.9 ± 101.2
Val	388.3 ± 131.4	512.3 ± 276.6	287.9 ± 21.3	372.1 ± 188.1	322.0 ± 55.7
Ile	165.6 ± 76.1	236.8 ± 145.7	120.3 ± 9.4	162.6 ± 106.6	141.4 ± 32.5
Leu	275.8 ± 109.4	371.2 ± 203.8	203.7 ± 17.4	266.7 ± 151.0	239.6 ± 54.2
Tyr	49.1 ± 8.2	69.6 ± 10.4*	72.9 ± 7.8*	64.4 ± 9.2	69.9 ± 10.3*
Phe	90.7 ± 15.1	115.7 ± 11.9*	111.8 ± 11.5	105.6 ± 10.4	104.9 ± 17.8
Cys	11.4 ± 2.3	13.0 ± 3.6	15.5 ± 2.0	12.3 ± 1.9	13.9 ± 1.9
Met	72.4 ± 8.5	82.2 ± 20.5	96.4 ± 9.5	80.6 ± 14.0	86.1 ± 6.7
Trp	49.0 ± 6.6	62.0 ± 3.7	64.2 ± 4.1*	54.1 ± 13.4	64.8 ± 7.6*
His	98.9 ± 21.7	121.4 ± 37.2	169.3 ± 19.3*	119.0 ± 34.0	163.4 ± 36.3
Pro	32.3 ± 7.8	43.6 ± 4.9	41.7 ± 3.4	34.2 ± 5.1	45.7 ± 9.4

	360 min		**1440 min**	
	50 mg/kg	**500 mg/kg**	**50 mg/kg**	**500 mg/kg**
Thr	184.6 ± 22.7	184.4 ± 38.5	291.7 ± 119.2	214.8 ± 74.0
Ser	113.6 ± 26.9	104.4 ± 48.1	146.8 ± 65.3	108.0 ± 46.7
Gly	360.9 ± 103.0	369.6 ± 143.5	503.5 ± 323.1	377.4 ± 303.5
Ala	407.1 ± 74.8	667.9 ± 250.8	328.2 ± 84.5	386.5 ± 216.3
Gln	215.4 ± 41.2	259.3 ± 95.6	302.4 ± 135.1	262.1 ± 150.0
Glu	89.3 ± 40.8	2309.0 ± 1729.4*	66.2 ± 16.5	61.3 ± 22.4
Asn	77.0 ± 14.8	77.4 ± 17.9	107.9 ± 33.2	85.2 ± 35.8
Asp	20.8 ± 7.9	86.0 ± 60.7	19.7 ± 5.9	14.2 ± 4.3
Arg	125.8 ± 23.6	89.7 ± 20.4	156.9 ± 55.4	91.7 ± 22.5
Lys	385.9 ± 145.9	287.9 ± 151.8	518.5 ± 152.3	308.7 ± 77.3
Val	405.4 ± 124.2	350.7 ± 108.2	499.9 ± 117.6	513.1 ± 144.9
Ile	183.5 ± 58.1	156.8 ± 45.7	217.1 ± 55.9	207.3 ± 65.1
Leu	308.6 ± 95.3	263.0 ± 75.7	348.2 ± 73.2	341.9 ± 103.5
Tyr	87.2 ± 15.3*	71.6 ± 7.4*	72.1 ± 13.3*	57.7 ± 6.1
Phe	120.3 ± 19.0*	104.6 ± 5.7	124.9 ± 15.5*	105.6 ± 15.1
Cys	14.0 ± 2.8	17.7 ± 2.9*	18.9 ± 3.5	18.3 ± 4.9
Met	88.7 ± 6.4	83.3 ± 11.1	98.0 ± 21.4	86.2 ± 16.1
Trp	65.1 ± 6.9*	60.0 ± 12.8	64.3 ± 10.7*	59.4 ± 5.7
His	130.1 ± 29.0	121.0 ± 24.4	252.5 ± 166.7*	126.4 ± 61.3
Pro	69.6 ± 16.0*	50.8 ± 15.0	61.7 ± 27.8*	43.1 ± 9.0

Variation in plasma amino acid concentrations after oral Glu administration. The blood were collected before (0 min; initial), 30, 120, 360 and 1440 (24 h) min after administration of 50, 500 mg/kg Glu

* $p < 0.05$ compared to initial group

Gene Expression

The expressions of genes related to nucleic acid synthesis in proximal and distal intestine are shown in Figs. 1 and 2 respectively. The expression of carbamoyl-phosphate synthetase2 (cps2), phosphoribosyl pyrophosphate amidotransferase (ppat), phosphoribosylformylglycinamidine synthase (pfas) were significantly increased 30 min after 50 mg/kg Glu administration compared to the initial group, returning to basal level after 120 min (Fig. 1). The expressions of those genes were not significantly different in the proximal intestine of 500 mg/kg Glu group and distal intestine of both 50 and 500 mg/kg Glu group compared to the initial group (Figs. 1,2).

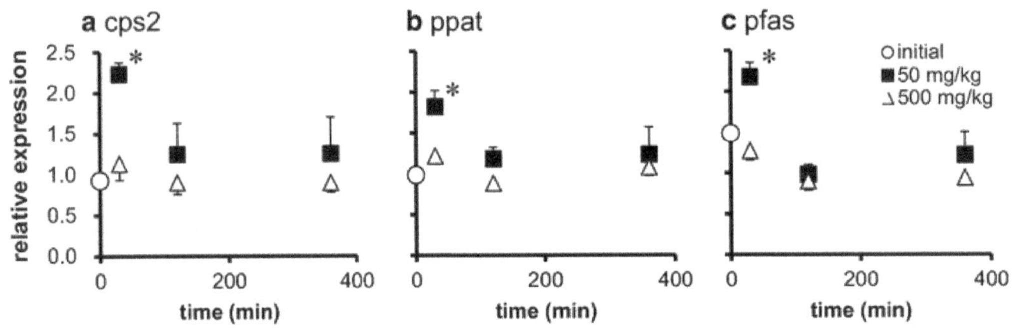

Figure 1: Alterations of gene expressions in proximal intestine by oral Glu intake. Relative expression of genes related to nucleic acid synthesis in proximal intestine. **a** cps2: carbamoyl-phosphate synthetase, **b** ppat: phosphoribosyl pyrophosphate amidotransferase, **c** pfas: phosphoribosylformylglycinamidine synthase. The tissues were collected before (initial), 30, 120, and 360 min after administration of 50 or 500 mg/kg Glu. The expression of the genes were estimated by quantitative RT-PCR and normalized by arbp. All values are expressed as the mean ± SEM (n = 6). *$p < 0.05$ as compared to initial group

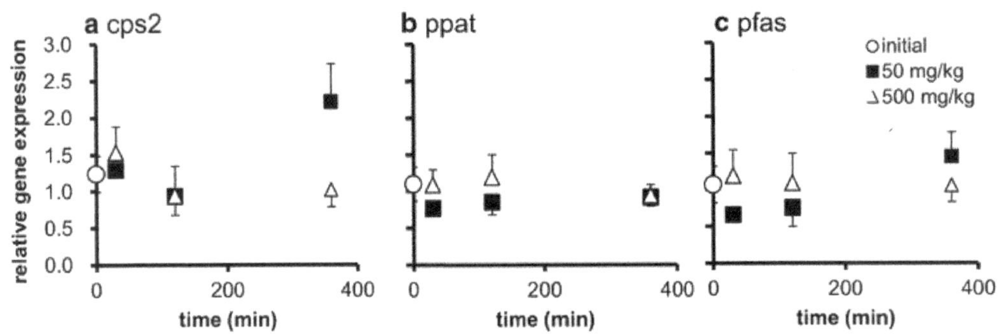

Figure 2 Alterations of gene expressions in distal intestine by oral Glu intake. Relative expression of genes related to nucleic acid synthesis in distal intestine. **a** cps2: carbamoyl-phosphate synthetase, **b** ppat: phosphoribosyl pyrophosphate amidotransferase, **c** pfas: phosphoribosylformylglycinamidine synthase. The tissues were collected before (initial), 30, 120, and 360 min after administration of 50, 500 mg/kg Glu p.o. The expression of the genes were estimated by quantitative RT-PCR and normalized by arbp. All values are expressed as the mean ± SEM (n = 6)

Histological Analysis

Figure 3 shows the number of BrdU positive cells in proximal and distal intestine. In proximal intestine of fish administered 500 mg/kg Glu, the number of BrdU positive cells significantly increased compared to 0 mg/kg group. In distal intestine there was no significant difference between each groups.

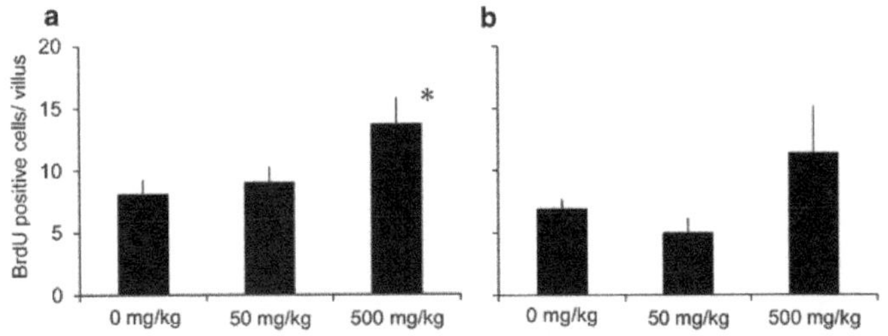

Figure 3 Effects of Glu intake on cell proliferation in the intestine. The number of BrdU positive cells per microvillus in **a** proximal intestine, **b** distal intestine. The tissues were collected 24 h after 0, 50, 500 mg/kg Glu (p.o.) and BrdU (i.p.) administration. Data are shown as the average of five villi from each fish. *$p < 0.05$ as compared to 0 mg/kg group

Experiment 2

Growth and Feed Performance

Weekly feed intake is shown in Fig. 4 and results of the growth and feed performance are shown in Table 5. The fish ate more FM-based feed than SBM-based feeds throughout the experimental period. Among three SBM-based feeds, there was no big difference in feed intake (Fig. 4). The fish fed SBM and Glu 1 % showed significantly lower final body weight compared to the fish fed FM. There was no significant difference between FM and Glu 2 % groups in the final body weight. Hepato-somatic index (HIS) and gallbladder-somatic index (GBSI) of the fish fed SBM, Glu 1 % and Glu 2 % were significantly lower compared to the fish fed FM.

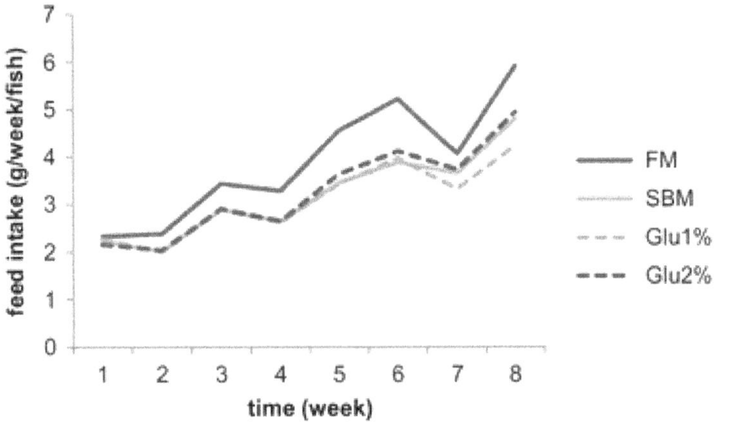

Figure 4: Weekly feed intake. The experimental diets were fed by hand to apparent satiation twice a day, 6 days a week for 8 weeks. Data are shown as week total intake per fish

Table 5: Fish growth and performance of four experimental feeds

	FM	SBM	Glu 1 %	Glu 2 %
Initial BW (g)	9.56 ± 0.51	9.73 ± 0.56	9.60 ± 0.49	9.70 ± 0.52
Final BW (g)	57.80 ± 5.43	41.44 ± 5.13*	40.24 ± 4.41*	47.22 ± 3.84
SGR (%/day)	2.57	2.07	2.05	2.26
Weight gain (%)	504.37	325.98	318.93	387.03
Feed efficiency ratio	1.03	0.85	0.85	0.87
DFC (%BW/day)	2.40	2.53	2.48	2.54
Mortality (%)	0.00	0.00	2.00	0.00
HSI (%)	1.29 ± 0.13	0.98 ± 0.18*	0.97 ± 0.09*	1.05 ± 0.17*
GBSI (%)	0.30 ± 0.06	0.17 ± 0.03*	0.17 ± 0.03*	0.21 ± 0.06*

The fish were challenged to 8 weeks of experiment. The final body weight and tissue weight were measured after fasting for the last 48 h of the experiment. The initial body weight was measured without fasting

BW body weight, *SGR* specific growth rate, *DFC* daily feed consumption, *HSI* hepato-somatic index (liver weight (g)/body weight (g)) × 100, *GBSI* gallbladder-somatic index (gallbladder weight (g)/body weight (g))

* $p < 0.05$ compared to FM group

Plasma Amino Acid Concentration

The free amino acid concentrations in plasma are shown in Table 6. Plasma Glu concentrations were not significantly different between groups. Plasma concentrations of Gln were significantly lower in Glu 1 % and Glu 2 % group compared to FM and SBM group.

Table 6: Variation in plasma Glu and Gln concentrations by 8 weeks of Glu supplementation in feed

	FM	SBM	Glu 1 %	Glu 2 %
Glu	2.92 ± 0.43	3.27 ± 0.35	3.82 ± 0.38	3.72 ± 0.44
Gln	9.33 ± 0.58	10.17 ± 0.86	3.56 ± 0.25	3.57 ± 0.43
Thr	162.2 ± 15.7	181.6 ± 10.4	179.0 ± 8.5	168.6 ± 21.7
Ser	180.0 ± 9.0	187.4 ± 20.5	163.0 ± 7.2	147.8 ± 13.2
Gly	1004.7 ± 69.0^{a}	1169.7 ± 83.4^{a}	695.1 ± 40.3^{b}	925.1 ± 69.1^{ab}
Ala	382.5 ± 22.1^{a}	599.2 ± 35.9^{b}	557.4 ± 55.5^{b}	673.0 ± 93.6^{b}
Gln	279.8 ± 17.4^{a}	305.2 ± 25.9^{a}	106.9 ± 7.4^{b}	107.2 ± 12.8^{b}
Glu	87.6 ± 12.9	98.0 ± 10.6	114.7 ± 11.3	111.5 ± 13.1
Asn	120.1 ± 12.1^{ab}	115.2 ± 17.5^{a}	157.2 ± 9.3^{ab}	178.2 ± 20.3^{b}
Asp	22.2 ± 1.7^{a}	68.5 ± 17.7^{b}	49.9 ± 9.6^{ab}	62.4 ± 7.3^{b}

Arg	103.0 ± 8.7	120.5 ± 11.8	126.7 ± 10.2	137.4 ± 14.7
Lys	220.3 ± 20.7	308.7 ± 47.4	286.2 ± 23.4	256.4 ± 28.9
Val	199.6 ± 7.4^{a}	291.4 ± 13.9^{b}	274.4 ± 15.1^{b}	264.7 ± 26.6^{ab}
Ile	82.4 ± 3.9^{a}	116.0 ± 5.2^{b}	117.2 ± 6.3^{b}	117.9 ± 14.9^{b}
Leu	146.5 ± 6.5^{a}	236.2 ± 10.7^{b}	257.6 ± 17.9^{b}	282.6 ± 42.0^{b}
Tyr	53.5 ± 2.6^{a}	60.9 ± 4.8^{ab}	73.3 ± 4.8^{b}	94.5 ± 10.5^{b}
Phe	108.5 ± 3.7	120.7 ± 5.4	116.7 ± 6.8	146.3 ± 16.2
Cys	13.0 ± 0.6	12.0 ± 0.6	14.0 ± 0.8	15.9 ± 2.2
Met	74.9 ± 2.8	92.7 ± 4.0	87.8 ± 3.7	95.2 ± 10.0
Trp	32.6 ± 0.8^{a}	26.4 ± 1.0^{ab}	20.5 ± 0.9^{b}	21.6 ± 2.4^{b}
His	188.8 ± 10.7^{a}	160.4 ± 10.7^{ab}	138.1 ± 5.1^{b}	136.8 ± 9.8^{b}
Pro	40.1 ± 2.5^{a}	82.4 ± 9.1^{b}	86.0 ± 8.7^{b}	136.7 ± 29.5^{b}

The fish were challenged to 8 weeks of experiment. The blood was collected after fasting during the last 48 h of the experiment. Different letters indicate significant difference between groups ($p < 0.05$)

Gene Expression

The expressions of genes related to nucleic acid synthesis in proximal and distal intestine are shown in Figs. 5 and 6 respectively. The expression of cps 2 significantly increased in Glu 1 % group compared to SBM group in proximal intestine. Neither expressions of ppat nor pfas showed significant difference between groups (Fig. 5). The expressions of cap 2 in distal intestine also significantly increased in Glu 1 % group compared to SBM group. The expressoin of ppat significantly decreased in the fish fed SBM-based feed (SBM, Glu 1 %, Glu 2 %) compared to the fish fed FM in distal intestine. There was no significant difference in the expression of pfas in distal intestine (Fig. 6).

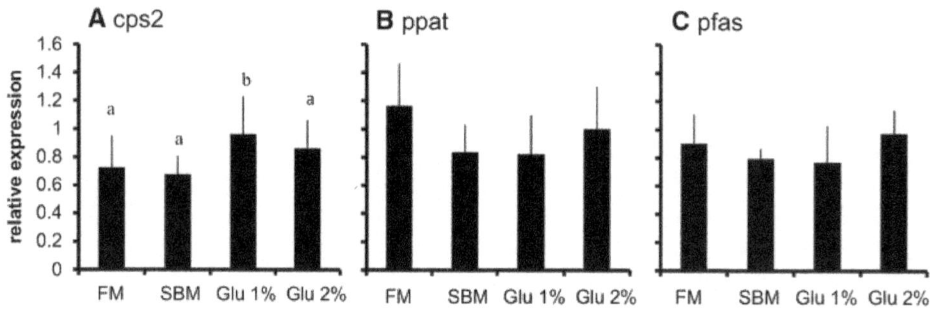

Figure 5: Alterations of gene expressions in proximal intestine by 8 weeks of Glu supplementation. Relative expression of genes related to nucleic acid synthesis in proximal intestine. **A**cps2: carbamoyl-phosphate synthetase, **A** ppat: phosphoribosyl pyrophosphate amidotransferase, **C** pfas: phosphoribosylformylglycinamidine synthase. The fish were challenged to 8 weeks of experiment. The tissues were collected after the 48 h-fasting. The expression of the genes were estimated by quantitative RT-PCR and normalized by arbp. All values are expressed as the mean ± SEM (n = 8). *Different letters* indicate significant difference between groups ($p < 0.05$)

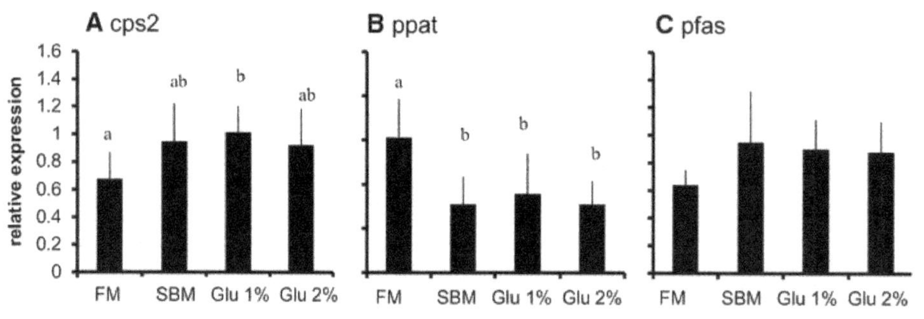

Figure 6: Alterations of gene expressions in distal intestine by 8 weeks of Glu supplementation in feed. Relative expression of genes related to nucleic acid synthesis in distal intestine. **A**cps2: carbamoyl-phosphate synthetase, **B** ppat: phosphoribosyl pyrophosphate amidotransferase, **C** pfas: phosphoribosylformylglycinamidine synthase. The fishes were challenged to 8 weeks of experiment. The tissues were collected after the 48 h-fasting. The expression of the genes were estimated by quantitative RT-PCR and normalized by arbp. All values are expressed as the mean ± SEM (n = 8). *Different letters* indicate significant difference between groups ($p < 0.05$)

Histological Analysis

Figure 7 shows the optical microscope images of transversal sections of distal intestine. The epithelial cells of mucosal folds in the distal intestine of fish fed SBM showed morphological changes such as absence of granulated pinocytotic vacuoles and thinner villus thickness. Neither the intestine of the fish fed Glu 1 % nor Glu 2 % improved these abnormalities (figure not shown). The villus thickness is shown in Fig. 8. Distal intestine of fish fed SBM feed showed significantly thinner villus. Though it was not significant difference, the villus thickness showed tendency to improve Glu dose-dependently in proximal intestine but not in distal intestine. The number of BrdU-positive cells per villi was counted 24 h after BrdU injection and shown in Fig. 9. There was no significant difference in the number of BrdU positive cells between groups in the proximal intestine though it showed tendency to increase Glu-dose dependently. In distal intestine, the number significantly increased in the SBM-based diet feed groups (SBM, Glu 1 %, Glu 2 %) compared to FM group.

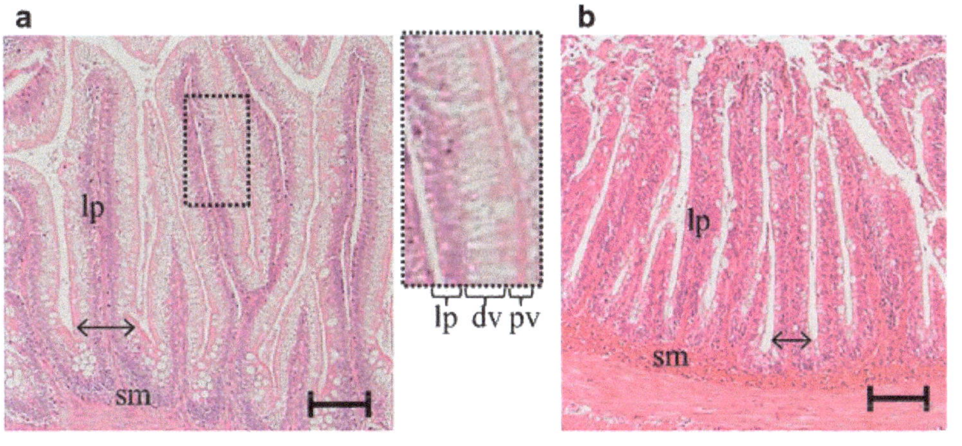

Figure 7: Morphological alteration in distal intestine by SBM feed. Transverse section of distal intestine mucosal fold **a** FM diet group, **b** SBM diet group. The fish were challenged to 8 weeks of experiment and fasted for the last 48 h of experiment. The *central panel* is an extended image of the area surrounded by a broken line in **a**. *dv* digestive vacuole, *lp*lamina propria, *pv* pinocytotic vacuoles, *sm* submucosa. *Scale bar* 100 μm

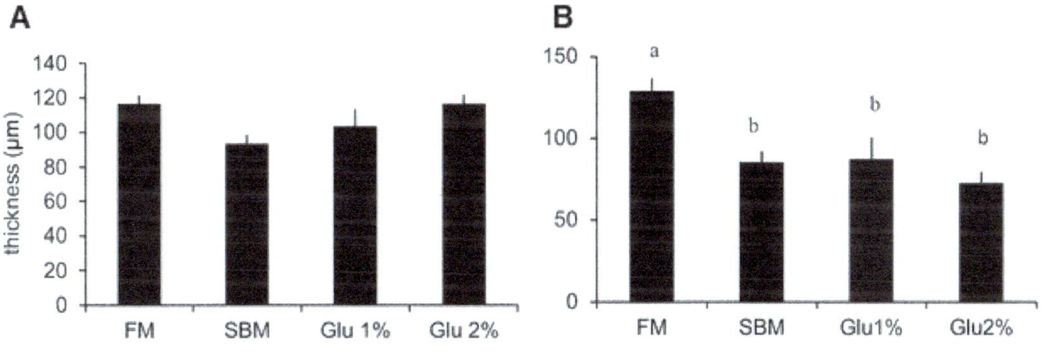

Figure 8: Effects of Glu intake on microvilli thickness in the proximal intestine. The thickness of microvillus of proximal intestine. The fish were fasted for the last 48 h of the experiment. All values are expressed as the mean ± SEM (n = 8). *Different letters* indicate significant difference between groups ($p < 0.05$)

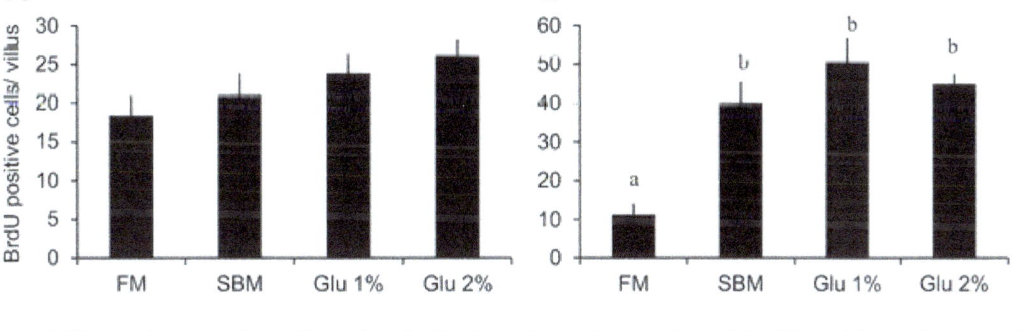

Figure 9: Effects of Glu intake on cell proliferation in the intestine. The number of BrdU positive cells per microvillus of **A** anterior intestine, **B** posterior intestine. The fish were challenged to 8 weeks of experiment. The fish were fasted for the last 48 h of the experiment. BrdU was administered i.p. 24 h before the sacrifice. All values are expressed as the mean ± SEM (n = 8). *Different letters* indicate significant difference between groups ($p < 0.05$)

DISCUSSION

This study demonstrated that single oral administration of glutamate (Glu) altered expressions of genes related to nucleotide synthesis in rainbow trout proximal intestine, and supplementation of 2 % of Glu in the SBM-based feed improved microvillus thickness in proximal intestine and ameliorated body weight gain. Though it was not significant difference, Glu tended to increase cell-proliferation in the proximal intestine dose-dependently. These data indicate that Glu has positive effect on rainbow trout fed SBM-based feed by reforming proximal intestine through altering cell-proliferation.

Previous studies have shown that supplementation of 1–4 % of Glu to the feed up-regulates gene expression of proliferating cell nuclear antigen (PCNA) and increases villus height in jejunum in weaning piglet, resulting in greater growth performance compared to non-supplemented feed (Rezaei et al. 2013; Wu et al. 2012). Glu in combination with glutamine (Gln) is also shown to increase villus height and growth performance in tilapia (Da Silva et al. 2010). In this study, we showed that the positive effect of Glu on intestine could be obtained by its single oral administration. Thirty minutes after 50 mg/kg Glu administration, the expressions of carbamoyl-phosphate synthetase 2 (cps 2), phosphoribosyl pyrophosphate amidotransferase (ppat) and phosphoribosylformylglycinamidine synthasegenes (pfas) were up-regulated (Fig. 1). These genes encode enzymes which catalyze the reactions of importing Gln to purine or pyrimidine base. Although plasma Gln concentrations did not increase significantly (Table 4), Glu can be converted into Gln in one enzymatic reaction (Newsholme et al. 2003) in the intestinal epithelium cells, and converted Gln can enhance up-regulation of these gene expressions. However, such mRNA up-regulations were not seen when fish was administered 500 mg/kg Glu. These results indicate the existence of an upper limit in intracellular Glu (or Gln) concentration which is suitable to up-regulate enzyme reaction. In mammal, majority of orally administered Glu is metabolized in intestinal epithelium cells before reaching portal vein (Reeds et al. 2000; Nakamura et al. 2013; Reeds et al. 1996), however, in our experimental condition, plasma Glu concentration increased sharply 120–360 min after 500 mg/kg Glu administration (Table 4). It is assumed that administered Glu was not metabolized in intestine because of its amount. On the other hand, the number of BrdU positive cells in proximal intestine showed tendency to increase 24 h after 500 mg/kg Glu administration (Fig. 3). Although the suitable intracellular concentration is still unclear, suitable administration concentrations would be exist in the range between 50 and 500 mg/kg.

In contrast, neither mRNA expressions nor BrdU-positive cell number in distal intestine were changed by Glu administration. Glu uptake by enterocyte depends on the existence of suitable transport systems. In mammal, it is reported that transporters responsible for amino acid uptake mainly exist proximal intestine (small intestine; Dave et al. 2004; Broberg et al. 2012). In fish, it is also reported that the absorption rates of amino acids are different from proximal to distal intestine, and proximal intestine uptakes more amino acids than distal intestine in rainbow trout (Bakke-McKellep et al. 2000; Santigosa et al. 2011). It is assumed that insufficient distribution of transporter responsible for Glu import is a reason for ineffectiveness of Glu on distal intestine.

The fish fed SBM diet showed significantly lower final body weight compared to the fish fed FM feed. Since daily feed consumption did not show any significant differences in this study, the direct cause of this growth failure seems to be the decline in feed conversion efficiency. Hepato-somatic index (HSI) and gallbladder-somite index (GBSI) were also significantly lowered in SBM group, suggesting impaired nutritional utilization. Lowered GBSI means decreased bile content because of anti-nutritional factors in SBM (Iwashita et al. 2008; Yamamoto et al. 2007,2008). Based on the histological observation, the area of lipid droplets in the liver decreased in the fish fed SBM diet (data not shown), similar to previous studies (Suzuki and Yamamoto2004). The decrease of bile is assumed to cause failure in lipid absorption, leading to less lipid content in the liver. Since lipid is one of major nutrients for fish, failure in its utilization supposedly lowers feed conversion efficiency. The epithelial cells of the distal intestine of fish fed FM diet showed well-developed microvilli and abundant digestive vacuoles. On the other hand, the distal intestine of SBM group showed thinner brush border and less digestive vacuoles, resulting in thinner villi (Fig. 6). Moreover, cell- proliferation in distal intestine was significantly increased in the fish fed SBM diet (Fig. 8), supposedly to recover from damage caused by SBM, as previously reported (Sanden et al. 2005). Neither these abnormalities nor cell-proliferation were altered by supplementation of Glu to SBM. Glu seems little effect on distal intestine even when administered for long term.

In this study, however, supplementation of 2 % of Glu on SBM-based feed improved the restrained weight gain caused by SBM (Table 5). The feed efficiency ratio slightly increased in Glu 2 % group (0.87) compared to SBM group (0.85), and feed intake was also increased in Glu 2 % group (35.8 g per fish) compared to SBM group (34.6 g per fish). Since the typical effects of anti-nutritional factors in SBM, such as lowered HSI/GBSI and abnormality in distal intestine morphologies, showed no improvement by Glu 2 % feed, Glu seems to act on other phenotype. In previous studies, no abnormalities have been reported in proximal intestine in the fish fed SBM-based diet (Suzuki and Yamamoto 2004; Buttle et al. 2001; Krogdahl et al. 2003; Burrells et al. 1999). In this study, however, the thickness of microvilli in proximal intestine not significantly but thinner in SBM group compared to FM group (Fig. 7). It is also reported that both proximal- and distal-intestine nutrient absorption is modified by replacement of fish meal by plant protein (Santigosa et al. 2011). Although there was no other abnormality in morphology like distal intestine, SBM is assumed to have some negative effect on proximal intestine, or the absorption of nutrients in proximal intestine was possibly lowered by SBM. Supplementation of Glu on SBM-diet dose-dependently improved the thickness of microvilli in proximal intestine (Fig. 7). Besides, though it was not significantly different, the number of BrdU positive cells in proximal intestine slightly increased in the fish fed SBM diet compared to FM diet and showed Glu dose-dependent tendency to increase (Fig. 8). It is indicated that Glu helps proximal intestine reform from some mal-effect of SBM. It is simply possible that Glu was absorbed as nutrients and expand the area of digestive vacuole, resulting in thicker microvilli. The absorbed Glu would be utilized as energy or converted into other molecules such as other amino acids, providing necessary elements to intestinal cells. The plasma Gln concentrations in Glu 1 % and 2 % group were significantly lower compared to SBM and FM group. Since plasma Gln is the second largest source of energy for intestine (Reeds et al. 1996, 2000), it indicates that intracellular metabolism in the intestine was activated by Glu supplementation to feed.

In conclusion, we showed that supplementation of 2 % of Glu to SBM-diet for 8 weeks improves growth of rainbow trout through reforming microvilli in proximal intestine. We also demonstrated that single oral Glu administration alters expressions of genes related to nucleotide synthesis in proximal intestine, and at least part of the long-term effect of Glu is indicated to result from up-regulation of cell-proliferation in the proximal intestine. Although further investigation is needed to clear the effects of Glu in fish, Glu has shown to have positive effect on the growth of rainbow trout fed SBM-based diet.

ACKNOWLEDGEMENTS

This study was financially supported (in part) by the Ministry of Agriculture, Forestry, and Fisheries of Japan.

REFERENCES

1. Bakke-McKellep AM, Nordrum S, Krogdahl Å, Buddington RK (2000) Absorption of glucose, amino acids, and dipeptides by the intestines of Atlantic salmon (*Salmo salar* L.). Fish Physiol Biochem 22:33–44
2. Berseth CL (1996) Gastrointestinal motility in the neonate. Clin Perinatol 23:179–190
3. Blachier F, Boutry C, Bos C, Tomé D (2009) Metabolism and functions of L-glutamate in the epithelial cells of the small and large intestines. Am J Clin Nutr 90:814S–821S
4. Broberg ML, Holm R, Tønsberg H, Frølund S, Ewon KB, Nielsen AL, Brodin B, Jensen A, Kall MA, Christensen KV, Nielsen CU (2012) Function and expression of the proton-coupled amino acid transporter PAT1 along the rat gastrointestinal tract: implications for intestinal absorption of gaboxadol. Br J Pharmacol 167:654–665
5. Burrells C, Williams PD, Southgate PJ, Crampton VO (1999) Immunological, physiological and pathological responses of rainbow trout (*Oncorhynchus mykiss*) to increasing dietary concentrations of soybean proteins. Vet Immunol Immunopathol 72:277–288
6. Burrin DG, Stoll B (2009) Metabolic fate and function of dietary glutamate in the gut. Am J Clin Nutr 90:850S–856S
7. Buttle LG, Burrells AC, Good JE, Williams PD, Southgate PJ, Burrells C (2001) The binding of soybean agglutinin (SBA) to the intestinal epithelium of Atlantic salmon, *Salmo salar* and Rainbow trout, *Oncorhynchus mykiss*, fed high levels of soybean meal. Vet Immunol Immunopathol 80:237–244
8. Cartmell J, Schoepp DD (2000) Regulation of neurotransmitter release by metabotropic glutamate receptors. J Neurochem 75:889–907
9. Chikwati EM, Venold FF, Penn MH, Rohloff J, Refstie S, Guttvik A, Hillestad M, Krogdahl A (2012) Interaction of soyasaponins with plant ingredients in diets for Atlantic salmon, *Salmo salar* L. Br J Nutr 107:1570–1590
10. Da Silva LCR, Furuya WM, Natali MRM, Schamber CR, Dos Santos LD, Vidal LVO (2010) Productive performance and intestinal morphology of Nile tilapia juvenile fed diets with L-glutamine and L-glutamate (*Desempenho e morfometria intestinal de juvenis de tilápia-do-nilo alimentados com dietas suplementadas com L -glutamina e L -glutamato*). Revista Brasileira de Zootecnia 39:1175–1179
11. Dave MH, Schulz N, Zecevic M, Wagner CA, Verrey F (2004) Expression of heteromeric amino acid transporters along the murine intestine. J Physiol 558:597–610
12. Hasebe M, Suzuki H, Mori E, Furukawa J, Kobayashi K, Ueda Y (1999) Glutamate in enteral nutrition: can glutamate replace glutamine in supplementation to enteral nutrition in burned rats? JPEN J Parenter Enteral Nutr 23:S78–S82
13. Iwashita Y, Yamamoto T, Furuita H, Sugita T, Suzuki N (2008) Influence of certain soybean antinutritional factors supplemented to a casein-based semipurified diet on intestinal and liver morphology in fingerling rainbow trout *Oncorhynchus mykiss*. Fish Sci 74:1075–1082
14. Kirchgessner AL (2001) Glutamate in the enteric nervous system. Curr Opin Pharmacol 1:591–596
15. Krogdahl Å, Bakke-McKellep AM, Baeverfjord G (2003) Effects of graded levels of standard soybean meal on intestinal structure, mucosal enzyme activities, and pancreatic response in Atlantic salmon (*Salmo solar* L.). Aquac Nutr 9:361–371
16. Lallès JP, Bosi P, Smidt H, Stokes CR (2007) Weaning—a challenge to gut physiologists. Livest Sci 108:82–93
17. Nakamura H, Kawamata Y, Kuwahara T, Torii K, Sakai R (2013) Nitrogen in dietary glutamate is utilized exclusively for the synthesis of amino acids in the rat intestine. Am J Physiol Endocrinol Metab 304:E100–E108
18. Newsholme P, Procopio J, Ramos Lima MM, Pithon-Curi TC, Curi R (2003) Glutamine and glutamate—their central role in cell metabolism and function. Cell Biochem Funct 21:1–9
19. Pluske JR, Hampson DJ, Williams IH (1997) Factors influencing the structure and function of the small intestine in the weaned pig: a review. Livest Prod Sci 51:215–236
20. Reeds PJ, Burrin DG, Jahoor F, Wykes L, Henry J, Frazer EM (1996) Enteral glutamate is almost completely metabolized in first pass by the gastrointestinal tract of infant pigs. Am J Physiol Endocrinol Metab 270:E413–E418
21. Reeds PJ, Burrin DG, Stoll B, Jahoor F (2000) Intestinal glutamate metabolism. J Nutr 130:978S–982S
22. Rezaei R, Knabe DA, Tekwe CD, Dahanayaka S, Ficken MD, Fielder SE, Eide SJ, Lovering SL, Wu G (2013) Dietary supplementation with monosodium glutamate is safe and improves growth performance in postweaning pigs. Amino Acids 44:911–923
23. Sanden M, Berntssen MHG, Krogdahl Å, Hemre GI,

Bakke-McKellep AM (2005) An examination of the intestinal tract of Atlantic salmon, *Salmo salar* L., parr fed different varieties of soy and maize. J Fish Dis 28:317–330

24. Santigosa E, García-Meilán I, Valentin JM, Pérez-Sánchez J, Médale F, Kaushik S, Gallardo MA (2011) Modifications of intestinal nutrient absorption in response to dietary fish meal replacement by plant protein sources in sea bream (*Sparus aurata*) and rainbow trout (*Onchorynchus mykiss*). Aquaculture 317:146–154
25. Suzuki N, Yamamoto T (2004) Histological observations of intestinal degeneration of defatted soybean meal diet and supplemental effect of soybean lecithin for fingerling rainbow trout,*Oncorhynchus mykiss*. J Sch Mar Sci Technol 2:17
26. Windmueller HG, Spaeth AE (1980) Respiratory fuels and nitrogen metabolism in vivo in small intestine of fed rats. Quantitative importance of glutamine, glutamate, and aspartate. J Biol Chem 255:107–112
27. Wu G (1998) Intestinal mucosal amino acid catabolism. J Nutr 128:1249
28. Wu X, Zhang Y, Liu Z, Li TJ, Yin YL (2012) Effects of oral supplementation with glutamate or combination of glutamate and *N*-carbamylglutamate on intestinal mucosa morphology and epithelium cell proliferation in weanling piglets. J Anim Sci 90:337–339
29. Yamamoto T, Shima T, Furuita H, Suzuki N (2002) Influence of feeding diets with and without fish meal by hand and by self-feeders on feed intake, growth and nutrient utilization of juvenile rainbow trout (*Oncorhynchus mykiss*). Aquaculture 214:289–305
30. Yamamoto T, Suzuki N, Furuita H, Sugita T, Tanaka N, Goto T (2007) Supplemental effect of bile salts to soybean meal-based diet on growth and feed utilization of rainbow trout*Oncorhynchus mykiss*. Fish Sci 73:123–131
31. Yamamoto T, Goto T, Kine Y, Endo Y, Kitaoka Y, Sugita T, Furuita H, Iwashita Y, Suzuki N (2008) Effect of an alcohol extract from a defatted soybean meal supplemented with a casein-based semi-purified diet on the biliary bile status and intestinal conditions in rainbow trout*Oncorhynchus mykiss* (Walbaum). Aquac Res 39:986–994
32. Yamamoto T, Iwashita Y, Matsunari H, Sugita T, Furuita H, Akimoto A, Okamatsu K, Suzuki N (2010) Influence of fermentation conditions for soybean meal in a non-fish meal diet on the growth performance and physiological condition of rainbow trout *Oncorhynchus mykiss*. Aquaculture 309:173–180
33. Zhang J, Yin Y, Shu XG, Li T, Li F, Tan B, Wu Z, Wu G (2013) Oral administration of MSG increases expression of glutamate receptors and transporters in the gastrointestinal tract of young piglets. Amino Acids 45:1169–1177

CHAPTER 12 FISH BIOLOGY AND BIOTECHNOLOGY FOR SUSTAINABLE AQUACULTURE

GUI Jian-Fang

The contribution of aquaculture to global food production and security has been widely acknowledged since the start of the 21st century [1–4]. Global aquaculture production has increased continuously over the last five decades, and particularly in China, aquaculture has become the fastest growing and most efficient agri-sector, with production accounting for more than 70% of the world's aquaculture output [5,6]. José Graziano da Silva, director-general of the United Nations Food and Agriculture Organization (FAO), stated in The State of World Fisheries and Aquaculture (2014), published by FAO, "Global fish production continues to outpace world population growth, and aquaculture remains one of the fastest-growing food producing sectors. If responsibly developed and practiced, aquaculture can generate lasting benefits for global food security and economic growth" [6]. The future of fish and aquaculture potentials have been the subject of extensive discussion [1–4].

Reviewing the process over the past 50 years, the rapid growth in aquaculture is clearly a result of developments in fish biology and biotechnology. As a multidisciplinary subject for studying the roles and mechanisms of fish phylogenesis, development, growth, reproduction and behavior, and thereby exploiting the biotechnology applicable to genetic breeding and disease prevention, fish biology and biotechnology has provided the innovation and technology for rapid development of the aquaculture industry. For example, the breakthrough in artificial propagation techniques for the "four important domestic fish"—silver, bighead, grass and black carp—in the early 1960s [7] established the scientific and technological basis for fry and fingerling production of other fish and aquatic species. Since the 1980s, biotechnological innovations and developments have provided sustained motivation for the establishment of genetic breeding, disease control and aquaculture seed programs [8]. Many novel aquatic varieties that have played a major role in aquaculture are the results of comprehensive and systematic studies on the biological characteristics of these species. For instance, most farmed varieties of the common and gibel carp were selected and developed on the basis of a thorough understanding of their

Citation: Gui JF. Fish biology and biotechnology for sustainable aquaculture. Sci China Life Sci, 2015, 58: 121–123, doi: 10.1007/s11427-015-4812-9

genetic background [9,10]. Over the last 20 years, significant progress has been made in fish biology and biotechnology, especially in the field of fish genetic breeding. To introduce these advances, the editorial committee has invited several scientists from the Institute of Hydrobiology of the Chinese Academy of Sciences, the College of Fisheries of Huazhong Agricultural University, Heilongjiang Fishery Institute of the Chinese Academy of Fishery Sciences, and the College of Life Sciences of Hunan Normal University to jointly contribute seven review articles to highlight a number of hot topics in this field.

Mei Jie and Gui Jian-Fang [11] provide a commentary review on the genetic basis and biotechnological manipulation of sexual dimorphism and sex determination in fish. They summarize recent breakthroughs and advances in this field, including several successful cases of biotechnological manipulation for sex control breeding, and consider some promising prospects for future research. Yin Zhan and his colleagues [12] collate recent work on the somatotropic endocrine axis of fish, and review some recent advances in the endocrine regulation of pituitary growth hormone (GH) and extrapituitary aspects of the GH/insulin-like growth factor (IGF) system. Xiao WuHan [13] summarizes novel progress in research of the hypoxia signaling pathway and its regulation, as well as the mechanism of hypoxia adaptation in fish. Based on their research experience of more than 20 years, Zhang QiYa and Gui Jian-Fang [14] review and compare the genomic data of diverse aquatic viruses, such as irido-, herpes-, reo- and rhabdoviruses, and outline some major advances on virus–host interactions in the infected aquaculture species.

Sun Yong Hua and Zhu ZuoYan, as well as their colleagues [15], review the history of fish-breeding methods based on classical genome manipulation, including polyploidy breeding and nuclear transfer, and emphasize the development and application of fish directional breeding based on transgenic technology and recently developed genome editing technologies. Tong Jin Gou and Sun XiaoWen [16] concisely summarize current genetic and genomic analyses for economically important traits, and discuss future directions and prospects of genomics and next-generation sequencing technology for fish molecular breeding. Liu ShaoJun and Liu Yun and their colleagues [17] appraise the biological foundations and applications of selective breeding technologies, including traditional selective, molecular marker-assisted, genome-wide selective and sex control breeding, and review and discuss the research status of and known problems in fish genetic breeding. Finally, I would like to cordially thank all the authors for contributing review articles on this topic, and wish that the advances in fish biology and biotechnology will become an inexhaustible source for the development of sustainable aquaculture and fisheries.

REFERENCES

1. Naylor RL, Goldburg RJ, Primavera JH, Kautsky N, Beveridge MC, Clay J, Folke C, Lubchenco J, Mooney H, Troell M. Effect of aquaculture on world fish supplies. Nature, 2000, 405: 1017–1024
2. Pauly D, Christensen V, Guénette S, Pitcher TJ, Sumaila UR, Walters CJ, Watson R, Zeller D. Towards sustainability in world fisheries. Nature, 2002, 418: 689–695
3. James HT, Geoff LA. Fishes as food: aquaculture's contribution. EMBO Rep, 2001, 21: 958–963
4. Worm B, Branch TA. The future of fish. Trends Ecol Evol, 2012, 27: 594–599
5. Gui JF, Zhu ZY. Molecular basis and genetic improvement of economically important traits in aquaculture animals. Chin Sci Bull, 2012, 57: 1751–1760
6. Food and Agriculture Organization of the United Nations. The State of World Fisheries and Aquaculture 2014. Rome, 2014
7. Wu HW, Zhong L. Progress and achievement on artificial proliferation of grass carp, black carp, silver carp and big head carp in China. Chin Sci Bull, 1964, 9: 900–907
8. Wu CJ, Gui JF. Fish Genetics and Breeding Engineering (in Chinese). Shanghai: Shanghai Scientific and Technical Publishers, 1999
9. Gui JF, Zhou L. Genetic basis and breeding application of clonal diversity and dual reproduction modes in polyploid Carassius auratus gibelio. Sci China Life Sci, 2010, 53: 409–415
10. Liu SJ. Distant hybridization leads to different ploidy fishes. Sci China Life Sci, 2010, 53: 416–425
11. Mei J, Gui JF. Genetic basis and biotechnological manipulation of sexual dimorphism and sex determination in fish. Sci China Life Sci, 2015, 58: 124–136
12. Dai XY, Zhang W, Zhuo ZJ, He JY, Yin Z. Neuroendocrine regulation of somatic growth in fishes. Sci China Life Sci, 2015, 58: 137–147
13. Xiao WH. The hypoxia signaling pathway and hypoxic adaptation in fishes. Sci China Life Sci, 2015, 58: 148–155
14. Zhang QY, Gui JF. Virus genomes and virus-host interactions in aquaculture animals. Sci China Life Sci, 2015, 58: 156–169
15. Ye D, Zhu ZY, Sun YH. Fish genome manipulation and directional breeding. Sci China Life Sci, 2015, 58: 170–177
16. Tong JG, Sun XW. Genetic and genomic analyses for economically important traits and their applications in molecular breeding of cultured fish. Sci China Life Sci, 2015, 58: 178–186
17. Xu K, Duan W, Xiao J, Tao M, Zhang C, Liu Y, Liu SJ. Development and application of biological technologies in fish genetic breeding. Sci China Life Sci, 2015, 58: 187–201

CHAPTER 13 FISH PROCESSING WASTES FOR MICROBIAL ENZYME PRODUCTION

Faouzi Ben Rebah, Nabil Miled

Keywords: Fish waste, Growth media, Protease, Lipase, Chitinolytic enzymes, Ligninolytic enzymes

INTRODUCTION

In recent years, there has been a constant increase in the exploitation of fish resources and the estimated quantity used for human consumption (105.6 million tons) is globally 75 % of the worldwide fish production. The remaining 25 % of the catch (34.8 million tons) are considered as wastes (FAO 2007). Furthermore, the commercial fish processing industry generates large quantities of solid waste and wastewater. Solid waste which represents 20–60 % of the initial raw material contains various kinds of residues (whole waste fish, fish head, viscera, skin, bones, blood, frame liver, gonads, guts, some muscle tissue, etc.) (Awarenet 2004). In some countries, these discards are not utilized, but incinerated or dumped at sea causing environmental problems (Bozzano and Sarda 2002). Recently, environmental regulations are becoming stricter, requiring new disposal methods based on the fact that fish wastes (solid waste and wastewater) may considered as an important source of protein, lipids and minerals with high biological value (Toppe et al. 2007; Kacem et al. 2011).

To avoid wasting by-products, various conventional disposal methods have been applied including ensilation and fermentation for the production of high-protein meals for animal feeds (Faid et al. 1997; Hassan and Heath 1986) as well as composting (Liao et al. 1997). Recent advances in industrial biotechnology process are exploited for an economic utilization of wastes in producing higher added value products. For example, fish oil with higher level of polyunsaturated fatty acids beneficial for human health (Kim et al. 2006; Zampolli et al. 2006; Chen et al. 2006) was produced and integrated into food products and beverages (Rubio-Rodriguez et al. 2010). Moreover, fish skin or cartilage from some species could be excellent raw materials for producing gelatin or chondroitin sulphate useful in food, cosmetic and pharmaceutical sectors (Blanco et al. 2006; Karim and Bhat 2009). By-products can be also hydrolysed by applying various treatments (heat, enzymatic and chemical treatments) and the hydrolysate quality can be also improved by fermentation (Yamamoto et al. 2005; Yano et al.2008; Xu et al. 2008). Generally, the obtained hydrolysates may have biological and functional properties

Citation: Faouzi Ben Rebah, Nabil Miled, Fish processing wastes for microbial enzyme production, DOI: 10.1007/s13205-012-0099-8

of interest in various sectors (human nutrition, cosmetology, aquaculture, microbiology, etc.) (Ben Rebah et al. 2008; Liaset et al. 2000; Martone et al. 2005). For example, proteolytic hydrolyses were reported to produce peptides with interesting biological activities (anti-hypertensers, immuno-modulators, antioxidants, anticoagulants, etc.) useful in the treatment of several diseases (osteoporosis, arthritis, diabetes, obesity, etc.) (Kim and Mendis2006).

Interestingly, fish by-products provide an excellent source for microbial growth, which can be exploited in producing various metabolites (lysine, enzymes, etc.) (Coello et al. 2000; Vazquez et al. 2006, 2008). A large number of microorganisms including bacteria, yeast and fungi produce different groups of enzymes (protease, lipase, etc.) having high biotechnological interest (food processing, detergent, textile, pharmaceutical products, medical therapy, etc.) (Underkofler et al. 1958). Microbial synthesis of enzymes has been reported to be influenced by various factors such as carbon sources, nitrogen sources, and operating parameters (temperature, pH, etc.) (Frost and Moss 1987; Gupta et al. 2004; Jacob and Prema 2006; Palaniyappan et al. 2009). Generally, agro-industrial residues mainly composed of complex polysaccharides are considered the best substrates for microbial growth and enzyme production (Pandey et al. 1999). The economies of this process are governed mostly by the cost and availability of a suitable carbon and nitrogen sources. Therefore, the investigation of low cost substrates easily available such as fish processing wastes became a necessity opening new avenues for their effective production and utilization. In this review, we will present the potential of different fish processing by-products tested as growth substrates for microbial enzyme production.

PROTEASES

Proteases are one of the most important groups of industrial enzymes, representing more than 65 % of the of global industrial enzyme market (Banik and Prakash 2004). Various microorganisms have been used to produce these enzymes (*Penicillium* sp. *Serratia marcescens*, *Streptomyces* sp., *Rhizopus oryzae*, *Pseudomonas*, *Bacillus* sp., *Vibrio*, etc.) (Banerjee et al. 1993; De Azeredo et al. 2004; Joo and Chang 2005; Vazquez et al. 2006).

Because of their vast diversity and their specificity of action, proteases have a large variety of biotechnological applications (Kumar and Takagi 1999; Gupta et al. 2002). Their use involves different industrial sectors, such as the detergent industries (Banik and Prakash 2004), bioremediation processes (Roberts et al. 2007), leather industry (Taylor et al. 1987), bioprocessing of used X-ray films for silver recovery (Fujiwara and Yamamoto 1987) pharmaceutical industries and protein hydrolysate production (Banik and Prakash 2004). Recently, the application of protease in producing bioactive peptide has received great attention as a viable alternative to chemical approach (Suetsuna 2000; Ma et al. 2007).

Numerous investigators have looked for ways of producing microbial proteases using inexpensive media. Interestingly, it was demonstrated that fish processing wastes (fish meat wastes, chitinous material from cephalopods and wastewater) offer good potential for this purpose (Triki-Ellouz et al. 2003; Vazquez et al. 2006; Wang and Yeh 2006; Haddar et al. 2010).

Fish processing by-products including heads and viscera were utilized for preparing microbial growth media. In some cases, fish wastes were cooked, pressed, minced and then dried (80 °C for 24–48 h) in order to obtain fish powder. In other cases, wastes were boiled in water and supernatants were recuperated. Moreover, protein hydrolysates obtained by acid, alkali, or enzymatic treatments of raw or defatted by-products were also used as a nitrogen source for protease production (Coello et al. 2000; Triki-Ellouz et al. 2003). Generally, fish powder or supernatants were added to the microbial basal medium (Table 1). For example, *Pseudomonas aeruginosa* MN7 and *Bacillus subtilis* were cultivated in media containing combined heads and viscera powder allowing an acceptable level of protease production (Ellouz et al. 2001; Triki-Ellouz et al. 2003). Similarly, marine peptones obtained from fish viscera of various marine species (rainbow trout "*Oncorhynchus mykiss*", swordfish "*Xiphias gladius*", squid "*Loligo vulgaris*" and yellowfin tuna "*Thunnus albacares*") allowed higher levels of protease activity for two vibrio species (*Vibrio anguillarum* and *Vibrio splendidus*) while compared to basal medium containing commercial peptones (Vazquez et al. 2006). According to Esakkiraj et al. (2009), a defatted tuna waste allowed higher level of protease activity by *Bacillus cereus*(134.57 U/ml) while compared to acid and alkali hydrolysates (60.37 and 65.96 U/ml, respectively) and to commercial peptone (124.90 U/ml). The enhancement of protease production in the defatted fish-based medium may be attributed to the lipid-free nature of product, which could support the protease synthesis by the microbial species than other nitrogen source preparations. A similar observation was notified for lipase production by*Rhizopus oryzae* (Ghorbel et al. 2005). More recently, a significant improvement of the alkaline protease production by *Bacillus mojavensis* A21 was obtained using *Sardinella* peptone (Haddar et al. 2010). However, is very important to note that optimal protease production is controlled by several growth conditions such as the medium composition (carbon and nitrogen sources, mineral salts, etc.), agitation, culture temperature and initial medium, pH, etc.

Table 1: Protease production by various microbial strains grown in fish processing waste based media

Fish raw materials	Preparation of the growth media[a]	Microbial strains	Activity (U/ml)	References
Heads and viscera of*Sardinella*	Raw materials cooked, pressed, minced and dried (80 °C, 24–48 h)	*Pseudomonas aeruginosa*MN7	7,800	Triki-Ellouz et al. (2003)
Heads and viscera of*Sardinella*	Raw materials cooked, pressed, minced and dried (80 °C, 24–48 h)	*Bacillus subtilis*	720	Ellouz et al. (2001)

Viscera from rainbow trout, swordfish, squid and yellowfin tuna	Raw materials ground with water and supernatants recovered by centrifugation and used to prepare the peptone	*Vibrio anguillarum*	35–68	Vazquez et al. (2006)
Viscera from rainbow trout, swordfish, squid and yellowfin tuna	Raw materials were ground with water and supernatants recovered by centrifugation and used to prepare the peptone	*Vibrio splendidus*	9–30	Vazquez et al. (2006)
Raw tuna waste	Raw materials cooked, bones removed, pressed to remove water and fat, pressed, minced and dried (80 °C, 24–48 h)	*Bacillus cereus*	74.77	Esakkiraj et al. (2009)
Defatted tuna waste	Extraction with chloroform/methanol	*Bacillus cereus*	134.57	Esakkiraj et al. (2009)
Acid-hydrolyzed tuna waste	Method described by Gao et al. (2006)	*Bacillus cereus*	60.37	Esakkiraj et al. (2009)
Alkali-hydrolyzed tuna waste	Method described by Batista (1999)	*Bacillus cereus*	65.96	Esakkiraj et al. (2009)

[a]The obtained fish powder was added to the basal medium

Several microbial strains grown in basal medium supplemented with different chitinous material preparations [shrimp shell powder (SSP), squid pen powder (SPP), chitin flake of shrimp shell (CFSS), chitin flake of crab shell (CFCS), shrimp and crab shell powder (SCSP)] (Table 2) were tested for protease activity. Generally, chitinous materials collected from marine food processing industry were treated (dried, milled, and sieved to powder with diameters of <0.053 mm) and added to the basal medium at different proportions as reported by Wang et al. (2002a). According to Table 2, protease activity was affected by the nature of the strain and the used carbon/nitrogen sources. For example, for both *Bacillus subtilis* TKU007 and *Bacillus* sp. TKU004, SSP was more suitable for protease production (Wang and Yeh 2006) than the other carbon/nitrogen sources. However, for the production of protease by *L. paracasei* subsp.*paracasei* TKU010, SPP (0.12 U/ml) was more suitable than SSP (0.08 U/ml) (Wang et al.2008a). Also, in the case of *Serratia ureilytica* TKU013, SPP (0.200 U/ml) was more suitable than SSP (0.190 U/ml) (Wang et al. 2009a). According to Wang et al. (2008b), higher protease activity might be controlled by the ratio of protein and chitin. In the findings of Wang et al. (2008b), the ratio of protein and chitin (48:38) in SSP is much closer than that of SPP (61:38) allowing a protease activity by *Chryseobacterium* sp. TKU014 of 0.010 and 0.007 U/ml with SSP and SPP, respectively. In addition to that, mineral salts may have a suppressed role on the protease productivity as reported for *L. paracasei* subsp. *paracasei* TKU010 and TKU012(Wang et al. 2008a).

Table 2: Protease production by various microbial strains grown in chitinous material based media

Chitinous materials (rate added to the basal medium in % w/v)	Microbial strains	Protease activity (U/ml)	References
Shrimp shell powder (SSP)[a]			
1	*Chryseobacterium* sp. TKU014	0.010	Wang et al. (2008b)
1.5	*Bacillus subtilis* TKU007	0.350	Wang and Yeh (2006)
1	*Bacillus subtilis* TKU007	0.350	Wang and Yeh (2006)
1	*Bacillus* sp. TKU004	0.090	Wang et al. (2006a)
1	*Lactobacillus paracasei*subsp *paracasei*TKU010	0.080	Wang et al. (2008a)
1	*Lactobacillus paracasei*subsp *paracasei* TKU012	0.130	Wang et al. (2008c)
1.5	*Serratia ureilytica*TKU013	0.190	Wang et al. (2009a)
2	*Bacillus cereus* TKU006	2.070	Wang et al. (2009b)
1	*Serratia* sp. TKU016	0.160	Wang et al. (2010)
Squid pen powder (SPP)[a]			
2	Bacillus sp. TKU004.	0.065	Wang et al. (2006a)
1	*Chryseobacterium* sp. TKU014	0.007	Wang et al. (2008b)
2	*Bacillus subtilis* TKU007	0.220	Wang and Yeh (2006)
1	*Bacillus subtilis* TKU007	0.220	Wang and Yeh (2006)
1	*Bacillus* sp. TKU004	0.060	Wang et al. (2006a)
1	*Lactobacillus paracasei*subsp *paracasei*TKU010	0.120	Wang et al. (2008a)

1	*Lactobacillus paracasei*subsp *paracasei* TKU012	0.140	Wang et al. (2008c)
1.5	*Serratia ureilytica*TKU013	0.200	Wang et al. (2009a)
1	*Serratia* sp. TKU016	0.090	Wang et al. (2010)
Chitin flake of shrimp shell (CFSS)			
2	*Bacillus subtilis* TKU007	<0.010	Wang and Yeh (2006)
1	*Bacillus subtilis* TKU007	<0.010	Wang and Yeh (2006)
1	*Bacillus* sp. TKU004	<0.010	Wang et al. (2006a)
1	*Lactobacillus paracasei*subsp *paracasei*TKU010	<0.010	Wang et al. (2008a)
1	*Lactobacillus paracasei*subsp *paracasei* TKU012	<0.010	Wang et al. (2008c)
Chitin flake of crab shell (CFCS)			
2	*Bacillus subtilis* TKU007	<0.010	Wang and Yeh (2006)
1	*Bacillus subtilis* TKU007	<0.010	Wang and Yeh (2006)
1	*Bacillus* sp. TKU004	<0.010	Wang et al. (2006a)
1	*Lactobacillus paracasei*subsp *paracasei*TKU010	<0.010	Wang et al. (2008a)
1	*Lactobacillus paracasei*subsp *paracasei* TKU012	<0.010	Wang et al. (2008c)
Shrimp and crab shell powder (SCSP)			
1	*Bacillus subtilis* TKU007	nd	Wang and Yeh (2006)
1	*Bacillus* sp. TKU004	0.060	Wang et al. (2006a)
1	*Lactobacillus paracasei*subsp *paracasei* TKU012	0.090	Wang et al. (2008c)
1	*Lactobacillus paracasei*subsp *paracasei*TKU010	0.040	Wang et al. (2008a)

nd not detectable

[a]Dried materials of SPP, SSP were prepared as described earlier (Wang et al. 2002a)

In order to enhance the protease production, shrimp and crab shell powder were treated using chemical reagents (HCl, NaOH and HCl/NaOH (Oh et al. 2000) and used for microbial growth. The chemical treatments and the rate of the preparation added to the microbial basal medium affect considerably the time of growth and the protease activity as reported by Liang et al. (2006) for *Monascus purpureus* CCRC31499. Interestingly, protease produced under optimal culture conditions can be exploited in the deproteinization of shrimp and crab shell wastes. For example, it was demonstrated that protease of *P. aeruginosa* K-187 allowed higher protein removal from crustacean wastes. The removal rates were about 72 % for shrimp and crab shell powder, 78 % for natural shrimp shell and 45 % for acid-treated shrimp and crab shell powder (Oh et al. 2000).

Wastewater from fish processing industry supplemented with cuttlefish by-products powder was also tested as growth media for microbial growth and protease production by five bacterial species (*Bacillus licheniformis*, *Bacillus subtilis*, *Pseudomonas aeruginosa*, *Bacillus cereus*BG1, and *Vibrio parahaemolyticus*). According to Souissi et al. (2008), all the tested strains can obtain their carbon and nitrogen source requirements directly from proteins of the cuttlefish by-products and, interestingly, the addition of the fishery wastewater improved the protease activity. For example, the protease activity of *Bacillus cereus* BG1, increased from 467 U/ml to reach 2771 U/ml while adding fishery wastewater to the cuttlefish by-product-based medium (Table 3). The positive effect on protease activity may be explained by the presence of very specific growth factors and amino-acids in the fishery wastewater.

Table 3: Proteases produced by some microbial strains cultivated in fishery waste based media (Souissi et al. 2008)

Strains	Type of protease	Growth media	Protease activity (U/ml)	Time of growth (h)
Bacillus cereus BG1	Metalloprotease	M[a]	487	48
		M + FWW[b]	756–2,771	48
Bacillus subtilis	Serine protease	CF[c]	178	16

		CF + FWW[d]	0–392	16
Bacillus licheniformis	Serine protease	CF	407	24
		CF + FWW	138–821	24
Pseudomonas aeruginosa	Metalloprotease	CF	1,680	24
		CF + FWW	160–1,694	24
Vibrio parahaemolyticus	Serine protease	CF	1,607	24
		CF + FWW	196–2,487	24

[a]M: 10 g /l of maltose and 10 g /l of cuttlefish by-products powder in artificial sea water (Krieg and Holt 1984)

[b]M + FWW: M mixed with FWW (fishery wastewater) at different concentrations

[c]CF: cuttlefish by-product medium (g /l of cuttlefish by-products powder in distilled water; Cuttlefish by-products powder was prepared from guts and stomachs (rinsed, heated, pressed, then, minced and dried at 80 °C for 24–48 h as reported by Souissi et al. 2008

[d]CF + FWW medium: CF prepared in crude and diluted FWW

The diversified utilization of fish processing wastes as potential media for microbial proteases production is expected to deliver an attractive and promising strategy for enzyme large-scale production. In commercial practice, the optimization of medium composition is done to maintain a balance between the various microbial growth nutrients. However, no defined medium has been established for the best production of proteases from different microbial sources. Generally, each microbial strain has its own special conditions for maximum enzyme production. Therefore, experiments should be conducted to elucidate the behaviour of each strain with each fish waste to identify inducers present in the fish waste. These researches would provide the incentive for commercial developments leading to large-scale and cost-effective production of proteases.

LIPASES

Lipases are a class of enzymes which catalyse the hydrolysis of long chain triglycerides and are of considerable commercial interest in various industrial applications (detergent, food, flavour industry, biocatalytic resolution of pharmaceuticals, esters and amino acid derivatives, fine chemicals, agrochemicals, use as biosensor, bioremediation, cosmetics, perfumery, etc.) (Hasan et al. 2006). With the rapid development of enzyme technology, lipases are currently receiving much attention involving various selected microorganisms especially from fungi, bacteria, and yeasts (Sharma et al. 2001).

Numerous investigators have looked for ways of producing microbial lipases using low cost media. A variety of fish processing by-products contain growth factors offering good potential as culture media, as shown by the high level of lipase activity produced by some microbial strains (Table 4). As reported in Table 4, various pre-treatment processes (heat treatment, chemical and enzymatic treatment, etc.) were applied on fish wastes before being used as growth media (Ben Rebah et al. 2008; Esakkiraj et al. 2010a). For example, lipase activity of *Staphylococcus xylosus* was evaluated in supernatants generated from boiled fish wastes (tuna, sardine, cuttlefish, and shrimp by-products) supplemented or not with the basal growth medium (containing 17 g/l casein peptone, 5 g/l yeast extract and 2.5 g/l glucose). Depending on the waste origin and the proportion of the added basal medium, lipase activity varied much among the tested samples and supernatants generated from shrimp and cuttlefish by-products exhibited the highest lipase activity (28 U/ml) (Ben Rebah et al. 2008). Similar experiments were reported while growing *Staphylococcus epidermidis* CMST Pi 2 (isolated from the intestine of shrimp *Penaeus indicus*) in a medium containing raw and treated (defatted, alkali and acid hydrolysates) tuna by-products (Esakkiraj et al. 2010a).

Table 4: Production of lipase by different microbial species grown in fish processing by-products

Fish raw materials	Preparation of the growth media	Microbial strains	Lipase activity (U/ml)	References
Defatted tuna by-products	Extraction with chloroform/methanol[a]	*Staphylococcus epidermidis*C-MST Pi 2	12.63	Esakkiraj et al. (2010a)
Defatted tuna by-products	Extraction with chloroform/methanol[a]	*Staphylococcus epidermidis*C-MST Pi 2	14.20	Esakkiraj et al. (2010a)
Tuna by-products	Raw materials were cooked, bones were removed, pressed to remove water and fat, pressed, minced and dried (80 °C, 24–48 h)[a]	*Staphylococcus epidermidis*C-MST Pi 2	8.17	Esakkiraj et al. (2010a)
Acid-hydrolyzed tuna waste	Method described by Gao et al. (2006) [a]	*Staphylococcus epidermidis*C-MST Pi 2	8.03	Esakkiraj et al. (2010a)
Alkali-hydrolyzed tuna waste	Method described by Batista (1999)[a]	*Staphylococcus epidermidis*C-MST Pi 2	8.13	Esakkiraj et al. (2010a)

Shrimp by-products	Raw materials were boiled (100 °C for 20 min) in water and supernatants were recuperated by centrifugation[b]	*Staphylococcus xylosus*	19–28	Ben Rebah et al. (2008)
Cuttlefish by-products	Raw materials were boiled (100 °C for 20 min) in water and supernatants were recuperated by centrifugation[b]	*Staphylococcus xylosus*	5–9.50	Ben Rebah et al. (2008)
Tuna by-products	Raw materials were boiled (100 °C for 20 min) in water and supernatants were recuperated by centrifugation[b]	*Staphylococcus xylosus*	0–4	Ben Rebah et al. (2008)
Sardine by-products	Raw materials were boiled (100 °C for 20 min) in water and supernatants were recuperated by centrifugation[b]	*Staphylococcus xylosus*	0–3	Ben Rebah et al. (2008)
Cod liver oil	1 % of fish oil added to the basal medium	*Staphylococcus epidermidis*C-MST Pi 1	14.8	Esakkiraj et al. (2010b)

[a]The obtained fish powder was added to the basal medium at different proportions

[b]The supernatant was used as a nutrient source for lipase production

A major concern associated with the use of fish wastes for microbial lipase production is the presence of lipids. A defatted fish meat preparation allowed a maximum lipase production (the optimum activity reached 14.20 U/ml while increasing the concentration of defatted fish meat) by *Staphylococcus epidermidis* CMST Pi 2 (Esakkiraj et al. 2010a). This observation confirmed the inhibitory effect of lipids on microorganism growth, as it was found when growing marine bacteria in fish peptone-based media (Vazquez et al. 2004) and *Staphylococcus xylosus* in tuna and sardine by-product-based media (Ben Rebah et al. 2008). In addition to that, regarding the effect of triglycerides on lipase production, cod liver oil showed as a suitable triglyceride to increase lipase production by *Staphylococcus epidermidis* CMST Pi 1 (Esakkiraj et al. 2010b).

Generally, microbial lipase production is highly influenced by medium components like nitrogen sources, carbon sources such as fatty acids, triglycerides and carbohydrates which can stimulate or repress lipase production. However, an adequate balance of nutrients and specific factors would ensure higher lipase activity (Ben Rebah et al. 2008; Ghorbel et al. 2005).

It appears, therefore, that there is a definite need for optimizing the microbial lipase production taking into consideration various factors especially the fish waste composition and microbial strain nutrient requirements. As pointed for proteases, the reported studies offer further exciting possibilities for the industrial use of the fish processing wastes.

CHITINOLYTIC ENZYMES

Chitinolytic enzymes are enzymes that degrade chitin and they are produced by various organisms such as viruses, bacteria, fungi, insects, higher plants and animals (Park et al. 1997). For example, these enzymes have been reported in several microorganisms such as *S. marcescens* (Suzuki et al. 2002), *Bacillus licheniformis* X-74 (Takayanagi et al. 1991),*Aeromonas* sp. No. 10S-24 (Ueda et al. 1995), *Streptomyces* sp. J. 13-3 (Okazaki et al. 1995),*Pseudomonas aeruginosa* K-187 (Wang and Chang 1997) and *Streptomyces griseus* HUT 6037 (Itoh et al. 2002). Recently, many novel chitinase/chitosanase producing microorganisms have been screened out from Taiwan soil (Wang et al. 2011).

Chitinolytic enzymes have various potential applications such as preparation of chitooligosaccharides and *N*-acetyl D-glucosamine which are known to have various biological activities (antimicrobial, antifungal, immunoenhancers, antitumor, etc.) (Tsai et al. 2000; Shen et al. 2009) with high interest in the pharmaceutical sector (Wen et al. 2002). Moreover, chitinases can be used for the control of pathogenic fungi in agriculture (Dahiya et al. 2005) and the degradation of crustacean chitinous waste in sea food industry. These enzymes are also useful for the preparation of single-cell protein, the isolation of protoplasts from fungi and yeast, etc. (Dahiya et al. 2006).

Microbial chitinase has been produced by liquid fermentation processes (batch, continuous and fed-batch fermentation) and is commercially available at a high cost (Dahiya et al. 2006). Generally, the production is controlled by physical factors (aeration, pH, and incubation temperature) and by the growth media components (Miyashita et al. 1991).

In order to increase the supply of active chitinase, it is necessary to reduce the production cost by using wastes for microbial growth. In this perspective, various chitinous materials from marine sources [shrimp shell powder (SSP), squid pen powder (SPP), shrimp and crab shell powder (SCSP)] have been utilized for chitinolytic enzyme production (Table 5) as alternative to waste disposal.

Table 5: Microbial production of chitinolytic enzymes using chitinous materials

Chitinous materials (rate added to the basal medium in % w/v)	Microbial strains	Enzyme type	Enzyme activity (U/mL)	References
Shrimp shell powder (SSP)				
0.5	*Pseudomonas* sp. TKU015	Chitosanase	0.026	Wang et al. (2008e)

0.5	*Pseudomonas* sp. TKU015	Chitinase	0.011	Wang et al. (2008e)
1	*Bacillus cereus*TKU018	Chitosanase	0.020	Wang et al. (2009d)
1.5	*Bacillus subtilis*TKU007	Chitosanase	0.030	Wang and Yeh (2008)
2	*Bacillus cereus*TKU006	Chitinase	0.089	Wang et al. (2009b)
1.5	*Serratia ureilytica*TKU013	Chitinase	0.032	Wang et al.2009a
0.5	*Serratia* sp. TKU020	Chitinase	0.180	Wang et al. (2009c)
0.5	*Serratia* sp. TKU020	Chitosanase	0.400	Wang et al. (2009c)
1	*Serratia* sp. TKU016	Chitosanase	0.022	Wang et al. (2010)
Squid pen powder (SPP)				
3	Bacillus sp. TKU004	Chitosanase	0.14–0.16	Wang et al. (2009e)
1.5	*Serratia ureilytica*TKU013	Chitinase	0.037	Wang et al. (2009a)
2	*Serratia marcescens*TKU011	Chitosanase	0.030	Wang et al. (2008d)
0.5	*Serratia* sp. TKU020	Chitosanase	0.420	Wang et al. (2009c)
1	*Serratia* sp. TKU016	Chitosanase	0.008	Wang et al. (2010)
Shrimp and crab shell powder (SCSP)				
3	*Pseudomonas aeruginosa* K-187	Chitinase	0.650	Wang and Chang (1997)
2	*Bacillus amyloliquefaciens*V656	Chitinase	0.017	Wang et al. (2002a)
2	*Bacillus cereus*YQ308	Chitinase	1.400	Chang et al. (2003)
3	*Bacillus subtilis*W-118	Chitinase	4.200	Wang et al. (2006b)
1	*Monascus purpureus*C-CRC31499	Chitinase	0.080	Wang et al. (2002b)

Dried materials of SPP SSP was prepared as described earlier (Wang et al. 2002a)

Interestingly, using shellfish chitin-containing wastes as the sole carbon/nitrogen source for chitinase/protease-producing strains can be exploited as potential biocontrol agents (Wang et al. 2006b). For example, SCSP of marine wastes induce the production of antimicrobial chitinases by *P. aeruginosa* K-187 (Wang and Chang 1997), *B. amyloliquefaciens* V656 (Wang et al. 2002a), *B. cereus* YQ308 (Chang et al. 2003), *B. subtilis* W-118 (Wang et al. 2006b) and *M. purpureus* CCRC31499 (Wang et al. 2002b). Moreover, *P. aeruginosa* K-187 isolated from the soil using SCSP as sole C/N source showed two kinds of bifunctional chitinases/lysozymes having antibacterial and cell lysis activities against many bacterial species (Wang et al. 1995; Wang and Chang 1997).

Silage obtained by lactic acid fermentation of shrimp head wastes containing chitin, proteins lipids and minerals was also used as substrate and inducer of β-*N*-acetylhexosaminidases of*Verticillium lecanii* in submerged fermentations (SF) and solid-state fermentations (SSF) (Shirai et al. 2001). The addition of sucrose or sugar cane pith bagasse reduces the growth time of *V. lecanii* (Matsumoto et al. 2001, 2004). Interestingly, a mixture of shrimp waste silage and sugar cane pith bagasse in SSF improved significantly the enzyme yield (Matsumoto et al.2004). Chitinous materials from marine sources can be considered as good inducers for microbial chitinase production.

LIGNINOLYTIC ENZYMES

Lignin is the most abundant natural aromatic polymer on earth and degradation of this recalcitrant aromatic polymer is caused by ligninolytic system from natural process in plants, animals, fungi and bacteria (Kirk and Farrell 1987). The ligninolytic system is an extracellular enzymatic complex including peroxidases, laccases and oxidases (Ruiz-Duenas and Martinez2009). Ligninolytic enzymes have capacities to remove xenobiotic substances (such as hydrocarbons, phenols, perchloroethylene, azo dyes, carbon tetrachloride aromatics, pesticides, lignin, humic substances, etc.) which are introduced into the environment by numerous industrial activities. The presence of these xenobiotics in the environment can pose dangerous and unstable situations, because of their possible harmful effects on many organisms (Danzo1997; Birnbaum 1994). Therefore, ligninolytic enzymes have potential applications in various sectors such as chemical, fuel, food, agricultural, paper, textile, cosmetic industrial sectors and in bioremediation purposes (Rodríguez and Toca 2006).

Generally, ligninolytic enzymes exhibit differential characteristics depending on various factors such as species, strains and culture conditions (Kirk and Farrell 1987). However, their production is very expensive and controlled by the raw material used for growth (Hacking1987). In this perspective, various wastes such as apple pomace (Vendruscolo et al. 2008; Gassara et al. 2010), brewery by-product (Bartolome et al. 2003; Gassara et al. 2010), municipal and industrial sludge (Gassara et al. 2010) were used to produce ligninolytic enzymes (pectinase, lignin peroxidase, manganese peroxidase, cellulase and xylanase). The only study reporting the production of ligninolytic enzymes using fishery waste was that of Gassara et al. (2010). In this study, the production of lignin peroxidase, manganese peroxidase and laccase by*Phanerochaete chrysosporium* BKM-F-1767 was investigated in fishery residues (from SAUMMOM Inc., Montreal, Canada) by solid-state cultures (Table 6). Compared to apple wastes, the poor results of *P. chrysosporium* in fishery residues may be related to the unavailability of nutrients and the absence of cellulose in these residues (Gassara et al.

2010). The study of Gassara et al. (2010) on the production of ligninolytic enzymes using fishery residues might offer an impetus for further research in this field. Nevertheless, other microbial strains should be tested using various fish processing wastes as growth media taking into consideration various factors (waste composition, the microbial strains, the strain nutrient requirements, fermentation parameters, etc.).

Table 6; Maximum enzyme activity during growth (expressed in units/gram dry substrate) by solid-state cultures of *Phanerochaete chrysosporium* BKM-F-1767 in fishery residues compared to apple waste with and without inducer (Gassara et al. 2010)

Enzymes	Without inducer	With veratryl alcohol	With copper sulphate
Fishery residues			
Manganese peroxidase	47.4	17	17.4
Lignin peroxidase	nd	nd	nd
Laccase	nd	nd	94.4
Apple waste			
Manganese peroxidase	243.7	631.25	213.5
Lignin peroxidase	nd	nd	nd
Laccase	nd	141.4	719.9

nd not detectable

CONCLUSIONS

The disposal of wastes generated by fishery processing industries represents an increasing environmental and health problem. However, these by-products have attracted considerable attention as an alternative feedstock and energy source, since they are abundantly available. Various microbes are capable of using these substances as carbon and energy sources beneficial in enzyme production process. A number of such substrates have been tested for the cultivation of microorganisms to produce several enzymes (protease, lipase, chitinases, peroxidases, laccases, oxidases, etc.). This may have numerous advantages for enzyme production process, such as superior productivity, simpler techniques, reduced energy requirements and reduced production costs. Generally, fish waste pre-treatments may be necessary to maximize microbial growth and enzyme production. However, each microbial strain has its own special conditions for maximum enzyme production. Therefore, it is of great significance to optimize the medium composition, taking into consideration the variability of fish waste composition, the nutrient requirements of microbial strain and fermentation parameters (pH, temperature, aeration, agitation, etc.). Nevertheless, the improvements in fish waste technology (pre-treatments, characterization, formulation, etc.) are still necessary before large-scale application of this new strategy can be realized.

CONFLICT OF INTEREST

The authors declare that they have no conflict of interest.

REFERENCES

1. Awarenet (2004) Handbook for the prevention and minimization of waste and valorization of by-products in European agro-food industries. Agro-food waste minimization and reduction network (AWARENET). Grow Programme, European Commission, pp 1–7
2. Banerjee UC, Agnihotri R, Bhattacharyya BC (1993) Purification of alkaline protease of*Rhizopus oryzae* by foam fractionation. Bioprocess Eng 9:245–248
3. Banik RM, Prakash M (2004) Laundry detergent compatibility of the alkaline protease from*Bacillus cereus*. Microbiol Res 159:135–140
4. Bartolome B, Gomez-Cordoves C, Sancho AI, Diez N, Ferreira P, Soliveri J, Copa-Patino J (2003) Growth and release of hydroxycinnamic acids from brewer's spent grain by*Streptomyces avermitilis* CECT3339. Enzym Microb Technol 32:140–144
5. Batista I (1999) Recovery of proteins from fish waste products by alkaline extraction. Eur Food Res Technol 210:84–89
6. Ben Rebah F, Frikha F, Kammoun W, Belbahri L, Gargouri Y, Miled N (2008) Culture of*Staphylococcus xylosus* in fish processing by-product-based media for lipase production. Lett Appl Microbiol 47:549–554
7. Birnbaum LS (1994) Endocrine effects of prenatal exposure to PCBs, dioxins, and other xenobiotics: implications for policy and future research. Environ Health Perspect 102:676–679
8. Blanco M, Sotelo CG, Chapela MJ, Perez-Martin RI (2006) Towards sustainable and efficient use of fishery resources: present and future trends. Trends Food Sci Technol 18:29–36
9. Bozzano A, Sarda F (2002) Fishery discard consumption rate and scavenging activity in the northwestern Mediterranean Sea. ICES J of Mar Sci 59:15–28
10. Chang WT, Chen CS, Wang SL (2003) An antifungal chitinase produced by *Bacillus cereus* with shrimp and crab shell powder as a carbon source. Curr Microbiol 47:102–108
11. Chen CC, Chaung HC, Chung MY, Huang LT (2006) Menhaden fish oil improves spatial memory in rat pups following recurrent pentylenetetrazole-induced seizures. Epilepsy Behav 8:516–521
12. Coello N, Brito L, Nonus M (2000) Biosynthesis of L-lysine by *Corynebacterium glutamicum*grown on fish silage. Bioresour Technol 73:221–225
13. Dahiya N, Tewari R, Tiwari RP, Hoondal GS (2005) Production of an antifungal chitinase from Enterobacter sp. NRG4 and its application in protoplast production. World J Microbiol Biotechnol 21:1611–1616
14. Dahiya N, Tewari R, Hoondal GS (2006) Biotechnological aspects of chitinolytic enzymes: a review. Appl Microbiol Biotechnol 71:773–782
15. Danzo BJ (1997) Environmental xenobiotics may disrupt normal endocrine function by interfering with the binding of physiological ligands to steroid receptors and binding proteins. Environ Health Perspect 105:294–301

16. De Azeredo LAI, Freire DMG, Soares RMA, Leite SGF, Coelho RRR (2004) Production and partial characterization of thermophilic proteases from *Streptomyces sp*. isolated from Brazilian cerrado soil. Enzym Microb Technol 34:354–358

17. Ellouz Y, Bayoudh A, Kammoun S, Gharsallah N, Nasri M (2001) Production of protease by*Bacillus subtilis* grown on sardinelle heads and viscera flour. Bioresour Technol 80:49–51

18. Esakkiraj P, Immanuel G, Sowmya SM, Iyapparaj P, Palavesam A (2009) Evaluation of protease-producing ability of fish gut isolate *Bacillus cereus* for aqua feed. Food Bioprocess Technol 2:383–390

19. Esakkiraj P, Austin Jeba Dhas G, Palavesam A, Immanuel G (2010a) Media preparation using tuna-processing wastes for improved lipase production by shrimp gut isolate *Staphylococcus epidermidis* CMST Pi2. Appl Biochem Biotechnol 160:1254–1265

20. Esakkiraj P, Rajkumarbharathi M, Palavesam A, Immanuel G (2010b) Lipase production by*Staphylococcus epidermidis* CMST-Pi 1 isolated from the gut of shrimp *Penaeus indicus*. Ann Microbiol 60:37–42

21. Faid M, Zouiten A, Elmarrakchi A, Achkari-Begdouri A (1997) Biotransformation of fish waste into a stable feed ingredient. Food Chem 60:13–18

22. FAO (2007) La situation mondiale des pêches et de l'aquaculture 2006. Département des pêches et de l'aquaculture, Rome

23. Frost GM, Moss DA (1987) Production of enzymes by fermentation. In: Rehm HJ, Reed G, Kennedy JF (eds) Biotechnology. vol 7a, VCH, Weinheim, pp 65–211

24. Fujiwara N, Yamamoto K (1987) Decomposition of gelatin layers on X-ray films by the alkaline protease from *Bacillus* sp. Hakkokogaku 65:531–534

25. Gao MT, Hirata M, Toorisaka E, Hano T (2006) Acid-hydrolysis of fish wastes for lactic acid fermentation. Bioresour Technol 97:2414–2420

26. Gassara F, Brar SK, Tyagi RD, Verma M, Surampalli RY (2010) Screening of agro-industrial wastes to produce ligninolytic enzymes by *Phanerochaete chrysosporium*. Biochem Eng J 49:388–394

27. Ghorbel S, Soussi N, Ellouz YT, Duffosse L, Guerard F, Nazri M (2005) Preparation and testing of *Sardinella* protein hydrolysate as nitrogen source for extracellular lipase production by*Rhizopus oryzae*. World J Microbiol Biotechnol 21:33–38

28. Gupta R, Beg QK, Khan S, Chauhan B (2002) An overview on fermentation, downstream processing and properties of microbial alkaline proteases. Appl Microbiol Biotechnol 60:381–395

29. Gupta R, Gupta N, Rathi P (2004) Bacterial lipases: an overview of production, purification and biochemical proprieties. Appl Microbiol Biotechnol 64:763–781

30. Hacking AJ (1987) Economic aspects of biotechnology. Cambridge University Press, Cambridge, p 317

31. Haddar A, Fakhfakh-Zouari N, Hmidet N, Frikha F, Nasri M, Sellami-Kamoun A (2010) Low-cost fermentation medium for alkaline protease production by *Bacillus mojavensis* A21 using hulled grain of wheat and *Sardinella* peptone. J Biosc Bioeng 110:288–294

32. Hasan F, Shah AA, Hameed A (2006) Industrial applications of microbial lipases. Enzym Microb Technol 39:235–251

33. Hassan TE, Heath JL (1986) Biological fermentation of fish waste for potential use in animal and poultry feeds. Agric Wastes 15:1–15

34. Itoh Y, Kawase T, Nikajdou N, Fukada H, Mitsutomi M, Watanabe T, Itoh Y (2002) Functional analysis of the chitin binding domain of a family 19 chitinase from *Streptomyces griseus*HUT6037: substrate-binding affinity and *cis*-dominant increase of antifungal function. Biosci Biotechnol Biochem 66:1084–1092

35. Jacob N, Prema P (2006) Influence of mode of fermentation on production of polygalacturonase by a novel strain of *Streptomyces lydicus*. Food Technol Biotechnol 44:263–267

36. Joo HS, Chang CS (2005) Production of protease from a new alkalophilic *Bacillus sp* I-312 grown on soybean meal: optimization and some properties. Proc Biochem 40:1263–1270

37. Kacem M, Sellami M, Kammoun W, Frikha F, Miled N, Ben Rebah F (2011) Seasonal variations of chemical composition and fatty acid profiles of viscera of three marine species from the Tunisian coast. J Aquat Food Prod Technol 20:233–246

38. Karim AA, Bhat R (2009) Fish gelatin: properties, challenges, and prospects as an alternative to mammalian gelatins. Food Hydrocolloids 23:563–576

39. Kim SK, Mendis E (2006) Bioactive compounds from marine processing by-products—a review. Food Res Int 39:383–393

40. Kim YJ, Kim HJ, No JK, Chung HY, Fernandes G (2006) Anti-inflammatory action of dietary fish oil and calorie restriction. Life Sci 78:2523–2532

41. Kirk TK, Farrell RL (1987) Enzymatic "combustion": the microbial degradation of lignin. Annu Rev Microbiol 41:465–501

42. Krieg RN, Holt JG (1984) Bergey's manual of systematic Bacteriology, vol 1. Williams and Wilkins Company, Baltimore, USA, pp 308–429

43. Kumar CG, Takagi H (1999) Microbial alkaline proteases from a bioindustrial viewpoint. Biotechnol Adv 17:561–594

44. Liang TW, Lin JJ, Yen YH, Wang CL, Wang SL (2006) Purification and characterization of a protease extracellularly produced by *Monascus purpureus* CCRC31499 in a shrimp and crab shell powder medium. Enzym Microb Technol 38:74–80

45. Liao PH, Jones L, Lau AK, Walkemeyer S, Egan B, Holbek N (1997) Composting of fish wastes in a full-scale in-vessel system. Bioresour Technol 59:163–168

46. Liaset BD, Lied E, Espe M (2000) Enzymatic hydrolysis of by-products from the fish-filleting industry; chemical characterisation and nutritional evaluation. J Sci Food Agric 80:581–589

47. Ma C, Ni X, Chi Z, Ma L, Gao L (2007) Purification and

characterization of an alkaline protease from the marine yeast *Aureobasidium pullulans* for bioactive peptide production from different sources. Mar Biotechnol 9:343–351

48. Martone CB, Borla OP, Sanchez JJ (2005) Fishery by-product as a nutrient source for bacteria and archaea growth media. Bioresour Technol 96:383–387

49. Matsumoto Y, Revah S, Saucedo G, Hall GM, Shirai K (2001) Chitinases production in solid state fermentation of shrimp waste silage. In: Muzzarelli RAA (ed) Chitin enzymology. Atec Edizioni, Italy, pp 381–389

50. Matsumoto Y, Saucedo-Castaneda G, Revah S, Shirai K (2004) Production of β-*N*-acetylhexosaminidase of *Verticillium lecanii* by solid state and submerged fermentations utilizing shrimp waste silage as substrate and inducer. Process Biochem 39:665–671

51. Miyashita K, Fujii T, Sawada Y (1991) Molecular cloning and characterization of chitinase gene from *Streptomyces lividans*. J Microbiol 137:2065–2072

52. Oh YS, Shih IL, Tzeng YM, Wang SL (2000) Protease produced by *Pseudomonas aeruginosa* K-187 and its application in the deproteinization of shrimp and crab shell wastes. Enzym Microb Technol 27:3–10

53. Okazaki K, Kato F, Watanabe N, Yasuda S, Masui Y, Hayakawa S (1995) Purification and properties of two chitinases from *Streptomyces* sp. J-13-3. Biosci Biotechnol Biochem 59:1586–1587

54. Palaniyappan M, Vijayagopal V, Renukal V, Viswanathan R, Viruthagiri T (2009) Screening of natural substrates and optimization of operating variables on the production of pectinase by submerged fermentation using *Aspergillus niger* MTCC 281. Afr J Biotechnol 8:682–686

55. Pandey A, Selvakumar P, Soccol CR, Nigam P (1999) Solid state fermentation for the production of industrial enzymes. Curr Sci 77:149–162

56. Park JK, Morita K, Fukumoto I, Yamasaki Y, Nakagawa T, Kawamukai M, Matsuda H (1997) Purification and characterization of the chitinase (ChiA) from *Enterobacter* sp. G-1. Biosci Biotechnol Biochem 61:684–689

57. Roberts DP, McKenna LF, Lakshman DK, Meyer SLF, Kong H, Souza JT, Lydon J, Baker CJ, Buyer JS, Chung S (2007) Suppression of dumping-off of cucumber caused by Pythium ultimum with live cells and extracts of *Serratia marcescens* N4–5. Soil Biol Biochem 39:2275–2288

58. Rodríguez S, Toca JL (2006) Industrial and biotechnological applications of laccases: a review. Biotechnol Adv 24:500–513

59. Rubio-Rodriguez N, Beltran S, Jaime I, de Diego SM, Sanz MT, Carballido JR (2010) Production of omega-3 polyunsaturated fatty acid concentrates: a review. Innov Food Sci Emerg Technol 11:1–12

60. Ruiz-Duenas FJ, Martinez AT (2009) Microbial degradation of lignin: how a bulky recalcitrant polymer is efficiently recycled in nature and how we can take advantage of this. Microb Biotechnol 2:164–177

61. Sharma R, Chisti Y, Banerjee UC (2001) Production, purification, characterization and applications of lipases. Biotechnol Adv 19:627–662

62. Shen KT, Chen MH, Chan HY, Jeng JH, Wang YJ (2009) Inhibitory effects of chitooligosaccharides on tumor growth and metastasis. Food Chem Toxicol 47:1864–1871

63. Shirai K, Guerrero I, Huerta S, Saucedo G, Castillo A, Gonzalez RO, Hall GM (2001) Effect of initial glucose concentration and inoculation level of lactic acid bacteria in shrimp waste ensilation. Enzym Microb Technol 28:446–452

64. Souissi N, Ellouz-Triki Y, Bougatef A, Blibech M, Nasri M (2008) Preparation and use of media for protease-producing bacterial strains based on by-products from Cuttlefish (*Sepia officinalis*) and wastewaters from marine-products processing factories. Microbiol Res 163:473–480

65. Suetsuna K (2000) Antioxidant peptides from the protease digest of prawn (*Penaeus japonicus*) muscle. Mar Biotechnol 2:5–10

66. Suzuki K, Sugawara N, Suzuki M, Uchiyama T, Katouno F, Nikaidou N, Watanabe T (2002) Chitinases, A, B and C1 of *Serratia marcescens* 2170 produced by recombinant *E. coli*: enzymatic properties and synergism on chitin degradation. Biosci Biotechnol Biochem 66:1075–1083

67. Takayanagi T, Ajisaka K, Takiguchi Y, Shimahara K (1991) Isolation and characterization of thermostable chitinases from *Bacillus licheniformis* X-74. Biochim Biophys Acta 1078:404–410

68. Taylor MM, Bailey DG, Feairheller SH (1987) A review of the use of enzymes in the tannery. J Am Leather Chem Assoc 82:153–165

69. Toppe J, Albrektsen S, Hope B, Aksnes A (2007) Chemical composition, mineral content and amino acid and lipid profiles in bones from various fish species. Comp Biochem Physiol B 146:395–401

70. Triki-Ellouz Y, Ghorbel B, Souissi N, Kammoun S, Nasri M (2003) Biosynthesis of protease by*Pseudomonas aeruginosa* MN7 grown on fish substrate. World J Microbiol Biotechnol 19:41–45

71. Tsai GI, Wu ZY, Su WH (2000) Antibacterial activity of a chitooligosaccharide mixture prepared by cellulose digestion of shrimp chitosan and its application to milk preservation. J Food Prot 63:747–752

72. Ueda M, Fujiwara A, Kawaguchi T, Arai M (1995) Purification and some properties of six chitinases from *Aeromonas* sp. No. 10S–24. Biosci Biotechnol Biochem 59:2162–2164

73. Underkofler LA, Barton RR, Rennert SS (1958) Microbial process report: production of microbial enzymes and their applications. Appl Microbiol 6:212–221

74. Vazquez JA, Gonzalez MP, Murado MA (2004) A new marine medium—use of different fish peptones and comparative study of the growth of selected species of marine bacteria. Enzyme Microb Technol 35:385–392

75. Vazquez JA, Docasal SF, Miron J, Gonzalez MP, Murado MA (2006) Proteases production by two *Vibrio* species

on residuals marine media. J Ind Microbiol Biotechnol 33:661–668

76. Vazquez JA, Docasal SF, Prieto MA, Gonzalez MP, Murado MA (2008) Growth and metabolic features of lactic acid bacteria in media with hydrolysed fish viscera. An approach to bio-silage of fishing by-products. Bioresour Technol 99:6246–6257
77. Vendruscolo F, Albuquerque PCM, Streit F, Esposito E, Ninow JL (2008) Apple pomace: a versatile substrate for biotechnological applications. Crit Rev Biotechnol 28:1–12
78. Wang SL, Chang WT (1997) Purification and characterization of two bifunctional chitinases/lysozymes extracellularly produced by *Pseudomonas aeruginosa* K-187 in shrimp and crab shell powder medium. Appl Environ Microbiol 63:380–386
79. Wang SL, Yeh PY (2006) Production of a surfactant- and solvent-stable alkaliphilic protease by bioconversion of shrimp shell wastes fermented by *Bacillus subtilis* TKU007. Process Biochem 41:1545–1552
80. Wang SL, Yeh PY (2008) Purification and characterization of a chitosanase from a nattokinase producing strain *Bacillus subtilis* TKU007 using shrimp shell powder as a medium. Process Biochem 43:132–138
81. Wang SL, Chang WT, Lu MC (1995) Production of chitinase by *Pseudomonas aeruginosa* K 187 using shrimp and crab shell powder as a carbon source. Proc Natl Sci Counc ROC B 19:105–112
82. Wang SL, Shih IL, Liang TW, Wang CH (2002a) Purification and characterization of two antifungal chitinases extracellularly produced by *Bacillus amyloliquefaciens* V656 in a shrimp and crab shell powder medium. J Agric Food Chem 50:2241–2248
83. Wang SL, Hsiao WJ, Chang WT (2002b) Purification and characterization of an antimicrobial chitinases extracellularly produced by *Monascus purpureus* CCRC31499 in a shrimp and crab shell powder medium. J Agri Food Chem 50:2249–2255
84. Wang SL, Kao TY, Wang CL, Yen YH, Chern MK, Chen YH (2006a) A solvent stable metalloprotease produced by *Bacillus* sp. TKU004 and its application in the deproteinization of squid pen for β-chitin preparation. Enzym Microb Technol 39:724–731
85. Wang SL, Lin TY, Yen YH, Liao HF, Chen YJ (2006b) Bioconversion of shellfish chitin wastes for the production of *Bacillus subtilis* W-118 chitinase. Carbohydr Res 341:2507–2515
86. Wang SL, Huang TY, Wang CY, Liang TW, Yen YH, Sakata Y (2008a) Bioconversion of squid pen by *Lactobacillusparacasei* subsp *paracasei* TKU010 for the production of proteases and lettuce growth enhancing biofertilizers. Bioresour Technol 99:5436–5443
87. Wang SL, Hsu WT, Yen YH, Wang CL (2008b) Purification and characterization of three novel keratinolytic metalloproteases produced by *Chryseobacterium indologenes* TKU014 in a shrimp shell powder medium. Bioresour Technol 99:5679–5686
88. Wang SL, Wang CY, Huang TY (2008c) Microbial reclamation of squid pen for the production of a novel extracellular serine protease by *Lactobacillus paracasei* subsp. *paracasei* TKU012. Bioresour Technol 99:3411–3417
89. Wang SL, Chen SJ, Wang CL (2008d) Purification and characterization of chitinases and chitosanases from a new species strain *Pseudomonas* sp. TKU015 using shrimp shells as a substrate. Carbohydr Res 343:1171–1179
90. Wang SL, Pen JH, Liang TW, Liu KC (2008e) Purification and characterization of a chitosanase from *Serratia marcescens* TKU011. Carbohydr Res 343:316–1323
91. Wang SL, Lin CL, Liang TW, Liu KC, Kuo YH (2009a) Conversion of squid pen by *Serratia ureilytica* for the production of enzymes and antioxidants. Bioresour Technol 100:316–323
92. Wang SL, Chao CH, Liang TW, Chen CC (2009b) Purification and characterization of protease and chitinase from *Bacillus cereus* TKU006 and conversion of marine wastes by these enzymes. Mar Biotechnol 11:334–344
93. Wang SL, Liou JY, Liang TW, Liu KC (2009c) Conversion of squid pen by using *Serratia* sp. TKU020 fermentation for the production of enzymes, antioxidants, and *N*-acetyl chitooligosaccharides. Process Biochem 44:854–861
94. Wang SL, Chen TR, Liang TW (2009d) Conversion and degradation of shellfish wastes by*Bacillus cereus* TKU018 fermentation for the production of chitosanase and bioactive materials. Biochem Eng J 48:111–117
95. Wang SL, Wu PC, Liang TW (2009e) Utilization of squid pen for the efficient production of chitosanase and antioxidant through prolonged autoclave treatment. Carbohydr Res 344:979–984
96. Wang SL, Chang TJ, Liang TW (2010) Conversion and degradation of shellfish wastes by*Serratia* sp. TKU016 fermentation for the production of enzymes and bioactive materials. Biodegradation 21:321–333
97. Wang SL, Liang TW, Yen YH (2011) Bioconversion of chitin-containing wastes for the production of enzymes and bioactive materials. Carbohydr Polym 84:732–742
98. Wen CM, Tseng CS, Cheng CY, Li YK (2002) Purification, characterization and cloning of a chitinase from *Bacillus* sp. NCTU2. Biotechnol Appl Biochem 35:213–219
99. Xu W, Yu G, Xue C, Xue Y, Ren Y (2008) Biochemical changes associated with fast fermentation of squid processing by-products for low salt fish sauce. Food Chem 107:1597–1604
100. Yamamoto M, Saleh F, Ohtsuka A, Hayashi K (2005) New fermentation technique to process fish waste. Anim Sci J 76:245–248
101. Yano Y, Oikawa H, Satomi M (2008) Reduction of lipids in fish meal prepared from fish waste by a yeast *Yarrowia lipolytica*. Int J Food Microbiol 121:302–307
102. Zampolli A, Bysted A, Leth T, Mortensen A, De Caterina R, Falk E (2006) Contrasting effect of fish oil supplementation on the development of atherosclerosis in marine models. Atherosclerosis 184:78–85

CHAPTER 14

VIABILITY OF SWEET POTATO FLOUR AS CARBON SOURCE FOR THE PRODUCTION OF BIO-FLOC IN FRESHWATER CULTURE OF TILAPIA, OREOCHROMIS SP.

Christopher Marlowe A. Caipang, Han Xuan Choo, Zemin Bai, Huilin Huang, Clara M. Lay-yag

Keywords: Aquaculture Biofloc technology BFT Carbon sources Water quality

INTRODUCTION

High-yield, high-density aquaculture is a rapidly growing industry and is accompanied with increased pollution of the aquatic environments (De Schryver and Verstraete 2009). Aquaculture animals usually absorb 20–25 % of protein in the feed, with the rest discharged into the water in the form of ammonia nitrogen, food remains and excrement, resulting in reduced water quality (Crab et al. 2007; Piedrahita 2003). High concentrations of ammonia–nitrogen and nitrite in aquaculture water also cause disease or even death of the farmed animals, which is a serious impediment to the development of the aquaculture industry (Yang and Qiu 2006). Several new methods have been proposed to solve the problems associated with high concentrations of ammonia–nitrogen and nitrite in the water. Biofloc technology (BFT) is one of these methods, which have shown numerous advantages (Wang et al. 2015). The use of BFT in aquaculture has received considerable attention because the adoption of this technique results in high production yields, feed protein recycling, better water quality, and bacterial infection control (Avnimelech 2006; Crab et al. 2007; Little et al. 2008; Choo and Caipang2015). The ideal candidates for BFT include shrimp, which are detritus feeders and tilapia, which are generally filter feeders (Azim and Little 2008). The basic principle of this technology is the generation of a nitrogen cycle in a stagnant aquaculture system by enhancing the growth of heterotrophic bacteria. This bacterial population utilizes nitrogenous wastes and subsequently serves as food for the cultured species (Avnimelech 1999; De Schryver et al. 2008). Biofloc systems have not only been shown to reduce nitrogenous wastes but they also improve the feed nutrient utilization efficiency and immunity of the cultured animals (Avnimelech 1999; Crab et al. 2009, 2012; Azim and Little 2008; Wang et al. 2015).

Maintaining an optimum carbon:nitrogen (C:N) ratio is crucial to the formation and production of bioflocs (Avnimelech 2006). High C:N ratio ranging 10–20 could promote the formation of biofloc as shown in previous studies (Avnimelech 1999; Hargreaves 2006; Asaduzzaman et al.2008; Ballester et al. 2010). However, the C:N ratio in most commercial feeds

Citation: Christopher Marlowe A. Caipang, Han Xuan Choo, Zemin Bai, Huilin Huang, Clara M. Lay-yag, Viability of sweet potato flour as carbon source for the production of biofloc in freshwater culture of tilapia, Oreochromis sp, DOI 10.1007/s40071-015-0117-7.

is approximately 10; thus, additional carbon is required to increase the ratio and maintain biofloc production (Avnimelech 1999; De Schryver et al. 2008). The C:N ratio in feed could be increased by the addition of extra carbohydrate to the basic feed, or by the reduction of the crude protein content in feed (Wang et al. 2015). Although bioflocs have the capacity to augment the nutritional requirements of fish and shrimp in aquaculture, a number of studies have indicated that the ability of bioflocs to control the water quality in the culture system and their nutritional properties are largely influenced by the type of carbon source that was used to produce the flocs (Crab 2010; Crab et al. 2010). There are several organic carbon sources that could favor the growth of specific bacteria, protozoa and algae, and hence influence the microbial composition of the bioflocs (Crab 2010). These differences in the composition of the bioflocs also affect their nutritional properties (Crab 2010; Ekasari et al. 2010, 2014).

Feeding experiments also demonstrated that the type of carbon source affected the availability, palatability and digestibility of these bioflocs (Crab 2010; Crab et al. 2010; Ekasari et al. 2010). For example, Ekasari et al. (2010) showed that regardless of the salinity, bioflocs that are produced using glycerol as a carbon source had higher total n-6 PUFAs than that of using glucose during the culture of tilapia. This was also shown by the studies done by Crab (2010). In shrimp culture, molasses is the most widely used carbon source for biofloc production (Baloi et al. 2013; Schveitzer et al. 2013; Correia et al. 2014). Previously, we compared wheat flour and corn flour as carbon sources for biofloc production in freshwater tilapia and showed that there were no significant effects on water quality between the biofloc groups and the control (Caipang et al.2015).

However, in terms of weight gain and feed conversion ratio (FCR), wheat flour performed better than corn flour, indicating that the former can be a good candidate as a source of carbon during biofloc production in freshwater tilapia culture. In spite of the popularity of molasses and wheat flour as carbon sources for biofloc production, further research should focus on the use of low-cost non-conventional agro-industrial products as carbon source. Different carbon sources stimulate the growth of the indigenous microbiota in various ways and would greatly influence the water quality, in situ feed production and utilization of the flocs by the cultured organisms. Various products or residues of the local industry can provide a low-cost external carbon source for application in biofloc technology in the culture of fish, but will need extensive research before implementation on a larger scale. Hence, this study evaluated the use of sweet potato (*Ipomoea batatas*) flour, an abundant and cheap source of carbon for biofloc production in terms of its effects on water quality, microbial population and yield during culture of tilapia in freshwater. The effects of sweet potato flour were compared against the performance of wheat flour to determine whether the former can be used as an alternative source of carbon for biofloc production in tilapia culture.

MATERIALS AND METHODS

Glass Aquaria and Experimental Design

The experiment was carried out at the Aquaculture Workstation of Temasek Polytechnic, Singapore. Glass aquaria (20 L capacity) were cleaned, disinfected and sundried prior to the experiment. Each aquarium was filled with 1 L of water taken from Bedok Reservoir, Singapore, scaled up to 15 L using dechlorinated freshwater and provided with aerators. After the addition of water, 100 mL of *Chlorella* (density of 10^6–10^8 cells mL^{-1}) was added to each aquarium and supplemented with 8.5 mL of the algal nutrient. The treatments consisted of A: biofloc production using wheat flour as carbon source; B: biofloc production using sweet potato flour as carbon source and C: no biofloc production with no addition of flour. Each treatment was done in duplicate.

Biofloc Production and Fish Stocking

For biofloc production in glass aquaria, the corresponding flour and commercial feed (crude protein of 35 %) combinations were added following a previously described procedure (Caipang et al. 2015). The amount of flour that was added together with the commercial feeds had a C:N ratio of 16:1 following the calculations of Crab et al. (2012). The use of the commercial feed for biofloc production in glass aquaria was to trigger the production of nitrogenous wastes and to be subsequently utilized by the heterotrophic bacteria that reside in the rearing water. Feeds and/or flour were added to each aquarium every morning at 9 AM. For the control group, 20 % of the total water volume was replaced every 2 days, while for the treatment groups, the water was replaced at the same volume as the control every 10 days or when the total suspended solids (TSS) exceeded 500 mg L^{-1}. At the 11th day of the experiment, tilapia fingerlings with average weight of 15 g were stocked at a density of 4 fish in each aquarium (4 g L^{-1}). The experiment lasted for 41 days.

Water Quality Monitoring

Water samples for determination of selected physico-chemical parameters were collected at the initial, 7th, 30th and 40th day of the experiment. Total ammonia–nitrogen (TAN), nitrite (NO_2-N) and nitrate (NO_3-N) from the water samples were determined using commercial water quality kits following the procedures of the manufacturer (Hach Company, Colorado, USA). TSS and biochemical oxygen demand (BOD) were analyzed using the procedures described in APHA (1989). The water samples for TSS and BOD determination were obtained and processed following previously described procedures (Caipang et al. 2015).

Bacterial Counts

Total heterotrophic bacteria and bacilli from the water in all glass aquaria were enumerated on the 20th and 40th day of the experiment. The enumeration of total heterotrophic bacteria provided an estimate of the culturable bacterial population in the rearing water during biofloc production, while the bacilli counts would likely indicate the presence of potential probiotic candidates that could be utilized in further studies. The water samples were serially diluted using 1 × PBS and 100 μL of the various dilutions was plated onto both nutrient agar plates and MRS agar plates for the enumeration of total heterotrophic bacteria and bacilli, respectively. Each dilution was performed in duplicate. The plates were then incubated at 28 °C for 2 days and bacterial counts were expressed as log10 colony-forming units (CFU) mL^{-1}.

Assessment of Fish Yield

At the end of the experiment, the weights of the tilapia were taken and recorded. The average body weight (ABW), average daily gain (ADG), specific growth rate (SGR) and feed conversion rate (FCR) were calculated from each group.

Data analyses

The data are represented as mean ± SD. One-way ANOVA was used to determine significant differences in the fish yield parameters among the different groups; thereafter, the differences between the groups were checked by Tukey's Multiple Comparison Test. All significance levels were set at $p < 0.05$.

RESULTS AND DISCUSSION

Physico-Chemical Parameters of the Rearing Water

The total ammonia–nitrogen concentration in the wheat flour treatment group was higher compared with the control and the sweet potato treatment group (Fig. 1a). There was a large degree of fluctuation in the NO_2-N concentration throughout the experiment for all the treatment groups with no observable trend among the treatment groups and the control (Fig. 1b). However, at the last sampling date, the NO_2-N concentration in the sweet potato flour group was significantly lower compared to the wheat flour group and the control. The levels of NO_3-N were generally lower in the sweet potato group than the wheat flour group and the control during the duration of the experiment (Fig. 1c).

These results suggest that sweet potato as a carbon source for biofloc production in a small-scale level might result in better water quality due to a reduction in the levels of the various forms of nitrogenous wastes within the system when this particular flour was used as a carbon source. In a previous study using wheat flour and corn flour as sources of carbon for small-scale biofloc production in freshwater tilapia showed variability in the levels of ammonia, nitrate and nitrite following addition of these carbon sources (Caipang et al. 2015). How these differences in the fluctuations of the nitrogenous wastes following the addition of these carbon sources related to the type of flour remain to be supported by further studies. Until such thorough experimentation is done, no firm conclusion can be advanced with regard to the best carbon source that could help mitigate the occurrence of high levels of nitrogenous wastes in freshwater BFT system for tilapia. However, the present study showed that the use of sweet potato in small-scale production of biofloc in freshwater tilapia culture seemed to be beneficial in lowering nitrogenous wastes in comparison with using either wheat or corn flour.

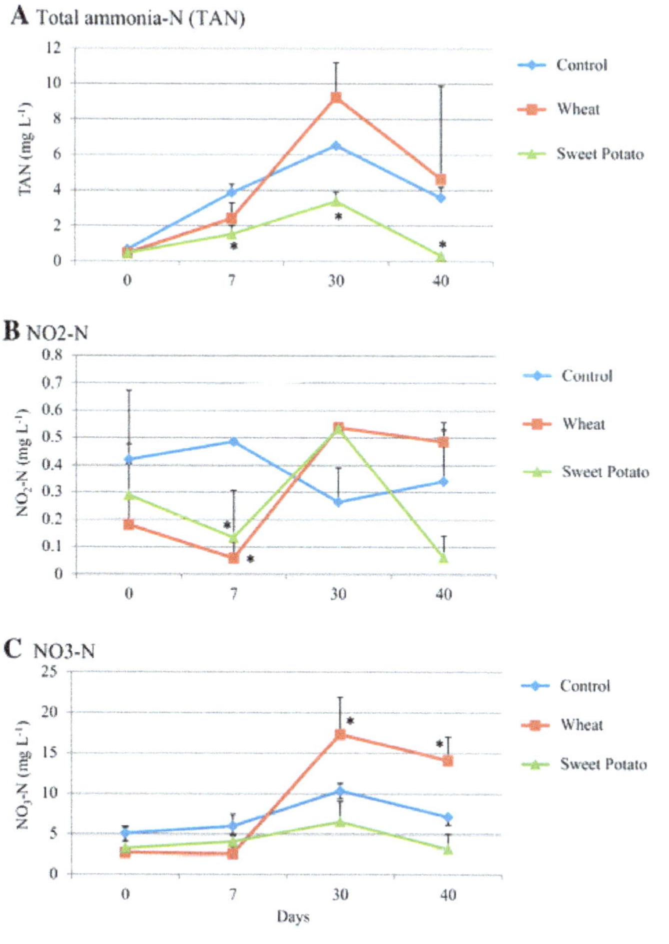

Figure 1: Levels of **a** total ammonia–nitrogen (TAN), **b** nitrite-N (NO_2-N) and **c** nitrate-N (NO_3-N) in the rearing water of freshwater tilapia with and without biofloc. *Asterisk* at each time point indicates significant difference at $p < 0.05$

The total suspended solids (TSS) in both the biofloc treatment groups were higher compared to the control (Fig. 2a). The levels of the biochemical oxygen demand (BOD) as shown in Fig. 2b corresponded to the TSS concentration, where higher readings for both parameters were obtained in both treatment groups compared to the control. The results obtained were similar to the findings of Azim and Little (2008), where TSS concentrations and BOD levels are higher in the BFT tanks compared to the control. Caipang et al. (2015) also obtained similar trends for the levels of TSS and BOD when they used wheat and corn flour as carbon sources for small-scale biofloc production in freshwater tilapia culture. Regardless of the type of carbon source, addition of such will result in higher BOD and TSS in BFT systems.

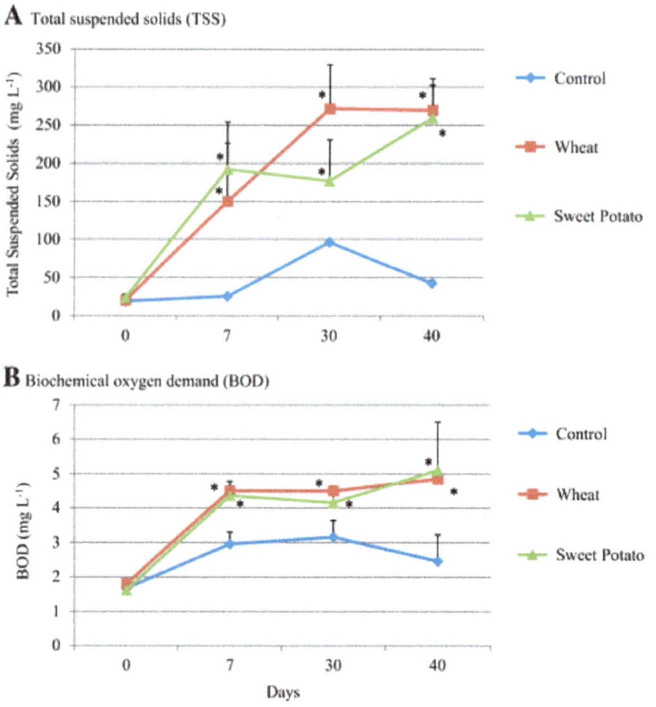

Figure 2: Levels of **a** total suspended solids (TSS) and **b** biochemical oxygen demand (BOD) in the rearing water of freshwater tilapia with and without biofloc. *Asterisk* at each time point indicates significant difference at $p < 0.05$

Bacterial Load

In Fig. 3a, it can be observed that total bacteria counts in the BFT groups were higher than the control throughout the duration of the experiment. This could be credited to the increased in bacteria population growth from the addition of carbon source in the BFT tanks (Avnimelech1999). Similar results were also obtained in our previous study using wheat and corn flour as carbon sources for biofloc production in freshwater tilapia (Caipang et al. 2015).

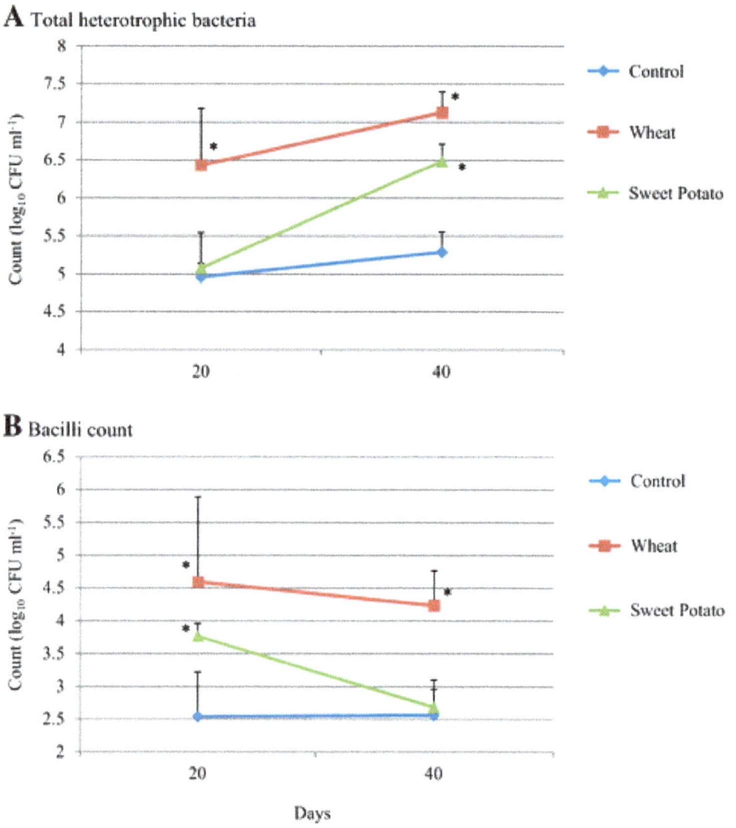

Figure 3: Total counts for **a** heterotrophic bacteria and **b** bacilli population in the rearing water of freshwater tilapia with and without biofloc. *Asterisk* at each time point indicates significant difference at $p < 0.05$

For the bacilli count, the wheat flour biofloc group has a consistently high bacilli population throughout the experiment period as shown in Fig. 3b. However, bacilli count was on a constant decrease in the sweet potato treatment, reaching a point where bacilli count was close to those of the control group at the last sampling date. An earlier study on the use of various carbon sources for biofloc production revealed that the organic carbon source of choice determines the composition of the flocs produced (De Schryver et al. 2008). This could have an effect on the bacterial populations that are closely associated with the flocs. The decreased population of bacilli in the group added with sweet potato flour could be an indication that this particular type of flour may not be a good source of carbon to maintain a stable population of bacilli compared to wheat flour. Additional research is needed to support the present study.

The presence of higher bacteria population in the BFT group as a result of the addition of carbon source could have resulted in the increased volume of the flocs. Higher biofloc production triggered the increase in TSS and a subsequent rise in respiration, which led to higher BOD levels. Previous studies on the use of bioflocs in tilapia culture are in agreement with the findings of the present study (Azim and Little 2008; Crab et al. 2010; Caipang et al.2015). It is interesting to find out whether a correlation exists between bacterial population and the levels of nitrogenous wastes in the rearing water as this relationship could provide insight on the possible role of these bacteria in the dynamics of nitrogenous waste in the water. Such relationship was not established in the present study because there were differences in the sampling dates for determination of nitrogenous wastes and the enumeration of bacterial counts. However, this link between the two parameters will be the point of focus in future studies involving the use of BFT in tilapia culture.

Fish Yield and Growth Parameters

Feed conversion rate (FCR) was significantly lower in the treatments compared to the control, with the FCR in the wheat flour treatment being lower compared to the sweet potato flour treatment (Table 1). The average daily gain (ADG; g day^{-1}) in the wheat flour BFT treatment was almost similar to the control, while the ADG in the sweet potato flour was lower compared to both the control and wheat flour BFT treatment. The specific growth rate (SGR) of tilapia in the sweet potato treatment was lower than the control and wheat flour BFT treatment. No fish mortality was observed in both the BFT and non-BFT groups during the monitoring period.

Table 1: Production of tilapia from biofloc and non-biofloc aquaria

Parameters	Control	Wheat	Sweet potato
Biomass (g)	75.8 ± 2.55	71.0 ± 2.55	54.2 ± 4.24*
Final weight (g)	33.5 ± 0.05	32.7 ± 0.66	28.5 ± 1.07*
FCR	0.79 ± 0.03	0.34 ± 0.01*	0.44 ± 0.03*
ADG (g day^{-1})	0.62 ± 0.01	0.59 ± 0.03	0.46 ± 0.04*
SGR (% d^{-1})	2.68 ± 0.01	2.61 ± 0.06	2.20 ± 0.10*

*Significant difference from the control at $p < 0.05$

The results obtained indicate that sweet potato flour as a carbon source could result in lower FCR compared to conventional tanks fed with commercial feed, suggesting the increased utilization of protein in these treatments. However, FCR in the sweet potato treatment is higher compared to the wheat flour treatment. Relatively slow growth rate was also observed in the sweet potato treatment compared to both control and wheat flour treatment. These could be credited to nutritional differences in flocs produced between the wheat flour treatment and sweet potato flour treatment, resulting in slower growth in fish feeding on flocs produced with sweet potato flour. This is supported by the study of Ekasari et al. (2010), which shows that there is a difference in nutritional content of biofloc produced by different carbon source. Further research on the nutritional difference in flocs between the two treatments is needed in this aspect.

CONCLUSION

In conclusion, small-scale studies showed that the bioflocs produced during the addition of sweet potato flour as a carbon source could lower nitrogenous waste and to a certain degree serve as food for tilapias in freshwater culture. However, to ensure that the fish are able to attain the desired growth and weight gain, efficient feeding management is needed when rearing fish in this particular type of system using sweet flour as carbon source for biofloc production.

ACKNOWLEDGMENTS

The data included in this paper are part of the Major Student's Project (Optimization Studies of Biofloc Technology (BFT) in Freshwater Tilapia Culture: Effects on Water Quality and Fish Growth) of H.X. Choo. The authors gratefully acknowledge the funding support provided by the School of Applied Science, Temasek Polytechnic.

Compliance with ethical standards

Conflict of interest

On behalf of the co-authors, I declare that there is no conflict of interest regarding the manuscript.

REFERENCES

1. APHA (American Public Health Association), American Water Works Association, and Water Pollution Control Association (1989) Standard methods for the examination of water and wastewater, 17th edn. American Public Health Association, Washington, DC

2. Asaduzzaman M, Wahab MA, Verdegem MCJ, Huque S, Salam MA, Azim ME (2008) C/N ratio control and substrate addition for periphyton development jointly enhance freshwater prawn*Macrobrachium rosenbergii* production in ponds. Aquaculture 280:117–123
3. Avnimelech Y (1999) Carbon and nitrogen ratio as a control element in aquaculture systems. Aquaculture 176:227–235
4. Avnimelech Y (2006) Bio-filters: the need for an new comprehensive approach. Aquac Eng 34:172–178
5. Azim ME, Little DC (2008) The biofloc technology (BFT) in indoor tanks: water quality, biofloc composition, and growth and welfare of Nile tilapia (*Oreochromis niloticus*). Aquaculture 283:29–35
6. Ballester ELC, Abreu PC, Cavalli RO et al (2010) Effect of practical diets with different protein levels on the performance of *Farfantepenaeus paulensis* juveniles nursed in a zero exchange suspended microbial flocs intensive system. Aquac Nutr 16:163–172
7. Baloi M, Arantes R, Schveitzer R, Magnotti C, Vinatea L (2013) Performance of Pacific white shrimp *Litopenaeus vannamei* raised in biofloc systems with varying levels of light exposure. Aquacult Eng 52:39–44
8. Caipang CMA, Choo HX, Bai Z, Huang H, Lay-Yag CM, Lim J (2015) Small-scale production of biofloc using various carbon sources for the freshwater culture of tilapia, *Oreochromis* sp. ABAH Bioflux 7:103–111
9. Choo HX, Caipang CMA (2015) Biofloc technology (BFT) and its application towards improved production in freshwater tilapia culture. AACL Bioflux 8:362–366
10. Correia ES, Wilkenfeld JS, Morris TC, Wei L, Prangnell DI, Samocha TM (2014) Intensive nursery production of the Pacific white shrimp *Litopenaeus vannamei* using two commercial feeds with high and low protein content in a biofloc-dominated system. Aquacult Eng 59:48–54
11. Crab R (2010) Bioflocs technology: an integrated system for the removal of nutrients and simultaneous production of feed in aquaculture. PhD thesis, Ghent University
12. Crab R, Avnimelech Y, Defoirdt T, Bossier P, Verstraete W (2007) Nitrogen removal techniques in aquaculture for a sustainable production. Aquaculture 270:1–14
13. Crab R, Kochva M, Verstraete W, Avnimelech Y (2009) Bio-flocs technology application in over-wintering of tilapia. Aquacult Eng 40:105–112
14. Crab R, Chielens B, Wille M, Bossier P, Verstraete W (2010) The effect of different carbon sources on the nutritional value of bioflocs, a feed for *Macrobrachium rosenbergii* postlarvae. Aquac Eng 41:559–567
15. Crab R, Defoirdt T, Bossier P, Verstraete W (2012) Biofloc technology in aquaculture: beneficial effects and future challenges. Aquaculture 365–357:351–356
16. De Schryver P, Verstraete W (2009) Nitrogen removal from aquaculture pond water by heterotrophic nitrogen assimilation in lab-scale sequencing batch reactors. Bioresour Technol 100:1162–1167
17. De Schryver P, Crab R, Defoirdt T, Boon N, Verstraete W (2008) The basics of bio-flocs technology: the added value for aquaculture. Aquaculture 277:125–137
18. Ekasari J, Crab R, Verstraete W (2010) Primary nutritional content of bio-flocs cultured with different organic carbon sources and salinity. HAYATI J Biosci 17:125–130
19. Ekasari J, Angela D, Waluyo SH, Bachtiar T, Surawidjaja EH, Bosier P, De Schryver P (2014) The size of biofloc determines the nutritional composition and the nitrogen recovery by aquaculture animals. Aquaculture 426–427:105–111
20. Hargreaves JA (2006) Photosynthetic suspended-growth systems in aquaculture. Aquac Eng 34:344–363
21. Little DC, Murray JF, Azim ME, LeschenW Grady K, Young J, Watterson A (2008) Warm-water fish production in the UK: limits to green growth? Trends Food Sci Technol 19:255–264
22. Piedrahita RH (2003) Reducing the potential environmental impact of tank aquaculture effluents through intensification and recirculation. Aquaculture 226:35–44
23. Schveitzer R, Arantes R, Costodio PFS, do Espirito Santo CM, Arana LV, Seiffert WQ, Andreatta ER (2013) Effect of different biofloc levels on microbial activity, water quality and performance of *Litopenaeus vannamei* in a tank system operated with no water exchange. Aquacult Eng 56:59–70
24. Wang G, Yu E, Xie J, Yu D, Li Z, Luo W, Qiu L, Zheng Z (2015) Effect of C/N ratio on water quality in zero-water exchange tanks and the biofloc supplementation in feed on the growth performance of crucian carp, *Carassius auratus*. Aquaculture 443:98–104
25. Yang SP, Qiu DQ (2006) Water quality in the high-density shrimp culturing ponds. Fisheries Sci 25:459–462

CHAPTER 15

SOCIAL AND ECOLOGICAL CHALLENGES OF MARKET-ORIENTED SHRIMP FARMING

Ngo Thi Phuong Lan

Keywords: Shrimp farming, the Mekong Delta of Vietnam, Market-oriented production, Risky business, Social and ecological challenges, Labor migration, Risk mitigation

BACKGROUND

Shrimp production has emerged as one important livelihood in the Mekong Delta of Vietnam. It promises to bring great rewards to farmers who have shifted from rice cultivation to shrimp farming. In the context of a sharp increase in shrimp farming area and the number of households engaging in this industry, I carry out this research to examine how this kind of livelihood affects the region and in the context of present practices of shrimp farming whether this is a sustainable livelihood for farmers.

INTRODUCTION

As a fertile delta having the greatest potential for agricultural production in Vietnam, in the context of increasing integration into the global market, the Mekong Delta has contributed greatly to the achievement of Vietnam's agricultural outputs. Under the impacts of various policies, trade liberation, and technological improvement, agricultural activities in the Delta are increasingly market-oriented. Farmers in the Delta have increasingly cultivated cash crops and become engaged in the global market. High-valued cash crops have increased in scale and intensity in a region where land is under de-facto private ownership. Consequently, although the Delta constitutes only 12% of the country's area; 30% of national agricultural land; and 21% of the total population, annually, the Mekong Delta contributes 50% of rice production; 90% of exported rice; 80% of aquaculture production; 60% of exported aquaculture products; 60% of total exported products for the country; and it contributes 18% towards the nation's GDP (Guidance Board of Vietnam's Southwest region, Ban Chỉ đạo Tây Nam Bộ và Trung tâm thông tin Sài Gòn 2005, p.17). Along with this economic growth at macro level, the Mekong Delta has undergone significant changes at the micro level, particularly with respect to livelihoods of the Delta's inhabitants. During recent decades, structural transformation in agriculture has become a prominent phenomenon in the Mekong Delta. Technological advances which provide high-productivity plant and animal breeds, new fertilizers and pesticides have played a crucial role in the transformation of the region. Further structural transformation in the Mekong Delta's agriculture has taken place with the shift from traditional rice cultivation to new agricultural products like fruit and aquaculture. In particular, the shift to export-oriented shrimp farming has been accelerated. For example, in Ca Mau province in the lower Mekong Delta, the shrimp cultivation area increased from 90,551 ha in 1999 to 239,398 ha in 2003 and 257,000 ha in

Citation: Ngo Thi Phuong Lan, Social and ecological challenges of market-oriented shrimp farming, DOI: 10.1186/2193-1801-2-675

2008 (Mai Trong Thong et al. 2006, p.7; information from Ca Mau province People's Committee). The rapid development of the shrimp industry in the Mekong Delta has had considerable influence on the Delta's economy, society and ecology.

Shrimp Farming In Vietnam: Market-Oriented Production

With 3,260 km of coastline, Vietnam is a suitable place for brackish aquaculture (Le Ba Thao2002, p.8). According to statistics from the Ministry of Agriculture and Rural Development (MARD) of Vietnam, Vietnam had 613,000 ha under shrimp production in 2010. Because of a wide discrepancy in temperature between seasons and low temperatures in winter, the northern part of Vietnam is not suitable for raising shrimp all year round. According to the Fisheries Informatic Center of the Vietnamese Directorate of Fisheries, sea and brackish water shrimp in the northern parts of Vietnam are raised in Hai Phong, Nam Dinh, Thai Binh, Ninh Binh, and Quang Ninh provinces. The central part of Vietnam is known both for shrimp farming and hatcheries. In this region, shrimp farms are concentrated in the provinces of Thanh Hoa, Nghe An, Ha Tinh, Quang Binh, Quang Tri, Ninh Thuan, Binh Thuan, Da Nang, Quang Nam, Quang Ngai, Binh Dinh, Phu Yen and Khanh Hoa. Khanh Hoa province is considered as a cradle of the shrimp farming movement thanks to its successful experiment in artificially producing shrimp fry in the 1980s. Presently, Khanh Hoa is still a well-known centre for shrimp hatcheries in the country.

In the south of Vietnam, shrimps are raised in the southeast region (Ho Chi Minh City, Dong Nai, Ba Ria – Vung Tau provinces), and in the Mekong Delta (Long An, Tien Giang, Ben Tre, Kien Giang, Tra Vinh, Soc Trang, Bac Lieu, Ca Mau provinces). Given its appropriate soil and climate conditions, the Mekong Delta, which is naturally bestowed with a dense network of rivers and ditches, fertile alluvial plains, and diversified mangrove forest, produces 75% of the country's shrimp products (GSO 2009, p.344). According to the Ministry of Agriculture and Rural Development, in 2008, with 539,607 ha, the Mekong Delta was the biggest shrimp producing region in Vietnam and accounted for approximately 89.3% of the country's shrimp area. Among eight provinces, Ca Mau has the largest area (50% of the region's total area) dedicated to shrimp raising. Results from the 2006 Rural, Agricultural and Fishery Census, conducted by the Vietnam General Office of Statistics, also confirm the importance of the Mekong Delta in terms of the quantity of shrimp farming households. Out of 337,614 households in Vietnam engaged in shrimp farming, 292,522 (86%) of households were from the Mekong Delta. The average size of shrimp farming land of 53% of these households was 0.5 ha, and only 16% of these households engaged in intensive and semi-intensive shrimp culture while 87% of shrimp farming households in the central part of Vietnam engaged in this type of shrimp culture (GSO 2007, p.436, p.437, p.448, p.449, p.453).

The years 2000 to 2002 were a landmark period for the shrimp industry in Vietnam as well as in the Mekong Delta with a dramatic increase in shrimp farming areas and production. Vietnam produces both fresh-water shrimp and brackish-water shrimp but the latter accounts for 97% of the total production. Vietnam's shrimp production increased from under 200 tons in 1976 to 158,755 tons per annum in 2001 (cited in EJF 2003, p.4). With this boom, since then Vietnam always entered the top five leading shrimp exporters in the world. This achievement is a result of various national policies, starting with the Doi Moi policy and various subsequent policies in support of shrimp farming. Current major export markets for shrimp from Vietnam are the United States and Japan.

Shrimp raising in the Mekong Delta is widely said to have emerged in Ca Mau province in the 1980s when farmers spontaneously let saline water into rice fields to raise shrimp naturally. However, some reports indicate that rice-shrimp rotation farming was applied in the 1960s (Vuong Do Quang and Lin 2001, p.4) and Soc Trang province was believed to be the first place to officially adopt this model in late 1980s (interviews with leaders of Soc Trang province). Although being officially permitted to raise shrimp in their rice fields by Resolution No. 09 in 2000 and Decision 173/2001/Qđ-TTg in 2001, shrimp farming in specific parts of the Mekong Delta provinces was given official permission at different times. For example, in Ca Mau province, shrimp were officially raised large-scale in mangrove forests as a commodity in 1992 (Vo Thanh Binh 2006, p.13); in Bạc Lieu province in 1998 (Guidance Board of Vietnam's Southwest Region Ban Chỉ đạo Tây Nam Bộ và Trung tâm thông tin Sài Gòn 2005, p.119), and in Long An province in 1994. Moreover, as a result of the locally specific implementation of national policies, each locality has its own trajectory of transformation from rice to shrimp cultivation. In Long An (the case of Tan Chanh commune), for example, the shift was gradual, while in Ca Mau (the case of Hoa My commune) it happened suddenly and simultaneously. Thanks to high value of shrimp, with favorable conditions, the Delta's farmers have rapidly shifted to this new livelihood. For example, when farmers were officially allowed to raise shrimp in rice fields, the shrimp-raising acreage of Cai Nuoc district, Ca Mau province increased by 8.5 times, reaching 65,000 ha in the period from 1997 to 2002 (Mai Trong Thong et al. 2006, p.8). In Can Duoc district of Long An province the shrimp-raising acreage increased by 5.6 times with 935 ha in the period from 1995–2002 (Tan Chanh's people committee, 2009).

In the Mekong Delta, the majority of farmers apply extensive and semi-extensive types of shrimp culture. According to the Ministry of Agriculture and Rural Development, in 2008, only 8% of the Delta's shrimp area was used for intensive shrimp culture while the rest of shrimp area consisted of extensive and advanced-extensive methods of shrimp raising. Research undertaken in 2009 and early 2010 in Hoa My commune of Ca Mau province showed 91% of the shrimp-farming households (n = 202) studied adopted extensive shrimp culture, the remaining 9% of households engaged in intensive and semi-intensive types of culture. In Tan Chanh, the biggest shrimp producing commune in Can Duoc district of Long An province, advanced extensive or semi-extensive culture was adopted by 100% of the households (n = 152) studied.

In principle, the density of shrimp post larvae is from three to five shrimps per m^2 under the advanced-extensive method of shrimp production. However, in Long An province, in order to have high productivity despite limited land, people release shrimp fry into their ponds with a much higher stocking density (from 10 to 20 fry per m^2) than the level recommended for extensive shrimp farming. Besides the two aforementioned types of shrimp culture, a minority of famers in both study sites also practice industrial shrimp farming which requires a much higher capital investment per ha per year, normally 40 times higher than the extensive type of farming and five times higher than the advanced-extensive type (calculated from research in the Ca Mau province study site). In contrast to other types of shrimp culture in which shrimp fry are released and harvested once per crop, in the extensive type, people usually release fry once every one to two months and harvest shrimp every day or every one to two weeks.

Presently, people mainly raise black-tiger shrimp (Penaeus Monodon), although white-legged shrimp (Penaeus Vannamei) have also been raised recently. The latter type of shrimp was introduced in Central Vietnam in 2000 and permitted to be raised in the Mekong Delta only in 2008. Although they grow more quickly and have a higher yield, white-legged shrimp usually cause extreme pollution to shrimp ponds due to their higher waste excretion. Thanks to their high effectiveness in terms of high yield, although being previously claimed to be a potential cause of disease outbreak and harmful to other species, white-legged shrimp are under consideration for removal from the list of harmful species and considered as an appropriate species for commercial production.

Along with this dramatic shifting in livelihoods, coastal areas in the Mekong Delta are witnessing emerging socio-economic and environmental challenges, for which people have to adopt various strategies to cope. Given these challenges, the promising shrimp industry becomes a risky business.

Ecological and Socio-Economic Challenges Amongst Shrimp Farming Communities in the Mekong Delta

The booming shrimp industry, which relies on saline water ecology, has been associated with serious environmental and social impacts. Some studies report that globally environmental problems associated with shrimp farming include destruction of mangroves and other wetland habitats; dispersion of chemicals and nutrients into the environment; pollution and salinity; and the depletion and biological contamination of wild fish and shrimp populations. Social impacts associated with shrimp farming include the increase of poverty and landlessness, food insecurity, and impacts on health and education (EJF 20032004; Barraclough and Finger-Stich,1996; Preston and Clayton 2003). In this part, drawing primarily on information from current research amongst the two aforementioned research study sites of Hoa My commune and Tan Chanh commune, the dynamics of these ecological and socio-economic challenges of shrimp farming in the Mekong Delta will be examine in detail.

Ecological Challenges

With the shift from rice to shrimp farming, local ecologies and landscapes of rural communities have experienced tremendous changes. In the two brackish-water study site communities that have now adopted the practice of rice-shrimp farming, inhabitants' past living patterns revolved around rice cultivation during the rainy season. Their leisure time was in the dry season. At present, the rhythm of farmers' lives is different. They mainly raise shrimp in the dry season and their leisure time is in the rainy season. In order to raise shrimp, people have converted their rice fields, residential land, or gardens into shrimp ponds. Commune landscapes are no longer characterized by green or yellow rice fields but treeless and spacious brackish shrimp ponds. The system of dikes and dams preventing saline movement was removed, while many canals were enlarged or newly dug. As a result, a dense and vast system of canals characterizes shrimp-farming communities, especially those engaging in extensive shrimp culture. Various fruit gardens no longer exist partly due to their clearance for shrimp ponds and partly due to saline intrusion. Nipa palms used to be grown across large areas of brackish-water communities to supply roofing materials for houses. Now, this kind of plant occupies only a small portion of the landscape, partly because of the space given over to shrimp ponds.

The diversity of local wild aquatic species is also affected by shrimp farming. At the two study sites of Hoa My and Tan Chanh communes, the practice of using poisonous chemicals to kill various fish competing with shrimp for food and killing shrimp fries in shrimp ponds contributes seriously to the decrease in wild species in these areas. Farmers report that many species of local fish have decreased dramatically, in type and quantity, owing to this practice.

Soil and water pollution is another serious problem associated with the shrimp industry. Besides the poor quality of shrimp fry, pollution is said to be an important cause of continuous shrimp losses. The practice of releasing high stocking densities of shrimp fry into ponds to maximize profits under the practice of intensive and semi-intensive types of shrimp culture, often leads to high levels of feeds, chemicals, and antibiotics being used in ponds. Effluence including chemical inputs and waste from shrimp farming ponds are often released directly into the natural environment without any treatment, even in the case of shrimp disease outbreaks. This is a direct source of contamination to soil, rivers and coastal habitats. Moreover, theoretically, it is good to raise shrimps in the dry season. However, because of repeated shrimp losses, farmers want to make up for their lost shrimp crops quickly and often expand the shrimp-raising time beyond the dry season. As a result, farmers raise shrimp throughout the year by keeping saline water in their ponds even in the rainy season. Unlike scientists who can figure out the scale of pollution by scientific measures, farmers measure this problem by observing their shrimp products. Farmers reported that in the past when only a few people engaged in shrimp culture and shrimp were raised using low stocking densities, the shrimp raising environment was good. Even though they did not pay much attention to taking care of shrimp or apply many technological measures, farmers could succeed remarkably with low investment. Now, with many farmers raising shrimp more intensively by increasing stocking density and the frequency of shrimp crops, despite being backed by advanced technology and aquatic engineers, farmers still frequently lose their crops. Moreover, the practice of using prohibited harmful chemicals that rapidly and effectively kill competing species or stimulate shrimp to shed their exoskeleton so that they grow more quickly, is said to cause certain damage to the ecosystem of shrimp farming areas. However, because of their low expense and lack of immediate impact to shrimp ponds, this practice of chemical use is still widely adopted by shrimp farmers in the Mekong Delta.

Farmers often blame their shrimp losses on the pollution caused by the intensification of shrimp farming. However, while they have limited land and suffer continuous shrimp losses, they still continue the practice of intensification because of their wish to increase the productivity. As a result, in order to raise shrimp intensively, farmers increasingly rely on technology in terms of food and chemicals that are direct sources of pollution to the shrimp environment. Shrimp disease outbreak has become a prominent and growing concern of shrimp farmers. Reports show that in the Mekong Delta failure rates in shrimp farming as high as from 70% to 80% started in 2001 (EJF 2003, p.17) when Vietnam began to promote large-scale shrimp production. The case of Tan Chanh commune in Long An province best illustrates this point. Although shrimp farming was adopted in 1994–1995, it was not until the period of 2000–2001 that this practice was widely adopted by the locals. The period of 2000–2001 was also considered a "golden age" for the shrimp industry with remarkable rewards to local shrimp farmers. Thanks to big profits from shrimp, the infrastructure of the community was being upgraded. Many concrete and durable houses were built in the community because of shrimp farming. However, along with the increase of shrimp farming land and intensive production, shrimp disease outbreak has since become more and more prevalent in the community. It affected 60% of shrimp farms in 2008. This number increased to 80-90% in 2009 and 2010 although farmers were reported to strictly observe instructions relating to the crop calendar, pond treatment and seed selecting. This was a huge loss because annually the shrimp industry brings about 20 billion VND (1 million USD) to farmers in Tan Chanh commune in a good harvest year.

Another ecological challenge to shrimp farmers is the effect of salinity. Attracted by extremely high profits of shrimp in comparison to rice, a majority of farmers engage entirely in shrimp farming instead of practicing rice-shrimp rotation which was encouraged by local governments as a friendly and sustainable livelihood. In Long An province, people have tried their utmost to use land for shrimp ponds. They have made use of every single acre of residential land to make shrimp ponds. Consequently, shrimp ponds are located very close to farmers' houses. At first, residents still cultivated rice to secure their food supply while raising shrimp. However, with the striking success of the first shrimp crops, farmers stopped cultivating rice to save time and energy for shrimp raising. Shrimp are raised year round, while rice fields are abandoned because rice is no longer considered as important as shrimp. As many farmers have not continued to grow rice, it has become difficult for other farmers to continue growing rice because of the invasion of rice crop destroyers such as insects, birds, and rodents. Gradually, having limited land and wishing to increase shrimp production, a majority of farmers removed the rice fields in their land to make the ponds deeper so that they could raise more shrimp. This means that it is now impossible for these farmers to grow rice anymore. Overall in the Long An province study site, after suffering serious repeated shrimp losses, farmers cannot return to rice cultivation for their food security. In Ca Mau province, because of much land is dedicated to extensive shrimp raising and the high expense of turning rice fields into ponds, farmers tended to keep saline water for longer periods in their ponds to raise shrimp in an effort to regain some of the outlay costs encountered from the initial conversion of their rice cultivation land into shrimp ponds. With the increase in salinity caused by the practice of keeping saline water in shrimp ponds, rice yields were not high and plants were even destroyed by saline water. Consequently, in Ca Mau province, rice farmers have to invest a great amount of labor and energy to keep their rice from crop destroyers and face a standing danger of salinity intrusion from canals or neighboring shrimp ponds where now the priority is keeping saline water in the fields.

The environmental challenges of shrimp production can also be seen from the aspect of the quality of shrimp. In the shrimp industry, intensive production leads, not only, to a polluted environment but also to the low quality of shrimp fry. The quality of post larvae stock plays an important role in the success of shrimp crops and farmers are unable to control this determinant. There is an upsurge in demand for shrimp fry with the boom in the shrimp industry while the infrastructure of shrimp hatcheries has not developed adequately. High demand and limited supplies have resulted in poor quality control systems. Broodstock ideally breed only once or twice and from parent shrimp taken from the open sea. However, in reality, to meet the high demand, parent shrimps are forced to overreproduce from six to eight broods. Later broods tend to be of poor quality with high mortality rates and increased susceptibility to disease. Due to limited natural parent shrimp, raised parent shrimp are chosen to produce broods which are also of poor quality.

In short, market-oriented production in the shrimp industry leads to serious ecological tensions which, in turn, negatively affect the industry itself and the livelihoods of inhabitants.

Socio-Economic Challenges of Shrimp Farming: Problems of Unequal Access and Household Production

The Problem of Unequal Access to Shrimp Farming

As a risky business, shrimp industry development is associated with newly emerging socio-economic tensions among shrimp farming communities in coastal areas of the Mekong Delta. With high returns, the shrimp industry has become an attractive business for many people who are capable of high capital investment. Therefore, although shrimp farming is the primary new livelihood practice amongst coastal communities of the Mekong Delta, studies show that this industry has not been viewed as an equal access industry for all local people (EJF 2003; Le Thi Van Hue and Scott 2008; Pham Van Khang 2008). Especially, in communities that adopt intensive shrimp culture, local people, mainly the poor, are excluded from this industry due to its requirement for high capital investment. Instead many local poor people have sold or leased their ponds instead of participating in shrimp farming themselves. Consequently, shrimp ponds are owned by outsiders, usually governmental officials or the rich (EJF 2003). In Soc Trang and Ben Tre provinces of the Mekong Delta, many local famers sell land to wealthier people who come from other places to invest in intensive shrimp culture (Pham Van Khang, 2008and researcher's own observations in Soc Trang province). It is reported that this phenomenon is also common in Ninh Thuan province of Central Vietnam, where only 96 out of 250 shrimp farmers are locals (Soo Kum Lin 2006, p.2).

Ironically, unequal access is not always bad for the locals. Reports show that due to the high risks of shrimp farming, the poor have became least indebted and relatively economically unaffected because of their lack participation. Meanwhile the rich, the upper middle and middle income households who are engaged in shrimp farming have lost their investment, plunging them into serious debt (Le Thi Van Hue and Scott 2008; Luttrell 2006). In Long An and Ca Mau provinces, despite knowing that intensive shrimp farming can bring great rewards, a majority of farmers prefer practicing extensive and semi-extensive types of shrimp culture rather than the intensive type because of its extremely high risk.

With the participation of outsiders who have enough capital to invest in the risky business, consequent landlessness is often considered as a negative result of the shrimp industry (Luttrell2006; EJF p.26; EJF 2003; Le Thi Van Hue and Scott p.17; Le Thi Van Hue and Scott 2008, p.70, Pham Van Khang 2008, p.45-47). However, outsider participation in shrimp farming is not always an issue. In the two research in Long An and Ca Mau provinces, where extensive and advanced-extensive types of shrimp culture are widely adopted, shrimp farmers are mostly local people who had previously cultivated rice. During the golden age of shrimp farming, land values in these localities increased. Outsiders who had capital and knowledge of raising shrimp (mainly aquaculture engineers) came to these sites to invest in shrimp farming. However, it was not easy for them to purchase land in these communities because most local landowners were reluctant to give up their land entirely due to their historical and cultural ties to the land. Therefore, those coming from elsewhere to invest in shrimp aquaculture had to rent ponds to raise shrimp. Often after several unprofitable crops, they then had to quit. In contrast to other studies, which claim landlessness as a negative impact of shrimp farming (EJF 2003; Luttrell2006; Le Thi Van Hue and Scott, 2008), this study shows that landlessness due to shrimp farming has not been a widely common phenomenon in shrimp farming communities, especially in those places where farmers adopt extensive and semi-extensive types of shrimp culture. To farmers, land is an economically and culturally valuable asset. Selling off their land is the last resort for farmers. During hard times, they resort to many strategies to keep their land. For example, at the beginning stage of shrimp farming, when the poor could not engage in shrimp farming, they either, rented their land out to others and took their land back when they could afford shrimp farming, or left their fields vacant while engaging in off-farm jobs to earn their living. In shrimp production, although incurring bad debt is a pervasive phenomenon, selling land, the only valuable asset of most farmers, to pay off their debt is not always the case. In the case that they cannot pay back loans or debts resulting from shrimp farming, they normally pledge their land (cầm cố) to their local neighbors who will give them some money (in gold equivalence) to use that plot of land until the land owners are able to pay them back that same amount of money (in gold equivalence with no interest charged). Failing shrimp farmers either migrate to other places to find jobs or rent back that same piece of land to raise shrimp in order to accumulate money to be able to get their land back. My study also shows that although loans from government banks with favorable interest are the most important source of capital that help people to take part in shrimp farming, other informal types of support from relatives, neighbors (in the form of cash, labor, and land) and people within the shrimp market network (buying food or shrimp post-larvae on credit) are also important, especially to the poor.

(Akerlof In shrimp farming, my study shows that the opportunity for access to high quality post-larvae stock is unequal. The poor quality control system and the nature of small-scale household production prevent a majority of shrimp farmers from accessing good quality shrimp fry. Only wealthy farmers who often adopt the intensive type of shrimp culture can afford the expenses of shrimp fry quality control. Other farmers, who often buy a small quantity of shrimp fry, are unable to demand high quality post-larvae stock. It is the fact that poor or small-scale farmers often buy post larvae stock that cannot be sold to wealthy farmers. Farmers have to rely on their network of brokers for their shrimp fry. In this type of market set-up, they have no power to negotiate and examine the quality of post-larvae shrimp and can actually be considered a lemon market1970).

Unequal access to shrimp production can be analyzed not only from the aspect of economic stratification but also from the standpoint of social class and ethnicity. In shrimp raising communities, because shrimp farming is a new practice that requires complicated technology, local governments have tried to help farmers by providing various forms of training. Training courses and broadcasting programs are the two most common channels of disseminating knowledge of raising shrimp to farmers. Due to the limitation of government resources to carry out these tasks, members of various unions (the farmers associations, the women associations and the veteran associations) are mainly selected to participate in these training courses. Moreover, in the Mekong Delta, the Kinh are not the only ethnic group engaging in shrimp farming. In the Delta, Khmer people make up a significant portion of the population. They are not good at Vietnamese language skills; language becomes a barrier for the Khmer in the Mekong Delta for accessing shrimp raising knowledge which is often provided in the Vietnamese language. Khmer shrimp-raising farmers in Cau Ngang district of Tra Vinh province reported that they did not receive any help from the local government in raising shrimp while, in reality, there were various shrimp-aid programs provided by the local government. Without access to shrimp-raising knowledge, many Khmer farmers in Cau Ngang district suffer serious loss which pushes them to quit shrimp farming, leaving their shrimp ponds vacant and migrating to other places to earn their living and to pay off their debt resulting from shrimp farming.

In the shift from rice cultivation to shrimp farming, unequal access to shrimp farming can be analyzed through the fact that farmers are not always free decision makers. In the decision-making process to raise shrimp, farmers play both an active and passive role. The phenomenon that all farmers in coastal areas are shrimp producers may imply that shrimp farming is an equal access opportunity to all local people. However, this study shows that although the potential to earn a high profit is a decisive factor to urge a majority of people to shift actively into shrimp farming, farmers are not always free-decision makers in this process. Some households were reluctant to shift to shrimp farming because they did not want to take on the risks involved with shrimp farming. Some of them were already living at the subsistence margin. Those reluctant to take on shrimp farming engaged in shrimp farming because they did not have any other alternative because, with the shift to shrimp farming, the irrigation system has favored saline water. They could not grow rice, the traditional crop from which they already had a stable income, while their neighbors were raising shrimp. Saline water intrusion from neighboring shrimp ponds was a standing danger for their rice crops. Results from the survey on structural transformation of agriculture-forestry-fisheries in the 1999–2005 period conducted by the Ca Mau People's Committee confirm this phenomenon. This report shows that in Cai Nuoc district of Ca Mau province, among the reasons for converting from rice to shrimp, aside from the reason to improve households' economic conditions, residing on planned areas for shrimp farming was also an important reason, accounting for 43% of the surveyed households (Ca Mau's People Committee 2005, p.49). Having other sources of income and the possibility of mobilizing social capital embedded in various social organizations in terms of financial capital, labor, and land helps poor farmers to participate in shrimp farming.

Household Production in Shrimp Farming: A Challenge to Sustainable Production

Households are both a social and economic unit where people are linked by kinship and marriage while performing economic activities together. Due to these overlapping characteristics, it is very difficult to represent the relationship between the household economy and market economy because these two concepts are often put as opposed to one another. The implication of this opposition is that because the household produces by itself what it needs, it does not need the market (Enrique 2005, p.405). In the case of shrimp farming in Vietnam, most shrimp farms are small-scale, relying on the labor source from members in the family. With the shift to shrimp farming, households are no longer associated with a subsistence economy but market-oriented production. In brackish areas of the Mekong Delta, before farming shrimp, farmers produced rice. However, due to unfavorable soil conditions, rice crop yields were not high. Therefore, a majority of farmers in these areas could not participate actively in market production. In Long An, for example, rice production was mainly for the household's subsistence. In the former rice-cultivation period, local inhabitants had practiced a subsistence economy in which they farmed rice, vegetables, fruits, pigs, poultry, and fish for their own consumption for the entire year. Thanks to the fresh-water ecology, people could also catch various wild fresh-water aquatic varieties, which were abundant in canals at that time. According to local government reports in Tan Chanh commune of Long An province, during the rice-cultivation period, 94% of households were involved in rice growing, 4% were involved in wild capture fisheries, and 30% were involved in fish-raising. Poultry, pigs and buffaloes were also raised to sell in the locality. Farmers could sell the excess volume of these products for cash to cover their families' other expenditures such as ceremonies, health care, and education. In the shrimp-production period, all household needs, including rice as the staple food, are met mainly through earnings made from market-oriented shrimp production. In coastal areas of the Mekong Delta, in the rice-cultivating era, rice was mainly produced for local inhabitants' consumption. In contrast to rice, at present, shrimp are mainly produced for the market. Shrimp farming has become the main agricultural activity with 80% of the studied households in Long An Province and 94% in Ca Mau Province involved in shrimp-raising. Due to the high risk of shrimp farming which cannot ensure a stable source of income, local inhabitants have experienced high levels of food insecurity. Without rice, other activities associated with rice such as animal husbandry also no longer exist. Going to markets (in Long An province) or buying from trading boats (in Ca Mau Province) for their meals have become a daily activity of shrimp farmers. Debt owed from buying food are still prominent in shrimp farming communities with local grocery stores still playing an important role in "helping" local inhabitants for their food security by selling food on credit and waiting for the harvest time of shrimp before receiving payment from customers. "Eating in advance, paying back later" is a pervasive motto in shrimp farmers' lives.

De facto private ownership of agricultural land proves to be a favorable factor that helps shrimp farmers to participate in market-oriented production. Although land remains the property of the state, usage rights can be legally transferred and exchanged, mortgaged and inherited. Farmers have the right to decide how to use their land for production. In shrimp farming, with the government's encouragement, farmers have eagerly converted their rice fields into shrimp ponds with the hope of improve their standard of living. Although private ownership of land is an advantageous condition in agricultural production, due to small-scale household production, shrimp farmers are finding sustainable production a difficult challenge to meet.

Having limited land is a tension that motivates shrimp farmers to raise shrimp more intensively. Farmers try to stock at higher densities and increase the frequency of their shrimp crops. Like other agricultural activities, shrimp farms are operated by households whose cultivated lands vary in size. Although it is reported that a majority of shrimp–farming households own cultivation land ranging from 0.5 to 2 hectares in size (50.58% for the whole country and 53.48% for the Mekong Delta) (GSO 2007, p.448-449), this survey in two provinces, one with the largest shrimp cultivation areas per household and the other with the smallest shrimp cultivation areas per household, shows a high variability in the size of shrimp farms. In Ca Mau province 55.3% of the households surveyed in Ca Mau province own over 1 hectare of shrimp farming land while the figure is only 6.9% in Long An provinces. 5 hectares and 0.068 hectares are the largest and the smallest shrimp areas of households studied in Ca Mau province while the equivalent measures are 1.3 hectares and 0.03 hectares in Long An province respectively. However, farm sizes may not inform the scale of capital investment because it depends on the type of culture. In my study, people mainly adopt extensive and semi-extensive shrimp farming in which shrimp acreage is a good indicator for the source of income. In Long An province, the average size farm is 0.4 ha while in Ca Mau the average size farm is 1 hectare. According to Vietnamese standards, these farm sizes are classified as a lack of cultivated land (Nguyen Dinh Huong 1999). It means that the income generated from these pieces of agricultural land cannot guarantee the household's subsistence. Income from a shrimp pond of a medium sized family with five people illustrates more details:

"My family has raised shrimps since 1996. My cultivation land is 5,000 m^2. In the first year, we earned nine million dong (equivalent to 750 USD), the next year 47 million dong (3,900 USD). This was the best crop that helped us to rebuild our house. From 2000 to 2003, we got on average 13 million dong (870 USD) annually. From 2004 to 2007, we suffered repeated losses or earned nothing. We lost from five to seven million dong (310–440 USD) each year. In 2008 and 2009, we earned 10 million dong (540 USD) annually." N.T.T.M, female, aged 35, Long An province, shrimp farmer

Profits from shrimp are much higher than from rice, normally from five to seven times with the same cultivated land. By raising shrimp, farmers can obtain large incomes that they "have never dreamed of when cultivating rice". However, due to the high risks in connection with price fluctuations, a changing environment, and disease outbreaks, this kind of production is no longer considered the main source of income for local inhabitants. A majority of farmers raise shrimp as one way to try their luck. In shrimp farming, it is very easy to make a loss or earn nothing. The case study of the above-mentioned household with an average-sized shrimp pond in Long An province shows that income generated from shrimp is not enough for a household with five people who at present have to rely mostly on the market for their daily consumption because their subsistence economy no longer exists. Farmers have various strategies to survive. In shrimp production, those engaged in advanced extensive types of shrimp culture have increased the shrimps' stocking density and raise shrimp throughout the year, while those engaged in extensive type of shrimp culture have raised various kinds of fish in shrimp ponds and try hard to adopt rice-shrimp rotation to ensure their food supply. When farmers cannot adapt locally, labor migration is their best solution to maintain their living. Labor migration has become a prominent phenomenon in shrimp farming communities especially after repeated shrimp losses. Both men and women migrate to industrial centers or other areas to find jobs. In Ca Mau province, while young people tend to travel far distances to Ho Chi Minh city, Dong Nai and Binh Duong provinces of the southeast region of Vietnam, other people tend to work locally with various manual jobs. In Long An province, a large wave of young rural women going to work in garment factories is partly a result of unsustainable production of shrimp farming.

Although local governments with various support policies have an important role in promoting shrimp cultivation, spontaneity is a common feature in shrimp production. There are no official or non-official organizations relating to shrimp farming at the community level. Households do their shrimp farming independently. As a result, individualist rationality is seemingly good at the household level but bad at the community level. For example, all farmers are instructed to release shrimp fry and to prepare their ponds at the same time in order to control disease outbreaks. However, farmers personally decide to release shrimp fry before or after the instructed cropping time. They do this because if all people raise shrimp at the same time, the prices of shrimp fry will be increased while shrimp products will decrease because of the interplay of supply and demand.

Moreover, in shrimp farming, water resources are a decisive factor for a successful shrimp crop. However, due to the unplanned system of water supply for shrimp ponds, this resource is easily contaminated by the practices of individual households. This implies an internal conflict among farmers regarding the use of water for shrimp farming. In Ca Mau, the conflict is between extensive and intensive shrimp farmers. Although there are regulations for intensive farmers regarding the dredging of their ponds, it is hard to check their compliance. Extensive shrimp farmers often claim that because of intensive farmers' lack of compliance, waste water leaks out from their intensive ponds. Likewise, in Long An Province, because people practice advanced shrimp culture they have to invest a lot in shrimp food. Although there are regulations on draining water, in their pursuit of individual benefits, people do not follow the regulations. For example, when someone's shrimp die out because of disease, according to regulations, that individual has to use chemicals to sterilize the pond before draining water out of the pond. However, people often do not do this due to high costs associated with purchasing the relevant chemicals. If neighboring farmers use this infected water, their shrimp would likely catch the disease. As another example, the waste water from a farmer's shrimp pond can damage his/her neighbors' ponds, especially of those which are farthest from canals. In both communities, the layout of shrimp ponds depends on individual households. Therefore, there are no separate channels for fresh water inflow and waste water outflow. People get water from and release it into the same canals. This condition is conducive to the outbreak of shrimp epidemics.

CONCLUSION

In the shrimp industry, it is obvious that shrimp farmers are vulnerable in their market production. Although various policies exist which aim to promote shrimp production, there is no policy relating to the output of shrimp products. Farmers have to rely on the network of middlemen who always try to make use of shrimp farmers in the whole process of production. With the shift from rice to shrimp, the Mekong Delta farmers are obviously risk-takers who readily engaged in market production. However, working in uncertain conditions, shrimp farmers are not rational actors who always take risks regardless of their conditions. Instead, in their production, shrimp farmers always try to mitigate or disperse their risks. For example, farmers do not strictly observe the local governments' technical instructions of shrimp farming, which require they invest much capital on shrimp production but do not prove to bring a remarkable success. This is why shrimp production is always judged with such terms as "spontaneous", "unprofessional" and "small-scale".

This paper attempts to present the social and ecological challenges of shrimp farming in Vietnam, which is becoming more and more market-oriented. Through the case of shrimp farming, it is obvious that market production could bring immediate benefits for farmers; however, it is also associated with new ecological and social tensions, which people cope with by relying on various strategies. Risk mitigation in shrimp production and labor migration are the two important strategies of local inhabitants for securing their livelihoods. Shrimp farming is a social process in which humans constantly shape and interact with their environment. Water pollution and poor quality post-larvae shrimp are direct consequences of market-oriented production. These results, in turn, lead to negative impacts on shrimp production. Moreover, in the context of poor management and weak enforcement mechanisms, although de facto land ownership is a favorable condition for market oriented shrimp farming, because of individualistic rationality, household production proves to be inappropriate for sustainable production.

Research Methods and Case Study Sites

This paper is based on ethnographic fieldwork undertaken in 2009–2010 in two shrimp raising communities of the Mekong Delta; one in the lower part (Hoa My commune, Cai Nuoc district, Ca Mau Province) and the other in the upper part of the Delta (Tan Chanh commune, Can Duoc district, Long An province). These two communities engage in extensive and advanced extensive shrimp culture, two major types of shrimp culture in the Mekong Delta. Extensive shrimp culture (colloquially called traditional extensive culture) is characterized by low capital investment in pond building, pond preparation, and shrimp raising techniques. Shrimps are left to grow naturally, without being fed processed food. Shrimp fry (post larvae) are released at low density, from one to two shrimps/ m^2. Shrimps rely completely on natural food. In the whole raising process, farmers only apply such simple pond-preparation measures as using derris elliptica to kill the various fish competing with shrimps for food or killing shrimp fry, using lime powder ($CaCO3$) to reduce alum, and occasionally using chemical fertilizers such as NPK to create algae as a natural food for shrimps in the pond. Having a large amount of land but limited capital, farmers apply these measures only to some extent, not as intensively as suggested by the government's aquaculture experts. In contrast to other types of shrimp culture in which shrimp fry are released and harvested once per crop, in the extensive type, people usually release fry every one or two months and harvest shrimps every day or every one or two weeks.

With a much improved technique over extensive shrimp culture, the advanced extensive type requires considerable investment in both capital and raising methods. Besides applying more intensively the measures as in extensive culture, farmers adopting the advanced extensive type also use processed pellets or self-processed food and vitamins to feed shrimps, as well as various chemicals such as sterilizer and bio-enzymes in pond preparation and shrimp raising. Pond dikes and ditches are prepared annually. In principle, the density of shrimp post larvae is from three to five shrimps/ m^2. However, in Long An, in order to have high productivity despite limited land, people release fry with a stock density much higher than the encouraged level, from 10 to 20 shrimps/ m^2. Besides the two aforementioned types of shrimp culture, a minority of famers also practice industrial shrimp farming which requires a much higher capital investment, normally 40 times higher than the extensive type and five times higher than the advanced extensive type per ha per year.

My research applies quantitative and qualitative methods to collect data. Quantitative methods are employed to generate general patterns and qualitative methods are used to get more detailed data, explanations for quantitative methods. Quantitative data from 354 households provide background information of socio-economic conditions and general picture of shrimp production in both study sites. Qualitative data from 72 interviews and participant observations reveal in-depth information on farmers' dynamics, analyses, and evaluation.

I apply random sampling method with the margin of sampling errors of 5% or less (Bernard1994, p.79) and random probability sampling method to select my survey samples. Accordingly, I choose 202 out of 409 households in Thi Tuong hamlet of Ca Mau province and 152 out of 245 households in Dinh hamlet of Long An province. On the basis of sex, age, economic background variables, I select household members, community elites, local authorities and other stakeholders in shrimp farming for in-depth interviews. Totally, we conduct 72 interviews and informal conversations relating to such topics as the shift from rice to shrimp, policies toward shrimp farming, market network, ecological and social changes with farmers who hold various positions in the community. Observations are made at shrimp production activities, rituals and meetings of study communities in order to check, to understand more and to have hints for further interviews.

In addition to the ethnographic fieldwork, this paper is also based on observations of shrimp farming in five other provinces, namely Ben Tre, Tra Vinh, Soc Trang, Bac Lieu, Kien Giang of the Mekong Delta since 2006.

ACKNOWLEDGEMENT

This paper was presented at the 2011 Vietnam Update, Australian National University "The Environment: Change, Challenges and Contestations," 17– 18 November, Australian National University. The author wishes to thank Dr. Philip Taylor, Olivia Dun, and Anh-Thu Ngo for having read and make useful suggestions of earlier drafts of this article.

Competing Interests

The author declares that I have no financial and non-financial competing interests with other parties.

REFERENCES

1. AkerlofGA: The market for 'Lemons': quality uncertainty and the market mechanism. Q J Econ1970, 84(3):488-500. (The MIT Press) 10.2307/1879431CrossRef
2. Ban Chỉ đạo Tây Nam Bộ và Trung tâm thông tin Sài Gòn (Guidance Board of Vietnam's Southwest region): Tây Nam Bộ tiến vào thế kỷ 21 (Southwest region moving forwards to the 21st century). Chính trị Quốc gia Publisher, Vietnam; 2005.
3. Barraclough S, Finger-Stich A: Some ecological and social implications of commercial shrimp farming in Asia. 1996.
4. Bernard RH: Research methods in anthropology: qualitative and quantitative methods. 3rd edition. Altamira Press, California; 1994.
5. Ca Mau's People Committee: Báo cáo chính thức kết quả khảo sát chuyển đổi cơ cấu nông – lâm – thủy sản giai đoạn 1999–2005 (Results of survery on structure transformation of agriculture – forestry – fishery period 1999–2005). Vietnam; 2005. Final report

6. EJF: Risky business: Vietnamese shrimp aquaculture – impacts and improvements. Environmental Justice Foundation, London, UK; 2003.
7. EJF: Farming the sea, costing the earth: why we must green the blue revolution. Environmental Justice Foundation, London, UK; 2004.
8. Enrique M: "Households and Their Markets in the Andes". In A handbook of economic anthropology. Edited by: Carrier JG. Edward Elgar Publishing; 2005:405-422.
9. GSO: Results of the 2006 rural, agricultural and fishery census. Volume 3 – agriculture, forestry and fishery. Thống kê Publisher, Hanoi; 2007.
10. GSO: Statistics handbook of 2009. Thống kê Publisher, Hanoi; 2009.
11. Hue LTV, Scott S: Coastal livelihood transitions: socio-economic consequences of changing mangrove forest management and land allocation in a commune of central Vietnam. Geogr Res2008, 46(1):62-73. 10.1111/j.1745-5871.2007.00492.xCrossRef
12. Huong ND: Sản xuất và đời sống của các hộ nông dân không có đất hoặc thiếu đất ở đồng bằng sông Cửu Long: thực trạng và giải pháp (Production and lives of landless or lack-of-land households in the Mekong Delta: present situation and solutions). Chính trị Quốc gia Publisher; 1999.
13. Luttrell C: Adapting to aquaculture in Vietnam: securing livelihoods in a context of change in two coastal communities. In Environment and livelihoods in tropical coastal zones Edited by: Hoanh CT, Tuong TP, Gowing JW, Hardy B. 2006, 17-29.
14. Preston N, Clayton H: Rice-shrimp farming in the Mekong Delta: Biological and Socioeconomic Issues, ACIAR (Australian Center for International Agricultural Research) Technical reports no. 52e. 2003.
15. Soo Kum Lin J: Sustainability of intensive shrimp farming: a case in Ninh Thuan, Vietnam. Master thesis. Lund University; 2006.
16. Tan Chanh's People Committee: Local report on shrimp production. Vietnam; 2009.
17. Thao LB: Thiên nhiên Việt Nam (Vietnam Nature). Vietnam: Giáo dục Publisher; 2002.
18. Thong MT, et al.: đánh giá mối quan hệ tự do hóa thương mại, đói nghèo nông thôn, môi trường trong nghề nuôi tôm (An Assessment of the linkage between trade liberalization, rural poverty, and environment in shrimp aquaculture in Cà Mau province, Vietnam). Vietnam; 2006. Project abstract
19. Van Khang P: Challenges to shrimp production in the Bentre Province. Master thesis in International Fisheries Management. Vietnam: University of TromsØ; 2008.
20. Vo Binh T: Potentials awakened for a more prosperous image of Ca Mau. In Ca Mau – New image in century XXI. National Political Publishing House, Vietnam; 2006.
21. Vuong Do Quang T, Lin CK: Rice – shrimp farming in the seawater intrusion zones of the Mekong Delta, Vietnam. ITCZM Monograph; 2001. No.6

CHAPTER 16 EVALUATION OF FLOCCULATED CONCENTRATES FROM INTENSIVE SHRIMP POND WATER AS A SUBSTITUTE FOR MICROALGAL CONCENTRATES IN THE NURSERY CULTURE OF JUVENILE OYSTER (CRASSOSTREA BELCHERI)

Suwat Tanyaros, Woraporn Tarangkoon, Tossaporn Klomkleing

Keywords: Flocculated concentrates Intensive shrimp pond Daily growth rate Crassostrea belcheri

INTRODUCTION

Highly nutritional microalgae are normally used to feed the larval and adult stages of bivalves in hatcheries. However, microalgae production requires a high level of investment and is very costly. Production of live microalgae has been seen as a major hatchery operating costs, accounting for 20–50 % of the total bivalve seed production costs (Coutteau and Sorgeloos1992). Hatchery-reared oysters in the nursery with a low mortality rate in the sea should be bigger than 5 mm in length (Tanyaros and Tarangkoon 2014), but feeding of spat from post settling up to this size result in high consumption of microalgae. It is at this stage of culturing that an alternative food source is most needed. Today, this problem can be solved using preserved microalgae which can be produced by concentrating algae from mass culture and preserving the products through refrigeration, freezing or drying (McCausland et al. 1999; Ponis et al. 2008). Initial cost estimates for supplement feed showed that it will add 2–3 % to juvenile oyster production costs during the nursery phase when oysters are 0.5–3 mm in size (McCausland et al. 1999). Successful partial replacement for a diet of live microalgae has been reported in several studies using microalgal concentrates (Nell and O'Connor 1991; McCausland et al. 1999; Heasman et al. 2000; Robert et al. 2001; Brown and Robert 2002; Ponis et al. 2003a, b, 2008; Knuckey et al. 2006).

Shrimp waste and uneaten feed are released directly into the water and settle at the bottom of the pond. Subsequently, the decomposition of wastes can produce excess inorganic nutrients which stimulate the production of algae in the ponds (Funge-Smith and Briggs 1998). Microalgal communities in intensive shrimp ponds have been studied (Tookwinas and Songsangjinda 1999; Yusoff et al. 2002; Rodrıguez and Paez-Osuna 2003; Islam et al. 2004; Casé et al. 2008; Shaari et al. 2011) and Bacillariophyceae constituted the largest number of species and accounted for the high densities in those reports. Bacillariophyceae often represents the main food source in marine ecosystems (Zhukova and Aizdaicher 1995) and many species have been used as food in oyster hatchery operation as summarized by Brown et al. (1989). As shrimp pond water typically contains elevated densities of microalgae, integrated culture of shrimp and oysters has been attempted to reduce

Citation: Suwat Tanyaros, Woraporn Tarangkoon, Tossaporn Klomkleing, Evaluation of flocculated concentrates from intensive shrimp pond water as a substitute for microalgal concentrates in the nursery culture of juvenile oyster (Crassostrea belcheri), DOI: 10.1007/s40071-016-0130-5

microalgal densities in shrimp culture units (Sandifer and Hopkins 1996; Jones and Preston 1999; Tanyaros 2000; Jones et al. 2001; Tanyaros et al. 2002). Manzi et al. (1988) and Jacob et al. (1993) showed the potential of*Crassostrea virginica* biculture using marine shrimp pond water as feed. High growth and survival rates for both shrimp and oysters were obtained in their studies. The results of these studies suggest the idea of harvesting microalgae from shrimp pond water in concentrates form may have potential for substituting microalgae in the nursing of juvenile oyster in a hatchery. In the past, flocculation of microalgae in marine shrimp culture system for improving the water quality has been investigated (Lertsutthiwong et al. 2009). However, the potential of flocculated concentrates from intensive shrimp pond water as a substitute for microalgae in the rearing of oysters has not been previously investigated. This article describes the preservation and use of flocculated concentrates from intensive shrimp pond in the hatchery diet for the nursery culture of the juvenile oyster, *C. belcheri*.

MATERIALS AND METHODS

Preparation of Flocculated Concentrates from Shrimp Pond Water

Water color is an important indicator of bio-optical properties as well as the general composition of microalgae. Types of microalgae which give the water a yellow–green or yellow–brown color are good for shrimp culture ponds. This color of water indicates a high probability that the water contains Bacillariophyceae (Reynolds 2006). Therefore, shrimp pond water of this desirable color from the shrimp farm at the Faculty of Science and Fisheries Technology, Rajamangala University of Technology Srivijaya, Trang campus, southern Thailand was selected and pumped into a 500 L fiberglass tank by submersible pump. The suspended materials in the pond water were flocculated by the method of Brown and Robert (2002). NaOH was added to the point of maximal mixing of the pond water to increase the pH of the water to between 10.0 and 10.6, then a white precipitate formed. At this stage, the polyelectrolyte [Polyacrylamide solution (Sigma-Aldrich)] was added to give a final concentration of 0.5 mg L^{-1}. When large flocs formed (~1 min), the mixing was stopped and the flocs were allowed to settle under gravity force (10–20 min). Surface water was then siphoned off and the floc collected. The suspension was deflocculated immediately by adjusting the pH of the floc slurry to pH 7.5–8.0 using concentrated HCl. The floc slurry was allowed to settle overnight at 4 °C and any supernatant was removed. The flocculated material was kept in a refrigerator at 4 °C prior to use in experiments. For each shrimp pond flocculation, three replicates of floc sample (20 mL) were collected and kept in a plastic bottle and then the samples were fixed in 10 % formaldehyde. The density of microalgae in the flocculated concentrates was determined using a compound microscope (Olympus CH2, Tokyo, Japan). The number of cells was calculated and cell densities expressed as the number of cells per milliliter. The nutrient concentrations in intensive shrimp pond increased with the cultured period and enhanced nutrient input affected phytoplankton density and composition (Casé et al. 2008). Therefore, two experiments with different of phytoplankton composition in flocculated concentrates from shrimp pond water were investigated to compare the effects on preservation and growth performance in juvenile oyster after substitute for microalgal concentrates. This study shows that unidentified nanocyanobacteria (99.2 %) constituted the greatest bulk of the microalgae population in the flocculated concentrates from shrimp pond water in experiment 1, while Bacillariophyceae (66.4 %) and unidentified nanocyanobacteria (33.3 %) were the dominant groups in experiment 2 (Table 1).

Table 1: Dominant group of phytoplankton in selected shrimp pond water

Group	**Experiment 1 ($n = 4$)**		**Experiment 2 ($n = 3$)**	
	Density (cell mL^{-1})	**%**	**Density (cell mL^{-1})**	**%**
Bacillariophyceae	$2.19 \times 10^5 \pm 3.73 \times 10^5$	0.8	$2.64 \times 10^7 \pm 4.71 \times 10^7$	66.4
	Nitzschia spp. (0.79)		*Chaetoceros* spp. (66.39)	
	Others (0.01)		Others (0.01)	
Dinophyceae			$1.20 \times 10^5 \pm 8.67 \times 10^4$	0.3
			Protoperidinium spp. (0.29)	
			Others (0.01)	
Cyanophyceae	$2.61 \times 10^7 \pm 1.57 \times 10^7$	99.2	$1.33 \times 10^7 \pm 1.09 \times 10^7$	33.3
	Unidentified nanocyanobacteria		Unidentified nanocyanobacteria	
Total		100		100

Microalgal CULTURES

A stock of microalgae (*Chaetoceros calcitrans* and *Tetraselmis suecica*) was grown in 1 L Erlenmeyer flasks (2 L Erlenmeyer flask) and then transferred to a 5 L Carboy using the Conwy medium (Walne 1974). Cultures were grown under continuous illumination by cool white fluorescent lights. Mass culture of microalgae was performed in 500 L open fiberglass tanks under ideal sunlight conditions. When the late logarithmic growth phase was reached, the microalgae were harvested by flocculation described by Brown and Robert (2002). Microalgal concentrates were preserved in a refrigerator at 4 °C prior been used in experiments.

Flocculated Concentrates Preservation Experiment

To study the effect of preservation on nutritional composition, three replicates of 500 mL of flocculated concentrates from shrimp pond water samples were collected and kept in plastic bottles. The samples were preserved in a refrigerator at 4 °C for 4 weeks. During preservation, 50 mL samples were taken weekly from each bottle for analysis of crude protein by the Kjeldahl method according to the Association of Official Analytical Chemists (2012). To analyze the pigment ratio for each preservation period, 10 mL of floc samples were taken and poured into 100 mL of filtered sea water. The samples were disaggregated to single cell suspensions by shaking prior to use for pigment determination. Chlorophyll *a* and pheophytin *a* concentration were determined using the spectrophotometric method (Lorenzen 1967), where the absorption of the extract was measured before and after acidification at 750 and 665 nm Chlorophyll *a* and pheophytin *a* (μg mL^{-1}) were given by the following formulae of Lorenzen (1967):

Chlorophylla(mg L−1)=[26.7v(A665na−A665a)]/(Vl)
Chlorophylla(mg L−1)=[26.7v(A665na−A665a)]/(Vl)

Pheophytina(mg L−1)=[26.7v(1.7A665a−A665na)]/(Vl)
Pheophytina(mg L−1)=[26.7v(1.7A665a−A665na)]/(Vl)

where v is the volume of 90 % acetone (mL); V is the volume of the filtered sample (mL); A_{665na} is the absorbance at 665 nm before acidification; A_{665a} is the absorbance at 665 nm after acidification; and l is the path length of the cuvette (cm).

At the beginning and at the fourth week of the experiments, flocculated concentrates samples were taken to determine fatty acid composition. The samples were extracted with chloroform–methanol (Folch et al. 1957) and portions of the total lipid extract were taken to determine fatty acid compositions. Fatty acids within the lipid extracts were derivatized to fatty acid methyl esters (FAME) using 14 % boron trifluoride–methanol (Van Wijngaarden 1967) before analysis of fatty acid composition by gas chromatography (model 6890 N; GC-FID, Agilent Technologies,Wilmington, NC, USA) using helium as the carrier gas and a flame ionization detector.

Substitution of Flocculated Concentrates for Microalgae in Rearing Juvenile Oysters

Experimental Diets and Juvenile Oysters

A mixed microalgal concentrates consisting of 50/50 % *C. calcitrans* and *T. suecica* was replaced by flocculated concentrates at different ratios (0 % (control), 25, 50, 75 and 100 %). The daily use of microalgal concentrates for both species in control (100 % microalgal concentrates) was calculated by the method of Helm et al. (2004). The volume for each species of microalgal concentrate was measured by cylinder and added to 10 L of filtered seawater. The suspension was thoroughly mixed by mildly agitation and then the percentage of transmission (%*T*) of the microalgal suspension was measured by spectrophotometer (biochrom Libra S11, Cambridge, England) and the value was recorded. In substitution treatments, the volume of microalgae concentrates were decreased according to the substitution rate, while the volume of flocculated concentrates from shrimp pond water was added to make the percentage of transmission similar in value to that of the control prior to use for the experiment.

Juvenile *C. belcheri* were produced in the hatchery of the Marine Shellfish Breeding Research Unit of the Faculty of Science and Fisheries Technology in Rajamangala University of Technology Srivijaya, Trang Campus, Thailand. The animals were graded by size to prevent growth retardation. Mean (±SD) shell widths (dorso-ventral) of 2.43 ± 0.15 and 1.86 ± 0.003 mm, shell lengths (antero-posterior) of 2.23 ± 0.09 and 1.82 ± 0.003 mm and individual wet weights of 2.78 and 1.50 mg were used in experiments 1 and 2, respectively.

Experimental System and Procedures

A semi-closed recirculation system was designed for nursing hatchery-reared juvenile oysters (see Fig. 1). Four sets were used, each consisting of a submersible pump, a rectangular 105 L (50 × 70 × 30 cm) fiberglass tank for placement of the nursing units, a cylindrical plastic 30 L container (60 cm diameter and 25 cm depth) for food storage, and three sets of cylindrical PVC pipe (15.2 cm diameter × 12 cm depth) used as nursing units. Each cylindrical PVC pipe was drilled 10 cm from the bottom for a 1.86 cm diameter overflow pipe. A screen using 600 μm mesh was fixed by a PVC clamp to the bottom of each cylindrical PVC section. The volume of water in each nursing unit was 1.82 L during system operation. Each nursing unit had a capacity of 754 juvenile oysters, equal to a stocking density of 4 juveniles cm^{-2} (Tanyaros et al. 2012). The water flow in each nursing unit was controlled by an adjustable valve at the overflow pipe to maintain a flow rate of 4 L min^{-1} (Tanyaros et al. 2012). The experiment was set up using a completely randomized design (CRD), and each treatment was conducted in triplicate.

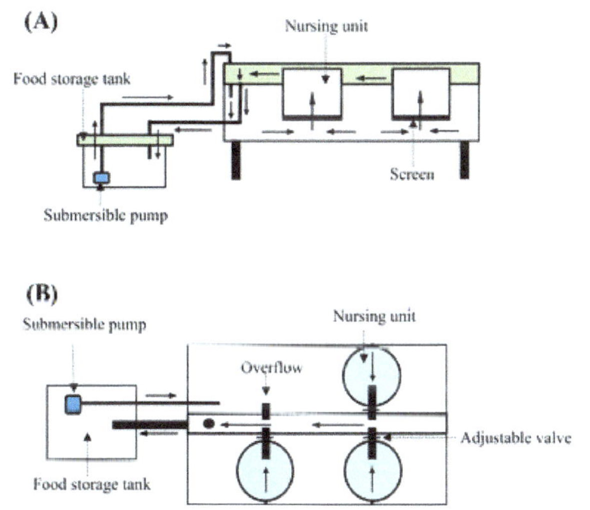

Figure 1: Diagram showing the experimental system: **a** *side view* and **b** *top view*

The water was totally renewed every 2 days, and food was added twice a day (morning and evening) at a rate calculated by the method mentioned above. The feeding rate for each nursing unit was adjusted weekly according to the wet weight of the oysters obtained each week. The number of surviving animals was ascertained at the end of the experiment. Samples from each replicate were taken weekly for measurement of shell width, shell length and wet weight. The daily growth rate (DGR) was then calculated using the method of Pérez Camacho et al. (1998). At the end of the experiment, all juvenile oysters from each experimental unit were counted. The survival rate was calculated and expressed as percentages.

Water salinity ranged from 31.6 to 33.2 and 29.7 to 30.3 ppt over the study period in experiments 1 and 2, respectively. Water quality parameters were as follows: dissolved oxygen $\geq$5.97 mg L^{-1}, total ammonia nitrogen $\leq$0.047 mg L^{-1}, water temperature 25.8–28.1 °C and pH $\geq$7.88 in experiment 1 and dissolved oxygen $\geq$6.68 mg L^{-1}, total ammonia nitrogen $\leq$0.53 mg L^{-1}, water temperature 29.7–31.6 °C and pH $\geq$7.92 in experiment 2.

Statistical Analysis

Data were analyzed statistically using one-way ANOVA to determine differences among treatments. If significant effects were present, the data were then subjected to Tukey's test to identify significant differences among treatment means.

RESULTS

Experiment 1

Flocculated Concentrates Preservation

Crude protein content in flocculated concentrates preserved in a refrigerator at 4 °C gradually decreased from the first week. The pheophytin *a*: chlorophyll *a* ratio suddenly increased during the third week of storage (Fig. 3). After 4 weeks of storage, most of the fatty acid composition in flocculated concentrates had completely degraded (Table 2).

Table 2: Fatty acid (FA) composition of fresh microalgae and flocculated concentrates from shrimp pond water at the outset (W0) and at the fourth week (W4) of storage at 4 °C

Fatty acid content (mg/100 g)				
Fatty acid compositions	**Experiment 1**		**Experiment 2**	
Monosaturate FA	406.9	769.6	63.1	ND
Polyunsaturate FA				
C18:2n6	27.2	63.5	8.3	ND
C18:3n6	52.5	ND	ND	ND
C18:3n3	6.3	ND	3.8	ND
C20:4n6	3.3	ND	ND	ND
C20:5n3 (EPA)	ND	ND	8.9	ND
C22:6n3 (DHA)	5.0	ND	9.6	ND

ND non detected

Effect of Flocculated Concentrates from Shrimp Pond Water Supplementation

The juvenile oysters were nursed in a semi-closed recirculation system for four weeks. There were significant differences ($P < 0.05$) in their mean absolute shell length and width. Substitution of 75 % of flocculated concentrates for microalgae concentrates in feed showed non-significant differences in both experiments in absolute shell growth compared with substitution of 25 and 50 % of flocculated concentrates or 100 % microalgal concentrates. Substitution with 100 % flocculated concentrates resulted in the lowest absolute shell growth (Fig. 4a). Mean absolute shell widths were 0.075 ± 0.01, 0.079 ± 0.02, 0.50 ± 0.01, 0.30 ± 0.01 and 0.019 ± 0.01 mm day^{-1} and mean absolute shell lengths were 0.037 ± 0.002 %, 0.040 ± 0.02, 0.033 ± 0.004, 0.017 ± 0.004 and 0.012 ± 0.002 mm day^{-1} for 0, 25, 50 and 75 % and 100 % substitution with flocculated concentrates, respectively. Mean daily growth rate differed significantly among the treatments over the study period ($P < 0.05$). Substitution of flocculated concentrates for 75 % of microalgae concentrates in feed resulted in no significant differences in daily growth rate in both the second and fourth week in comparison with 25 and 50 % substitution with flocculated concentrates or 100 % microalgal concentrates, while replacing 100 % of microalgal concentrates with flocculated concentrates showed the lowest daily growth rate (Table 3). No differences among the mean percentages for survival at the end of the experiment ($P < 0.05$). Mean survival rates were 90.5 ± 6.5, 97.2 ± 2.1, 98.7 ± 1.3, 96.0 ± 4.0 and 94.4 ± 4.7 % for 0, 25, 50, 75 and 100 % substitution with flocculated concentrates, respectively (Fig. 5a).

Table 3: Effect of partial replacement of microalgae with flocculated concentrates from shrimp pond water on the daily growth rate at the second (DGR^2) and at the fourth week (DGR^4) of the juvenile oyster *C. belcheri*

Substitution rate (%)	**Experiment 1**				**Experiment 2**			
	DGR^2 (% day^{-1})	**$\%^A$**	**DGR^4 (% day^{-1})**	**$\%^A$**	**DGR^2(% day^{-1})**	**$\%^A$**	**DGR^4 (% day^{-1})**	**$\%^A$**
0	203.0 ± 9.9	100^a	236.7 ± 5.7	100^{ab}	35.1 ± 4.1	100^c	113.3 ± 9.0	100^b
25	211.1 ± 5.4	104^a	251.2 ± 9.0	106^a	55.6 ± 3.8	160^b	183.1 ± 13.8	163^a
50	213.0 ± 1.6	105^a	243.3 ± 3.7	102^{ab}	64.1 ± 5.4	183^{ab}	205.5 ± 13.5	181^a

75	200.7 ± 1.5	99[a]	228.5 ± 5.1	97[b]	67.8 ± 6.1	195[a]	203.8 ± 4.3	180[a]
100	184.8 ± 2.1	91[b]	205.0 ± 1.7	87[c]	28.9 ± 1.3	83[c]	44.8 ± 3.3	40[c]

Different superscripts (vertical only) denote significant differences among means ($P < 0.05$)

[A]DGR expressed as a percentage of DGR obtained with 100 % microalgae

Experiment 2

Flocculated Concentrates Preservation

A slight decrease in protein content appeared after the third week (Fig. 2). The pheophytin*a*:chlorophyll *a* ratio was increased at the third week of preservation (Fig. 3). Most of the fatty acid composition in flocculated concentrates had completely degraded after four weeks (Table 2).

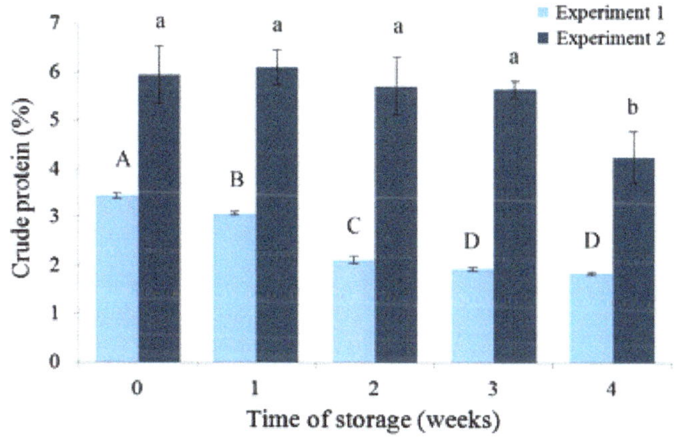

Figure 2: Total protein content (average ± SD; $n = 3$) in flocculated concentrates from shrimp pond water during different weeks of storage at 4 °C. *Bars* labeled with *different letters* are significantly different ($P < 0.05$)

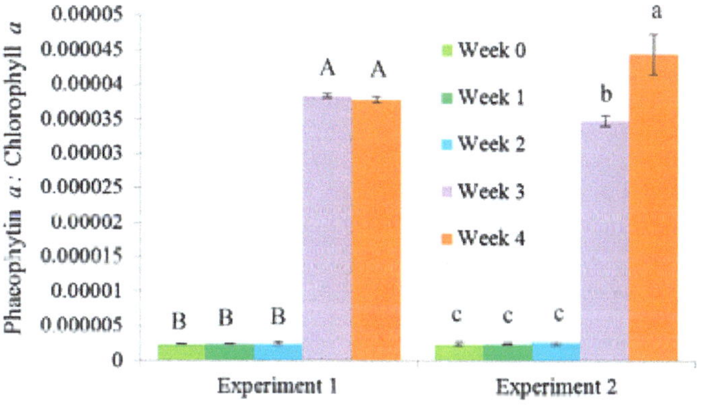

Figure 3: Ratio of pheophytin *a*: chlorophyll *a* (average ± SD; $n = 3$) in flocculated concentrates from shrimp pond water during different weeks of storage at 4 °C. *Bars* labeled with*different letters* are significantly different ($P < 0.05$)

Effect of Flocculated Concentrates from Shrimp Pond Water Supplementation

No differed on substitution of 75 % of flocculated concentrates for microalgal concentrates in feed on growth performance compared with substitution of 25 and 50 % of flocculated concentrates or 100 % microalgal concentrates. Substitution with 100 % flocculated concentrates resulted in the lowest absolute shell growth as same as in experiment 1 (Fig. 4b). Mean absolute shell widths were 0.077 ± 0.003, 0.105 ± 0.01, 0.135 ± 0.01, 0.113 ± 0.02 and 0.027 ± 0.002 mm day^{-1} and mean absolute shell lengths were 0.047 ± 0.001, 0.064 ± 0.003, 0.085 ± 0.01, 0.064 ± 0.01 and 0.015 ± 0.003 mm day^{-1} for 0, 25, 50 and 75 % and 100 % substitution, respectively. Mean daily growth rate differed significantly among the treatments ($P < 0.05$). Replacing 75 % of microalgae in feed with flocculated concentrates showed significant differences in survival rates in comparison with 25 and 50 % replacement or 100 % microalgal concentrates. Mean survival rates were 88.4 ± 2.9, 86.3 ± 1.5, 90.7 ± 2.4, 97.5 ± 1.8 and 91.8 ± 1.9 % for 0, 25, 50, 75 and 100 % substitution with flocculated concentrates, respectively (Fig. 5b).

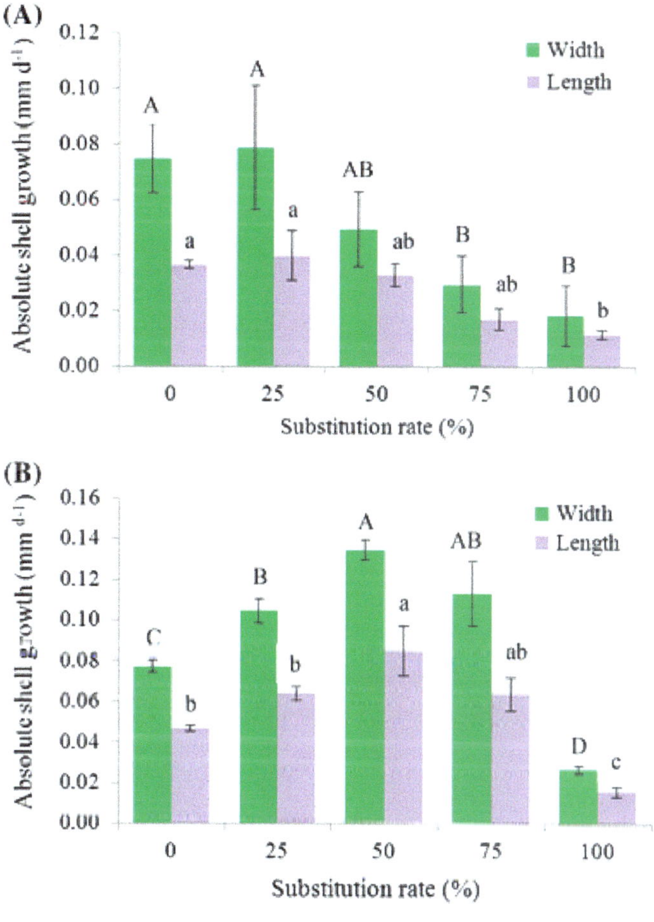

Figure 4: Mean (±SD) absolute shell length and width of hatchery-reared juvenile oysters fed on microalgae and algal concentrates from shrimp pond water at different ratios at the end of the 4 weeks experimental duration. *Bars* labeled with *different letters* are significantly different ($P < 0.05$) (**A** = Experiment 1 and **B** = Experiment 2)

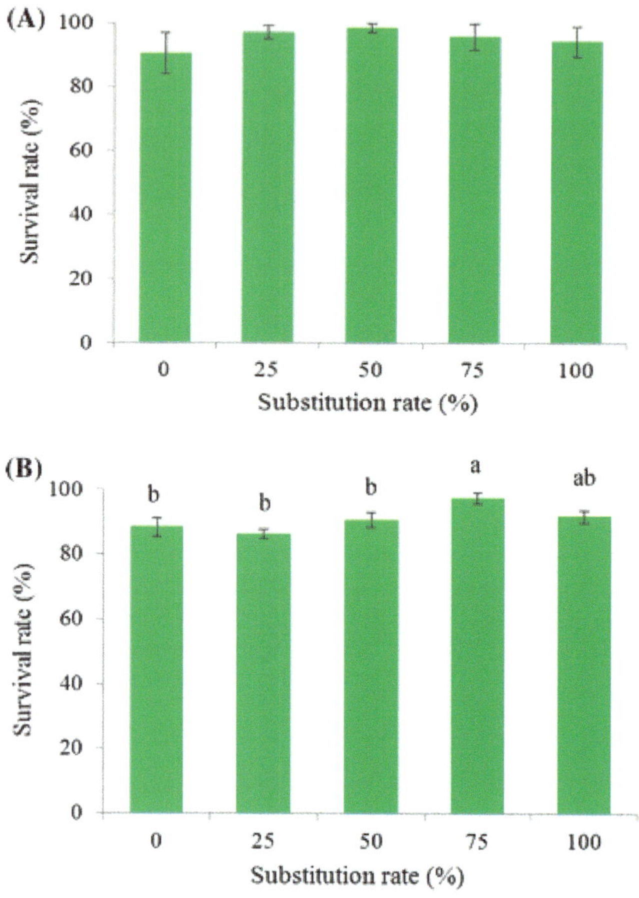

Figure 5: Mean (±SD) survival rate of hatchery-reared juvenile oysters fed on microalgae and different rations flocculated concentrates from shrimp pond water at the end of the 4 weeks experimental duration. *Bars* labeled with *different letters* are significantly different ($P < 0.05$) (**A** = Experiment 1 and **B** = Experiment 2)

DISCUSSION

Flocculated Concentrates Preservation

Flocculation is presented as a cost-effective alternative for preparing microalgal concentrates for the bivalve diet and is suited for on-site production (Knuckey et al. 2006). Flocculated concentrates from shrimp pond water contains a high concentration of microalgae, but the species compositions vary throughout the culture period. Changes in nitrogen-to-phosphorus ratios and ammonia concentrations coincided with changes in microalgal community structure (Burford 1997). Once the flocculated concentrate is obtained, suitable preservation is required to maintain nutritional value for a reasonable time. Previously, several methods have been used to preserve microalgal concentrates from on-site production, such as adding additives or preservatives (Tzovenis et al. 2004), freezing (Heasman et al. 2000) and refrigeration (Nunes et al. 2009). In this study, the simple technique of storage in a refrigerator at 4 °C was selected as this method is easy and cost-effective for use in a hatchery. The results show the composition of microalgae in flocculated concentrates affected the nutritional value during the storage period. Flocculated concentrates containing unidentified nanocyanobacteria showed a reduction in crude protein content from the first week of storage, a very short shelf life. This is similar to findings in an experiment conducted by Brown and Robert (2002); there were dramatic decreases in protein content in microalgae concentrates during the first week of storage. However, the crude protein content in flocculated concentrates containing the groups Bacillariophyceae and unidentified nanocyanobacteria in this experiment decreased in the third week of storage. This finding is similar to that of a study involving concentrated *C. calcitrans*; the chemical composition of cells remained stable during the first 3 weeks of storage but the organic matter decreased significantly after that period (Ponis et al. 2003a). Generally, the maximum time the microalgal pastes can be kept and still retain its nutritional value equivalent to fresh microalgae ranged from about 1 to 4 weeks, depending upon the species of algae (Heasman et al. 2000). The pheophytin *a*: chlorophyll *a* ratio has been used as indicator of the chemical status of microalgal concentrates over the period of storage. In this study, the flocculated concentrates had greatly increased values in the pheophytin *a*: chlorophyll *a* ratio in the third week of the storage period in both experiments, as a consequence of deterioration of the stored cells. Several experiments have been investigated pheophytin *a*: chlorophyll *a*evolution in microalgal concentrates during the storage period (Brown and Robert 2002; Ponis et al. 2003a). Those results have shown variations in chlorophyll *a* degradation, depending on species. For instance, for *Pavlova lutheri*, the value for pheophytin *a*:chlorophyll *a* increased remarkably during the second week of storage (Ponis et al. 2003a), while concentrates of *Thalassiosira pseudonana* showed no sign of degradation of chlorophyll *a* after around 21 days storage at 4 °C (Knuckey et al. 2006). Furthermore, after a 1-month storage period, the algae concentrate of *C. calcitrans* forma *pumilum*, *Chaetoceros* sp., *Skeletonema costatum* obtained by flocculation showed a loss in chlorophyll *a* of approximately 20–30 % (Brown and Robert2002). Therefore, unsuitable storage methods may have negatively influenced the nutritional quality and directly affected the growth of juvenile oysters.

The gross chemical composition of food can be a defining factor when choosing diets as substitutes for microalgae in the nursery culture of bivalves in hatcheries. Therefore, preservation methods must be devised that keep the fatty acid content unchanged or that involve modifications that do not decrease the nutritional value of the biomass. In our study, the HUFA concentration in flocculated concentrates preserved in a refrigerator at 4 °C decreased to an undetectable level at 4 weeks of storage time. Degeneration of fatty acids in flocculated concentrates depends on the species of microalgae and storage techniques. For example, studies have found that concentrates of *Isochrysis galbana* stored at 4 °C with continuous illumination had fatty acids located in storage lipids which decreased over time while polyunsaturated fatty acids (EPA, DHA) remained unaltered (Grima et al. 1994), while spray-dried *T. suecica* can cause a loss by oxidation of highly unsaturated fatty acids (Robert et al. 2001). The poor performance reported by operators of dried microalgae was mostly associated with the difficulty of keeping cells in suspension without disintegrating them (Laing and Millican 1992). Types with a good nutritional value and the best shelf life include the *C. calcitrans*, *S. costatum* and *Tetraselmis* spp. concentrates prepared by centrifuge, whereas the flagellates *P. lutheri* and *Isochrysis* sp. are easily damaged and deteriorate rapidly (McCausland et al. 1999; Heasman et al. 2000). Preservation of flocculated concentrates from shrimp pond water in a refrigerator at 4 °C in this study has been shown to be an unsuitable method.

Assessment of Flocculated Concentrates from Shrimp Pond Water as a Supplementary Diet for Juvenile Oysters

Flocculated concentrates of microalgae can be used effectively as supplementary diets in bivalve hatcheries. Substitution of microalgal concentrates for fresh microalgae have been used successfully as part of mixed or complete diets for larval or juvenile bivalves (Nell and O'Connor1991; McCausland et al. 1999; Heasman et al. 2000; Knuckey et al. 2006). Flocculated concentrates from selected shrimp pond water contain a high proportion of microalgae, but are composed of different species. Previous studies have shown that 1–2 week old preparations of microalgal concentrates (*C. calcitrans* and *S. costatum*) gave growth equivalent to live microalgae when presented as part of a mixed diet to juvenile *Crassostrea gigas* (McCausland et al. 1999) and in the same species, the use of the same concentrates for a longer period (4 weeks) gave significantly lower growth compared to corresponding diet of fresh microalgae (Ponis et al. 2003a). Those results and our findings in the preservation study indicate that flocculated concentrates produced from fresh microalgae and from intensive shrimp pond water used as a diet in this study could be kept in a refrigerator at 4 °C for a period of less than one week, a reasonable time, and maintain their nutritional value without effect on the growth of experimental oysters. From experiments, a partial substitution in the diet of 75 % of flocculated concentrate from intensive shrimp pond water for microalgal concentrates produced higher growth in comparison with oysters fed on 100 % microalgal concentrates in both experiments. In contrast, replacement with 100 % flocculated concentrates from shrimp pond gave significantly lower growth compared to replacements of below 75 % or the use of 100 % microalgal concentrates. In this case, complete replacement probably causes nutritional deficiency because of an imbalance of nutrients. This finding is similar to the results of a study by Brown and Robert (2002), who demonstrated that flocculated concentrates of microalgae (*Chaetoceros* sp.) can be used effectively as major dietary components for juvenile oysters. In follow-up experiments, concentrates prepared from *C. muelleri* cultures were effective as a supplementary diet and improved the growth of juvenile *C. gigas* and the scallop *Pecten fumatus* reared under commercial conditions, though not as effectively as corresponding live microalgae (Knuckey et al. 2006). In general, alternative diets have been tested as a partial replacement for fresh food, whereas bivalves are normally fed with a multispecies diet that allows a better balancing of essential compounds, with particular reference to polyunsaturated fatty acids (PUFAs), and assures high growth and survival (Enright et al. 1986; Laing and Millican 1986). Nevertheless, when used as a complete replacement for fresh microalgae, these diets generally do not adequately feed juveniles. The nutritional value of these concentrates and pastes of different microalgal species has been evaluated with juvenile oysters (Nell and O'Connor 1991; McCausland et al. 1999; Heasman et al. 2000; Robert et al. 2001) with some promising results. It has been shown that a mixed diet of two or more algal species provides a better balance of nutrients, because certain algae differ in their lipid (Brown et al. 2001) and highly unsaturated fatty acids (HUFAs) content (Renaud et al. 1999 and Leonardos and Lucas2000). A similar result was found in this study, as the inclusion of the microalgal concentrates (*C. calcitrans* and *T. suecica*) component in a mixed diet may have compensated for any nutritional deficiency in the concentrates. The presence of HUFAs, specifically EPA (eicosapentaenoic acid), DHA (docosahexaenoic acid) and ARA (arachidonic acid) has been seen as an indication of high nutritional value (Knauer and Southgate 1999) that promotes good growth and a high survival rate in hatchery and nursery production of larval and juvenile oysters. The best diet for culturing European oyster juveniles was a mixture of microalgal diets which led to better growth rates and higher levels of PUFAs (Ronquillo et al. 2012). In addition, flocculated concentrates from shrimp pond water not only contain microalgae, but also clay minerals. This may enhance efficiency in ingestion and assimilation in oysters. Oysters live and grow in estuarine habitats characterized by high concentrations of suspended inorganic material. Several experiments have been carried out to determine how suspended clay particles affect the amount of food ingestion by the oysters *C. virginica* (Urban and Kirchman 1992) and*C. gigas* (Sornin et al. 1988) and the mussels *Mytilus edulis* (Bayne et al. 1987). Those studies concluded that a suitable threshold concentration of clay particle mixed in food facilitates ingestion and increases the time food remains in the intestine, leading to increase in the efficiency of absorption. The role of clay minerals in supporting growth has been reported in the oyster *C. virginica* (Nell and Jordan 1983). High protein content in flocculated concentrates used in experiment 2 might support growth performance in juvenile oysters in comparison with experiment 1.

Various studies have been carried out on the diet of different species of hatchery-reared bivalves in which 40–50 % of the microalgae diet has been successfully replaced by different types of cereal meals and macroalgae (Pérez Camacho et al. 1998; Albentosa et al. 1999, 2002; Tanyaros and Chuseingjaw 2014), Our findings indicate that substituting flocculated concentrates of shrimp pond water for 75 % of microalgae led to better growth of juvenile oysters, but using flocculated concentrates as a mono-diet is unlikely to be suitable for feeding juvenile oysters. Further research is needed to find the best way to minimize problems with the use of flocculated concentrates, such as methods and storage time to best maintain the nutritional value and also to prevent clumping in the water. In addition, more research is needed on the toxicological effects of synthetic polymer flocculates for safe use in a bivalve hatchery.

ACKNOWLEDGMENTS

The authors thank Supatcha Chusiengjaw and Pacharee Kewprakan for sample collection. We also wish to thank William Martin for assistance in editing the manuscript. This study was funded by the Rajamangala University of Technology Srivijaya, annual budget year 2014.

REFERENCES

1. Albentosa M, Fénandez-Reiriz MJ, Pérez Camacho A, Labarta U (1999) Grow performance and biochemical composition of *Ruditapes decussatus* spat fed on microalgal and wheatgerm flour diets. J Exp Mar Biol Ecol 232:23–37CrossRef

2. Albentosa M, Pérez Camacho A, Fénandez-Reiriz MJ, Labarta U (2002) Wheatgerm flour in diets for Manila clam, *Ruditapes philippinarum*, spat. Aquaculture 212:335–345CrossRef
3. Bayne BL, Hawkins AJS, Navarro E (1987) Feeding and digestion by the mussel *Mytilus edulis* L. (Bivalvia: Mollusca) in mixtures of silt and algal cells at low concentrations. J Exp Mar Biol Ecol 111:1–22CrossRef
4. Brown M, Robert R (2002) Preparation and assessment of microalgal concentrates as feeds for larval and juvenile Pacific oyster (*Crassostrea gigas*). Aquaculture 207:289–309CrossRef
5. Brown MR, Jeffrey SW, Garland CD (1989) Nutritional aspects of microalgae used in mariculture: a literature review. CSIRO Marine Laboratories Report 205
6. Brown MR, McCausland MA, Kowalski K (2001) The nutritional value of four Australian microalgal strains fed to Pacific oyster *Crassostrea gigas* spat. Aquaculture 165:281–293CrossRef
7. Burford M (1997) Phytoplankton dynamics in shrimp ponds. Aquacult Res 28:351–360CrossRef
8. Casé M, Leca EE, Leitão SN, Anna EES, Schwamborn R, de Moraes T Jr (2008) Plankton community as an indicator of water quality in tropical shrimp culture ponds. Mar Poll Bull 56:1343–1352CrossRef
9. Coutteau P, Sorgeloos P (1992) The use of algal substitutes and the requirement for live algae in hatchery and nursery rearing of bivalve molluscs: an international survey. J Shellfish Res 11:467–476
10. Enright CT, Newkirk GF, Craigie JS, Castell JD (1986) Growth of juvenile *Ostrea edulis* L. fed*Chaetoceros gracilis* Schutt of varied chemical composition. J Exp Mar Biol Ecol 96:15–26CrossRef
11. Folch AJ, Less M, Sloan-Stanley GH (1957) A simple method for isolation and purification of total lipids from animal tissues. J Biol Chem 226:497–509
12. Funge-Smith SJ, Briggs MRP (1998) Nutrient budgets in intensive shrimp ponds: implications for sustainability. Aquaculture 164:117–133CrossRef
13. Grima EM, Sánchez Peŕez JA, García Camacho F, Acién Femández FG, López Alonso D, Segura del Castillo CI (1994) Preservation of the marine microalga *Isochrysis galbana*; influence on the fatty acid profile. Aquaculture 123:377–385CrossRef
14. Heasman MP, Diemar J, O'Connor W, Sushames T, Foulkes L (2000) Development of extended shelf-life microalgae concentrate diets harvested by centrifugation for bivalve molluscs: a summary. Aquacult Res 31:637–659CrossRef
15. Helm MM, Bourne N, Lovatelli A (2004) Hatchery culture of bivalves: a practical manual. FAO Fisheries Technical Paper No. 471. Food and Agriculture Organization of the United Nations Publishing, Rome, Italy
16. Islam S, Sarker J, Yamamoto T, Wahab A, Tanaka M (2004) Water and sediment quality, partial mass budget and effluent N loading in coastal brackish water shrimp farms in Bangladesh. Mar Poll Bull 48:471–485CrossRef
17. Jacob GS, Pruder GD, Wang JK (1993) Growth trial with the American oyster *Crassostrea virginica* using shrimp pond water as feed. J World Aquacult Soc 24:344–554CrossRef
18. Jones AB, Preston NG (1999) Sydney rock oyster, *Saccostrea commercialis* (Iredale & Roughley), filtration on shrimp farm effluent: the effects on water quality. Aquacult Res 30:51–57CrossRef
19. Jones AB, Dennison WC, Preston NP (2001) Integrated treatment of shrimp effluent by sedimentation, oyster filtration and macroalgal absorption: a laboratory scale study. Aquaculture 193:155–178CrossRef
20. Knauer J, Southgate PC (1999) A review of the nutritional requirements of bivalves and the development of alternative and artificial diets for bivalve aquaculture. Rev Fish Sci 7:241–280CrossRef
21. Knuckey MR, Brown MR, Robert R, Frampton DMF (2006) Production of microalgal concentrates by flocculation and their assessment as aquaculture feeds. Aquacult Eng 35:300–313CrossRef
22. Laing I, Millican PF (1986) Relative growth and growth efficiency of *Ostrea edulis* L. spat fed various algal diets. Aquaculture 54:245–262CrossRef
23. Laing M, Millican PF (1992) Indoor nursery cultivation of juvenile bivalve molluscs using diets of dried algae. Aquaculture 102:231–243CrossRef
24. Leonardos N, Lucas IAN (2000) The nutritional value of algae grown under different culture conditions for *Mytilus edulis* L. larvae. Aquaculture 182:301–315CrossRef
25. Lertsutthiwong P, Sutti S, Powtongsook S (2009) Optimization of chitosan flocculation for phytoplankton removal in shrimp culture ponds. Aquacult Eng 41:188–193CrossRef
26. Lorenzen CJ (1967) Determination of chlorophyll and pheopigments: spectrophotometric equations. Limnol Oceano 12:343–346CrossRef
27. Manzi JJ, Rourke CB, Bobo MY, Steele GH, Smiley RA (1988) Result of an oyster-shrimp pond biculture study. J Shellfish Res 7:205–206
28. McCausland MA, Brown MR, Barrett SM, Diemar JA, Heasman MP (1999) Evaluation of live microalgae and microalgal pastes as supplementary food for juvenile Pacific oysters (*Crassostrea gigas*). Aquaculture 174:323–342CrossRef
29. Nell RIE, Jordan SJ (1983) Preferential ingestion of organic material by the American oyster*Crassostrea virginica*. Mar Ecol Prog Ser 13:81-A
30. Nell JA, O'Connor WA (1991) The evaluation of fresh algae and stored algal concentrates as a food source for Sydney rock oyster, *Saccostrea commercialis* (Iredale and Roughley). Aquaculture 99:277–284CrossRef
31. Nunes M, Pereira A, Ferreira JF, Yasumaru F (2009) Evaluation of the microalgae paste viability produced in a mollusk hatchery in southern Brazil. J World Aquacult Soc 40:87–94CrossRef
32. Pérez Camacho A, Albentosa M, Fernández-Reiriz MJ, Labarta U (1998) Effect of microalgal and inert (cornmeal and cornstarch) diets on growth performance

and biochemical composition of *Ruditapes decussatus* seed. Aquaculture 160:89–102CrossRef

33. Ponis E, Robert R, Parisi G (2003a) Nutritional value of fresh and concentrated algal diets for larval and juvenile Pacific oysters (*Crassostrea gigas*). Aquaculture 221:491–505CrossRef
34. Ponis E, Robert R, Parisi G, Tredici M (2003b) Assessment of the performance of Pacific oyster (*Crassostrea gigas*) larvae fed with fresh and preserved *Pavlova lutheri* concentrates. Aquacult Inter 11:69–79CrossRef
35. Ponis E, Parisi G, Chini Zittelli G, Lavista F, Robert R, Tredici MR (2008) *Pavlova lutheri*: production, preservation and use as food for *Crassostrea gigas* larvae. Aquaculture 282:97–103CrossRef
36. Renaud SM, Thinh LV, Parry DL (1999) The gross chemical composition and fatty acid composition of 18 species of tropical Australian microalgae for possible use in mariculture. Aquaculture 170:147–159CrossRef
37. Reynolds CS (2006) The ecology of phytoplankton. Cambridge University Press, New YorkCrossRef
38. Robert R, Parisi G, Rodolfi L, Poli BM, Tredici MR (2001) Use of fresh and preserved*Tetraselmis suecica* for feeding *Crassostrea gigas* larvae. Aquaculture 192:333–346CrossRef
39. Rodrıguez RA, Paez-Osuna F (2003) Nutrients, phytoplankton and harmful algal blooms in shrimp ponds: a review with special reference to the situation in the Gulf of California. Aquaculture 219:317–336CrossRef
40. Ronquillo JD, Fraser J, McConkey A-J (2012) Effect of mixed microalgal diets on growth and polyunsaturated fatty acid profile of European oyster (*Ostrea edulis*) juveniles. Aquaculture 360–361:64–68CrossRef
41. Sandifer PA, Hopkins JS (1996) Conceptual design of a sustainable pond-based shrimp culture system. Aquacult Eng 15:41–52CrossRef
42. Shaari AL, Surif M, Latiff FA, Omar WMW, Ahmad MN (2011) Monitoring of water quality and microalgae species composition of *Penaeus monodon* ponds in Pulau Pinang, Malaysia. Trop Life Sci Res 2011:51–69
43. Sornin J, Deslouspaoli J, Hesse O (1988) Experimental study of the filtration of clays by the oyster *Crassostrea gigas* (Thunberg): adjustment of particle size for best retention. Aquaculture 69:355–366CrossRef
44. Tanyaros S (2000) Use of the oyster *Crassostrea belcheri* (Sowerby) as a biofilter in intensive shrimp pond water. Dissertation. Asian Institute of Technology. Bangkok, Thailand
45. Tanyaros S, Chuseingjaw S (2014) A partial substitution of microalgae with single cell detritus produced from seaweed (*Porphyra haitanensis*) for the nursery culture of tropical oyster (*Crassostrea belcheri*). Aquacult Res 2014:1–9. doi:10.1111/are.12662
46. Tanyaros S, Tarangkoon W (2014) Water quality, growth and mortality of tropical oyster*Crassostrea belcheri* spat in the conserved natural oyster beds of Trang province, southern Thailand. In: 2nd International Conference on Fisheries Science. 30–31 July, 2014. Colombo, Sri Lanka, pp 55–59
47. Tanyaros S, Choopunth P, Kurokura H (2002) A basic design and operating for shrimp-settling-oyster integration system to reduce discharge problem. Fish Sci 68:843–846
48. Tanyaros S, Pattanatong T, Tarangkoon W (2012) Effect of water flow rate and density on nursing hatchery-reared juvenile oysters, *Crassostrea belcheri* (Sowerby 1871) in a semi-closed recirculation system. J Appl Aqua 24:364–373
49. The Association of Official Analytical Chemists (2012) Official methods of analysis of AOAC international, 19th edn, Virginia. AOAC 2001.11: 2012
50. Tookwinas S, Songsangjinda P (1999) Water quality and phytoplankton communities in intensive shrimp culture ponds in Kung Krabaen Bay, eastern Thailand. J World Aquacult Soc 30:36–45CrossRef
51. Tzovenis I, Triantaphyllidis G, Naihong X, Chatzinikolaou E, Papadopoulou K, Xouri G, Tafas T (2004) Cryopreservation of marine microalgae and potential toxicity of cryoprotectants to the primary steps of the aquacultural food chain. Aquaculture 230:457–473CrossRef
52. Urban ER Jr, Kirchman DL (1992) Effect of kaolinite clay on the feeding activity of the eastern oyster *Crassostrea virginica* (Gmelin). J Exp Mar Biol Ecol 160:47–60CrossRef
53. Van Wijngaarden D (1967) Modified rapid preparation of fatty acids esters from lipids for gas chromatographic analysis. Anal Chem 39:848–849CrossRef
54. Walne PR (1974) Culture of bivalve molluscs: 50 years' experience at Conwy. Fishing News (Books) Ltd, West Byfleet
55. Yusoff FM, Zubaidah MS, Matias HB, Kwan TS (2002) Phytoplankton succession in intensive marine shrimp culture ponds treated with a commercial bacterial product. Aquacult Res 33:269–278CrossRef
56. Zhukova NV, Aizdaicher NA (1995) Fatty acid composition of 15 species of marine microalgae. Phytochem 39:351–356

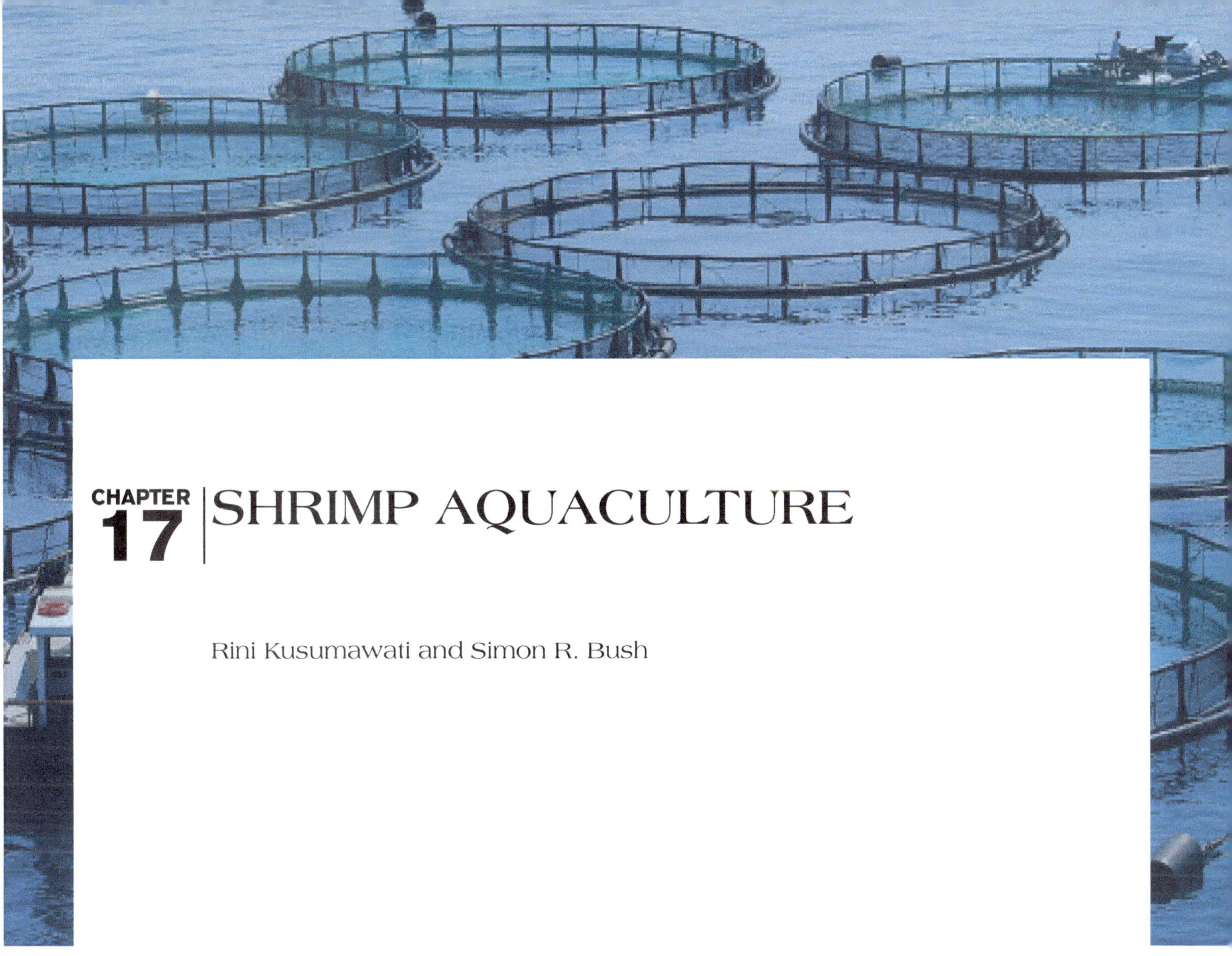

CHAPTER 17 | SHRIMP AQUACULTURE

Rini Kusumawati and Simon R. Bush

INTRODUCTION

Better management practices (BMPs) for shrimp farming are designed to standardize on-farm practices and reduce the economic risk associated with disease and environmental degradation. Similar to third party certification standards, BMPs set out a range of technical indicators that producers can use as targets to improve their production practices (Vandergeest, 2007; Anh et al. 2011; Padiyar et al. 2012). However, unlike private certification standards, such as those of the Aquaculture Stewardship Council (ASC), Global Aquaculture Alliance (GAA) and Naturland, BMPs are in most cases not formally monitored by either governments or by third party auditors. Instead governments and NGOs alike use them to disseminate technical practices to producers in a more structured and formalized fashion in order to standardize production inputs and practices.

BMPs are based in large part on the Principles for Sustainable Shrimp Farming, developed by a consortium of international agencies including WWF (The Consortium 2006) and the FAO Code of Conduct for Responsible Fisheries (CCRF) (FAO 1995). When further defined at the national level, these international principles and codes are adapted to suit the specific demands of intensive and/or extensive production systems (e.g. Anh et al 2011; Padiyar et al. 2012). In geographically diverse countries such as Indonesia, these national level BMPs are further adapted to the range of production systems found in different regions or provinces by both government departments and NGOs. In doing so these actors enact a process of translating a generic set of global concerns into local contexts with the intention of increasing the likelihood of adoption and subsequently performance of shrimp aquaculture.

Translating BMPs into local shrimp farming contexts has proven to be a challenging process. Despite their involvement, locally defined BMPs have been viewed with suspicion by governments and farmers as an attempt to supplant state regulation, and render social and political conflicts surrounding shrimp farming as technical problems with technical solutions (Ha and Bush 2010; Islam 2008; Li 2007; Vandergeest and Unno 2012). As argued by (Béné2005), the perceived threat posed by BMPs is that advocates are able to "depoliticis[e] the problem of shrimp farming and refram[e] it into a neutral, bio-physical problem, where only technical - and not structural or political - solutions are required" (p. 611). Others have

Citation: Rini Kusumawati and Simon R. Bush, shrimp aquaculture, DOI 10.1186/s40152-015-0039-4

shown how BMPs fit into a wider technical system of non-state regulation by setting producers on the path towards more stringent private standards that facilitate access to international markets, such as ASC or GAA (Belton et al. 2009; Anh et al. 2011; Ha et al. 2012). With few exceptions (e.g. Konefal and Hatanaka 2010), these studies have largely ignored the capacity of NGOs or governments to facilitate this process of translation, and as such the extent to which the knowledge and interests of global and local actors influence their successfully co-production.

This paper analyses fills this gap in the literature by providing a detailed case study of a NGO-led multi-stakeholder process of co-producing BMPs and implementation in the northern part of East Kalimantan, Indonesia. In this case, co-production is not linked to any single event, but is instead seen as an outcome of networked places and processes stretching beyond the local and link to the global places and historical events (Goldman 2010). In order to understand this process, we first examine the extent to which NGO-led BMPs are able to capture and mitigate both technical and social challenges of production. We then examine the extent to which these BMPs (de)politicize social and environmental issues around shrimp aquaculture by including or excluding different knowledge and practices. In doing so we analyse the transfer of knowledge embodied in global principles to local actors and environments, the composition and content of multi-stakeholder meetings and the degree to which the final BMPs incorporated the practices of local producers.

We examine the process of translation in three parts. First, we review the process of developing the international principals in Aceh by WWF in partnership with other national and international organisations following the 2004 tsunami (ACIAR et al. 2007). We then analyse how WWF transferred these Aceh BMPs to two districts of East Kalimantan - Tarakan and Bulungan – through a series of multi-stakeholder meetings and a farm-level pilot programme. Based on these results we reflect on the position and role of WWF, as an international NGO, in mediating the politicization of shrimp aquaculture through technical BMP standards.

CO-PRODUCING STANDARDS

Research on aquaculture standards have focused on a range of questions related to the extent of expert and industry capture during their definition, the poor representation of producers from developing countries (Vandergeest 2007; Belton et al. 2010; Bush et al. 2013), and the effects of transferring globally derived knowledge into a national to local settings (Béné 2005; Konefal and Hatanaka 2011; Bush and Belton 2012; Ha et al. 2013). A common theme running through these studies is how NGO-state-industry networks 'translate' global norms, knowledge and policy goals into local knowledge and practices, with questions focusing on the inclusion/exclusion of actors, consensus seeking in local settings, and the extent to which they lead to changes in production practices.

Translation can be understood as a process of knowledge co-production, emphasizing the creative friction that emerges between the interaction of different knowledge, each with their own grounds for belief, procedures for validation (Jasanoff 2004; Long 2001; Tsing 2005). Similar to findings into multi-stakeholder initiatives in other sectors, the technical definition of these standards does not occur in a vacuum. Instead the formation of the networks of actors that constitute these multi-stakeholder initiatives is as equally important to understanding which knowledge is translated into measurable standards and which knowledge is excluded (see Cheyns 2011; Anh et al. 2011; Ponte and Cheyns 2013). Put differently, co-production draws attention to how content is decided upon, as well as who is able to contribute to that content.

Co-production is seen in two distinct ways. From a managerial perspective it implies a collaborative process of stakeholder engagement designed to "address a defined problem and build an integrated system-oriented understanding of [a] problem" (Armitage et al. 2011: 996). Based on the notions of resource co-management, this perspective opens up the possibility to actively steer a process of co-production which results in a shared understanding of environmental problems that in turn results in conformity and effective management. However, from a critical perspective, and reminiscent of Cooke and Kothari's (2001) 'tyranny of participation', co-production can also reveal the difference between the ingenuous inclusion of certain actors and meaningful collaborative process that lead to substantive stakeholder inclusion and empowerment. In this sense, the uniformed and formalized knowledge codified in BMPs is representative of the relations of power that negotiate the inclusion of different knowledges (Konefal and Hatanaka 2011; Ponte and Cheyns 2013). The conditions of including the knowledge and interests of actors in the final standards are therefore as fundamentally political as the conditions of exclusion (Cheyns 2011). In this sense, the final standards do not always represent the aims, perceptions, values, interests and relationships of actors with different technical and political capacities to contribute.

Co-production can also be observed through producer practices, which represent the outcomes of the discursive practice inherent in standard definition and implementation. In the case of BMPs, exploring practices and their outcome help to uncover how co-production continues beyond standard definition to standard implementation. We argue that by understanding these practices and their material outcomes, it is possible to distinguish how responsive standard setters are to producer innovation or localization. Following on from what Tsing (2005) labels 'friction', our analysis explores the difference between discursive and material practice by focusing on the "awkward, unequal, unstable, and creative qualities of interconnection across difference" (p. 4). In the case of BMPs, this means identifying how the creative qualities of friction are expressed in producer practices, while also reflecting the willingness of the standard setters to re-negotiate the content of standards and, therefore, the collective knowledge they embody. This moves us beyond studies on the formalized standard setting processes in networks and multi-stakeholder initiatives to explore the possibility of incorporating alternative practices and knowledge in standard formation and revision.

SITE SELECTION AND METHODOLOGICAL APPROACH

The methodology adopted for this study is structured around what Marcus (1995) labels a 'multi-sited ethnography'; investigating the context within which decisions over standards for East Kalimantan were made, how different actors influenced these decisions, and how the different knowledge that constitute these standards have been communicated across space (also see Falzon 2009). This is methodology is also complimentary to Tsing's (2005) claim that the connections between actor networks spanning global sites requires a heterodox approach, taking into account the spaces in which these actors operate (e.g. farms and workshops) as well as the connections between them based on material and informational flows.

Although based on a single 'case', the process of defining standards and implementing and the friction between these processes is thought to be representative of the steps and processes common to multi-stakeholder processes in general. However, it the goal of our analysis is not necessarily to provide generalizable results. Instead we seek to abstract our specific empirical findings (see Lund 2014) into the wider conceptual precepts of knowledge co-production, and in doing so contribute to a wider understanding of how knowledge used by who and how.

Fieldwork for this study was conducted from 2009 to 2013 in Tarakan and Bulugan districts of East Kalimantan. Tarakan has developed as a major centre for shrimp processing companies thanks to its existing infrastructure for international trade (Ilman et al. 2009), and because of the value of the surrounding demersal fishery, both of which has attracted investment from Malaysian, Taiwanese and Japanese investors. Farmed shrimp are primarily produced in the Sesayap Estuary of Tarakan, Bulungan, Kabupaten Tanah Tidung and Nunukan districts (see Fig. 1). It is precisely the location of shrimp ponds in these remote islands that have given the shrimp from the north coast East Kalimantan a reputation of the high quality desired by Japanese shrimp importers, and subsequently the market for entrepreneurs to invest in super-extensive ponds of up to 100 ha in size.

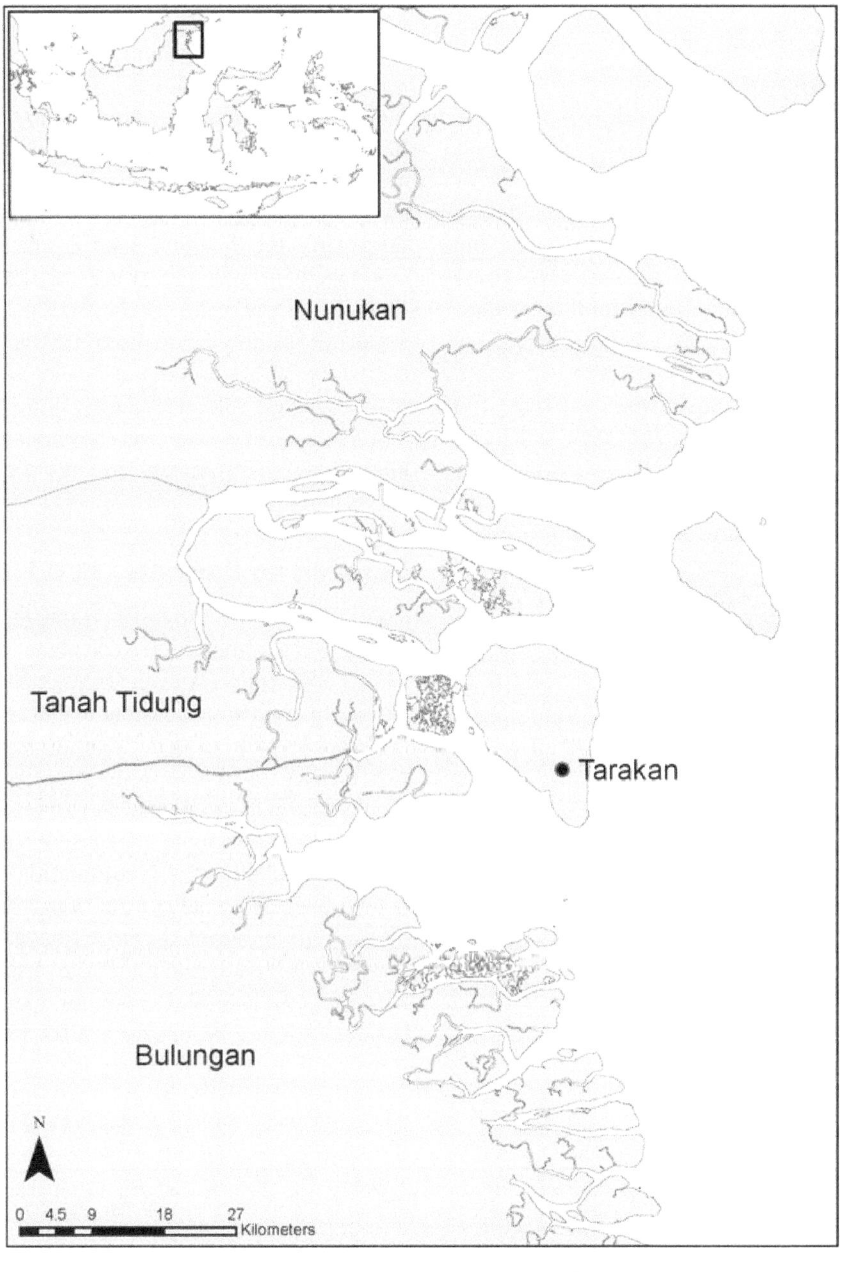

Figure 1: Map of districts involved in the WWF BMPs program in East Kalimantan

Primary data was collected through in-depth interviews with main actors working in shrimp production and trade, as well as those working in regulation, policy and extension. The first group of actors consisted of eight pond owners, three middlemen and three cold storage owners, including those both included and excluded in definition of BMP for Tarakan. The second group were consisted of government functionaries of the Marine Affairs and Fisheries (MAF) office in Tarakan, as well as three officials from the same department at the provincial level of East Kalimantan,[1] and four at the national level. In addition, three Indonesia-based NGOs were interviewed both in Tarakan and Jakarta. Observations were also made during several visits to shrimp ponds, hatcheries, nurseries, and processing companies, as well as at meetings and seminars on sustainable shrimp production in both Indonesia and Europe – for example, during the WWF shrimp aquaculture dialogue meeting in Jakarta in March 2010. Secondary data was gathered by reviewing and examining data on both shrimp farming and better management practices from websites, newsletters, reports and newspaper articles.

BUILDING THE BMPS NETWORK

From Principles to Standards

The first BMPs in Indonesia were developed for Aceh through a collaboration of government departments with international and national NGOs, and development institutions as part of the of a wider reconstruction programme for shrimp aquaculture after the 2004 tsunami (ADB et al. 2007). Many international agencies offered assistance for the rehabilitation of coastal regions, including the aquaculture sector that was one of the sources of livelihood of the inhabitants. However, as Rimmer et al. (2012) argue, many of the agencies involved did not have adequate technical capacity to rehabilitate shrimp ponds and their supporting infrastructure. Organisations such as WWF therefore saw the development of BMPs as a means of providing guidelines, based on the international principles for sustainable shrimp aquaculture, for these organisations to either reduce the risk of poor reconstruction or at best develop technically improved shrimp ponds. Although made difficult by the disintegration of community and farming structures following the tsunami, the Aceh BMPs have been criticised for not taking consideration local perspectives on reconstruction, and for reproducing previous practices instead of new, potentially more sustainable farming approaches; a pattern noted by McGregor (2007) in other initiatives for rebuilding Aceh.

The Aceh BMPs represent the first step of translating the international principles into the Indonesian context, and WWF-Indonesia saw them as an opportunity to create locally specific standards for other regions in country. As stated by a staff member of WWF-Indonesia, the international collaboration that went into the development of the Aceh BMPs provided considerable credibility within Indonesia. WWF-Indonesia planned to invest further in the development of other locally specific BMPs through the "development of pilot projects enriched with literature studies" (Interview, Jakarta, February 2013). WWF-Indonesia then entered a new phase of technical translation of the BMPs aimed at advocating environmental improvement by district level governments. Part of a their wider advocacy strategy – including BMPs for 11 other aquaculture species – the translation process intended to address a widely perceived (see Smith et al. 2003; Satria and Matsuda 2004) lack of capacity for environmental policy and management at the district level.

The process of translating BMPs into 'new' local contexts in Indonesia is therefore not insulated from networks of global experts. It is instead part of WWFs wider strategy to facilitate different organizations to input their experience and expertise into the promotion of responsible aquaculture production. However, unlike the development of the Aceh BMPs, WWF-Indonesia argued that the new round of BMPs translation should focus directly on local government, industry and shrimp producers. Nonetheless, the association of WWF-Indonesia with the global WWF network did mean that the BMPs they facilitated were also designed to assist shrimp producers in Indonesia to move towards international third-party certification, such as the ASC shrimp standards; developed through the WWF-facilitated shrimp aquaculture dialogue (ShAD). As we now go onto illustrate through the Tarakan case, these 'localised' BMPs remain closely linked to the wider ShAD process and, according to WWF-Indonesia, play a strategic role in preparing producers to be ready for ASC certification. They can therefore be seen as a bridge between knowledge of producers and the experts within the global environmental regulatory networks of which they are a part.

Building on Business - WWF and MMA

The decision to move to Tarakan, with vastly different production systems to Aceh (see Table 1), was led by Mustika Minanusa Aurora (MMA); the largest processing and exporting company operating in that region. MMA was already actively developing their own environmental programme, in part because of a new found environmental concern of the company's owner, but also in response to demand for 'green' shrimp from their main Japanese buyer Nicherei. The general manager explained the incorporation of environmental issues in their business strategy by arguing: "If we want our product to hit the market, we need a story [that can promote the product]. If our customers believe that our products are environmental friendly, organic, and pose no harm to the environment, [I believe] there would be a price increase to our products" (Interview, Tarakan, February 2009).

Table 1: Main technical differences of production system between Aceh and Northern part of East Kalimantan

Pond characteristics	Aceh (Source: Zainun et al. 2007)	East Kalimantan (Source: Ilman et al. 2009)
Size	2 – 5 ha/pond	5 – 25 ha/pond
Type	Mainly traditional, but there are also semi intensive and intensive ponds	Traditional extensive

Stocking	1000—20,000/ha (traditional)	1 - 2/m2 – 10,000 - 20,000/ha
	20,000—60,000/ha (semi-intensive)	
	100,000—600,000/ha (intensive)	
Feed	Algal growth in ponds, rice bran, pellet for semi-intensive; and for intensive farming system, feed is given as recommended.	Natural algal growth. No additional feed.

The demand for green products by Nicherei was originally focused on the reforestation in Kalimantan province in order to support their 'green shrimp' label. The first phase of collaboration focused on a 'environmental responsibility programme' which contributed US$ 1 per box of shrimp sold to fund a reforestation programme in Kutai Kartanegara district of East Kalimantan. Recognising a mismatch between inland reforestation and their claim of 'green shrimp', MMA then proposed that Nicherei shift their funding to mangrove rehabilitation in Tarakan and the surrounding region. The resulting 'Forest of Life' project started in 2006 providing Nichirei and MMA the chance to claim mangrove reforestation as a marketable quality of their shrimp. Using the claim of a 'green made farm' on their packaging with further explanation that the shrimp were sourced from extensive systems that are 'close to the natural environment', they established what they perceived as a Japanese consumer driven scheme for mangrove rehabilitation.

The 'green made farm' label was successful in establishing a flow of funding, however after some poor technical results with their reforestation programme MMA requested the assistance of WWF to provide assistance on mangrove planting. In doing so WWF also became a key partner, providing legitimacy to the environmental claims of the company in return for establishing their own network in Tarakan. They were given office space at the MMA factory and access to the companies' suppliers, as well as a basis from which to develop relations with the local district government. In April 2008, WWF Indonesia, MMA, and Environment and the Tarakan Natural Resources Agency (*Badan Lingkungan Hidup* - BLH) signed an MoU stating that MMA, with assistance from WWF, had a responsibility to plant 150 ha of mangrove in five years. For WWF the document represented a formalization of their collaboration with the MMA, the enrolment of the government into their activities, and provided a statement of political support for introducing their BMPs programme.

Translation into a Local Setting

The collaboration with MMA provided a platform on which WWF could access and enrol local actors into the process of translating the Aceh BMPs to the local context of Tarakan. WWF supported at least three 'public' meetings to introduce the concept of BMPs, outline their experience from Aceh, and ultimately to develop a new set of BMPs for East Kalimantan. At the first consultation, held in August 2009, WWF invited government agencies from different sectors: the Marine and Fisheries Agency (MFA), Forestry Agency, and Environmental Agency from five districts: Tarakan, Bulungan, Kabupaten Tanah Tidung, Nunukan and Berau. The aim of the meeting was to enrol the five district governments by introducing the BMPs from Aceh and the process of developing BMPs suitable for East Kalimantan. In the second meeting, held in January 2010, WWF reduced the number of government departments invited and brought in a selected group of local industry representatives. MFA were invited to chair the meeting and provide 'technical input' to the standards. The meeting was scheduled in Tarakan at the MFA office, with other departments who attended the first meeting (such as environment and forestry), excluded. WWF also invited representatives from fisher and shrimp producer organisations selected on the basis of their perceived ability to 'speak' in public. They were also key suppliers of MMA, who was the only one of seven export processors in Tarakan to be invited. On justifying this exclusive selection policy, the WWF officer reflected on close personal ties with MMA; "[WWF] involves MMA in the BMPs development because we have worked with them since the beginning. I would feel uncomfortable if we invited other cold storages to our project" (WWF field staff, Tarakan, January, 2010).

The exclusive and targeted selection of participants was reflected in the content and discussion of the meeting, as well as the subsequent strategy for trialling the draft BMPs. The head of MFA stated in her opening address that MFA give full support to WWF-Indonesia in the development of BMPs for Tarakan. She also argued that the BMPs provide a clear solution to the decline of shrimp production caused by producer mismanagement. According to her, BMPs also provide a solution to help the government support shrimp producers to improve the quality of farming systems in the Tarakan area.

However, the praise given by the head of MFA also reveals a subtext against which the BMPs were being introduced. In follow up interviews MFA argued that a 'third party' like WWF-Indonesia was needed to intervene in the management of shrimp farming across Northeast Kalimantan. While local government recognised key issues associated with producer practices, mangrove clearance and overfishing of broodstock from surrounding waters, they were unable to unilaterally control these activities. The exclusion of other government departments also avoided a potential institutional clash between forestry agency and other technical agencies around the illegality of farming on forestry lad in some coastal area. Their attempts to draw in the provincial governor had failed because of the lack of jurisdiction of the province in district affairs and the relative isolation of Tarakan – on the periphery of Kalimantan and the Indonesian state (see Visser and Adhuri 2010). For the government, the definition of BMPs by a select group of actors represents the contestation over coastal resources.

The selection of pond owners by WWF also reflects the complexity of pond ownership and decision making over shrimp farming in the region. As described elsewhere, there is a clear distinction between producers, who are in practice farm managers, and the pond owners who attended the meeting

(Kusumawati et al. 2013). WWF expected pond owners to be resource persons; providing technical knowledge into the formation of the standards and, in doing so, legitimizes the BMPs in the wider farming community. But on reflection, the group selected was more representative of the patronage networks that control ownership of the shrimp industry than it was of technical knowledge on farming. The pond owners fulfilled their 'speaking' role as hoped for by the WWF, but as commented by participants in later interviews, were better at expressing their opinions (*bisa bicara*) than grounded technical knowledge of shrimp farming.

The political positioning of MFA and the selection of pond owners rather than farm managers to represent producer interests affected the outcome of the meeting. Instead of presenting the Aceh BMPs, receiving input, and translating them into BMPs for farming systems in Tarakan, WWF was met with resistance. During the meeting the pond owners refuted the knowledge of WWF because they perceived changes to the infrastructure of the their ponds as capital intensive, and fundamentally opposed to the techniques they used on their farms and, perhaps most importantly, unrelated to the immediate problems they are facing in terms of falling productivity. This was captured by one pond owner who stated during the meeting that "If you show these standards to producers, they will shake their heads." He went on to argue that producers and pond owners alike"... never think about what good [pond] construction is like. We only know how to stock, maintain shrimp, and harvest them. That's it. If shrimp disease spreads, we would like to know how to deal with it and the cure" (Pond owner, Tarakan, January, 2010). Although the advice from WWF to change the design of ponds was backed up by the expert knowledge on which they are based, it created an immediate friction because of its abrupt introduction and radical shift from current practices.

Recognizing the resistance from the producers in the formal meetings, WWF changed their strategy to enrol the producers into the BMPs process through interviews and demonstration. The interviews were designed to collect the experiences and opinions of pond owners and their workers on issues related to production. The inclusion of workers was an important step given their absence at the formal meetings. However, the selection of producers remained narrow, continuing with suppliers to MMA. It was only in 2012, after the BMPs for Tarakan were issued, that this strategy was revised and new pilot farms in Tarakan were built to demonstrate the new set of BMPs to non-MMA related producers (WWF-Indonesia 2012). We now turn our attention to the experiences of producers prior to 2012 who were involved in the pilot project run by WWF and MMA to test the applicability of the Aceh BMPs in Northeast Kalimantan while at the same time from and contributed to the design of the Tarakan BMPs.

TRANSLATING BMPS INTO PRACTICE

The field-testing by WWF in partnership with MMA offered an opportunity to test the Aceh BMPs, translate them to the local conditions of Tarakan. In doing so WWF hoped to avoid the politicised debate of the meetings and overcome the producer's critique that the BMPs were of no relevance to the challenges they faced. The goal was to develop a shared or co-produced set of BMPs that balanced the expert knowledge embodied in the Aceh BMPs and the practice knowledge of the producers. We now turn to an analysis of our observations of WWF's testing of three specific BMPs from Aceh, the feedback received by WWF from producers and the reasoning behind rejection, acceptance and adaptation of these standards. We compare the BMPs from Aceh, producers' interpretation of and objection to the standards, and the ways in which the standards accommodated the wider interests of shrimp production in contested coastal land.

Pond Location and Mangrove Rehabilitation

The location of ponds is a fundamental parameter in both the international principles for sustainable shrimp farming and the Aceh BMPs standards. The standard requires local government to establish rules for the construction of ponds in and adjacent to coastal mangrove forest. The original manual developed for the Aceh BMPs stipulated a minimum buffer zone of 150 m along the seaward facing edges of ponds (ADB et al. 2007). In the Tarakan BMPs this seaward buffer was maintained, but reflecting the location of ponds in the Seyasap delta, added a requirement for ponds to maintain a 50 m buffer along river facing edges of their ponds and to re-establish mangrove in and directly around the ponds in order to "increase the environmental carrying capacity" of the extensive shrimp systems (WWF-Indonesia, 2011). To comply with this adjusted BMPs, the producers in the farm trial were asked to plant mangroves along the dyke and the raised, shallow central areas inside the ponds (see Fig. 2).

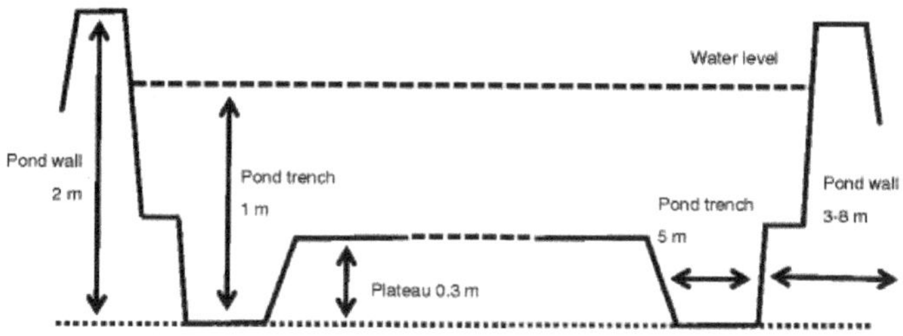

Figure 2: Pond design as recommended by BMPs for Tarakan and Bulungan (WWF Indonesia2011)

The scientific justification for planting mangroves in and around the ponds is to increase the micronutrient load of the ponds, which in turn increases the production of bacteria and plankton upon which shrimp in extensive systems feed (Alongi et al 1999; Gatune et al. 2012). Other benefits include the provision of cryptic habitat for shrimp, increased shade cover to prevent algal blooms, and improved nutrient cycling (Paez-Osuna 2001; Primavera 1997; Robertson and Phillips 1995). All of these benefits are scientifically supported and are listed on the MMA website developed in partnership with WWF (seehttp://www.shrimp.co.id/eng/index.php, accessed on 15 November 2013). However, despite the scientific motivation for implementing these measures producers expressed technical concerns about increasing the mangrove cover of the ponds.

The producers did not accept the rehabilitation of the mangroves, arguing that the ecology of the ponds would be disturbed and the productivity of the ponds would decline. But instead of rejecting the proposal outright, they proposed an alternative approach for planting mangroves on the dykes and inside the ponds at a wider spacing. Instead of the one to two meters spacing between trees they successfully negotiated a 10 m gap between trees. Their rationale for doing so was to protect the production of plankton, the main source of food in extensive shrimp systems, which require adequate sunlight to for photosynthesis. A secondary concern was that if mangroves become too dense dangerous wildlife would encroach on the 'forested' pond.

Although technically framed, the concerns of these producers were underscored by the wider politics of access and ownership to coastal land in Northeast Kalimantan. Under national law, the mangrove forest is part of production forest are (*Kawasan Budidaya Kehutanan*) falls under the ownership of the state through the Ministry of Forestry who issues user permits (Ilman et al. 2009). However, the administrative decentralization of the Indonesian state has reduced the enforcement of national forest regulation (e.g. Smith et al 2003). Once a forest is cleared, usufruct rights are granted by village leaders with customary rights over the forested land. Pond owners then need a letter from the sub-district head to acknowledge this permission before making a formal request (often in retrospect) to clear the forest and develop a pond to the district level MFA. This formal permission is often not sought because it enables MFA to levy a tax over production (*Surat Ijin Usaha Perikanan* or SIUP in Indonesian). As a result access rights remain ambiguous; as long as the producers do not convert the land back to forest, they are able to maintain their customary access agreement and avoid any legal enforcement.

The case demonstrates that although technically framed, resistance by producers to the BMP standard for pond location and mangrove rehabilitation is part of a wider conflict around access and use of coastal land. Planting trees in and around their ponds changes the status of that land and disqualifies them from continuing shrimp aquaculture. The final version of the BMPs for Tarakan and Bulungan (2011) maintained that mangrove needs to be planted around the dike, inside the ponds and around the water gate. But unlike the standards in Aceh they do not specify the exact distance between trees – leaving this up to the farmers. The standards therefore provide room for interpretation by producers, but in doing so feed into a wider politics of land-use.

Extensive Traditional Aquaculture Practices

The BMPs standards also set out guidelines on the design and construction of ponds to create a productive farming system that reduced the incidence of disease and promote biodiversity. The both the meetings and the pilot programme gave considerable attention to the characteristics of the extensive system found in the Tarakan region in mind; open tidal water flows, variable pond size and shape, and low density in seed stocking (WWF-Indonesia 2011, p.1; see Fig. 3). However, despite the focus on these local conditions, nearly all changes to pond design and management was met with resistance.

The first concern of the producers focused on the economic rationality for large pond sizes. Capital costs for pond construction are based on distance from Tarakan and a cost per square meter, making little difference for the size of pond being built. However, other pond infrastructure is more cost sensitive. One of the biggest capital costs is the construction of the gate through which water is exchanged and shrimp recruited to the pond and harvested. Each gate has one or two caretakers who live at the gate and manage the pond. Dividing larger ponds requires further capital investment and operational cost. But perhaps more importantly, the reduction in pond size also moves production to a higher intensity of production. The traditional extensive system is based on limited stocking of shrimp supplemented with wild shrimp recruited through water exchange (see Ilman et al. 2009). Smaller ponds mean that farmers move from a relatively less to more extensive form of production, which increases capital outlay and operational costs, as well as increasing production risk.

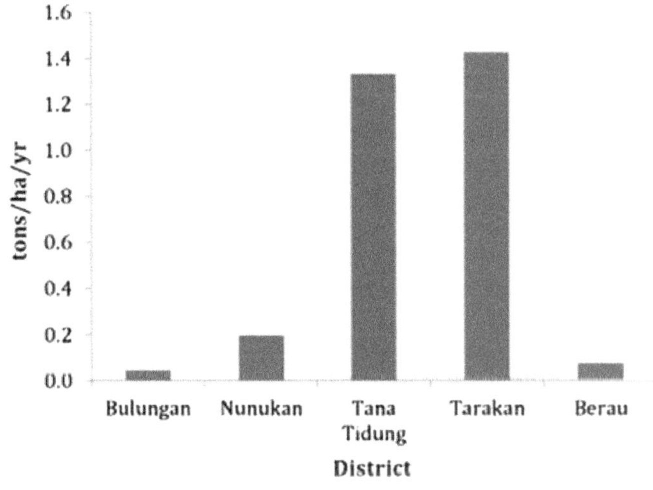

Figure 3: Stocking densities in study districts (Source: Badan Pusat Statistik BPS Provinsi Kalimantan Timur 2012)

The proposals for pond management were based on the smaller size of ponds; including drying and liming the soil to reduce soil acidity, fertilizing the pond and managing water input (WWF-Indonesia 2011). These are well established measures for more intensive forms of production aimed at mitigating the incidence of disease by reducing the incidence of virus carrying organisms, reducing excessive nutrient loads and reducing the acidity of pond soil – all of which are also major factors in reducing the overall productivity of ponds (e.g. Tho et al 2011; Walker and Mohan 2009). However, producers state that none of these measures are feasible in their extensive systems. Contrary to the BMPs they argue that drying their ponds will trigger a higher level of acidity by drawing pyrite out of the soil. WWF countered that this can be overcome by washing, turning and liming the soil. However, this process becomes difficult because of the difficulty in ensuring that all points of the ponds are dry (Ilman et al. 2009). Other farmers do not empty their ponds at any point in the production cycle, instead keeping water in the canals in the pond during harvesting to maintain the stock of wild and juvenile shrimp.

The extensive production systems are also low cost, requiring minimal maintenance and input. The wider experience of shrimp aquaculture indicates that implementing BMPs measures on pond design and management would lead to higher stocking densities, feed and ultimately labour. Although incurring greater costs in the short-term, these measures are expected to improve the overall efficiency of production in the long-term. However, the farmers see a range of risks associated with implementing the BMPs and increasing the intensity of production. First, the measures trialled in the pilot ponds by WWF failed – survival rates and yield per hectare were significantly lower and the quality of shrimp poorer (see Table 2).

Table 2: Observed and target metrics of MMA-WWF trial ponds based on BMPs (Source:http://www.shrimp.co.id/eng/index.php accessed on March 2012)

Metric	Observed	Target
Survival rate	< 20 %	30 %
Yield/ha	< 75 kg	100 kg
Average shrimp size	40 g	40 g
Quality of shrimp	Variable	Stable

Second, increasing the intensity of production is recognised by producers as increasing the risk of disease. This is supported by observations that open extensive systems have a higher incidence of viruses such as White Spot Syndrome Virus (WSSV) but with a lower virulence than in intensive systems (Dieu 2010). Producers are therefore faced with the prospect of following the BMPs under conditions of uncertainty and with a potentially higher degree of risk. They are faced with redesigning their ponds and changing pond management practices with little assurance that these practices would increase productivity.

REFLECTIONS ON CO-PRODUCED BMPS

BMPs are designed to combine global experiences on better management in shrimp farming, consultation with the local stakeholder and field experience of those initiating the translation process (Padiyar et al. 2012). But like any globally derived standards they are not introduced and translated in a social vacuum. The translation of the standards is a function and outcome of social relations between global actors embedded in expert networks and local actors embedded in politicised places and environments. Reflecting the findings of others, we concur that the particularities of place disrupt and transform the implementation of globally conceived standards (Foley and Hébert 2013; Roth and Dressler 2012). Our results demonstrate this on three levels: the formation of the networks that support the translation of the standards to Tarakan, the design of the multi-stakeholder meetings, and the farm-level BMPs pilot program.

The International Principles were published with the expectation that public and/or the private sector actors would use them to develop national policy or BMPs. The attempt to develop BMPs can therefore not be separated from the wider advocacy strategies of WWF as an international environmental NGO, especially given their role in the development of the BMPs for Aceh, the International Principles for Sustainable Shrimp Farming and the Shrimp Aquaculture Dialogue. Even though the adoption of the global standards should be suitable to the social, economic, and environmental conditions of the targeted area, the translation from principles to standards has led to some diversification in the content of the BMPs. However, as outlined by Mohan et al. (2010) most, if not all BMPs are closely related in content. In this sense, and supporting the findings of others (Anh et al. 2010; Belton et al. 2010; Cheyns 2011; Fransen and Kolk 2007), BMPs are co-produced in a network dominated by global expert actors with interests and agendas that transcend those of producers embedded in local places. The multi-stakeholder meetings in Tarakan were designed to enrol local actors and manage a 'system-oriented' collaborative process of defining and codifying problems associated with shrimp farming into local BMPs. However, these meetings were unable to create a shared understanding of how aquaculture practices in extensive systems lead to environmental problems that in turn lead decreases the productivity of ponds. We argue this is because the inclusion of actors in the meetings was politicized in two different ways. First, the close relationship of WWF with MMA, that was instrumental for WWF to gain access to Tarakan, also restricted the inclusion of other processors and producers. Although no stipulation was placed on WWF by MMA, the relationship imposed a perceived obligation set by a 'pre-existing social arrangement' (McCarthy 2002) to restrict the programme to those involved in MMA's supply chain. Second, the active exclusion of pond owners and government actors in the second meeting avoided debate on the (il) legality of land use. Instead of creating consensus, the composition and content of the meetings fed into a politics of regulation and control over the environment by reinforcing global expert knowledge (cf. Cheyns 2011; Ponte and Cheyns 2013). The difficulties WWF faced in fostering consensus over the content of the BMPs reflected the relations of power within Tarakan that determine who is heard and who is not in the governance of the shrimp industry.

The politicization of BMPs during the meetings transferred over to the farm-level pilot program. The counter arguments of producers to BMPs for pond design, location and management were largely technical, but underscored by the politics of access and control over land and the distribution of economic risk to farmers rather than regulators. However, the pilot program did offer the producers an opportunity to enter into a technical dialogue with WWF over the design of the BMPs that was not possible during the meetings. The co-produced BMPs also represent the fine division between a technical outcome and the reification of highly contested political relations of aquaculture production. The examples we illustrate explore these divides. WWF acknowledges that the aquaculture practices in the region have emerged in response to the local challenges of production. However, with the goal of increasing the productivity of ponds, or at least mitigating current declines, they sought to adapt these practices to increase the relative intensity of production. But by not stipulating the distance on mangrove replantation around and inside the pond, WWF tried to minimize the issue of land ownership by 'rendering' (Li2007) land access issues into technical terms.

Comparing the meetings and the pilot program questions the extent whether the co-production of knowledge is built upon a collective dialogue, or simply remains a 'transactional' form of co-production: 'I'll trade you this for that' (Needham 2008). It is not only the inclusion/exclusion from a process of co-production but, returning to Cooke and Kothari's 'tyranny of participation', also the quality of knowledge contributed. Who is included, how and when, then become important questions to determine why and how those participating contribute to the BMPs process at different moments and at events. The pond owners at the meetings enabled WWF to develop a meaningful dialogue and buy-in to the BMPs process because they are the primary decision makers. However, the technical focus of the meetings forced them to comment on production practices that are better captured by the experiential knowledge of farm managers. In contrast, the pilot program provided technical input that reflects a transactional form of co-production, but provided by farm managers who are not responsible for subsequent decisions over major changes in infrastructure or investment in the ponds.

WWF has been successful in accessing complex local social networks responsible for shrimp farming in East Kalimantan. However, the results indicate that the BMPs are not only a tool to negotiate the technical dimensions of shrimp aquaculture, but create an extended 'site' of advocacy (cf. De Vos and Bush 2011). By expanding their networks, WWF can gradually extend their influence by acting as a conduit for international principles and debates around shrimp aquaculture to pond owners and agencies that have until now not been included in discussions around sustainability (Kusumawati et al. 2013). Despite the apparent failure of the multi-stakeholder process and farm trials in reaching a consensus based set of BMPs, the embeddedness of WWF in these local networks will enable on-going engagement and advocacy for conservation and sustainable production. But their ongoing success will be reliant on their ability to balance the technical and political dimensions of the BMPs.

CONCLUSION

Our study identifies a range of sites and events led by WWF-Indonesia, where the Tarakan BMPs were co-produced with processors, government and shrimp producers. The goal of translating BMPs into the local context through this multi-stakeholder process was to ensure both technical relevance and a higher rate of adoption and impact. Despite being supported by an international network of experts, we demonstrate that the resistance WWF faced in both the meetings and the pilot program indicates that the formulation of BMPs is far from a linear process of translation. Instead we observe that the activities implemented by WWF represent a process of co-production, through which attempts are made to incorporate and represent different knowledge in the content of the standards.

The results show that through this process of co-production, BMPs represent and perpetuate existing social relations of production, as well as a politics of environmental control and degradation. Despite the neutral scientific and depoliticized terms used in BMPs (Béné 2005), we therefore conclude that BMPs have the potential to contribute to the politicization of social and environmental issues around shrimp farming. Recognizing the political role of BMPs also reflects on the precarious role of NGOs such as WWF in leading the process of co-production. Despite the international label attributed to such NGOs, they need to engage with networks embedded in complex local political economies to co-produce locally relevant and effective standards. Close involvement in local networks may well lead to the formulation of relevant standards that will end up having a higher rate of adoption. However, the impacts of these co-produced standards may be compromised by how inclusive or exclusive the process of co-production is and the degree to which they justify existing farming practices and resource access.

Groups like WWF hope that BMPs will provide a stepping stone for farmer compliance with third party standards, such as ASC and GAA. For such hopes to be realized further research is needed to understand who benefits and who doesn't from their implementation. Doing so will make clearer the link between these different sets of standards and over the long term may help avoid many of the criticisms of developing world and small holder exclusion leveled at third party certification schemes.

Footnotes

1 When we started the research and during the data collection, Tarakan and Bulungan were part of East Kalimantan province. During the writing this paper, Tarakan and Bulungan became part of newly established North Kalimantan province. However, we will still refer those districts as part of East Kalimantan.

ACKNOWLEDGEMENTS

Prof. Leontine Visser provided comments on an earlier draft of this paper. Funding for this research was provided by the Interdisciplinary Research Fund of Wageningen University.

Competing Interests

The authors declare that they have no competing interests.

REFERENCES

1. ADB/ACIAR/AwF/BRR/DKP/FAO/GTZ/IFC/MMAF/NACA/WWF. 2007. *Practical Manual of better management practices for tambak farming in Aceh*. Banda Aceh: Asian Development Bank ETESP, Australian Centre for International Agriculture Research, Food and Agriculture Organization of the United Nations, International Finance Corporation of the World Bank Group.
2. Alongi, D.M., F. Tirendi, and L.A. Trott. 1999. Rates and pathways of benthic mineralization in extensive shrimp ponds of the Mekong delta, Vietnam. *Aquaculture* 175(3–4): 269–292.
3. Anh, P.T., C. Kroeze, S.R. Bush, and A.P.J. Mol. 2010. Water pollution by intensive brackish shrimp farming in South east Vietnam: causes and option for control. *Agricultural Water Management* 97: 872–882.
4. Anh, P.T., S.R. Bush, A.P.J. Mol, and C. Kroeze. 2011. The multi-level environmental governance of Vietnamese aquaculture: global certification, national standards, local cooperatives. *Journal of Environmental Policy and Planning* 13(4): 373–397.
5. Armitage, D., F. Berkes, A. Dale, E. Kocho-Schellenberg, and E. Patton. 2011. Co-management and the co-production of knowledge: learning to adapt in Canada artic. *Global Environmental Change* 21: 995–1004.
6. Badan Pusat Statistik (BPS) Provinsi Kalimantan Timur. 2012. *East Kalimantan in Number 2012*. Samarinda: Badan Pusat Statistik (BPS).
7. Belton, B., D. Little, and K. Grady. 2009. Is responsible aquaculture sustainable aquaculture? WWF and the eco-certification of tilapia. *Society and Natural Resources* 22(9): 840–855.

8. Belton, B., F. Murray, J. Young, T. Telfer, and D.C. Little. 2010. Passing the Panda Standard: a TAD off the mark? *Ambio* 39(1): 2–13.

9. Béné, C. 2005. The good, the bad and the ugly: discourse, policy controversies and the role of science in the politics of shrimp farming development. *Development Policy Review* 23(5): 585–614.

10. Bush, S.R., B. Belton, D. Hall, P. Vandergeest, F.J. Murray, S. Ponte, P. Oosterveer, M.S. Islam, A.P.J. Mol, M. Hatanaka, F. Kruijssen, T.T.T. Ha, D.C. Little, and R. Kusumawati. 2013. Certify sustainable aquaculture? *Science* 341(6150): 1067–1068.

11. Bush, S.R., and B. Belton. 2012. Out of the factory and into the fish pond: can certification transform Vietnamese Pangasius? In *Food practices in transition: changing food consumption, retail and production in the age of reflexive modernity*, ed. G. Spaargaren, P. Oosterveer, and A. Loeber, 257–290. London: Routledge.

12. Cheyns, E. 2011. Multi-stakeholder initiatives for sustainable agriculture: limits of the 'inclusiveness' paradigm. In *Governing through standards: origins, drivers and limitation*, ed. S. Ponte, P. Gibbon, and J. Vestergaard, 210–235. Basingstoke: Palgrave MacMillan.

13. Cooke, B., and U. Kothari. 2001. The case for participation as Tyranny. In *Participation: the New Tyranny*, ed. B. Cooke and U. Kothari, 1–5. London: Zed Books.

14. De Vos, B., and S.R. Bush. 2011. Far more than market-based: questioning the impact of the Dutch Viswijzer (Good Fish Guide) on fisheries governance. *Sociologia Ruralis* 51(3): 284–303.

15. Dieu, B.T.M. 2010. *On the epidemiology and evolution of white spot syndrome virus of shrimp*. Wageningen: Wageningen University.

16. Falzon, M.A. 2009. Introduction: multi-sited ethnography: Theory, Praxis and locality in contemporary research. In *Multi-sited ethnography: Theory, Praxis and Locality in contemporary research*, ed. Falzon Mark-Anthony, 1–24. Surrey, England: Ashgate Publicy Limited.

17. FAO. 1995. *Code of conduct for responsible fisheries*. Rome: FAO.

18. Foley, P., and K. Hébert. 2013. Alternative regimes of transnational environmental certification: governance, marketization, and place in Alaska's salmon fisheries. *Environment and Planning A* 45(11): 2734–2751.

19. Fransen, L.W., and A. Kolk. 2007. Global rule-setting for business: a critical analysis of multi-stakeholder standards. *Organization* 14(5): 667–684.

20. Gatune, C., A. Vanreusel, C. Cnudde, R. Ruwa, P. Bossier, and M. De Troch. 2012. Decomposing mangrove litter supports a microbial biofilm with potential nutritive value to penaeid shrimp post larvae. *Journal of Experimental Marine Biology and Ecology* 426–427: 28–38.

21. Goldman, M.J. 2010. Circulation of environmental knowledge networks, expertise, and science in practice. In *Knowing Nature: Conservations at the intersection of political ecology and science studies*, ed. M.J. Goldman, P. Nadasdy, and M.D. Turner, 203–208. USA: The University of Chicago Press.

22. Ha, T.T.T., S.R. Bush. 2010. Transformations of Vietnamese shrimp aquaculture policy: empirical evidence from the Mekong Delta. Environment and Planning C: Government and Policy 28: 1101 - 1119.

23. Ha, TTT., H. van Dijk, S.R. Bush. 2012. Mangrove conservation or shrimp farmer's livelihood? The devolution of forest management and benefit sharing in the Mekong Delta, Vietnam. *Ocean and Coastal Management*, Vol. 69: 185–193.

24. Ilman, M., D. Wiharyanto, and D. Desyana. 2009. *Study on Shrimp Farming in Northern part of East Kalimantan*. Indonesia: WWF.

25. Islam, M.S. 2008. From Pond to plate: towards a twin-driven commodity chain in Bangladesh shrimp aquaculture. *Food Policy* 33: 209–223.

26. Konefal, J., and M. Hatanaka. 2011. Enacting third-party certification: A case study of science and politics in organic shrimp certification. *Journal of Rural Studies* 27 (2): 125–133.

27. Jasanoff, S. (ed.). 2004. *States of knowledge: the co-production of science and the social orders*. London: Routledge.

28. Kusumawati, R., S. R. Bush, L. E. Visser. 2013. Can patron be bypassed? Frictions between local and global regulatory networks over shrimp aquaculture in East Kalimantan. Society and Natural Resources 26: 898–911.

29. Li, T.M. 2007. *The will to improve: governmentality, development, and the practice of politics*. USA: Duke University Press.

30. Long, N. 2001. Development sociology: Actor perspective. London: Routledge.

31. Lund, C. 2014. Of what is this a case?: analytical movements in qualitative social science research. *Human Organization* 73(3): 224–234.

32. Marcus, G.E. 1995. Ethnography in/of the world system: the emergence of multi-sited ethnography. *Annual Review of Anthropology* 24: 95–117.

33. McCarthy, J.F. 2002. Turning Circle: district governance, illegal logging, and environmental decline in Sumatra, Indonesia. *Society and Natural Resources* 5 (10): 867–886.

34. McGregor, A. 2007. Development, foreign aid and post-development in Timor-Leste. *Third World Quarterly* 28(1): 155–170.

35. Mohan, C.V., and S.S. De Silva. 2010. Better Management Practices (BMPs) – gateway to ensuring sustainability of small scale aquaculture and meeting modern day market challenges and opportunities. *Sustainable Aquaculture XV* 1: 9–14.

36. Needham, C. 2008. Realising the potential of co-production: negotiating improvements in public services. *Social Policy and Society* 7(2): 221.

37. Padiyar, P.A., M.J. Phillips, B. Ravikumar, S. Wahju, T. Muhammad, D.J. Currie, K. Coco, and R.P. Subasinghe. 2012. Improving aquaculture in post-tsunami Aceh,

Indonesia: experiences and lessons in better management and producer organizations. *Aquaculture Research* 43(12): 1787–1803.

38. Paez-Osuna, F. 2001. The environmental impact of shrimp aquaculture: causes, effects, and mitigating alternatives. *Environmental Management* 28(1): 131–140.

39. Ponte, S., and E. Cheyns. 2013. Voluntary standards, expert knowledge and the governance of sustainability networks. *Global Networks*. doi:10.1111/glob.12011.

40. Primavera, J.H. 1997. Socio-economic impacts of shrimp culture. *Aquaculture Research* 28 (10): 815–827.

41. Rimmer, M.A., M.J. Phillips, P.A. Padiyar, C. Kokarkin, S. Raharjo, S. Bahrawi, and C. Desyana. 2012. Cooperation in aquaculture rehabilitation and development in Aceh, Indonesia.*Development in Practice* 22(1): 91–97.

42. Robertson, A., and M. Phillips. 1995. Mangroves as filters of shrimp pond effluent: predictions and biogeochemical research needs. *Hydrobiologia* 295(1-3): 311–321.

43. Roth, R.J., W. Dressler. 2012. Market-oriented conservation governance: the particularities of place. Geoforum 43 (3): 363–366.

44. Satria, A., and Y. Matsuda. 2004. Decentralization of fisheries management in Indonesia.*Marine Policy* 28(5): 437–450.

45. Smith, J., K. Obidzinski, S. Subarudi, and I. Suramenggala. 2003. Illegal logging, collusive corruption and fragmented governments in Kalimantan, Indonesia. *International Forestry Review* 5(3): 293–302.

46. The Consortium. 2006. *International principle for responsible shrimp farming. Shrimp farming and the environment*. Bangkok, Thailand: Network of Aquaculture centre in Asia-Pacific - NACA, Bangkok, Thailand.

47. Tho, N., V.N. Ut, and R. Merckx. 2011. Physico-chemical characteristics of the improved extensive shrimp farming system in the Mekong Delta of Vietnam. *Aquaculture Research*42 (11): 1600–1614.

48. Tsing, A. 2005. *Friction: an ethnography of global connection*. USA: Princeton University Press.

49. Vandergeest, P. 2007. Certification and communities: alternatives for regulating the environmental and social impacts of shrimp farming. *World Development* 35(7): 1152–1171.

50. Vandergeest, P., A. Unno. 2012. A ne extraterritoriality? Aquaculture certification, soveregnty, and empire. *Political Geography* 31 (6): 358–367.

51. Visser, L.E., and D.S. Adhuri. 2010. Territorialization re-examined: transborder marine resources exploitation in Southeast Asia and Australia. In *Transborder governance of forest, rivers and seas*, ed. W. de Jong, D. Snelder, and N. Ishikawa, 83–98. London: Earthscan.

52. Walker, P.J., and C.V. Mohan. 2009. Viral disease emergence in shrimp aquaculture: origins, impact and the effectiveness of health management strategies. *Reviews in Aquaculture* 1(2): 125–154.

53. WWF-Indonesia. 2011. *Better Management Practices. Guidebook for small-scale fisheries: No feed no aeration tiger shrimp farming*. Fromhttp://www.wwf.or.id/tentang_wwf/upaya_kami/marine/publication/bmp/?24804/BMP---Budidaya-Udang-Windu-tanpa-Pakan-tanpa-Aerasi retrieved at 24 June 2012.

54. WWF-Indonesia. 2012. *WWF: environmental friendly tiger shrimp farming*. Press release posted on 01 August 2012. From http://www.wwf.or.id/berita_fakta/pressrelease/?25742/WWF-Budidaya-Udang-Windu-Tradisional-Bisa-Ramah-Lingkungan. Accessed 17 Dec 2012.

55. Zainun, I., S. Budidarsono, Y. Rinaldi, and M.C. Adek. 2007. *Socio-economic aspect of Brackish water aquaculture (tambak) production in Nanggroe Aceh Darussalam: Integrated natural resource management and livelihood paradigms in recovery from the tsunami in Aceh*, vol. ICRAF Working Paper Number 46. Bogor: ICRAF

CHAPTER 18

CO-CULTURE WITH CHAETOMORPHA SP. ENHANCED GROWTH PERFORMANCE AND REDUCED FEED CONVERSION RATIO OF THE GIANT TIGER PRAWN, PENAEUS MONODON

Isao Tsutsui, Jaruwan Songphatkaew, Chonlada Meeanan, Dusit Aue-umneoy, Halethichanok Sukchai, Piyarat Pinphoo, Sirimas Klomkling, Monthon Ganmanee, Hiroyuki Sudo, Kaoru Hamano

Keywords: Cladophoraceae Feed conversion ratio Growth rate Intensive aquaculture Penaeidae Proximate analysis

INTRODUCTION

Aquaculture has risen in importance worldwide over the past two decades because of increasing global demand for fisheries products. World aquaculture production has rapidly increased from approximately 13 million tons in 1990 to approximately 66 million tons in 2012 (FAO 2014). In 2012, the total world aquaculture production reached almost 42 % of total fishery production (FAO 2014).

Reduction of the amount of commercial feed used in intensive aquaculture is useful both from an environmental aspect and for cost management. Residual commercial feed causes eutrophication in intensive aquaculture; therefore, reduction of commercial feed helps prevent eutrophication inside and outside of aquaculture environments. From another point of view, commercial feed represents about 43–68 % of the total culture cost, which makes it the highest production cost in intensive aquaculture (Chandra et al. 2010; Hung and Quy 2013). However, it is first necessary to find a no or low-cost alternative to enable reduction of commercial feed usage.

Trials using several species of filamentous green algae (*Ulva*, *Chaetomorpha*, and *Cladophora*) as raw or dried feed for aquacultured crustaceans and fish have recently been successfully conducted. Integrated aquaculture of brackish water prawns or freshwater herbivorous fish from tropical zones with species from the genus *Ulva* has resulted in increased growth of aquaculture species (Limsuwan et al. 2008; Cruz-Suárez et al. 2010; Gamboa-Delgado et al.2011; Siddik and Anh 2015). Although other filamentous green seaweeds including*Chaetomorpha* and *Cladophora* may be as easily applied in intensive aquaculture systems as*Ulva* species because of their euryhalinity, so far only few studies have investigated their application as raw feed for aquaculture species (Tsutsui et al. 2010).

An unidentified species of *Chaetomorpha* grows abundantly in the central Thailand. This alga has extremely wide euryhalinity and eurythermy (Tsutsui et al. 2015), making it of potentially higher benefit to intensive shrimp aquaculture

Citation: Tsutsui I, Songphatkaew J, Meeanan C, Aue-umneoy D, Sukchai H, Pinphoo P, Klomkling S, Ganmanee M, Sude H, Hamano K (2015). Co-culture with Chaetomorpha sp. enhanced growth performance and reduced feed conversion ratio of the giant tiger prawn, Penaeus monodon. International Aquaculture Research, v. 7, 193-199. DOI: 10.1007/s40071-015-0103-0.

ponds than *Ulva* species; however, the use of*Chaetomorpha* sp. in aquaculture has not yet been tested. This study aims to investigate the effect of *Chaetomorpha* sp. as raw feed on the growth of giant tiger prawns, and assess*Chaetomorpha* sp. as a potential alternative for reducing feeding cost in intensive shrimp culture. This study will also detail the proximate analysis of *Chaetomorpha*, which has not yet been reported.

MATERIALS AND METHODS

All research activities in this study were permitted by the National Research Council of Thailand (NRCT) (Project ID 2011/005). Care and handling of the experimental animals were conducted in accordance with institute regulations and Japanese law (Kurosawa 2008). The experiment was performed in a semi-outdoor experimental facility with a transparent roof and natural temperature at the Shrimp Co-culture Research Laboratory (SCORL), King Mongkut's Institute of Technology Ladkrabang (KMITL), Thailand.

Experimental Shrimp and Seaweed

A batch of 15-day-old giant tiger prawn post-larvae (PL_{15}) was purchased from a commercial nursery in Chon Buri Province, Thailand. After transportation to SCORL, the PL_{15} were acclimated in a 2000-L fiberglass tank with seawater of salinity 20 PSU and gentle aeration until becoming 20-day-old juveniles (J_{20}). Commercial grow-out pellets with >40 % crude protein (Vital prawn pellets, Higashimaru Co. Ltd., Kagoshima, Japan) were provided three times a day.

The *Chaetomorpha* sp. used in this study have been allocated accession numbers AB759541 and AB819998 in the DNA Data Bank of Japan (DDBJ) (Tsutsui et al. 2015). It was originally collected at a salt swamp in Samut Sakhon Province, Thailand, and cultured at SCORL in 70-L plastic tanks at a seawater salinity of 20 PSU.

Experimental Design

Three replications each were made of *Chaetomorpha* sp. in monoculture (control) and in co-culture of giant tiger prawns and *Chaetomorpha* sp. (experimental treatment). The 70-L plastic control and treatment tanks (40 cm long × 57 cm wide × 30 cm high) were randomly arranged, and connected to each other by polyvinyl chloride pipes. The water was circulated through all tanks with a flow rate of about 30 L h^{-1} to maintain the same water conditions, and was passed through an external filter to improve water quality. Water was adjusted to an approximate salinity of 20–25 PSU. Water temperature was recorded but not adjusted, and varied from 26.7 to 31.5 °C during the experiment.

Five J_{20} individuals (body weight 0.39 ± 0.03 g) were released in each tank, creating a density of about 20 individuals per m^2. Commercial pelleted feed (vital prawn) was fed three times daily in each tank. Feeding quantity was recorded every time for each tank; quantity was adjusted according to the appetite of the juveniles, which was shown as feed remaining on the bottom of the tank. Excreta and remaining shrimp feed were removed from the tanks every morning.*Chaetomorpha* sp. was maintained to cover about 30–50 % of the water surface or bottom of the co-culture tanks. Seaweed was freely grazed by shrimp in co-culture tanks. Any seaweed, including diatoms, growing in the monoculture tanks (controls) was inhibited and scraped off from the walls with a brush every week.

Measurement of Growth and feed Conversion ratio (FCR)

The individual body weight (wet weight) of each juvenile in each tank after 10 weeks was measured and recorded using a digital balance (PL602-S, Mettler-Toledo, Schwerzenbach, Switzerland). Specific growth rate (SGR) and FCR were calculated by the following formulae:

$$\text{SGR } (\%\ d^{-1}) = (\ln W_t - \ln W_i) \times D^{-1} \times 100 \quad (1)$$

where W_t is terminal wet weight, W_i is initial wet weight, and D is culture duration

$$\text{FCR} = F/(W_t - W_i) \quad (2)$$

where F is feed given, W_t is terminal wet weight, and W_i is initial wet weight

Analysis of Proximate Composition and Energy

Gross energy and proximate composition of *Chaetomorpha* sp. cultured at SCORL were analyzed by standard methods based on AOAC (2012) by Central Laboratory (Thailand) Co. Ltd., Bangkok, Thailand. Protein content was measured by Kjeldahl method on AOAC official method 981.10 using a factor of 6.25. Ash was measured after incineration at 600 °C for 2 h based on AOAC official method 942.05. Fat was extracted with ether using a Soxhlet apparatus and weighed following AOAC official method 920.39. Crude fiber was measured by the loss on ignition of the residue after digestion with 0.313 N sulfuric acid and 0.255 N sodium hydroxide based on AOAC official method 978.10. Carbohydrates were calculated by subtracting the protein, lipid and ash levels from 100 %. Gross energy was initially calculated as calories (protein, 4 kcal g^{-1}; fat, 9 kcal g^{-1}; carbohydrate, 4 kcal g^{-1}), and then converted to joules.

Statistics Analysis

Data were shown as mean ± standard deviation (SD). Differences in SGR and FCR between control and treatment tanks were tested via the Tukey–Kramer honestly significant difference (HSD) test using JMP 10 software (SAS Institute, Cary, NC, USA); $P < 0.05$ was considered to denote significant difference.

RESULTS

Growth

Mean body weight of shrimp from co-culture treatment tanks at the end of 10 weeks was 11.20 ± 0.652 g, which was significantly higher and approximately 50 % heavier than shrimp from monoculture tanks (7.15 ± 1.278 g) (Table 1). The size frequency of shrimp in the co-culture treatment tanks showed that the ratio of larger sizes was higher than that in the monoculture tanks at the end of the experimental period (Fig. 1). SGR in the co-culture treatment tanks was 4.79 ± 0.08 % day^{-1}, which was also significantly higher than that in control tanks (4.14 ± 0.27 % day^{-1}) (Table 1).

Table 1: Growth performance, SGR, and FCR of giant tiger prawn juveniles in monoculture and in co-culture with *Chaetomorpha* sp.

	Monoculture	Co-culture
Mean initial weight (g) of individual shrimp	0.39 ± 0.026[a]	0.39 ± 0.034[a]
Mean terminal weight (g) of individual shrimp	7.15 ± 1.278[a]	11.20 ± 0.652[b]
Final weight difference compared with control (%)	NA	156.6
SGR (% day^{-1})	4.14 ± 0.27[a]	4.79 ± 0.08[b]
FCR	2.39 ± 0.28[a]	1.46 ± 0.62[b]
FCR reduction rate compared with control (%)	NA	38.9

Values are shown as mean value ± standard deviation from triplicate data

Different superscript labels within the same row indicate significant difference between means (Tukey–Kramer HSD test, $P < 0.05$)

NA not applicable, *SGR* specific growth rate, *FCR* feed conversion ratio

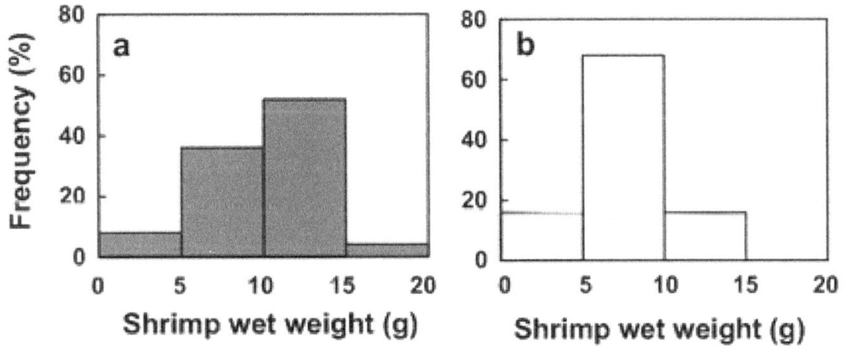

Figure 1: Size variation of giant tiger prawns at 10 weeks in (**a**) co-culture with *Chaetomorpha* sp. and in (**b**) monoculture

Fcr

FCR was significantly lower in co-culture (1.46 ± 0.62) than in monoculture tanks (2.39 ± 0.28) (Table 1). FCR reduction rate in co-culture tanks was 38.9 % compared with control tanks (Table 1).

Proximate Composition

The proximate composition of wind-dried *Chaetomorpha* sp. per 100 g dry weight (DW) was: crude protein 20.4 ± 0.16 g, fat 0.8 ± 0.03 g, ash 14.0 ± 0.05 g, carbohydrate 64.8 g (including crude fiber 21.8 ± 0.06 g) (Table 2). Gross energy in 100 g DW of *Chaetomorpha* sp. was 1455.0 ± 0.35 kJ (Table 2).

Table 2: Gross energy and proximate composition of *Chaetomorpha* sp. and related species (per 100 g DW)

Species	Present species (*Chaetomorpha* sp.)	*Ulva clathrata*	*Ulva prolifera*	*Ulva* sp.
Energy (kJ)	1455.0 ± 0.35	1045.6[a]	1381.3[a]	1312.1[a]
Protein (g)	20.4 ± 0.16	20.7 ± 3.1	20.0	26.6
Fat (g)	0.8 ± 0.03	1.5 ± 0.3	0.3	0.7
Ash (g)	14.0 ± 0.05	38.4 ± 5.5	17.9	22.5
Carbohydrate[b](g)	64.8	39.4	61.8	50.8
Crude fiber[c](g)	21.8 ± 0.06	5.6 ± 0.5		
Reference	Present study	Cruz-Suárez et al. (2010)	MEXT (2010)	MEXT (2010)

Values are shown as mean value ± standard deviation from triplicate data. Proximate composition was re-calculated minus moisture content from original data

DW dry weight, *MEXT* Ministry of Education, Culture, Sports, Science and Technology, Japan

[a]For comparison, these values were re-calculated from the original data by applying the same gross energy coefficient (protein, 4 kcal g^{-1}; fat, 9 kcal g^{-1}; carbohydrate, 4 kcal g^{-1})

[b]Carbohydrate = 100 − (protein + fat + ash); therefore, no SD applied

[c]Crude fiber is treated as a component of carbohydrate

DISCUSSION

Research into improving the growth and/or feed efficiency of various aquaculture species using filamentous green seaweeds as raw or dried feed has been increasing since around 2010 (Tsutsui et al. 2007; Limsuwan et al. 2008; Ergün et al. 2009; Cruz-Suárez et al. 2010; Tsutsui et al. 2010; Gamboa-Delgado et al. 2011; Siddik and Anh 2015; Siddik et al. 2014). For species related to the seaweed investigated in the present study, *Ulva clathrata* improved growth of whiteleg shrimp by 107–164 % (Cruz-Suárez et al. 2010; Gamboa-Delgado et al. 2011), *U. intestinalis* improved growth of whiteleg shrimp and giant tiger prawns (without specific data) (Limsuwan et al. 2008), and *Chaetomorpha ligustica* increased SGR of giant tiger prawns by up to 157 % at the early juvenile stage (Tsutsui et al. 2010). Although direct comparisons with the studies cited above cannot be made because of differences in co-cultured species, our results suggest that the *Chaetomorpha* sp. is able to improve shrimp growth by almost the same amount as that of the other filamentous green seaweeds under co-culture conditions.

Previous studies have found that the filamentous green seaweed *Ulva* spp. can reduce FCR for shrimp and herbivorous fish (Cruz-Suárez et al. 2010; Siddik and Anh 2015). For example, *U. clathrata* reduced the FCR for whiteleg shrimp by approximately 47.5–62.1 % (Cruz-Suárez et al. 2010), and *U. intestinalis* reduced the FCR for spotted scat, red tilapia, and giant gourami by 26–58 % (Siddik and Anh 2015). Our study found that *Chaetomorpha* sp. as a raw feed reduced the FCR for giant tiger prawns by approximately 40 % under co-culture conditions, indicating that *Chaetomorpha* sp. can reduce artificial feed expenses as efficiently as *Ulva* species. This is the first report of FCR reduction for a farmed species by the genus *Chaetomorpha* under co-culture conditions, as previous reports have only described improvement in shrimp growth (Tsutsui et al. 2010).

In Table 2, the extremely high content of ash in *U. clathrata* (38.4 %) is believed to be salt contamination on the surface of the thalli. Considering the difference in ash content, proximate composition analysis showed that the protein content of *Chaetomorpha* sp. was the same or a little less than the closely related *Ulva* species. Conversely, *Chaetomorpha* sp. had higher energy and fiber content, but lower ash level (Table 2). Artificial commercial feed containing about 30–60 % protein is generally used in intensive shrimp aquaculture (Shigeno et al. 1972; Cruz-Suárez et al. 2010; Tsutsui et al. 2010; Hung and Quy 2013). The protein content of*Chaetomorpha* sp. and several species of *Ulva* are much lower than that of artificial feeds. However, despite its lower protein content, the enhancement of shrimp growth and reduced FCR showed that the *Chaetomorpha* sp. contributed as effectively as commercial feed to shrimp growth.

It has previously been reported that mysids and *Acetes* shrimp living in mangrove estuaries have cellulose-degrading capabilities (Niiyama et al. 2012). Although it has never been reported whether Penaeidae species have similar cellulose-degrading abilities, there is a high probability of this, as Penaeidae belong to the same taxonomic class (Malacostraca) as mysids and *Acetes*shrimp, and, like mysids and *Acetes* shrimp, are omnivorous. If the giant tiger prawn does have cellulose-degrading abilities, the *Chaetomorpha* sp. seaweed is more likely to contribute to the growth of this shrimp, as this alga has a higher fiber content than *Ulva* species. In addition,*Chaetomorpha* sp. as a supplemental raw feed may be more advantageous to growth enhancement and FCR reduction of giant tiger prawns owing to its higher gross energy content compared with *Ulva* species. However, the possible contribution of cellulose to growth enhancement and FCR reduction in giant tiger prawn requires future, detailed dietary studies.

Almost all filamentous green seaweeds currently used for growth enhancement of aquaculture species as raw or dried dietary feed are from the genus *Ulva*; species used for this application include *U. clathrata* for whiteleg shrimp, (Cruz-Suárez et al. 2010; Gamboa-Delgado et al.2011), *U. intestinalis* for whiteleg shrimp (Limsuwan et al. 2008), *U. intestinalis* for spotted scat, red tilapia, and giant gourami (Siddik and Anh 2015), *U. rigida* for Nile tilapia (Ergün et al. 2009), and *Ulva* sp. for Nile tilapia (Siddik et al. 2014). The frequent use of *Ulva* spp. may be due to their euryhalinity and eurythermy (Msuya and Neori 2002; Kakinuma et al. 2004; McAvoy and Klug 2005; Tsutsui et al. 2012), which facilitates their application to variable aquaculture environments with a wide range of salinities, and to their widespread availability. The *Chaetomorpha* sp. investigated in this study has a wider salinity tolerance range than *Ulva*species; *Chaetomorpha* sp. can grow in salinities of 0.2–80 PSU and has a high growth rate maximum of <60 % day^{-1} (Tsutsui et al. 2015). Therefore, *Chaetomorpha* sp. may be better and more suitable than *Ulva* species in intensive shrimp aquaculture ponds.

CONCLUSION

The *Chaetomorpha* sp. that grows abundantly in the central Thailand enhanced the growth of giant tiger prawns and reduced the FCR when used as a raw supplemental feed under co-culture conditions. The protein content of this species is almost the same or a little less than other filamentous green seaweeds, such as *Ulva* species; however, it has the advantage of higher fiber and gross energy levels. Thus, *Chaetomorpha* sp. has the potential to reduce commercial feed costs in the shrimp aquaculture industry through co-culturing. This species of *Chaetomorpha*may be more suitable and better for shrimp co-culture systems worldwide than *Ulva* species, which are currently used as raw or dried supplemental feeds.

ACKNOWLEDGMENTS

The authors thank the National Research Council of Thailand (NRCT) for providing useful advice on conducting this study in Thailand. We also greatly appreciate the helpful advice of Dean Sakchai Choochote, and the Head of Fisheries group, Dr. Paveena Taveekijakarn, Faculty of Agricultural Technology,

King Mongkut's Institute of Technology Ladkrabang (KMITL). We thank Dr. Masayoshi Saito, Program Director, Japan International Research Center for Agricultural Sciences (JIRCAS) and Dr. Yukio Maeno, ex-Director, Fisheries Division, JIRCAS for their constructive comments. Our appreciation also goes to Mr. Prachara Manoromchatchawan and Ms. Patumwadee Reuangsawang, Faculty of Agricultural Technology, KMITL for their devoted support of the experiment.

Conflict of interest

The authors declare that they have no competing interests.

REFERENCES

1. AOAC (2012) Official methods of analysis, 19th edn. AOAC International, Maryland
2. Chandra KJ, Sarker D, Khaleque MA, Das DR (2010) Economic analysis of floodplain aquaculture at Daudkandi upazilla in Comilla. J Bangladesh Agril Univ 8:323–332
3. Cruz-Suárez LE, León A, Peña-Rodríguez A, Rodrgíuez-Peña G, Moll B, Ricque-Marie D (2010) Shrimp/*Ulva* co-culture: a sustainable alternative to diminish the need for artificial feed and improve shrimp quality. Aquaculture 301:64–68
4. Ergün S, Soyutük M, Güroy B, Güroy D, Merrifield D (2009) Influence of *Ulva* meal on growth, feed utilization, and body composition of juvenile Nile tilapia (*Oreochromis niloticus*) at two levels of dietary lipid. Aquacult Int 17:355–361
5. Food and Agriculture Organization of the United Nations (FAO) (2014) FAO year book 2012 Fishery and Aquaculture Statistics, Rome
6. Gamboa-Delgado J, Peña-Rodríguez A, Ricque-Marie D, Cruz-Suárez LE (2011) Assessment of nutrient allocation and metabolic turnover rate in pacific white shrimp *Litopenaeus vannamei*co-fed live macroalgae *Ulva clathrata* and inert feed: dual stable isotope analysis. J Shellfish Res 30:969–978
7. Hung LT, Quy OM (2013) On farm feeding and feed management in whiteleg shrimp (*Litopenaeus vannamei*) farming in Viet Nam. In: Hasan MR, New MB (eds) On-farming feeding and feed management in aquaculture. FAO Fisheries and aquaculture technical paper No. 583. Rome, FAO, pp 337–357
8. Kakinuma M, Kuno Y, Amano H (2004) Salinity stress responses of a sterile mutant of *Ulva pertusa* (Ulvales, Chlorophyta). Fish Sci 70:1177–1179
9. Kurosawa TM (2008) Japanese regulation of laboratory animal care with 3Rs. In: Proceedings of 6th world congress on alternatives and animal use in the life sciences. Japanese Society for Alternatives to Animal Experiments 14, pp 317–321
10. Limsuwan C, Kawsuralikhit C, Choocherd N, Hongrat P, Banlerdate D (2008) Co-culture shrimp and *Enteromorpha* sp. National Research Council of Thailand, Bangkok (in Thai)
11. McAvoy KM, Klug JL (2005) Positive and negative effects of riverine input on the estuarine green alga *Ulva intestinalis* (syn. *Enteromorpha intestinalis*) (Linnaeus). Hydrobiologia 545:1–9
12. Ministry of Education, Culture, Sports, Science and Technology (MEXT), JAPAN (2010) Standard tables of food composition in Japan (2010) MEXT, Tokyo
13. Msuya FE, Neori A (2002) *Ulva reticulata* and *Gracilaria crassa*: macroalgae that can biofilter effluent from tidal fishponds in Tanzania. Western Indian Ocean J Mar Sci 1:117–126
14. Niiyama T, Hanamura Y, Tanaka K, Toyohara H (2012) Occurrence of cellulose activities in mangrove estuarine Mysids and *Acetes* shrimp. JIRCAS Work Rep 75:35–39
15. Shigeno K, Kumada K, Deshimaru O, Aramaki T, Kuroki K, Kitaue K (1972) Studies on the artificial diets of prawn-I. Relationship between the feed efficiency and crude protein in the diets. Bull Jpn Soc Sci Fish 38:101–106
16. Siddik MAB, Anh NTN (2015) Preliminary assessment of the Gut weed *Ulva intestinalis* as food for herbivorous fish. Int Aquat Res. 7:41–46
17. Siddik MAB, Nahar A, Rahman NM, Anh NTN, Nevejan N, Bossier P (2014) Gut weed,*Enteromorpha* sp. as a partial replacement for commercial feed in Nile tilapia culture. World J Fish Mar Sci 6:267–274
18. Tsutsui I, Aue-umneoy D, Srisapoome P, Hamano K (2007) Possible implications of the co-cultivation of black tiger shrimp and Cladophoraceae species on a Southeast Asian shrimp farm. Abstract of 21th International Seaweed Symposium, p 77, International Seaweed Symposium, Kobe
19. Tsutsui I, Kanjanaworakul P, Srisapoome P, Aue-umneoy D, Hamano K (2010) Growth of giant tiger prawn, *Penaeus monodon* Fabricius, under co-culture with a discarded filamentous seaweed, *Chaetomorpha ligustica* (Kützing) Kützing, at an aquarium-scale. Aquacult Int 18:545–553
20. Tsutsui I, Hamano K, Aue-umneoy D, Songphatkaew J, Srisapoome P, Ruangsomboon S, Klomkling S, Ganmanee M, Taveekijakarn P, Maeno Y (2012) Common underwater plants in coastal areas of Thailand. JIRCAS International Agriculture Series 21, JIRCAS, Tsukuba
21. Tsutsui I, Miyoshi T, Aue-umneoy D, Songphatkaew J, Meeanan C, Klomkling S, Sukchai H, Pinphoo P, Yamaguchi I, Ganmanee M, Maeno Y, Hamano K (2015) High tolerance of*Chaetomorpha* sp. to salinity and water temperature enables survival and growth in stagnant waters of central Thailand. Int Aquat Res 7:47–62

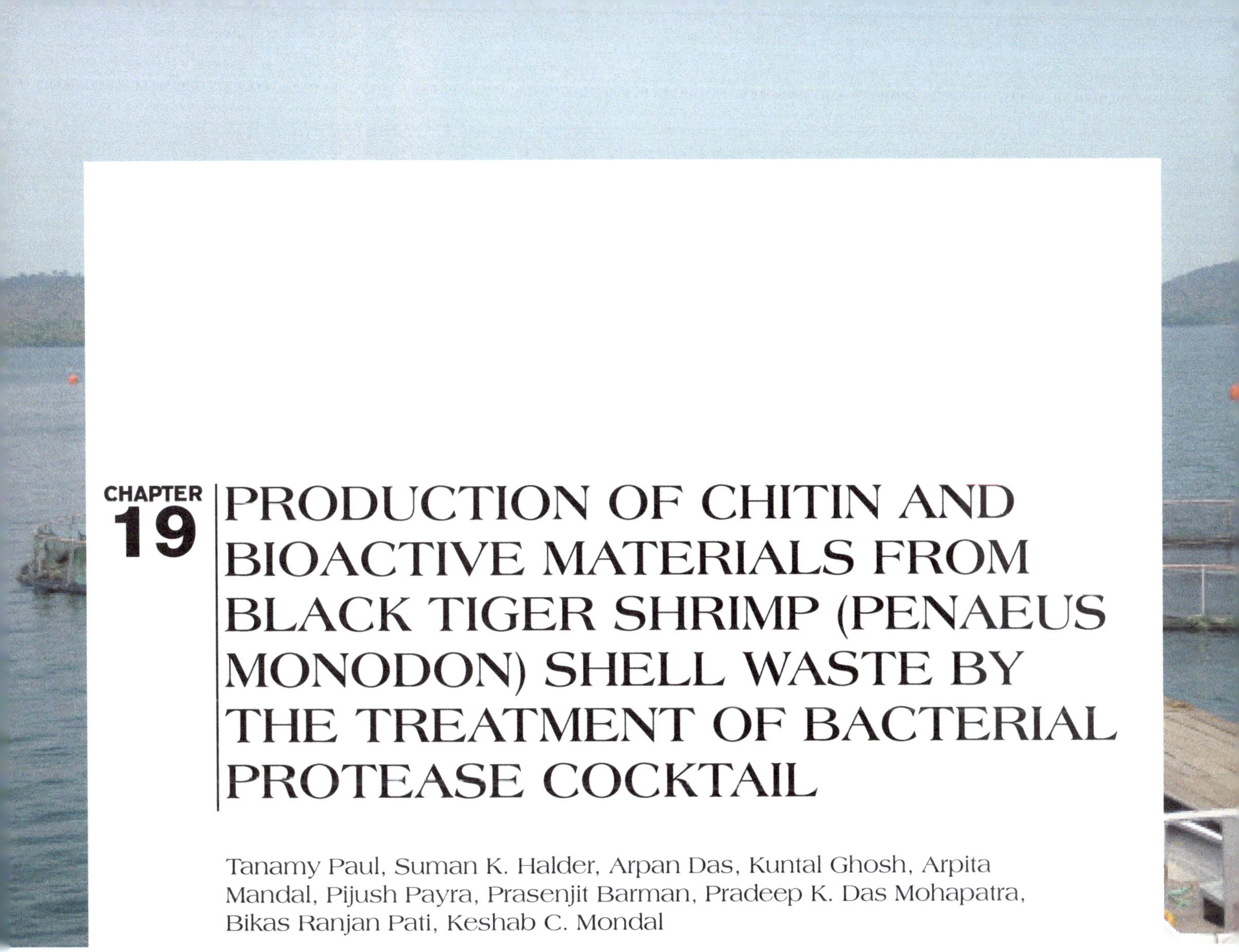

CHAPTER 19

PRODUCTION OF CHITIN AND BIOACTIVE MATERIALS FROM BLACK TIGER SHRIMP (PENAEUS MONODON) SHELL WASTE BY THE TREATMENT OF BACTERIAL PROTEASE COCKTAIL

Tanamy Paul, Suman K. Halder, Arpan Das, Kuntal Ghosh, Arpita Mandal, Pijush Payra, Prasenjit Barman, Pradeep K. Das Mohapatra, Bikas Ranjan Pati, Keshab C. Mondal

Keywords: Black tiger shrimp shell wastes Paenibacillus woosongensis Cocktail protease Chitin Amino acid

INTRODUCTION

Chitin, one of the most abundant renewable biopolymers on earth, is a linear chain molecule composed of several hundred units of (1-4)-2-acetamido-2-deoxy-β-D-glucan. Among the natural chitinous resources, fishery wastes especially shrimp and crab shells have the highest content followed by the exoskeleton of arthropods and the fungal cell walls. Except that some are processed to make cheap feeds on shrimp and crab shell powder, most wastes are discarded at will (Wang et al. 1997). In consideration of the amount of chitin produced annually all over the world, it is the most copious renewable natural resource after cellulosic substrate (Rinaudo2006). Chitin has many unique functional properties including biocompatibility, biodegradability and non-toxicity for which it has been widely applied in the sectors of food, agriculture, medicine, and materials (Muzzarelli et al. 2012; Franco and Peter 2011; Ling et al.2011). Now a days, chemical (Abdou et al. 2008; Chandumpai et al. 2004), enzymatic (Valdez-Pena et al. 2010; Nakagawa et al. 2011) and microbiological methods (Ghorbel-Bellaaj et al.2011; Jung et al. 2007) have been applied to prepare chitin from shrimp shell powders (SSPs). However, the use of these chemicals can seriously pollute the ecological environment, produce abundant waste, and damage human health. Moreover, the addition of acid and alkali can hydrolyze the polymer, resulting in inconsistent physiological properties of the final product. Along with increased demands on environment-friendly society and rapid development of fermentation technology, more eco-friendly processes using enzymatic and microbiological methods for producing chitin have attracted great interests. The enzymatic method includes the use of trypsase, papain, and pepsase (Valdez-Pena et al. 2010). After enzymatic treatment, the experimental liquor is found to contain high amounts of protein and amino acids, indicating a high nutritional value which can be used for food, feed or as a nitrogen source in growth media for microorganisms (Gildberg and Stenberg 2011). In fact, the potentiality of chitin as a storehouse of proteins and amino

Citation: Tanamy Paul, Suman K. Halder, Arpan Das, Kuntal Ghosh, Arpita Mandal, Pijush Payra, Prasenjit Barman, Pradeep K. Das Mohapatra, Bikas Ranjan Pati, Keshab C. Mondal, Production of chitin and bioactive materials from Black tiger shrimp (Penaeus monodon) shell waste by the treatment of bacterial protease cocktail, DOI: 10.1007/s13205-014-0245-6.

acids has prompted the present researchers to undertake the study. As mentioned above, the objectives of this work were to investigate the influence of several operating parameters such as, enzyme/substrate ratio, temperature, pH, agitation speed and incubation time on their efficacy of deproteinization (DP) and demineralization (DM) degree of shrimp shells by non-commercial *P. woosongensis* TKB2 crude cocktail enzyme. RSM has employed to optimize these process parameters better than classical one-variable-at-a-time experimentation. Physicochemical and structural properties of extracting chitin were evaluated by scanning electron microscopy (SEM), Fourier transform infrared spectrometer (FT-IR), X-ray diffraction (XRD) and ^{13}C CP/MAS-NMR analysis. In this work, we have given an HPLC analysis of free amino acids present in lyophilized protein fraction from shrimp shell waste hydrolysate obtained by enzymatic process. To the best of our knowledge, no studies have reported the preparation of chitin from BTSHW using crude protease. That is why, emphasis has been laid on enzymatic procedure on lieu of treatment with acid or alkali which is at all not eco-friendly practices when it is discarded in the surroundings.

EXPERIMENTAL

Raw Materials and Microorganism

The black tiger shrimp (*Penaeus monodon*) shells were obtained in fresh condition from the local market of Midnapore, West Bengal, India. Prior to use, the black tiger shrimp shells were washed thoroughly with distilled water. The shells were ground in a mixture grinder (Bajaj-Bravo, India) and the resulting small pieces of shells were used as raw materials (BTSHW). Casein, Bovine serum albumin and commercial chitin were supplied by HiMedia Co., India. All the other chemicals were of analytical and HPLC grade.

Paenibacillus woosongensis TKB2, a keratinase producing strain (Paul et al. 2013), was used for this study.

Culture Medium and Enzyme Production

Paenibacillus woosongensis TKB2 was grown in 50 ml of liquid medium in a 250 ml Erlenmeyer flask containing 0.75 % (wv^{-1}) chicken feather, 0.05 % K_2HPO_4, 0.025 % $MgSO_4$, 5 % NaCl, 0.020 % $CaCO_3$, and 0.015 % $FeSO_4$. The incubation temperature and pH were 30 °C and 8.5, respectively. After the culture was incubated for 48 h, the culture broth was centrifuged at 4 °C for 20 min at 10,000 rpm followed by the membrane filter (0.45 µm, Millipore, USA). The filtrate containing the crude enzyme was used in tests for Black tiger shrimp shell waste deproteinization.

Protein Estimation and Zymogram Analysis

Protein content in fermented liquor of BTSHW and solid BTSHW after fermentation from different periods of time was measured according to Lowry et al. (1951). SDS–PAGE 10 %, polyacrylamide containing 1 % casein (HiMedia, India) was performed for zymogram analysis of crude enzyme extract. The gel was stained with Coomassie brilliant blue R-250 (Mark, India).

Separation of Carotenoid from BTSHW

Extraction and separation of carotenoid from shrimp shell waste were investigated by application of the methods of Sindhu and Sherief (2011) using hexane: isopropanol 3:2 (vv^{-1}) as organic solvent.

A known weight of homogenized wet and dry shrimp shell waste (1 g) was extracted with 10 ml of solvent to assess carotenoid recovery. The carotenoid extract was filtered using Whatman No. 2 filter paper. Recovered shrimp waste was repeatedly extracted with fresh solvent until the filtrate was colorless, to a maximum of three times. The pooled extracts were phase separated with equal volume of 1 % (w/v) NaCl solution. The appeals were collected and dehydrated with anhydrous sodium sulfate, and then evaporated to dryness under vacuum and the residue was dissolved in a 5 ml of hexane. The carotenoid was quantified in the Black tiger shrimp shell extract of the present study at 470 nm in hexane. The yield of the carotenoids was calculated as astaxanthin (Simpson and Haard 1985) using the following equation:

$$\text{Carotenoid yield } (\mu\text{g astaxanthin/g sample}) = \frac{A_{468\,\text{nm}} \times V_{\text{extract}} \times \text{Dilution factor}}{0.2 \times W_{\text{sample}}} \quad (1)$$

where, A is absorbance, V is volume of extract, 0.2 is the A_{468} of 1 µg ml^{-1} of standard astaxanthin and W is weight of sample in grams.

After separation of carotenoid from BTSHW, solid residual BTSHW was used for deproteinization experiments.

Deproteinization (DP) of Black tiger shrimp shell Waste (BTSHW)

Deproteinization was expressed as percentage and computed by the following equation as described by Rao et al. (2000).

$$\%\ \text{Deproteinization} = [(P_O \times O) - (P_R \times R)] \times 100/P_O \times O \quad (2)$$

where, P_O and P_R are protein concentrations (%) before and after hydrolysis; while, O and Rrepresent the mass (grams) of initial sample and hydrolyzed residue on dry weight basis, respectively.

Acid treatment of BTSHW

BTSHW was mixed with 2 N HCl solution at a ratio of 1:8 (wv^{-1}). The mixture was stored at 50 °C temperature for 4 days (Wang et al. 1997), and then filtered. After filtration, the solid residues were washed with de-ionized water until a neutral pH was obtained.

Alkali treatment of BTSHW

BTSHW was mixed with 2 N NaOH solution at a 1:8 (wv^{-1}) ratio. The mixture was allowed to react at 50 °C for 4 days (Wang et al. 1997) and then it was centrifuged. After centrifugation, the supernatant was used for analysis of protein concentration.

Crude enzymatic treatment of BTSHW

The culture supernatant containing crude proteases (1.57 mg

ml^{-1} of 71.4 U ml^{-1}) obtained from the liquid-phase fermentation described above was tested for deproteinization. BTSHW was mixed with an enzyme solution at a 1: 8 (w/v) ratio. After the substrates were incubated with the protease solution at 50 °C for 4 days, the solid residues were isolated and washed, followed by deproteinization analysis of fermented liquor. The protein content was analyzed for the calculation of the protein removal rate.

Commercial enzyme treatment of BTSHW

For comparison, one other commercially available protease (Sigma-Aldrich, USA) was also studied for shrimp shell waste deproteinization capabilities. The same experimental procedure was employed for this protease with same incubation temperature for 4 days.

Demineralization (DM)

Demineralization was carried out in a dilute 1.25 M HCl solution. Solid fractions obtained after hydrolysis with different processes were treated with 1.25 M HCl in 1:10 (wv^{-1}) ratio for 6 h at room temperature under constant stirring. The chitin product was filtered through four layers of gauze with the aid of a vacuum pump and washed to neutrality with de-ionized water and then freeze-dried. The residual minerals were estimated by HPLC (Agilent-1200, USA).

Optimization of deproteinization process

To investigate the most contributing factors and the suitable level of each contributing factor in the deproteinization using a crude enzyme, response surface experimental design (RSM) was employed. The Box–Behnken design with three factors and three levels (+1, 0, −1), including three replicates at the centre point, was used for fitting a second-order response surface. A polynomial quadratic equation was applied to evaluate the effect of each independent variable to the response:

$$Y = \beta_0 + \sum \beta_i x_i + \sum \beta_{ij} x_i x_j + \sum \beta_{ii} x_i^2 \quad (3)$$

where, Y is the predicted response, x_i and x_j are the independent input variables in coded values that influence the response variable Y, β_0 the offset term, β_i represents the linier effect of x_i, β_{ij} represents the interaction effect between x_i and x_j, β_{ii} represents the quadratic effect of x_i. For statistical calculations, the variables X_i were coded as x_i according to the following equation:

$$x_i = (X_i - X_0)/\,\delta X \quad (4)$$

where, x_i is the dimensionless coded value of the variable X_i, X_0 is the value of the X_i at the centre point and δX is the step change. Each experimental design was carried out in triplicates.

Analysis of Chemical Properties of Sample

The moisture and ash content were estimated by the standard AOAC (2005) methods. Total nitrogen present in shrimp shell protein hydrolysate was determined using the Kjeldahl method. Total protein was estimated by multiplying total nitrogen content in chitin by the factor of 6.25. The BTSHW powder was analyzed for the selected minerals (Ca, Na, K and Mg) by high-performance liquid chromatography (Agilant-1200, USA).

Scanning Electron Microscopic Analysis

SEM (JEOL JSM5800, Japan) was used to clarify the superficial characteristics of the samples at 2,000× magnification. The samples were fixed on a sample holder and coated with a thin gold layer of 3 mm by a sputter coater (POLARON-SC7620) for conductivity. Shrimp shell wastes from different treatments were used for surface analysis through SEM and comparative study with commercial chitin.

Fourier Transform Infrared Spectroscopy (FT-IR) Analysis

Fourier transform infrared spectroscopy (PerkinElmer-L160000A, USA) was used to record infrared spectra of samples between 4,000 and 500 cm^{-1}. The degree of acetylation (DA) was determined using the following equation (de Vasconcelos et al. 2011).

$$DA\,(\%) = 115 \times \left(A_{1,655}/A_{3,450}\right) \quad (5)$$

where, $A_{1,655}$ and $A_{3,450}$ were the absorbances of samples at wave numbers of 1,655 and 3,450 cm^{-1}, respectively. All samples were analyzed three times and DA values were reported as mean ± standard deviation.

X-ray diffraction

The wide-angle X-ray diffraction (WAXD) analysis was applied to detect the crystallinity of commercial chitin in comparison with chitin prepared by crude protease treatment and their patterns were recorded using a Rigaku Americas, 9009 (Rigaku Co., USA). 2θ was scanned from 5 to 50° at a coating time of 2 s with an angle step width of 0.05°. The crystallinity index (Cr_{Ipeak}) was calculated as (de Vasconcelos et al. 2011).

$$Cr_{Ipeak} = (I_{110} - I_{am})/I_{110}, \quad (6)$$

where I_{110} was the maximum intensity (arbitrary units) of the (110) lattice diffraction pattern at 2θ = 20° and I_{am} was the intensity of amorphous diffraction in the same units at 2θ = 16°.

^{13}C CP/MAS-NUCLEAR MAGNETIC RESONANCE (NMR) STUDY

Chitin structural analysis was carried out by ^{13}C NMR with CP/MAS technique (cross-polarization, magic-angle spinning) by BRUKER-ASX300 instrument. NMR spectra were recorded at a ^{13}C frequency of 75.5 MHz (field of 7.04 T). CP/MAS sequence was used with the following conditions: the ^{13}C spin lattice relaxation time was 5 s; powdered samples were positioned in an alumina rotor used for the double air-bearing-type MAS system and spun as fast as 8 kHz. Contact time was 8 ms.

Oligopeptides and Amino Acid Quantification of Black Tiger Shrimp Shell Hydrolysate

After incubation for 4 days, the shrimp shell waste was centrifuged to collect the chitin fraction (sediment) and the protein-rich liquor (liquor). The protein liquor was filtered through 5 kDa Biomax polyethersulfone (PES) cutoff membrane using Amicon Stirred Ultrafiltration Cells (MA, USA), lyophilized. The samples were desalted by Zip Tip pipette chips containing C18-reverse-phased media (Millipore, Billerica, MA, USA) according to the supplier's protocols before applying to a matrix-assisted laser desorption ionization-time of flight (MALDI-TOF) spectrometer. α-Cyano-4-hydroxycinnamic acid (10 mg ml^{-1}) was prepared in 50 % acetonitrile/0.05 % trifluoroacetic acid as the matrix solution. Oligopeptides estimation was carried out according to Kumar et al. (2008) using Bovin serum albumin as a standard.

To evaluate free amino acids were extracted first from the dry samples with methanol. The free amino acids were quantified with ninhydrin method (Sun et al. 2006). Briefly, Amino acid solution (1 ml) and ninhydrin solution (1 ml) were put in screw-capped test tubes and heated in a boiling water bath for a pre-determined period of time. The schematic representation of the chemical reaction between amino acids and ninhydrin to develop purple color is represented in Fig. 1. After heating, the tubes were immediately cooled in an ice bath. Then, 5 ml of 50 % alcohol was added into each tube and thoroughly mixed with a Vortex mixer for 15 s. The absorbance (570 nm) of the reaction mixture was measured with a spectrophotometer (UV-2300, Japan). Two independent experiments each with duplicate samples were conducted and the reported values were the means of the two experiments. Linear regression was performed with commercial statistical software on a personal computer.

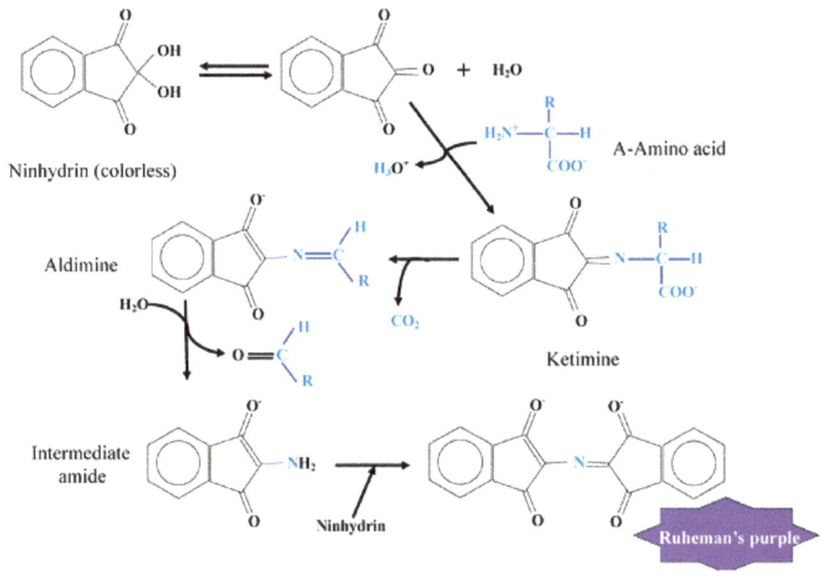

Figure 1: Schematic representation of biochemical estimation of amino acids using ninhydrin

Statistical Analysis

All experiments were carried out in triplicate, and average values with standard deviation are reported. Mean separation and significance were analyzed using the Sigmaplot software package (Sigmaplot-11, USA). Correlation and regression analysis was carried out using the EXCEL program.

RESULTS AND DISCUSSION

We have earlier reported the production of a keratinase from *P. woosongensis* TKB2 when grown in media containing only chicken feather as a sole nitrogen source (Paul et al. 2013). This indicated that the enzyme is a protease in nature and can deproteinize shrimp shell wastes.

Zymogram Analysis of Crude Enzyme

Crude enzyme showed four bands all identically protease activity (Fig. 2). This complied that the bacterial strain produced four protease enzymes that hydrolyzed the casein to produce clear zones on gel slab. These protease cocktails were used for deproteinization of Black tiger shrimp shell waste.

De-Carotenoid the BTSHW

The solvent extracted carotenoid was in the form of a paste with an orange–red color. The highest carotenoid yield (48.9 μg g^{-1} waste) from waste was obtained after 24 h of contentious agitation with hexane.

Protein Removal from BTSHW

Once the enzyme production was obtained, deproteinization of BTSHW was studied by incubation of these substrates with the crude cocktail protease of *P. woosongensis* TKB2. The percent of protein removal for BTSHW after 4-day incubation was 80 and 73.33 % for cocktail protease and commercial protease, respectively, whereas that of alkali- and acid-treated BTSHW was 91.66 and 61.89 %, respectively. By contrast, before optimization, 74.09 % deproteinization was observed with crude protease from BTSHW. The possible reason for the low protein removal for BTSHW may be a Maillard reaction associated with the cooking and high-temperature drying process. The Maillard reaction is known to render proteins (due to cross-linking with sugars and other compounds) resistant to protease treatment (Sikorski et al.2009). The fact that acid-treated BTSHW did not yield a favorable result is possibly due to the destruction of many of the proteins during the treatment (Wang and Chio 1998).

In the present study, chitins obtained by deproteinization of shrimp shell wastes with crude enzyme or by chemical extraction were carried out and chemical compositions of chitins were compared (Table 1). High enzymatic deproteinization of BTSHW was achieved which reached 80 ± 0.4 % with an E/S ratio of 1:8 (wv^{-1}) after 72 h incubation. Chemical deproteinization by NaOH (2 N) and HCl (2 N) of BTSHW was investigated at different incubation times (data not shown).

Demineralization of BTSHW

In the recovery of chitin from shrimp shell waste, associated minerals should be removed as a second stage. The demineralization was completely achieved within 6 h at room temperature after treatment with 1.25 N HCl solutions at a ratio of 1:10 (wv^{-1}). One of the factors determining the good quality of chitin is the low mineral content (Tolaimate et al. 2003). The four types of chitins obtained in this work contain metals significantly lower than those reported in other works (Percot et al. 2003; Chandumpai et al. 2004). The residual contents of Ca, Na, Mg, and K in four types of chitins were about 0.03, 0.012, 0.013, and 0.02 %, respectively.

The treatments employed to extract chitin from the shrimp shell wastes shown in Table 1produced a mass of water insoluble white fibrous material, which indicated that a good yield for the chitin extraction was attained and no pigments were present in the chitin (Canizares et al.2002). In fact, the discoloration of the chitin prepared by inversion of the two steps (i.e., deproteinization followed by demineralization) was less important. In addition, the HCl treatment may probably remove the residual acid-soluble proteins.

Attempts to compare the deproteinization effect found in this study with other reports were made. We found that the variation in analytical methods, calculation approach and demineralization made comparatively impractical. Nevertheless, in view of the shortage associated with the chemical treatment using a strong acid and base, the use of proteolytic enzymes to hydrolyze proteins associated with Black tiger shrimp shell waste to recover chitin was very promising.

Optimization of Process Parameters

A response surface design is applied when the optimal region for running the process has been identified (Jain and Vigneshwaran 2012). RSM using Box–Behnken design was applied to determine the optimal levels of the three selected variables which significantly influenced deproteinization process. The respective coded (−1, 0, +1) and un-coded values of the variables at various levels are as follows—pH −8.5, 9.0, 9.5; temperature (°C)—40, 50, 60 and agitation speed (rpm) of 80, 100, 120. To study the combined effect of these variables, experiments were performed using different combinations. Table S1 summarizes the Box–Behnken experimental design along with the experimental and predicted responses from each individual experiment. By applying multiple regression analysis on the experimental data, the following second-order polynomial equation was found to describe deproteinization experiment:

$$\begin{aligned} Y_{\text{Deproteinization}\,(\%)} = {} & +74.08 + 4.76 \times A + 0.63 \times B \\ & + 11.07 \times C - 4.40 \times A \times B - 7.76 \\ & \times A \times C + 0.54 \times B \times C - 24.13 \\ & \times A^2 - 27.48 \times B^2 - 4.95 \times C^2 \end{aligned} \quad (7)$$

The statistical significance of the Eq. 7 was checked by the *F* test. The analysis of variance (ANOVA) for response surface quadratic model is summarized in Table S2. The model *F* value of 113.27 for deproteinization process implies that the model was significant, and the coefficient of variation (R^2 value) ensured a satisfactory adjustment of the quadratic model to the experimental data and indicated that approximately 95 % of the variability in the dependent variables (responses) could be explained by this model.

The *P* value denoting the significance of the coefficients was also important in understanding the pattern of the mutual interactions between the variables. In this case, independent variables (A, B, C) and the quadratic terms of the factors (A^2, B^2, D^2) were significant model terms.

The predicted sum of squares (PRESS) which is a measure how a particular model fits each point in the design was 844.24 for response. The adequate precision value which measures the 'signal-to-noise ratio' was found to be >4 which indicates an adequate signal. Thus, it can be concluded that the model was statistically sound and can be used to navigate the design space.

To determine the optimal levels of the variables, 3D contour plots (Fig. 3) for deproteinization process were constructed according to the Eq. 4. To depict the interactive effects of independent variables on responses, two variables were kept constant while the other two variables varied within certain ranges. These contour plots provided a visual interpretation of the interaction between two factors, namely, pH and agitation speed followed by pH and temperature to facilitate the location of optimum experimental conditions (Jain and Vigneshwaran 2012). In this experiment the plots had elliptical contours enclosing the region of maximum activity within the experimental range.

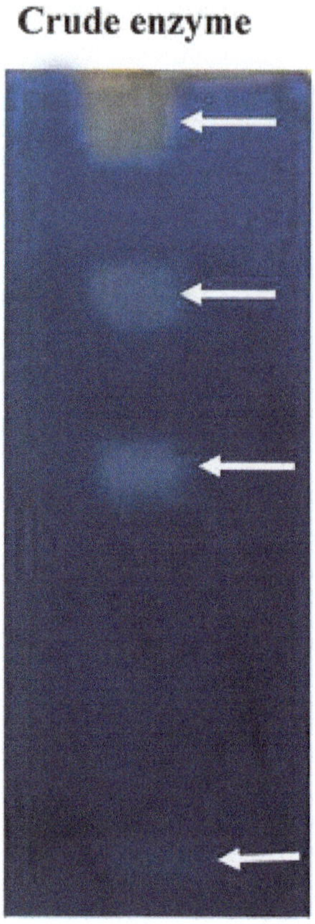

Figure 2: Zymogram of crude protease showed four distinguishable hydrolytic bands. *Arrows*indicate the protease hydrolytic bands

The optimized variables were found using a desirability objective function that assigns relative importance to the responses. Solutions with higher desirability gave an optimum pH (8.82), temperature (50.05 °C) and agitation speed (100.98 rpm). Under these conditions, confirmation experiments were conducted in three replicates. The maximum protein yield of 79.93 % was observed. The statistical model produced 1.1-fold increased deproteinization than that in un-optimized state, which also showed the acceptability of the model for better deproteinization by enzymatic protease cocktail yield. The result indicates better deproteinization than some current reports (Oh et al. 2000; Wahyuntari and Junianto Setyahadi2005), though very few higher yields had been also reported (Younesa et al. 2012). Our reports envisage a shorter period of treatments in which further treatments such as solicitation (de Vasconcelos et al. 2011) are not required.

Chitin Characterization

The characteristics of the raw BTSHW and chitins that were prepared by enzymatic and chemical (acid and alkali) treatments are shown in Table 1. The ground BTSHW before treatment contained a relatively high content of protein (48.16 ± 0.20 %), chitin (34.23 ± 0.25 %) and ash (30.66 ± 0.35 %). These results are comparable with those reported by previous studies (Manni et al. 2010; Younesa et al. 2012). The demineralization experiments were used in this study to reduce the mineral content to permissible limits in the chitin. Indeed, the ash content was reduced to about 0.83 %. This was lower than that found by Sini et al. (2007) and Younesa et al. (2012). Further, low-ash content for both chitins indicated the suitability of removal of calcium carbonate and other minerals from the raw BTSHW. There were no significant differences in the moisture content and ash among the four chitins prepared ($p > 0.05$). In contrast, the protein content was significantly higher in the chitin isolated with enzymatic deproteinization both by commercial and cocktail protease ($p < 0.05$) than chemical treatments. It is time that the enzymatic process we have adopted does not completely remove the residual protein associated with BTSHW. However, the residual protein on the instant case is much lower than usually found in the literature.

Table 1: Properties of the chitins obtained by deproteinization with crude proteases, commercial protease, alkali and by acidic methods

Percentage (%)	Raw BT-SHW	Chitin 1 (cocktail protease)	Chitin 2 (commercial protease)	Chitin3 (acid treatment)	Chitin 4 (alkali treatment)
Moisture	65.06 ± 0.35	4.05 ± 0.05	4.58 ± 0.34	5.70 ± 0.35	3.51 ± 0.37
Ash	30.66 ± 0.35	0.83 ± 0.06	1.18 ± 0.34	3.68 ± 0.53	1.06 ± 0.02
Chitin	34.23 ± 0.25	26.43 ± 0.44	21.21 ± 0.21	10.19 ± 0.17	27.28 ± 0.27
Protein	48.16 ± 0.20	20.07 ± 0.07	26.69 ± 0.17	38.11 ± 0.12	8.29 ± 0.27
Lipid	5.2 ± 0.20	–	–	–	–
Appearance	–	White flakes	Slight yellowish flakes	Yellowish flakes	Yellowish flakes
Chitin recovery (%)	–	77.23 ± 0.21	62.06 ± 0.12	29.93 ± 0.05	79.83 ± 0.17

Although the deproteinization percentage is lower than that of chemical treatment, enzymatic deproteinization helps to avert many drawbacks of chemical treatment such as heavy metal residues, over-hydrolysis, break down of chitin, etc. Needless to mention, heavy metals referred are potential bio-hazards.

Analysis of Superficial Characteristics by Scanning Electron Microscopy (SEM)

Scanning electron microscopy images of BTSHW after DP by enzymatic and chemical treatments are shown in Fig. S1. The smashed BTSHW displayed smooth microfibrillar crystalline structure and left layer structure largely intact (Fig. S1a), while the SEM images of BTSHW after DP showed a slight fracture (Fig. S1b–e). The HCl treated BTSHW showed a hazy layer on the surface of BTSHW, which may be due to denatured proteins and minerals (Fig. S1b). The DP under the optimal conditions using alkali treatment and commercial enzyme treatment left layers stacked and showed signs of perforation with some depositions of mineral compounds (Fig. S1c, d). The BTSHW processed by cocktail enzyme and commercial chitin was morphologically similar (Fig. S1e and f). However, the Black tiger shrimp shell fragments became highly perforated and spongy after crude enzymatic treatment. It was apparent that the crude enzymatic treatments improved DP and DM efficacy and produced chitin polymer (Liu et al. 2012).

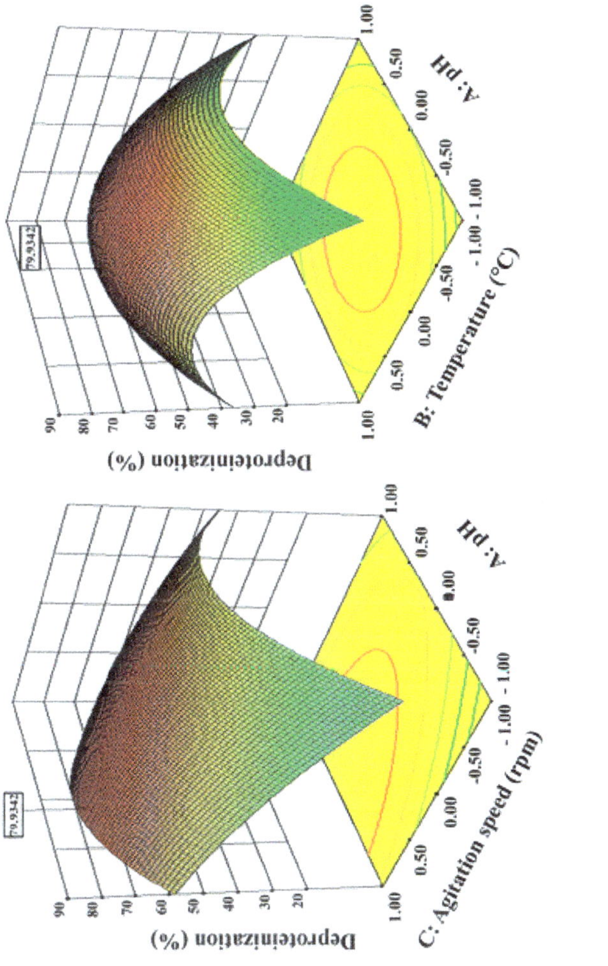

Figure 3: Box–Behnken response surface design curves for deproteinization with combination of agitation speed vs pH and temperature vs pH

FT-IR analysis of BTSHW

The differences in the FT-IR scans among all samples are displayed in Fig. 4. The spectra were characterized by three bands at 1,377, 1,654, and 2,932 cm^{-1}, which corresponded to the vibrations of –NH, –C–O, and –CO–CH$_3$ group, respectively (Liu et al. 2012). The bands between 890 and 1,156 cm^{-1} represented polysaccharide structures. It was notable that the bands at 1,654 and 1,377 cm^{-1} for other samples were more intense than that of commercial chitin (Fig. 4a), which confirmed the existence of chitin (Moreira et al. 2011). Compared to the commercial chitin, the band observed at 2,932 cm^{-1} demonstrated an intensification of the peak (Fig. 4b–e), and suggested the occurrence of deacetylation. Extent of deacetylation of the samples after cocktail enzymatic treatment, commercial protease, acid and alkali (chemical treatment) was 82.25, 80.34, 55.25 and 84.46 %, respectively (Table 2).

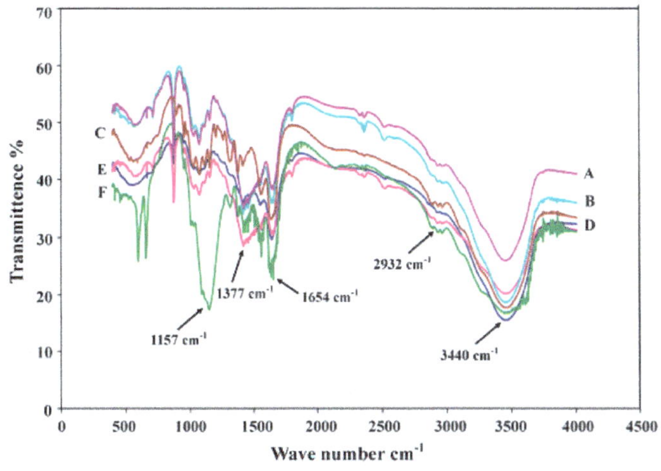

Figure 4: FT-IR spectra of commercial chitin (**a**), chitin from commercial protease (**b**), chitin from crude protease (**c**), from NaOH treatment (**d**), raw BTSHW and treatment with HCl (**e**) in this study

Table 2: DA and crystallinity index of commercial chitin, chitin obtained from crude protease, commercial protease, alkali and acid treatments

Samples	DA (%)	Crystallinity index (%)
Commercial chitin	88.10 ± 0.11	97.9
Chitin obtain from cocktail protease treatment	82.25 ± 0.07	88.0
Chitin obtain from commercial protease	80.34 ± 0.11	–
Chitin obtain from acid treatment	55.25 ± 0.06	–
Chitin obtain from alkaline treatment	84.46 ± 0.14	–

X-Ray Diffraction Analysis

The crystallinity indexes of commercial chitin and chitin extracted by crude protease treatment were determined from the scattering intensity at two angles, one at 2θ = 9–10° and another at 2θ = 19–20° (Fig. 5). The results were in agreement with the literature, in which the purified chitin had a wide-angle X-ray diffraction pattern and showed two crystalline peaks at 2θ = 9.3 and 19.1° (Yen and Mau 2004). Similarly, Yen and Mau reported that fungal chitin displayed two crystalline reflections at 5.4 5.6 and 19.3 19.6° (Yen and Mau 2007). The crystallinity indices of commercial chitin and chitin obtained from cocktail protease treatment were 97.9 and 81.0 %, respectively (the baseline at 2θ = 16°). Overall, it was found that the application of protease cocktail reduced the crystallinity of chitin from 97.9 % in commercial chitin to 88.0 % in enzymatic treatment (Table 2). A lower crystallinity of polysaccharides indicates disruption of intra- and inter-molecular hydrogen bonds, in turn provides the possibility for more efficient chemical modifications in subsequent processing steps (Yen et al. 2009). The X-ray diffractograms of chitin powder obtained from crude protease treatment showed narrowed peak areas than the commercial chitin, confirming that further purification is necessary to obtain satisfactory chitin extractive.

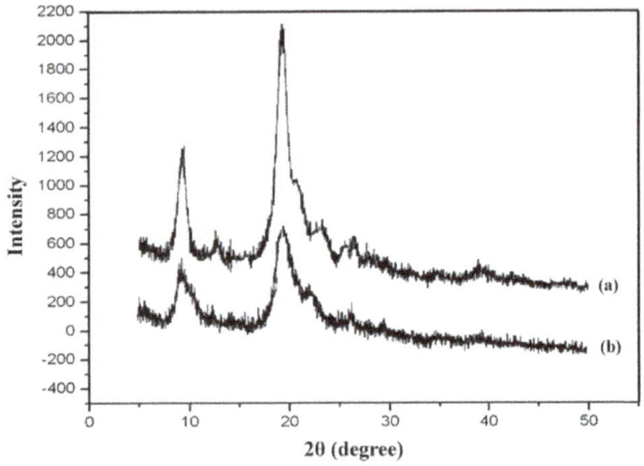

Figure 5: X-ray diffractogram of commercial chitin (**a**) and chitin from crude protease treatment (**b**) in this study

Detection of Chemical Shifts in BTSHW and Comparison with Chitin

NMR is one of the most reliable instruments in the examination of polysaccharide composition and sequential structure. NMR is a non-destructive method ensuring in retained structure and conformation of the polysaccharide, making it possible to monitor reactions and other structural and physical properties under different solvent conditions. Solid state ^{13}C CP/MAS-NMR is known to be very sensitive to changes in the local structure. ^{13}C CP/MAS-NMR spectrum of the chitin prepared by enzymatic deproteinization and commercial chitin is shown in Fig. S2. NMR analysis of the treated shrimp shell waste chitin gave similar peak pattern with that of commercial chitin. There are eight signals for the eight carbon atoms of chitin. The C_1–C_6 carbons of *N*-acetylglucosamine monomeric unit are shown between 50 and 110 ppm, indicating the high structural similarity. The carbonyl group is around 173 ppm, while the methyl group of the acetyl group produced a peak at around 23 ppm. The lesser intensity of the peaks of carbonyl and acetyl groups suggests that the deacetylation reaction is complete. Fig. S2 shows the chitin spectrum, in which the deacetylation of chitin is evident, since there are tight peaks at 23 and 173 ppm that correspond to the CH_3 and C=O groups, respectively. The other peaks correspond to C_1 (δ104.37), C_2 (δ60.71), C_3 (δ73.69), C_4 (δ83.40), C_5 (δ75.88), and C_6(δ55.33). As we have no other picks, it may be argued that the product was significantly pure and the removal of protein during extraction was almost total (Younesa et al. 2012). The presence of some background noise may be accounted for the presence of a possible by-product or some impurity in the sample, especially in commercial chitin (Fig. S2). The low-protein content is considered as a good indicator for purity of the final product.

Oligopeptides and Amino Acid Separation and Estimation

In response to the increase in the amount of soluble protein in the culture, oligo or small peptide contents of the supernatants were analyzed to determine whether the increase in soluble proteins was derived from BTSHW. The culture supernatants were subjected to MALDI-TOF mass spectrometric analysis to survey the distribution of the molecular masses of the solubilized products. As shown in Fig. 6, numerous but specific signals were observed in the sample after fermentation for 4 days revealed multiple peaks from 1,015 to 3,182 *m/z*, indicating the presence of peptides generated by protein hydrolysis. For example, the thirteen sharply signal peaks (*m/z* = 1,015.41, 111.30, 1,561.26, 1,927.64, 1,965.05, 1,995.40, 2,099.90, 2,183.08, 2,357.30, 2,679.77, 2,834.90, 3,136.05, and 3,182.96) were observed. From this mass analysis, it is clear that molecular masses must be derived from the degradative products of shrimp shell.

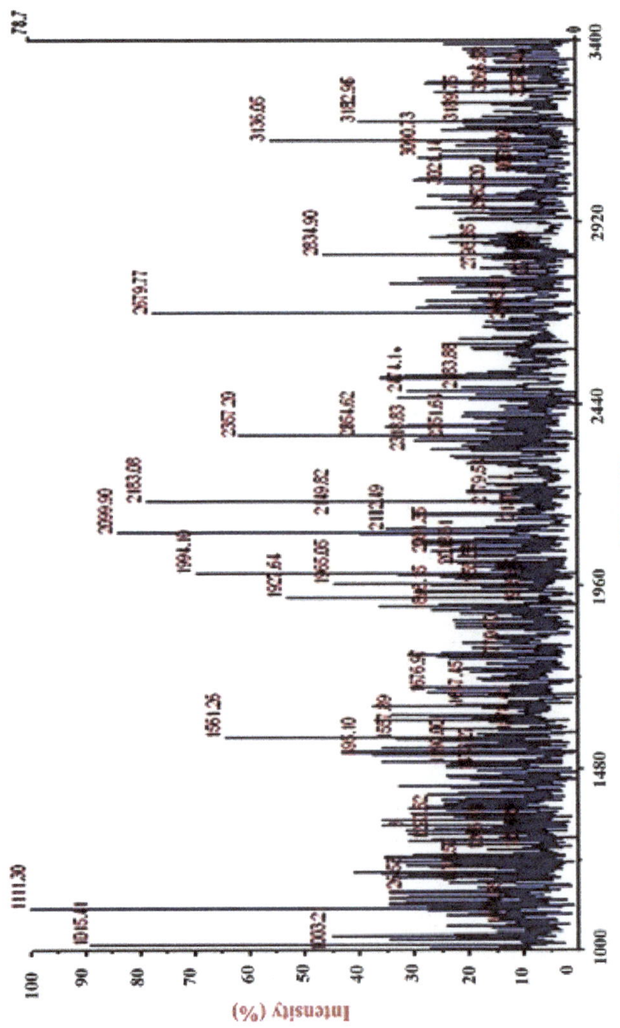

Figure 6: MALDI-TOF mass spectrum for oligopeptides present in treatment spent hydrolysate

The colorimetric method proposed allows simultaneous analysis of free amino acids in fermented broth. The concentration of the amino acids standard solution injected was 16.9 μg ml^{-1}.

CONCLUSIONS

This study has established that fermentation using a protease cocktail of *P. woosengensis* TKB2 alone gives chitin with satisfactory DM and DP efficacy and the resultant chitin has similar physicochemical and structural properties to commercial chitin. The oligopeptides and amino acid-rich deproteinized liquor should be used for feed and medium supplement. The enzyme method is a relatively simple profitable and environment-friendly alternative to the chemical method. This study at shaking flask under laboratory conditions may be suitable for large-scale operations also. RSM model significantly influences the DP process and enhances the DP rate of 1.1-fold. Moreover, this study gives yields satisfactory result for DP for chitin production at shake flask level. The issue of large-scale production of chitin through DP may be taken up for future studies.

ACKNOWLEDGMENTS

This work was financially supported by the University grant Commission (UGC, New Delhi, India) in RFSMS scheme [Sanction No: F.11-114/2008 (BSR)]. The paper has been meticulously revised by Prof. Nandadulal Bhattacharyya, Vidyasagar Institute of Health, Midnapore was greatly acknowledged.

Conflict of interest

Authors were declared that there is no conflict of interest.

REFERENCES

1. Abdou ES, Nagy KSA, Elsabee MZ (2008) Extraction and characterization of chitin and chitosan from local sources. Bioresour Technol 99:1359–1367
2. AOAC (2005) Official methods of analysis, 18th edn. Association of Official Analytic Chemists, Washington, DC
3. Canizares E, Gonzalez Jose A, Hau L, Osorno H (2002) Evaluation of two chemical treatment alternatives for obtaining chitin from shrimp solid waste. Alimentaria 331:31–34
4. Chandumpai A, Singhpibulporn N, Faroongsarng D, Sornprasit P (2004) Preparation and physic-chemical characterization of chitin and chitosan from the pens of the squid species,*Loligo lessoniana* and *Loligo formosana*. Carbohydr Polym 58:467–474
5. de Vasconcelos CL, Bezerril PM, Pereira MR, Ginani MF, Fonseca JLC (2011) Viscosity–temperature behavior of chitin solutions using lithium chloride/DMA as solvent. Carbohydr Res 346:614–618
6. Franco TT, Peter MG (2011) Advances in chitin and chitosan research. Polym Int 60:873–874
7. Ghorbel-Bellaaj O, Jellouli K, Younes I, Manni L, Ouled-Salem M, Nasri M (2011) A solvent-stable metalloprotease produced by *Pseudomonas aeruginosa* A2 grown on shrimp shell waste and its application in chitin extraction. Appl Biochem Biotechnol 164:410–425
8. Gildberg A, Stenberg E (2011) A new process for advanced utilization of shrimp waste. Proc Biochem 36:809–812
9. Jain P, Vigneshwaran N (2012) Effect of Fenton's pretreatment on cotton cellulosic substrates to enhance its enzymatic hydrolysis response. Bioresour Technol 103:219–226
10. Jung WJ, Jo GH, Kuk JH, Kim YJ, Oh KT, Park RD (2007) Production of chitin from Black tiger shripm shellwaste by successive fermentation with *Lactobacillus paracasei* KCTC 3074 and*Serratia marcescens* FS-3. Carbohydr Polym 68:746–750
11. Kumar AG, Swarnalatha S, Gayathri S, Nagesh N, Sekaran G (2008) Characterization of an alkaline active–thiol forming extracellular serine keratinase by the newly isolated *Bacillus pumilus*. J Appl Microbiol 104:411–419
12. Ling SLF, Yee CY, Eng HS (2011) Removal of a cationic dye using deacetylated chitin (chitosan). J Appl Sci 11:1445–1448
13. Liu SF, Sun J, Yu LN, Zhang CS, Bi J, Zhu F, Qu MJ, Jiang C, Yang QL (2012) Extraction and characterization of chitin from the beetle *Holotrichia parallela* motschulsky. Molecules 17:4604–4611
14. Lowry OH, Rosebrough NJ, Farr AL, Randall RJ (1951) Protein measurement with the folin phenol reagent. J Biol Chem 193:267–275
15. Manni L, Ghorbel-Bellaaj O, Jellouli K, Younes I, Nasri M (2010) Extraction and characterization of chitin, chitosan, and protein hydrolysates prepared from shrimp waste by treatment with crude protease from *Bacillus cereus* SV1. Appl Biochem Biotechnol 162:345–357
16. Moreira RM, Pereda M, Marcovich NE, Roura SI (2011) Antimicrobial effectiveness of bioactive packaging materials from edible chitosan and casein polymers: assessment on carrot, cheese, and salami. J Food Sci 76:54–63
17. Muzzarelli RAA, Boudrant J, Meyer D, Manno N, DeMarchis M, Paoletti MG (2012) Current views on fungal chitin/chitosan, human chitinases, food preservation, glucans, pectins and inulin: a tribute to Henri Braconnot, precursor of the carbohydrate polymers science, on the chitin bicentennial. Carbohydr Polym 87:995–1012
18. Nakagawa YS, Oyama Y, Kon N, Nikaido M, Tanno K, Kogawa J, Inomata S, Masui A, Yamamura A, Kawaguchi M, Matahira Y, Totani K (2011) Development of innovative technologies to decrease the environmental burdens associated with using chitin as a biomass resource: mechanochemical grinding and enzymatic degradation. Carbohydr Polym 83:1843–1849
19. Oh YS, Shih IL, Tzeng YM, Wang SL (2000) Protease produced by *Pseudomonas aeruginosa* K-187 and its application in the deproteinization of shrimp and crab shell wastes. Enzyme Microb Technol 27:3–10
20. Paul T, Halder SK, Das A, Bera S, Maity C, Mandal A, Das PS, Mohapatra PKD, Pati BR, Mondal KC (2013) Exploitation of chicken feather waste as a plant growth promoting agent using keratinase producing novel isolate *Paenibacillus woosongensis* TKB2. Biocatal Agric Biotechnol 2:50–57
21. Percot A, Viton C, Domard A (2003) Optimization of chitin extraction from shrimp shells. Biomacromolecules 4:12–18
22. Rao MS, Munoz J, Stevens WF (2000) Critical factors in chitin production by fermentation of shrimp biowaste. Appl Microbiol Biotechnol 54:808–813
23. Rinaudo M (2006) Properties and applications. Prog Polym Sci 31:603–632
24. Sikorski P, Hori R, Wada M (2009) Revisit of α-chitin

crystal structure using high resolution X-ray diffraction data. Biomacromolecules 10:1100–1105

25. Simpson BK, Haard NF (1985) The use of enzymes to extract carotenoprotein from shrimp waste. J Appl Biochem 7:212–222
26. Sindhu S, Sherief PM (2011) Extraction, characterization, antioxidant and anti inflammatory properties of carotenoids from the shell waste of arabian red shrimp *Aristeus alcocki,* ramadan 1938. Open Conf Proc J 2:95–103
27. Sini TK, Santhosh S, Mathew PT (2007) Study on the production of chitin and chitosan from shrimp shell by using *Bacillus subtilis* fermentation. Carbohydr Res 342:2423–2429
28. Sun S, Lin Y, Weng Y, Chen M (2006) Efficiency improvements on ninhydrin method for amino acid quantification. J Food Compost Anal 19:112–117
29. Tolaimate A, Desbrieres J, Rhazi M, Alagui A (2003) Contribution to the preparation of chitins and chitosans with controlled physico-chemical properties. Polymer 44:7939–7952
30. Valdez-Pena AU, Espinoza-Perez JD, Sandoval-Fabian GC, Balagurusamy N, Hernandez-Rivera A, De-la-Garza-Rodriguez IM, Contreras-Esquivel JC (2010) Screening of industrial enzymes for deproteinization of shrimp head for chitin recovery. Food Sci Biotechnol 19:553–557
31. Wahyuntari B, Junianto Setyahadi S (2005) Process design of microbiological chitin extraction. Microbiol Indones 5:39–45
32. Wang SL, Chio SH (1998) Deproteinization of shrimp and crab shell with the protease of*Pseudomonas aeruginosa* K-187. Enz Microb Technol 22:629–633
33. Wang SL, Chio SH, Chang WT (1997) Production of chitinase from shellfish waste by*Pseudomonas aeruginosa* K-187. Proc Natl Sci Counc ROC (B) 21:71–78
34. Yen MT, Mau JL (2004) Physicochemical properties of chitin from shiitake stipes and crab shells. Ann Tainan Woman's Coll Arts Technol 23:229–240
35. Yen MT, Mau JL (2007) Selected physical properties of chitin prepared from shiitake stipes. LWT–Food Sci Technol 40:558–563
36. Yen MT, Yang JH, Mau JL (2009) Physicochemical characterization of chitin and chitosan from crab shells. Carbohydr Polym 75:15–21
37. Younesa I, Ghorbel-Bellaaja O, Nasria R, Chaabounib M, Rinaudoc M, Nasri M (2012) Chitin and chitosan preparation from shrimp shells using optimized enzymatic deproteinization. Process Biochem 47:2032–2039

CHAPTER 20 | COMMUNITY-BASED SHRIMP AQUACULTURE

Eranga K Galappaththi and Fikret Berkes

Keywords: Community-based management, Common-pool resources, Multi-level institutions, Shrimp aquaculture Sri Lanka

INTRODUCTION

Can shrimp aquaculture be carried out through community-based institutions? And if so, can community-based aquaculture be considered a potential alternative to large-scale, company-owned aquaculture? Such questions have theoretical significance for commons management, and practical significance for sustainable aquaculture. We explore these questions based on coastal shrimp farming communities on the Indian Ocean in the northwestern part of Sri Lanka. The most common shrimp species cultured in these communities is the giant tiger prawn or Asian tiger shrimp (*Penaeus monodon*), a prime export species. Sri Lanka is not a big player in the global shrimp trade, and the operations in question have received little attention even in Sri Lanka itself (Galappaththi [2010]). In terms of production volume and technology, Sri Lankan shrimp aquaculture sector is relatively small compared to other Asian countries (MFARD [2011]; Anon. [2007]).

Citation: Eranga K Galappaththi and Fikret Berkes, Community-based Shrimp Aquaculture, DOI: 10.1186/s40152-014-0013-6

In most of South and Southeast Asia, shrimp aquaculture has a mixed history of social (Nayak and Berkes [2010]) and environmental impacts (Bhatta and Bhat [1998]; Primavera [1997]; Huitric et al. [2002]). Commercial operations are typically large-scale and company-owned. They have resulted in the destruction of coastal vegetation, including mangroves, that act as a buffer to protect coastlines from such disasters as tsunamis (Adger et al. [2005]). Historically, shrimp aquaculture operations have often collapsed after a few years of profit-taking, often the victim of shrimp diseases as a consequence of concentrating such large numbers of one species. For example, Huitric et al. ([2002]) tracked the rise and fall of large-scale shrimp farming in Thailand and around the Gulf of Thailand, leaving behind its wake collapsed farms and damaged land and waterscapes. However, in some areas, the collapse of large-scale farms creates the opportunity for the start of small, locally owned operations at the abandoned locations.

Such is the case in the northwestern part of Sri Lanka, where the country's large-scale shrimp farms used to be located and where the small-scale operations now exist. Sri Lanka's shrimp production reached a peak in 1998 (Munasinghe et al. [2010]), and then declined after a major outbreak of two virus diseases. In the place of large farms, numerous small-scale operations appeared over time, mostly family-level farms run by producers who had learned their trade by working earlier for the large companies. These producers patterned the new farms along the lines of agricultural operations, to take advantage of economies of scale afforded by community-based associations. These were associations that helped coordinate production, reduce pollution loads in the shared water resource (the key common-pool resource in this system), and functioned as a conduit of information between the farmers and the state.

Community associations have been known to play a major role in managing small-scale fisheries throughout the world (Digby [1975]; McCay et al. [2011]). Although there is a growing literature on such associations in finfish aquaculture (Haque et al. [2011]; Von Essn et al. [2013]), there is very little literature on their role in shrimp aquaculture. Even within Sri Lanka, the small-scale operations that replaced the collapsed industrial shrimp aquaculture are little known, much less appreciated (U. Amarasinghe, pers. comm.). How does small-scale aquaculture work? How can individual ownership be combined with community level coordination? As Bush et al. ([2010]) pose the question, how can small-scale shrimp farming techniques be integrated into a landscape level approach to protect mangroves and control water-borne shrimp diseases?

In the context of aquaculture, the commons problem occurs due to overuse of environmental resources as inputs and the addition of pollutants to the environment as waste (Pa´ez-Osuna [2001]). Solving the problem should start by addressing the two main issues related to commons management, the excludability problem and the subtractability problem (Ostrom et al. [1999]). Based on these two characteristics, Ostrom et al. ([1999]) define common pool resources as a class of resources for which exclusion is difficult and joint use involves subtractability. Controlling access to resources (the excludability problem) is addressed by the fact that the farms are privately owned. However, the private property regime is not ideally suited to deal with the subtractability problem when individual holdings are interconnected. Hence, rules are needed to reduce the impact of users on one another (Feeny et al. [1990]). Where community institutions exist, common property regimes may be suitable for users as an arena in which to make and enforce their own rules. Such community-based management works in many coastal resource contexts in Asia and elsewhere (Pomeroy and Rivera-Guieb [2006]; Berkes et al. [2001]). It is based on the idea that local resource users have a greater interest in the sustainable use of local resources than do large companies and other non-local players (Ostrom [1990]).

For common pool resources, collective management can work better than pure government management or private management (Ostrom et al. [1999]; Ostrom [2005]). Where users are interdependent, collective action is more appropriate than individual action and more effective for problem solving (Ostrom [2005]). This is where associations such as cooperatives come in. The International Cooperative Alliance defines cooperative as "an autonomous association of persons united voluntarily to meet their common economic, social, and cultural needs and aspirations through a jointly-owned and democratically-controlled enterprise" (ICA [2014]; Amarasinghe and Bavinck [2011]: 389). They work best in societies which have traditions of collective management (Yamamoto [1995]; Makino and Matsuda [2005]; Amarasinghe and De Silva [2001]). A cooperative functions as a local institution that can coordinate the response to the exclusion problem of commons, limiting access rights to the resource. It also deals with the subtractability problem, helping make and enforce local rules among the parties that have access rights, coordinating with government agencies and market institutions, and acting as an information broker (Beem [2007]).

Sri Lanka has a history of managing resources collectively in many sectors, including fisheries, aquaculture and agriculture (Amarasinghe and Bavinck [2011]; Amarasinghe [2006]). Historically, the Sri Lankan government has recognized and promoted "bottom-up" management through legislation aimed at developing and uplifting certain economic sectors (Anon. [2007]). Many of these resources are managed by cooperatives that facilitate collective decision making (Degen [1998]; Samarakoon et al. [2011]). Even though not all associations discussed in this paper are formally constituted as cooperatives, they function like cooperatives, despite varying designations.

However, the Sri Lankan literature on collective management does not include shrimp aquaculture carried out by community associations and cooperatives. The paper addresses this gap by, first, attempting to explore and understand how shrimp aquaculture is carried out by community-based institutions in northwestern Sri Lanka. Second, the paper discusses the evolution of small-scale shrimp aquaculture governance and the role of existing commons institutions. Finally, the paper explores the importance of community-based, small-scale aquaculture as a potential alternative to large-scale aquaculture.

METHODS AND STUDY AREA

This paper is based on a qualitative research study conducted during 120 days of fieldwork in three coastal communities of northwestern Sri Lanka (Figure 1). The communities were *Ambakandawila*, *Koththanthive*, and *Karamba,* hereafter referred to as communities A, B, and C respectively. These three communities were selected to capture the entire process of shrimp farming operation within the available limited timeframe. During the field work period, the three communities were in different stages of the shrimp farming process. For example, during the month of May, community A was in the middle of farming; community B was in the harvesting stage, and community C was in the pond preparation stage. The management activities and focus of the community-level shrimp farmers' associations varied according to the production stages of farmers in the community. For example, the main concern of community C was to finish postlarvae (shrimp "seed") stocking prior to the deadline. Community A— which was in the middle of the shrimp farming stage—focused on protecting their farms from shrimp diseases. Community B was in the process of shrimp harvesting and preparing ponds for the next culture season.

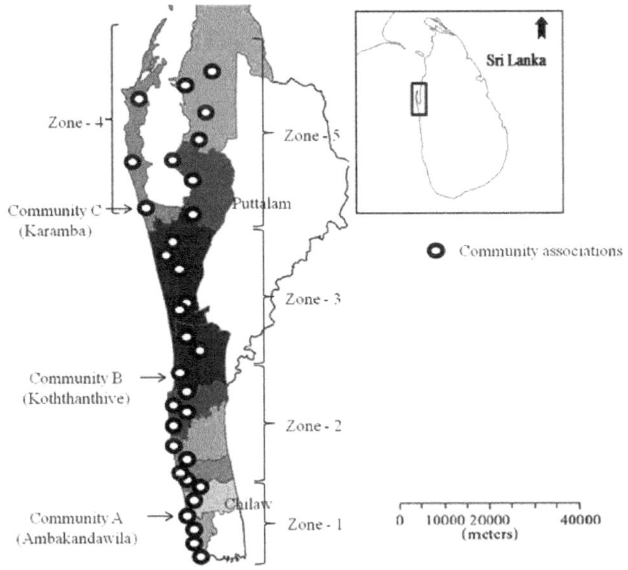

Figure 1: Study area: Three coastal communities (A, B, and C) in the northwestern Sri Lanka and the distribution of communal institutions (Community associations/ *Samithies*).

The research strategy adopted was the case study approach (Yin [2009]). Primary data were collected through: a) participant observation; b) semi-directive interviews with shrimp farmers and the leaders of shrimp farming community associations; c) focus group discussions; and d) key informant interviews with influential people involved in shrimp farming. Data collection took place from April to August in 2012. The sample of farmers was selected using a snowball sampling technique. Thirty eight (38) shrimp farmers and community associations (A: 13; B: 11; and C: 14) were interviewed. Further, the data were validated through seven key informant interviews and three focus group discussions. The data from participant observations helped to obtain a better contextual understanding of the shrimp farming areas. During the field work, a couple of community associations, which did not belong to the study sample, also invited the researcher to visit and observe their meetings.

Data gathered through participant observations and semi-directive interviews were documented immediately in the field. Analysis of findings began with the labeling and coding of data to help to distinguish differences and similarities. Mind maps and diagrams on concepts and, management processes, were developed based on the contextual understanding developed during field work. Insights and experiences gained through participant observation also helped the researcher to simplify the analysis process. Descriptive statistics (e.g., frequencies, percentages, average values) were used in assessing the background of community-based shrimp aquaculture in northwestern Sri Lanka. Microsoft Excel was used for data processing.

Community A is located in the southern part of northwestern coastal area, close to *Chilaw* lagoon. This is a rural community with people of Sinhalese origin. At the time of research there were about 150 households involved in small motorized craft fishing, shrimp breeding (hatchery), and shrimp feed selling in addition to shrimp farming. Community B is located in the middle of northwestern coastal area on the *Mundel* lagoon. This is also a rural isolated community but with Tamil ethnic backgrounds. The size of community B is about 100 households. The main household income generating activities are: fishing; paddy farming; and government employment. Community C is located in the northern part of the northwestern coastal area, close to the *Puttalam* lagoon. This community has a mix of people with Sinhalese, Tamil, as well as Muslim origins. Salt extraction and coconut cultivation were the main household activities, other than shrimp farming.

DESCRIPTION OF THE AQUACULTURE OPERATIONS

Almost all of the existing shrimp aquaculture operations in northwestern Sri Lanka are small-scale. The number of ponds operated in a farm can be considered as an indicator of the extent of shrimp farming. In 2011–12, 37 percent of the sampled farmers operated two to five ponds, whereas 11 percent operated just one pond. Twenty-nine percent and 18 percent of shrimp farmers operated five to ten and ten to fifteen ponds, respectively. Only five percent of farmers had more than 15 ponds. The size of the smallest pond was about 0.2 hectares whereas the largest was about 0.8 hectares. As of 2013, there were many abandoned shrimp ponds in the northwestern coastal area, and only about 45 percent of ponds is operational. Small-scale shrimp farmers do not invest money in the construction of new ponds; rather they use existing abandoned farms to reduce costs. Another indicator of the scale is the total land area of the farm (i.e., the full extent or the size of the farm), including the ponds which were being operated and rested. Fifty five percent of shrimp farmers had farming areas between one and three hectares in size. Thirteen percent of farmers had less than one hectare of land. Sixteen percent and 11 percent of farmers had farm areas between three and five and five and seven hectares, respectively. The recorded percentage of farmers with more than seven hectares of farm lands was five percent. Only the farmers from community C used more than seven hectares of farm land. Multiple reasons were given by the farmers for resting some of the ponds in their farms: lack of labor; relatively low profit margins and increasing cost; limited shrimp stocking capacity issued by the farmers' co-operative; and pond disinfecting to reduce disease risk. Draining a pond and exposing it to sunlight presumably reduces disease risk.

Shrimp farms were owned and operated by individuals, not by community associations or large private corporations. However, 53 percent of these farms were on government-owned lands without any compensation paid for using the lands for commercial purposes. There were unresolved legal issues mostly related to land ownership, and it seemed political influences were in play. Thirty-seven percent of farmers used their own private lands. The remaining 11 percent of farmers rented lands from shrimp farmers who had discontinued farming previously. Due to the high input cost of farming, the general practice of farmers was to minimize the number of employees as a way to reduce costs. In most cases, family members provided the labor needed. Involvement of family members in the farming operation was common. Seventy-four percent of shrimp farmers said they receive the support and involvement of their family members and/or relatives. Sixty four percent of them received support from female family members --wives, daughters, mothers, and/or other close relatives. These women were involved in shrimp farming in diverse ways: labor (shrimp harvesting); book keeping; day-to-day management activities; financial contributions towards farming; and community association activities. Complementing the individual/family-level farm ownership, community level farming activities were governed by community-level rules designed and enforced by the community institutions as a part of a multi-layered management structure.

MULTI-LAYERED MANAGEMENT STRUCTURE

Shrimp aquaculture operations are managed under a commons regime consisting of three kinds of institutions: private; communal; and governmental. At the heart of this management regime is the shrimp farmers' (or community) association, called *samithiya* in Sinhala. The distribution of community level associations in the northwestern part of the island is shown in Figure 1. These institutions are self-organized and formulate and implement their own rules. For example, community-level rules specify that a pond should not be harvested when shrimp are less than one month old or less than five grams in average weight (community A); water should not be released or pumped until the two month culture period is completed (community A; community B); and partial harvests are not allowed (community A; community C). Obtaining community association membership is a requirement for any person starting-up a shrimp farm. Community associations have collective authority and a high level of involvement in regulating and managing shrimp aquaculture. Elected officers of the community associations enforce the rules and assess penalties for those not conforming. Penalties vary from fines to denial of permit—a document issued by the community association required for carrying out shrimp farming during a particular crop season.

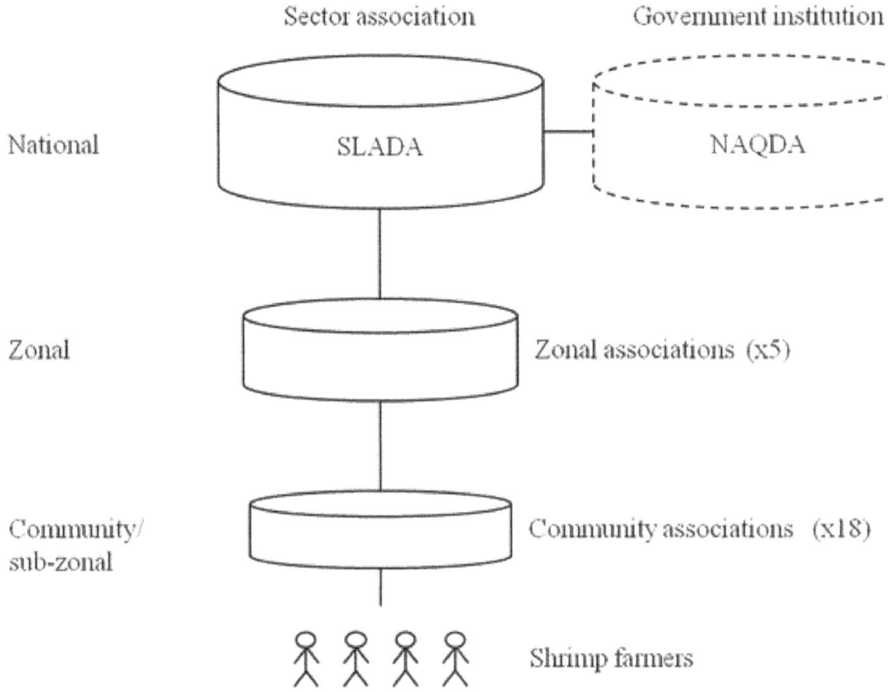

Figure 2: Evolution of small-scale shrimp aquaculture.

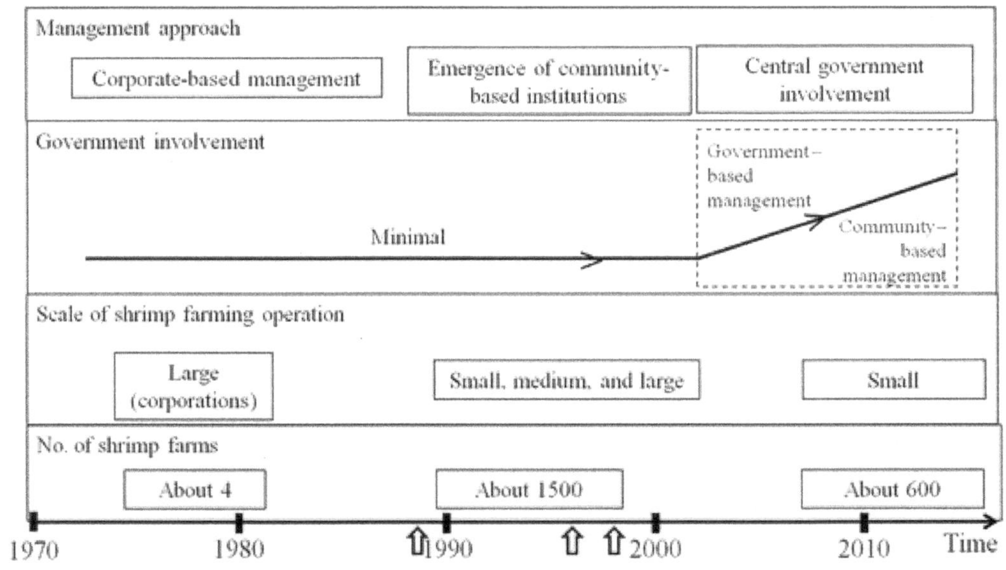

Figure 3: Multi-level structure of commons institutions.

Most of the community associations consist of 20–60 members. Members pay an annual membership fee. The elected officials of a community association include: president, vice president, secretary, treasurer, and assistant treasurer. These officers have the authority to make decisions on behalf of the membership—a power granted through the collective agreement. However, the officials are also responsible for elaborating and adequately justifying such decisions during the upcoming meetings. The community associations collectively make decisions on the day-to-day regulatory operations related to community level shrimp aquaculture. For example, monitoring the application of proper management practices for shrimp disease prevention; inspection of ponds prior to issuing post larvae bill; and supporting members to obtain electricity connection to farms, are some of the major activities. Most of the associations have government aquaculture extension officers working closely with them to ensure that practices comply with national level regulations. However, the extension officers are merely observers and are not part of the decision-making. Collective decision making is one of the important characteristics of community associations. Moreover, community associations act as hubs for sharing information required for operations. These community institutions are inter-connected through zonal-level associations. Table 1 shows characteristics of the three community associations studied.

Table 1: Characteristics of community associations

Characteristics	Community A	Community B	Community C
Year of establishment	2006	1996	1995
Nature of origin (farmer-initiated or government-initiated)	Government-initiated	Farmer-initiated	Farmer-initiated
Registered year	2012	2002	1998
Registered under	NAQDA	Divisional secretariat office	Cooperative Act
Current membership	30	61	30
Number of officers	7	7	11
Availability of a constitution	Yes	Yes	Yes
Bank account/Letter head/Official rubber stamp	Yes	Yes	Yes
Meeting frequency	Monthly, special, and annual meetings	Monthly and special meetings	Monthly and special meetings
Governance structure	Mostly flat	Hierarchical	Flat
Cohesiveness	High	Moderate	Very high

One basic feature of shrimp aquaculture in the study area is the zonal system design introduced by the SLADA[a] (Sri Lanka Aquaculture Development Association) for the main purpose of managing disease. The shrimp farming area is divided into five main zones and 32 sub-zones. Some zones include just one sub-zone whereas the others represent up to eight sub-zones. If the number of farmers in a particular sub-zone is too small, then adjacent sub-zones are amalgamated to form a single association to represent a cluster of sub-zones. Leaders of sub-zonal associations represent the zonal association. All zonal and sub-zonal associations are collectively represented in SLADA with at least one representative from each zone and sub-zone.

SLADA is the national level association, which represents all the direct stakeholders in the shrimp farming sector—farmers; hatchery owners; feed suppliers; and shrimp processors. This association was established in 2005 with an initial membership of 16 people. Later, zonal and sub-zonal community associations were allowed to become members, resulting in a total membership of about 50 (about 20 institutional members). SLADA is an independent body. Even though the main objective of SLADA is to develop the shrimp aquaculture sector, initially the mandate was not restricted to shrimp aquaculture. Development and planning of ornamental fish aquaculture and inland aquaculture were also included in the mandate. In developing the shrimp farming sector, SLADA appointed six committees targeting different subject areas, such as environmental protection; hatchery development; and shrimp farm development. Moreover, SLADA and NAQDA (National Aquaculture Development Authority)—the government line institution responsible for aquaculture development and management, which falls under the of Ministry of Fisheries and Aquatic Resource Development in Sri Lanka —jointly developed a technical advisory committee representing expertise from local universities and research institutes. In 2006, SLADA was legally registered as a corporation by the members themselves (Figure 2).

EVOLUTION OF SMALL-SCALE SHRIMP AQUACULTURE GOVERNANCE

Evolution of small-scale shrimp aquaculture in northwestern Sri Lanka is related to the collapse of large-scale shrimp farming operations, the management approach, government's involvement and impact, and the management of the key common-pool resource, water. Figure 3 illustrates the evolution of shrimp aquaculture from the 1970s to 2012.

Collapse of Shrimp Farming Operations

In the late 1970s, shrimp aquaculture in Sri Lanka started off as a commercial-scale venture. Initially, the farms were geographically limited to the eastern part of the country (Batticaloa District) and only four companies (Carson, Enris, Lever Brothers and Andrew's) were in the industry. To begin with, all the operations were done in an integrated fashion, whereby shrimp breeding (hatcheries), farming, and processing, as well as exporting activities were managed by the same company. According to the respondents, some farms were as large as 150 ha in area with hundreds of ponds. At the time, these companies reportedly earned attractive profits. The information related to culture systems, input supply, and marketing were accessible only to a few stakeholders attached to these companies. During the early 1980s, the profit making potential of this new industry attracted investors such as politicians, bank managers, and wealthy business owners. At the same time, some of the employees of these pioneer companies started their own farms in eastern and northwestern provinces.

Later (in the 1990s), the government intervened by promoting shrimp farming as a profitable self-employment opportunity among locals, leading to many other community-level small-scale farmers entering the industry. As a result, farming area expanded over the northwestern part of the country. Small-scale farmers conducted their operations within their community areas, while large and medium-scale farmers shifted from place to place by converting mangrove forests (Gunawardena and Rowan [2005]) and coconut plantations into shrimp farms (Cattermoul and Devendra [2002]). However, the expansion of aquaculture into the eastern part of the country was restricted due to the civil war that was taking place. During this time, there were outbreaks of three viral diseases. The first major disease outbreak Monodon Baculo Virus (MBV) came in 1988/9. The second major disease outbreak was the White Spot Syndrome Virus (WSSV) in 1996, followed by a third outbreak in 1998 with both WSSV and Yellow Head Virus (YHV). Most of the pioneer companies shut down operations a result of these outbreaks.

As of 2012, almost all the 600 shrimp farms in the northwestern Sri Lanka are small-scale operations. Large or medium-scale shrimp operations have not survived under disease conditions. As documented earlier, the average size of a shrimp farm is small compared to the operations that took place at the beginning of the industry.

Changes in Management Approach

The type of management system found in the 1970s and 1980s was a corporate-based management with four large multi-national companies controlling the industry. However, the decline of corporate shrimp farming during the 1990s created space for small-scale operations with their own associations. This is the era of emergence of community-based institutions across northwestern Sri Lanka. Community associations were registered under different legislations/authorities such as the Cooperative Act, divisional secretariat offices, and most recently with government institutions like NAQDA.

Before 2003, there were no governmental, technical or other controls over shrimp farming activities. Resource management activities were mostly limited to self-managed community-level associations, even though these associations were not recognized by the government. There was neither any proper linkage between farmers and government nor proper management over shrimp aquaculture production and quality. For instance, there was no specific timeframe for stocking shrimp postlarvae. Shrimp farmers and hatcheries continued production throughout the year, and people were free to enter and leave shrimp farming whenever they wished.

Government's Involvement and Impact

Until 2003, government involvement in shrimp aquaculture was minimal and was mostly restricted to divisional and provincial level general administration, and research on water quality of the northwestern lagoons. The direct involvement of the central government in the shrimp aquaculture sector started in 2003/4 through NAQDA. This action led to many changes in management, resulting in a collaborative management approach. NAQDA's approach was to work in collaboration with SLADA and community-level shrimp farmer and breeder (hatchery owner) associations. Initially, it was not easy for the independent-minded actors in the sector (especially shrimp farmers and breeders) to adjust to the new management system, as they were skeptical of the purposes behind government involvement.

Design and introduction of the zonal crop calendar system in 2005/06 was one of the initial tasks of SLADA and the individual community institutions. The crop calendar system emerged out of a study conducted by a few of the founding members of SLADA with previous experience and knowledge of aquaculture. The study report recommended management that involved the development of a zonal crop calendar system. This crop calendar system was initiated by SLADA in collaboration with NAQDA. The calendar system was designed to limit farmers accessing common water bodies through a schedule or timetable leading to a major change in the way farming operations were conducted. Under this system, farmers were not allowed to access the water bodies at any time but only during a particular period when all of the community farmers of a subzone could access the same water body. The initial reaction of shrimp farmers and hatchery owners was to resist, mainly because the crop calendar limited the production cycle (number of crops per year). Moreover, the introduction of new standards and quality improvement measures targeting shrimp farms and hatcheries led to social conflicts between shrimp farmers/hatchery owners and NAQDA. Adoption of these measures required costly physical infrastructure changes such as construction of chlorine baths at farm and hatchery entrances to control disease spread. NAQDA field extension officers were tasked with monitoring these standards by visiting farms and hatcheries. However, these officers at times were not allowed to enter farms and hatcheries—and there were reported incidences of serious verbal conflicts and even physical harassment.

During the first few crop cycles under the new zonal crop calendar system, some relatively large-scale shrimp farmers and large-scale hatchery owners in Sri Lanka left the industry as it was not profitable to maintain their farms and hatcheries in an idle state during parts of the year as they still had to incur costs in terms of salaries and fixed costs such as maintenance. The overall effect of these changes was a decrease in national shrimp aquaculture production after 2003. However, farmers and hatchery owners who remained in the sector experienced a very successful crop year after the introduction of the zonal crop calendar. Farmers had a good harvest due to reduction in disease (fewer infected ponds), and hatchery owners received a good price for postlarvae. After realizing the success, some farmers who had left earlier, joined again in the following season; however, the next season was not as successful, in part due to lower market prices.

During the period, 2006 to 2009/10, there were ups and downs in terms of shrimp production. People started to think that the white spot disease had been eradicated, but it returned from time to time. Most of the medium-scale farmers also left farming due to the uncertain nature of production and significant increases in costs of labor, fuel, and chemicals such as chlorine. Although small-scale farms were affected by the same factors, small operations tended to have less operational and financial risks, mainly due to use of family labor.

By this time, the majority of farmers and hatchery owners had adapted to the new crop calendar system. NAQDA, with the support of SLADA, further strengthened the existing community associations while establishing new associations in certain other communities. NAQDA's aquaculture field extension officers were able to visit farming communities without any problems, and NAQDA itself became more flexible after their experience. Some illegal activities such as shrimp postlarvae stocking and releasing disease-infected water into the natural water body, still occurred, but NAQDA and SLADA are able to take action against these activities with the support of community-level associations. Shrimp aquaculture management system became well established as a result of these changes.

Based on the SLADA report recommending a crop calendar system, different tasks were assigned to different levels of management. Community (primary) associations regulate community-level farming practices with the support of NAQDA field extension officers. For this purpose, there is a shrimp farm monitoring and extension unit in *Battulu Oya* established in 2008. It is the responsibility of second-level zonal associations to develop the infrastructure of shrimp farming areas in collaboration with divisional secretariat offices and provincial councils. Maintaining links (mainly communication) between the different management levels is also the role of zonal associations. Finally, the role of top-level management (SLADA and NAQDA) is to develop an effective national-level shrimp aquaculture system.

Water: Managing The Key Common-Pool Resource

Shrimp farms take water from a common source at the beginning of the production cycle and discharge used water at the end of the cycle, after shrimp have been harvested. Almost all the shrimp farms in the northwestern area are directly or indirectly connected to single water bodies, mainly via a canal that connects them (the Dutch canal) and the lagoons. *Puttalam* lagoon (28,000 ha), *Mundal* lagoon (3,600 ha), and *Chilaw* lagoon (700 ha) are the three main lagoons connected by the Dutch canal. *Kala Oya*, *Mi Oya*, *Daduru Oya*, and *Maha Oya* are the four main rivers connected to this system. The interconnected nature of this common water body makes the shrimp farms susceptible to shrimp diseases. Hence, one of the major tasks of the aquaculture management system is to control the spread of disease. Wetland ecosystem biodiversity and ecosystem services such as water filtration through mangroves may also have a role to play in this regard but their importance has not been investigated.

Regarding the common water supply as the key resource, the excludability problem (Ostrom et al. [1999]) is addressed by community associations with their own rules to exclude outsiders from shrimp farming in their communities. This is accomplished in two ways. First, the community association organizes the membership and oversees how and when the members access the water source at the given time period, as specified by the zonal crop calendar (main aim is to control the spread of any diseases). Second, if an outsider wants to start shrimp farming in a particular community, attending association meetings for a minimum period of six months is mandatory to be eligible for membership. This rule is in place to discourage community dwellers/outsiders, who intend to start shrimp farming, from moving from one farming location to another within a short period of time, i.e. relatively irresponsible shrimp farming. This rule also ensures that the prospective members understand how shrimp farming operations are conducted within the community in a collective manner. The subtractability problem is associated with the release of waste water into the common water body. If a particular pond is disease infected and if the farmer releases contaminated pond water to the common water body, there is a high probability for disease spread into other farms. Disease cannot be completely eradicated but it can be controlled. Constant vigilance and monitoring by farmer families themselves make early detection possible to deal with shrimp disease. If an outbreak is caught early, shrimp can still be harvested quickly and sold.

Large-scale shrimp disease problems, such as those that led to the demise of the aquaculture companies in the 1990s, mainly occur when uncontrolled amounts of disease-infected water are released into a common water body. By limiting the amounts of waste water and staggering the times of release, the zonal crop calendar system reduces the incidence of disease, and helps farmers to control infections. Compliance is monitored and managed by community level associations making use of their own rules. These rules are implemented by the collectively elected officers of community associations, and enforced by higher level authorities as appropriate. According to Gazette number 1536/12-(February 13, 2008), shrimp can only be farmed in compliance with the crop calendar. To operate a shrimp farm, farmers must obtain a permit from NAQDA (as specified in Gazette number 1677/7-October 25, 2010) issued under the consent of the community association. Punishments for free riders have been specified depending on the severity of the violation and range from fines paid to the community association to court cases filed by NAQDA.

DISCUSSION AND CONCLUSIONS

Most of the commercially managed aquaculture operations

in South and Southeast Asia are large-scale, corporate-based industries that often produce negative social and environmental impacts. Unsustainablity of large-scale aquaculture is well documented, but there have been some major attempts to make them more sustainable (Primavera [2006]; Bush et al. [2010]). Many large-scale aquaculture operations are known to move from one area to another, in order to avoid disease. Large-scale aquaculture tends to make its profit during the first few years, with the expectation that they can move onto another area, rather than aiming to establish sustainable practices and remain in one area (Huitric et al. [2002]).

By contrast, small-scale community-based aquaculture farmers do not have the option of relocating, and have strong incentives to seek ways to make sustainability work. They cannot afford to act in ways that will undermine their livelihoods, if they have attachment to place and intend to continue to live in that area as part of the local community. These considerations are well documented by commons scholars for a range of resource types in many parts of the world (Ostrom [1990]; [2005]). Thus, by design, small-scale aquaculture has built-in incentives to be sustainable (or else go out of business). Based on the Sri Lanka experience, the survival of small-scale aquaculture (as opposed to large-scale aquaculture which has not) seems to be related to a number of factors:

A mixed governance regime that combines private ownership with community-level commons institutions, which deal with exclusion and subtractability problems.

A management system that is able to limit and control the release of waste water, minimizing its environmental impact and the transmission of shrimp disease. Small-scale operators have low overheads and low operating costs, as family members work in the farm.

A close relationship with aquaculture ponds, allowing for constant monitoring by farmer families themselves for shrimp disease. There are relatively small net losses even under disease conditions, since disease detection is fast and the potential for limiting economic damage is relatively high. This is because the shrimp can still be sold at market prices, if disease is detected at the initial stage.

The social enterprise nature of small-scale aquaculture, whereby operations are based on an economic model in which resources provide for broader goals, such as improving human and environmental well-being, rather than merely maximizing profits for external shareholders (Ridley-Duff and Bull [2011]).

Competition over scarce resources is a common phenomenon throughout the world, and interactions among users tend to be competitive. However, in the context of small-scale shrimp aquaculture in Sri Lanka, farmers co-operate to overcome common challenges, such as disease. Such cooperation is not common in the world of free enterprise, and as Bruno Dyck (pers. comm.) pointed out; it can be considered as a different way of doing things, even a different worldview. There seem to be two major factors that shape the operations of Sri Lanka's small-scale aquaculture producers.

First, shrimp farmers are empowered by their experience having worked for the old (and failed) aquaculture companies; they constantly apply this knowledge (stocking, growing and harvesting shrimp) to the practice of their small-scale operations. Second, starting at the local level, multi-level commons institutions have been developed by shrimp farmers in cooperation with the government, for commons management (Berkes and Davidson-Hunt [2008]). These commons institutions function collaboratively, using a participatory approach. There are no formal co-management arrangements with the government. However, small-scale aquaculture, as practiced in Sri Lanka, functions like co-management because it involves shared decision-making between government and producers. The overall system is a hybrid regime of private, communal, and state commons management, in which government devolves certain powers to other parties.

The evolution of this commons management model, which dates back to 1990s, has been relatively rapid and has taken place under certain historical conditions. The community-based shrimp aquaculture management system in northwestern Sri Lanka seems to have evolved to its current state under the influence of two major drivers. The first is the tendency of large aquaculture operations in Sri Lanka in the 1970s and 1980s to develop disease. Shrimp disease problem is a key factor in the demise of large operations in general (Bush et al. [2010]). Disease spreads through the interconnected water bodies, which tend to be controlled centrally in large-scale operations, but separately in small-scale operations through the zonal crop calendar system. The second driver is the ability of Sri Lankan producers to use collective approaches. There is a historical and cultural tendency in Sri Lanka towards a collective approach in managing and sharing resources (Gunawardena and Steele [2008]).

Backbone of the community-based shrimp aquaculture management system is multi-level commons institutions. These institutions have been developed by shrimp farmers themselves, in cooperation with the government. This bottom-up multi-level institutional structure is effective in managing small-scale shrimp aquaculture management. Small-scale shrimp aquaculture owners are families. Their operations are owner-managed under community-level institutions, with government supervision and coordination. This private-communal-state mixed regime appears to be a viable alternative to "big aquaculture".

The current (as of 2013) small-scale shrimp aquaculture systems in Sri Lanka are by no means perfect. Shrimp farmers deal with new problems on a daily basis and constantly argue among themselves in local associations. As new challenges appear, they debate new approaches. The present management system will likely continue to evolve in the face of numerous changes in the social-ecological system into the future. However, the recent historical experience with northwestern Sri Lanka shrimp aquaculture systems suggests four lessons: (1) Small-scale shrimp aquaculture operations have been able to survive in an environment in which large-scale ones have not; (2) this has been mainly due to the ability to control the spread of disease under the recent management regime for small-scale shrimp aquaculture; (3) the mixed regime (private-communal-state) commons management model with multi-level institutional linkages has been effective in providing private incentives, community-level rules and government coordination and oversight; and (4) continuous learning and adaptation, such as adjusting the crop calendar from year to year and combining farmer knowledge with government technical knowledge, has provided a pathway to economic sustainability. The present study is significant for several reasons. Even though small-scale community-based shrimp aquaculture practices exist elsewhere in the world (for example in Thailand, Berkes, unpublished field notes), it seems that they have not been documented or studied. The present study attempts to shed light on this area by confirming the existence of community-based operations and by providing evidence that such operations seem to be viable. Further, the study suggests that the comparatively lower environmental and social impacts of these operations make them highly attractive as an alternative approach. Principles and practices of Sri Lanka's community-based management system provide lessons in multi-level governance that may be applicable to other aquaculture operations.

DECLARATIONS

Acknowledgement

We thank the shrimp farmers of northwestern Sri Lanka, and gratefully acknowledge Rohan Paaris from the Sri Lanka Aquaculture Development Association for his invaluable support and guidance extended during the field work period. We thank Upali Amarasinghe (University of Kelaniya, Sri Lanka), Bruno Dyck (University of Manitoba), and Iroshani Tennakoon (Manitoba Health) for help and for sharing valuable thoughts and experiences in this field.

Competing interests

The authors declare that they have no competing interests.

REFERENCES

1. Adger WN, Hughes TP, Folke C, Carpenter SR, Rockström J: Social-ecological resilience to coastal disasters. *Science*2005, 309: 1036–1039. 10.1126/science.1112122
2. Amarasinghe US: Culture-based fisheries development in Sri Lanka: a case study. In *Better-practice approaches for culture-based fisheries development in Asia*. Edited by: Silva SS, Amarasinghe US, Nguten TTT. ACIAR Monograph 120, Canberra, Australia; 2006:50 72.
3. Amarasinghe O, Bavinck M: Building resilience: fisheries cooperatives in southern Sri Lanka. In *Poverty mosaics: realities and prospects in small-scale fisheries*. Edited by: Jentoft S, Eide A. Springer, Dordrecht; 2011.383–406.
4. Amarasinghe US, De Silva SS: Sri Lankan reservoir fishery: a case for introduction of a co-management strategy. *Fisheries Management and Ecology* 2001, 6(5):387–400. 10.1046/j.1365-2400.1999.00170.x
5. *Ten year development policy framework of the fisheries and aquatic resources sector (2007–2016), Ministry of fisheries and aquatic resources, March 2007, 4*. 2007.
6. Beem B: Co-management from the top? The roles of policy entrepreneurs and distributive conflict in developing co-management arrangements. *Marine Policy* 2007, 31: 54–549. 10.1016/j.marpol.2006.12.001
7. Berkes F, Davidson-Hunt I: Commons in a multi-level world. *International Journal of the Commons* 2008, 2(1):1–6.
8. Berkes F, Mahon R, McConney P, Pollnac R, Pomeroy R: *Managing small-scale fisheries: alternative directions and methods*. International Development Research Centre, Ottawa; 2001.
9. Bhatta R, Bhat M: Impacts of aquaculture on the management of estuaries in India. *Environmental Conservation* 1998, 25(2):109–121. 10.1017/S0376892998000162
10. Bush SR, van Zwieten PAM, Visser L, Van Dijk H, Bosma R, De Boer WF, Verdegem M: Scenarios for resilient shrimp aquaculture in tropical coastal areas. *Ecology and Society* 2010, 15(2):15.
11. Cattermoul N, Devendra A: *Effective management for biodiversity conservation in Sri Lankan coastal wetlands, A measurement of the ecological footprint of shrimp farming in the Chilaw lagoon area*. University of Portsmouth Cemare, Portsmouth, UK; 2002.
12. Degen P: Fisheries co-management in Sri Lanka: learning from the experiences of the others. *Mekong Fisheries Network Newsletter* 1998, 4(1):1–2.
13. Digby M: *Organisation of Fishermen's Cooperatives*. Plunkett Foundation, Oxford, UK; 1975.
14. Feeny D, Berkes F, McCay BJ, Acheson JM: The tragedy of the commons: twenty-two years later. *Human Ecology* 1990, 18(1):1–19. 10.1007/BF00889070
15. Galappaththi EK: *Exploratory study on supply chains of cultured prawns in Sri Lanka*. Postgraduate Institute of Agriculture, University of Peradeniya, Sri Lanka; 2010.
16. Gunawardena W, Rowan JS: Economic valuation of a mangrove ecosystem threatened by shrimp aquaculture in Sri Lanka.*Environmental Management* 2005, 36(4):535–550. 10.1007/s00267-003-0286-9
17. Gunawardena A, Steele P: The stake-net fishery association of Negombo lagoon, Sri Lanka: why has it survived over 250 years and will it survive another 100 years? In *Promise, trust and evolution: managing the commons of South Asia*. Edited by: Ghate R, Jodha N, Mukhopadhyay P. Oxford Scholarship, New York; 2008:144–154. 10.1093/acprof:oso/9780199213832.003.0007
18. Haque ABMM, Visser LE, Dey MM: Institutional arrangements in seasonal floodplain management under community-based aquaculture in Bangladesh. *Asian Journal of Agriculture and Development* 2011, 8(1):1–18.
19. Huitric M, Folke C, Kautsky N: Development and government policies of the shrimp farming industry in Thailand in relation to mangrove ecosystems. *Ecological Economics* 2002, 40: 441–455. 10.1016/S0921-8009(02)00011-3
20. *International Co-operative Alliance*. 2014.
21. Makino M, Matsuda H: Co-management in Japanese coastal fisheries: institutional features and transaction costs. *Marine Policy* 2005, 29: 441–450. 10.1016/j.marpol.2004.07.005
22. McCay BJ, Weisman W, Creed C: Coping with environmental change: Systematic responses and the roles

of property and community in three fisheries. In *World Fisheries: A Social-Ecological Analysis, eds. Ommer, RE, RI Perry, K Cochrane, and P Cury*. Wiley-Blackwell, Oxford; 2011:381–400. 10.1002/9781444392241.ch23

23. MFARD. 2011. Ministry of Fisheries and Aquatic Resource Development of Sri Lanka. Sri Lanka:. Accessed 14 May 2014., [http://www.fisheries.gov.lk/].
24. Munasinghe MN, Stephen C, Abeynayake P, Abeygunawardena IS: Shrimp farming practices in the Puttallam district of Sri Lanka: implications for disease control, industry sustainability, and rural development. *Veterinary Medicine International*2010, 2010: 1–6. 10.4061/2010/679130
25. Nayak PK, Berkes F: Whose marginalization? Politics around environmental justice in India's Chilika lagoon. *Local Environment* 2010, 15(6):553–567. 10.1080/13549839.2010.487527
26. Ostrom E: *Governing the commons: the evolution of institutions for collective action (11–13)*. Cambridge University Press, New York; 1990.
27. Ostrom E: *Understanding institutional diversity*. Princeton University Press, Princeton; 2005.
28. Ostrom E, Burger J, Field CB, Norgaard RB, Policansky D: Revisiting the commons: local lessons, global challenges.*Science* 1999, 284: 278–282. 10.1126/science.284.5412.278
29. Pa´ez-osuna F: The environmental impact of shrimp aquaculture: causes, effects, and mitigating alternatives. *Environmental Management* 2001, 28(1):131–140. 10.1007/s002670010212
30. Pomeroy RS, Rivera-Guieb R: *Fishery co-management: a practical handbook*. International Development Research Centre, Ottawa; 2006.
31. Primavera JH: Socio-economic impacts of shrimp culture. *Aquaculture Research* 1997, 28: 815–827. 10.1111/j.1365-2109.1997.tb01006.x
32. Primavera JH: Overcoming the impacts of aquaculture on the coastal zone. *Ocean and Coastal Management* 2006, 49: 531–545. 10.1016/j.ocecoaman.2006.06.018
33. Ridley-Duff RJ, Bull M: *Understanding Social Enterprise: Theory and Practice*. Sage Publications, London; 2011.
34. Samarakoon JI, Nightingale M, Hermes R, Joseph BL, Salagrama V: *Review of Community-based Integrated Coastal Management: Best Practices and Lessons Learned in the Bay of Bengal, South Asia*. Ecosystems and Livelihoods Group, Asia, IUCN, Colombo; 2011.
35. Von Essn LM, Ferse SCA, Glaser M, Kunzmann A: Attitudes and perceptions of villagers toward community-based mariculture in Minahasa, North Sulawesi, Indonesia. *Ocean and Coastal Management* 2013, 73: 101–112. 10.1016/j.ocecoaman.2012.12.012
36. Yamamoto T: Development of a community-based fishery management system in Japan. *Marine Resource Economics*1995, 10: 21–34.
37. Yin, RK. 2009. Case study research, design and methods (4th ed.). In Applied social research method series 5. Sage Publications, Los Angeles.

INDEX